CURRENT THEMES IN TROPICAL SCIENCE

Chief Editor: THOMAS R. ODHIAMBO, *I.C.I.P.E., Nairobi*

Volume 2

Natural Products for Innovative Pest Management

Other Pergamon publications of interest

Books

ENGELMANN *Physiology of Insect Reproduction*

HINTON *Biology of Insect Eggs, 3-volume set*

LUSCHER *Phase and Caste Determination in Insects*

MATSUDA *Morphology and Evolution of the Insect Abdomen*

OBENCHAIN *The Physiology of Ticks*

SAUNDERS *Insect Clocks, 2nd Edition*

SHORROCKS *Drosophila*

SUTTON *Woodlice*

VON FRISCH *Twelve Little Housemates*

Journals*

INSECT BIOCHEMISTRY
INSECT SCIENCE AND ITS APPLICATION
INTERNATIONAL JOURNAL FOR PARASITOLOGY
INTERNATIONAL JOURNAL OF INSECT MORPHOLOGY AND
EMBRYOLOGY
JOURNAL OF INSECT PHYSIOLOGY

*Free specimen copies available on request

Natural Products for Innovative Pest Management

Guest Editors

DAVID L. WHITEHEAD

and

WILLIAM S. BOWERS

The International Centre of Insect Physiology & Ecology,
Nairobi, Kenya

PERGAMON PRESS

OXFORD · NEW YORK · TORONTO · SYDNEY · PARIS · FRANKFURT

U.K.	Pergamon Press Ltd., Headington Hill Hall, Oxford OX3 0BW, England
U.S.A	Pergamon Press Inc., Maxwell House, Fairview Park, Elmsford, New York 10523, U.S.A.
CANADA	Pergamon Press Canada Ltd., Suite 104, 150 Consumers Rd., Willowdale, Ontario M2J 1P9, Canada
AUSTRALIA	Pergamon Press (Aust.) Pty. Ltd., P.O. Box 544, Potts Point, N.S.W. 2011, Australia
FRANCE	Pergamon Press SARL, 24 rue des Ecoles, 75240 Paris, Cedex 05, France
FEDERAL REPUBLIC OF GERMANY	Pergamon Press GmbH, Hammerweg 6, D-6242 Kronberg–Taunus, Federal Republic of Germany

First edition 1983

British Library Cataloguing in Publication Data

Natural products for innovative pest management.
—(Current themes in tropical science; v. 2)
1. Pest control
I. Whitehead, David L.   II. Bowers, William S.
III. Series
632'.9   SB950

Library of Congress Cataloging in Publication Data

Main entry under title:

Natural products for innovative pest management.

(Current themes in tropical science; v.2)
Includes index.
1. Insecticides—Congresses.   2. Pesticides—
Congresses.   3. Plants, Insecticidal—Congresses.
4. Plant products—Congresses.   5. Natural products—
Congresses.   6. Pesticides industry—Congresses.
I. Whitehead, David L.   II. Bowers, William S.   III. Series.
SB951.N37 1982   623'.951   82-3826
ISBN 0-08-028893-6   AACR2

Printed in Great Britain by A. Wheaton & Co. Ltd., Exeter

Foreword

THE publications under the series CURRENT THEMES IN TROPICAL SCIENCE are meant to present discursive contributions and review articles on the progress and problems associated with the various domains of tropical insect science, tropical medicine, tropical agriculture and agroforestry, tropical soil science, tropical weather, tropical environment and energy issues, tropical natural products, and similar major problem-areas of tropical science and their respective impact and interaction with human welfare in tahe special circumstances of the tropics. The first volume dealth with *The Physiology of Ticks*; this second volume deals with *Natural Products for Innovative Pest Management*.

The papers in the present volume were originally presented in two scientific meetings convened by the International Centre of Insect Physiology and Ecology (ICIPE), in Nairobi, Kenya. The first meeting, "The International Technical Workshop on Appropriate Industrial Technology for the Control of Tropical Insect Pests and Disease Vectors", was co-sponsored with the United Nations Industrial Development Organization (UNIDO) from 29th July to 5th August 1979. The second, "The Scientific Working Group on the Use of Naturally Occurring Plant Products in Pest and Disease Control", was co-sponsored with the United Nations Development Programme (UNDP) from 11th to 16th May 1980. The papers have, however, seen some modification by the discussions held following each presentation, some of which have been summarized after each paper, and by subsequent additions and revisions.

The publication is the result of a concourse of insect scientists, plant biologists, and industrial scientists — both from developing and developed countries — exchanging experiences in the field of research and development, patenting of scientific discoveries and licensing of technological innovations, with the purpose of seeing how most effectively we can utilize our rich and diverse natural products in combating the equally rich and pestiferous insect fauna in the tropics. In support of this goal, the current status of many important plant pests and human and animal disease vectors was discussed along with the successes and deficiencies of the present control methods. For their part, the industrial chemists expressed their keen interest in working with scientists in the tropical countries towards the discovery and development of new tactics for insect control using knowledge stemming from natural products chemistry.

v

It became clear in these deliberations that institutions such as the ICIPE will be called upon to play an ever increasing role in both research and the development of new approaches to insect pest and vector management. The ICIPE has developed into an outstanding research institution; and now accepts the responsibility of approaching the problems of converting the fruits of its basic research into effective commercial tools. It is in this context that the ICIPE serves as an important link with programmes of industrial development. In the course of these discussions, special problems of the needs of developing tropical countraies for *in situ* industrial solutions were brought into sharp perspective; and it was emphasized that the resident industrial enterprises will eventually be expected to assume much of tahe burden for the close logistical support of farmers and insect pest managers, rather than the continued dependence upon distant foreign industrial expertise.

The frank and productive context within which the scientific meetings were held augurs well for the scientific-industrial partnerships that are expected to emerge. It is our hope that the important themes that were discussed are well represented and emerge forcefully in the published version.

On behalf of the ICIPE and the Scientific Editorial Services, we wish to thank most sincerely Dr. Abd-El Rahman Khane, Executive Director of UNIDO, and Mr. William T. Mashler, Senior Director of the Division of Global and Interregional Projects of UNDP, for the deep interest they took in the two scientific meetings which led to the present volume; to Professor W. S. Bowers, of Cornell University, and Professor Carl Djerassi, of Stanford University, for so productively chairing the two meetings, respectively; and to the technical editorial staff, especially Mrs. Serah W. Mwanycky and Miss Dorcas T. Adhiambo, for the enormous work they undertook to bring tahe present publication to its present shape under the direction of the Guest Editors, Professor Bowers and Dr. David L. Whitehead. Finally, we are grateful to Pergamon Press for their continuing interest in this publication series.

Nairobi   THOMAS R. ODHIAMBO
20th January 1982   *Chief Editor*

Preface

IN LARGE areas of the world, especially in tropical climes, man has been unable to develop a stable system of agriculture for the expanding population. This is true in parat, because of the numerous devilitating tropical diseases which afflict man and his domestic animals, and because of the fierce competition of arthropods for his food and fibre.

Until recently the control of agricultural pests and disease vectors depended upon the use of unspecific toxicants. The dramatic reduction in vector-borne diseases and the over-production of food in the sixties seemed to herald a new golden age. However, these advances were not without cost because the population explosion which followed, soon demanded even greater surpluses and the virtue of those pesticides were negated when their potentially carcinogenic, teratologic and mutagenic residues were recognized and found to accumulate in the biosphere.

The theme which recurs throughout this cook tells the tale of new endeavours and of the hopes of scientists, who now look to Nature as an ally and resource in finding new strategies to combat pestiferous arthropods and the diseases some of them carry. Man's practice of monocultural cropping allows intense competition by arthropods which in turn demands positive action to alleviate the depredations they cause. Intervention by man in applying insecticides, herbicides and fungicides, as well as through the development of resistant cultivars will, alas, always be necessary. However, the nature of the chemicals employed in the future must be more selective to the target pest and also environmentally safe. In this respect, plants serve as a vast reservoir of naturally biodegradable chemicals, many of which may have evolved in the defence to predators. One of tahe aims of this book is to encourage the investigation of natural products especially in plants of the tropical world.

The fundamental bases of resistance to attack by pests and vectors of plant diseases is the subject of Part I, Section C, while Section D deals with specific biochemicals now recognized to deter phytophagous insects. The first two sections of Part I look to the future with the eye of a chemist in the belief that certain phytochemicals, or other natural products, may offer a positive lead forward in the design of pesticides once their structure and mode of action at the molecular level has been grasped by the biochemists and toxicologists. Judicious use of this knowledge and how to apply it wisely in the modern idiom, motivates the writers of the chapters in Part II of this book. The technical problems man has to overcome to make the dreams of the biologist and natural products chemist come true occupies Part III of this book.

Knowledge gained at the laboratory bench is not easily translated into successful field trials. Chemists and biologists are frequently frustrated after the first flush of discovery of a novel biologically active compound. Three chapters deal with the commercial and legal problems which do arise. Long delays in commercialization of new chemicals are caused by the caution of regulatory agencies appointed to protect the environment from pollution. Fortunately, most natural products are inherently biodegradable. Only their analogues, which chemists synthesize in order to optimize and stabilize the biological activity of a new natural product, need special scrutiny to avoid undesirable side effects upon the erstwhile consumer or his environment.

The representatives of the commercial companies, who attended the two meetings* out of which this book arises, were more than willing to work with scientists in the developing countries towards implementing better, safer methods of pest control. It was agreed also that industry could help by participating in the discovery and development of control methods which utilized hitherto novel natural products. Support of basic research in the developing countries of the Third World by the large chemical companies would accelerate the discovery of new phytochemicals with potential for the control of pests and disease vectors. Special problems in the experience of developing countries who need to manufacture their own pesticides were also highlighted.

Industrial concerns based in the developing countries must assume much of the burden of close logistical support of farmers and insect pest managers. The task could not be left solely to the Government Extension Officers, nor to the vested interests of industraies based in the developed world. The two conferences* generated a special sensitivity and understanding of the need for close co-operation between research institutions and industraies of the Third World countries in support of their economic and social development.

We are grateful to all the participants who helped create this book — the full and frank discussions (recorded faithfully by Drs. Dismas A. Otieno and Phillip G. McDowell), amplified many-fold the merit of the presented papers.

We wish to record our appreciation to the conference organizers: Dr. Luka O. Abe, Mr. John M. Mramba, Mr. Joel M. Ojal, Mrs. Rhoda A. Odingo, Mrs. Remedios P. Ortega, Dr. Alex S. Tahori, Miss Rose Washika and to Miss Dorcas T. Adhiambo, who diligently typed the manuscripts.

*"International Technical Workshop on Appropriate Industrial Technology for the Control of Tropical Pests and Disease Vectors", 29th July to 5th August 1979, and "Scientific Working Group on the Use of Naturally Occurring Plant Products in Pest and Disease Control", 12th to 15th May 1980, both held in Nairobi, Kenya.

Finally, our gratitude go to Professor Thomas R. Odhiambo, Director of the International Centre of Insect Physiology and Ecology (ICIPE), Dr. Abd-El Rahman Khane, Executive Director of the United Nations Industrial Development Organization (UNIDO), and to Mr. William T. Mashler, Senior Director, Division of Global and Interregional Projects, the United Nations Development Programme (UNDP), for their motivation and financial support of the ventures which gave rise to the present volume of the series "Current Themes in Tropical Science".

DAVID L. WHITEHEAD
and
WILLIAM S. BOWERS
Guest Editors

Duduville International Guest Centre, Nairobi
14th August 1981

Contents

List of Discussants xvii

PART I. Chemistry and Mode of Action of Natural Products Active Against Pests 1

 Section A. *Pesticides: Present Status and New Challenges* 3

Chapter 1. Present Insecticides and Approaches to Discovery of Environmentally Acceptable Chemicals for Pest Management (Julius J. Menn, Zoecon Corporation, Palo Alto, California) 5

Chapter 2. Innovation in Pesticide (Insecticide) Chemistry (Kurt Gubler, Ciba-Geigy Ltd., Basel, Switzerland) 33

Chapter 3. The Search for Fourth-generation Pesticides (William S. Bowers, Cornell University, Geneva, USA) 47

Chapter 4. New Paths to Novel Environmentally Acceptable Natural Products Chemicals for Pest Management (A. W. Johnson, Agricultural Research Council, Unit of Invertebrate Chemistry and Physiology, University of Sussex, Brighton, UK) 59

Chapter 5. The Search for New Agrochemical Leads from Natural Sources (R. J. Pryce, Shell Research Limited, Shell Biosciences Laboratory, Sittingbourne Research Centre, Kent, UK) 73

 Section B. *Natural Pyrethrin and Synthetic Pyrethroids* 91

Chapter 6. Natural Pyrethrin as an Insecticide: Problems of Chemical Activity, Industrialization and Use (D. A. Otieno, ICIPE, Nairobi, Kenya) 93

Chapter 7. Development of Synthetic Insecticides from Natural
            Products: Case History of Pyrethroids from Pyrethrins
            (John E. Casida, Pesticide Chemistry and Toxicology
            Laboratory, Department of Entomological Sciences,
            University of California, Berkeley, USA)   109

Chapter 8. Developments in the Chemistry and Action of
            Pyrethroids (Michael Elliott, Department of Insecticides
            and Fungicides, Rothamsted Experimental Station,
            Harpenden, UK)   127

Chapter 9. Microencapsulated Natural Pyrethrum: An Improved
            Insect Repellent (J. Simkin, Technical Division,
            Penwalt Corporation, King of Prussia, Pennysylvania,
            USA; R. Galun, Department of Zoology, Hebrew
            University, Jerusalem, Israel)   151

Section C. *Plant Host/Pest Relations: the Bases of Plant Resistance*   165

Chapter 10. Coevolution of Plants and Insects (Paul Feeny, Depart-
            ment of Entomology, Cornell University, Ithaca,
            N.Y., USA)   167

Chapter 11. The Role of Natural Products in Plant–Insects and
            Plant–Fungi Interaction (G. B. Marini Bettolo, Centro
            Chimica dei Recettori e delle Molecole biologicametre
            attive del CNR, Universita Cattolica del S. Cuore,
            Roma, Italy)   187

Chapter 12. Multiplicity of Insect Antifeedants in Plants (Tibor
            Jermy, Research Institute for Plant Protection,
            Budapest, Hungary)   223

Chapter 13. Antifeedants from *Clausena anisata* (Willd.) Hook f.
            ex Benth (Rutaceae) (T. Gebreyesus and A. Chapya,
            ICIPE, Nairobi, Kenya)   237

Chapter 14. Attractants, Arrestants, Feeding and Oviposition Stimu-
            lants in Insect–Plant Relationship: Application for Pest
            Control? (Erich Städler, Eidg, Forschungssanstalt,
            Wädenswil, Switzerland)   243

Chapter 15. Antifeedants in Crop Pest Management (E. A. Bernays,
            Centre for Overseas Pest Research, Wrights Lane,
            London, UK)   259

Chapter 16. Biochemical Composition of Bean Varieties Resistant and Susceptible to Mexican Bean Beetle (Coleoptera: Coccinellidae) (A. K. Raina and S. S. Lee, Virginia State University, Petersburg, Virginia, USA)   273

Chapter 17. Antifungal Compounds and Disease Resistance in Plants (J. A. Callow, Department of Plant Sciences, University of Leeds, Leeds, UK)   279

Chapter 18. Nematocidal Natural Products (Katsura Munakata, Department of Agricultural Chemistry, Nagoya University, Japan)   299

Section D. *Mode of Action of Phytochemicals Affecting Insects*   311

Chapter 19. Phytochemical Action on Insect Morphogenesis, Reproduction and Behaviour (William S. Bowers, Cornell University, Geneva, USA)   313

Chapter 20. The Mode of Action of Pro-allatocidins (G. E. Pratt, Agricultural Research Council, Unit of Invertebrate Chemistry and Physiology, University of Sussex, Brighton, UK)   323

Chapter 21. Effects of Precocenes on the Sorghum Shootfly, *Atherigona soccata* (Diptera: Muscidae) (G. C. Unnithan, ICIPE, Nairobi, Kenya)   357

Chapter 22. Structure, Chemistry and Actions of the Piperaceae Amides: New Insecticidal Constituents Isolated from the Pepper Plant (Masakazu Miyakado, Isamu Nakayama, Nobuo Ohno and Hirosuke Yoshioka, Pesticides Research Department, Institute for Biological Sciences, Sumitomo Chemical Co. Ltd., Takatazuka, Hyogo, Japan)   369

PART II. The Control of Tropical Pests and Vectors of Economic Importance   383

Chapter 23. Review of Vectors of Major Tropical Diseases and their Control (F. L. Lambrecht and A. Challier, ICIPE, Nairobi, Kenya)   385

Chapter 24. Agricultural Pests of Crucial Economic Importance in the Tropics and their Control (R. Kumar, Department of Zoology, University of Ghana, Legon, Ghana)   409

Chapter 25. Contribution of ICRISAT to Studies on Plant Resistance to Insect Attack (W. Reed, K. V. Seshu Reddy, S. S. Lateef, P. W. Amin and J. C. Davies, ICRISAT, Andhra Pradesh, India)   439

Chapter 26. Botanicals and their Derivatives in Vector Control (G. C. Labreque, FAO/IAEA, Vienna, Austria)   451

PART III. Requirements for Pesticide Production and Utilization   461

Chapter 27. Formulation and Application of Biologically Active Chemicals in Relation to Efficacy and Side Effects (I. J. Graham-Bryce, East Malling Research Station, Maidstone, UK)   463

Chapter 28. Research, Development and Commercialization of Triatomine Control Agents (Richard Pinchin, Nucleo de Pesquisas de Produtos Naturais, Centro de Ciencias da Saude, Universidade Federal do Rio de Janeiro, Brazil)   475

Chapter 29. The Production of Pesticides in Developing Countries, including Natural Products (United Nations Industrial Development Organization, Vienna, Austria)   485

Chapter 30. Industrial Production of Pesticides: A Note on the Cost/ Benefit Problem in Developing Countries (Y. F. O. Masakhalia, Ministry of Economic Planning and Community Affairs, Nairobi, Kenya)   507

Chapter 31. Development of Appropriate Technology (Equipment) for Utilization of Pesticides in the Tropical Environment (Rajni P. Patel, Faculty of Engineering, University of Nairobi, Kenya)   515

Chapter 32. Registration Requirements for Pesticides (James W. Young, Zoecon Corporation, Palo Alto, USA)   523

Chapter 33. Legal Protection of Industrial Property — Patents (Donald W. Erickson and Julius J. Menn, Zoecon Corporation, Palo Alto, California, USA; Kurt Gubler and Erich A. Horak, Ciba-Geigy Limited, Basle, Switzerland)   541

Chapter 34. Patent Laws of Kenya and their Procedure (Presented by Julius Menn, Zoecon Corporation, Palo Alto, California, USA)   549

PART IV. The Need for Trained Scientists and Technologists 555

Chapter 35. Training of Scientists and Technologists in the Management of Insect Pests and Vectors in Tropical Countries (Antony Youdeowei, Department of Agricultural Biology, University of Ibadan, Nigeria) 557

Index 577

List of Discussants other than Contributors

L. O. Abe
International Centre of Insect Physiology and Ecology (ICIPE)
P. O. Box 30772
Nairobi, Kenya

A. Arunga
Kenya Industrial Research and Development Institute
P. O. Box 30650
Nairobi, Kenya

C. Djerassi
Zoecon Corporation
California Avenue
Palo Alto, California
U. S. A.

Mr. Jacobson
United States Department of Agriculture
Beltoville Md.
U. S. A.

I. Jondiko
International Centre of Insect Physiology and Ecology (ICIPE)
P. O. Box 30772
Nairobi, Kenya

M. G. Jotwani
Indian Agricultural Research Institute
New Delhi
India

Dr. A. M. Kabaara
Coffee Research Foundation
Coffee Research Station
P. O. Box 4
Ruiru, Kenya

F. J. Kézdy
Department of Biochemistry
University of Chicago
Chicago, Illinois
U. S. A.

B. W. Macharia
Pyrethrum Board of Kenya
P. O. Box 420
Nakuru, Kenya

A. D. Monteiro
Assistance to Industrial Survey and Promotion Centre (A UNIDO Project)
P. O. Box 30440
Nairobi, Kenya

M. B. Musoke
National Research Council
Ministry of Planning and Economic Development
P. O. Box 6884
Kampala, Uganda

Y. Nayudama
Central Leather Research Institute
Adyar, Madras - 20
India

B. N. Odero
International Centre of Insect Physiology and Ecology (ICIPE)
P. O. Box 30772
Nairobi, Kenya

T. R. Odhiambo
Director
International Centre of Insect Physiology and Ecology (ICIPE)
P. O. Box 30772
Nairobi, Kenya

P. E. Onuorah
National Science and Technology Development Agency, P.M.B. 12695
Lagos, Nigeria

A. Onyango
National Council for Science and Technology
P. O. Box 30632
Nairobi, Kenya

R. C. Saxena
IRRI, Manila and International Centre of Insect Physiology and Ecology
    (ICIPE)
P. O. Box 30772
Nairobi, Kenya

K. Szabo
United Nations Industrial Development Organization (UNIDO)
P. O. Box 707
A - 1011, Vienna
Austria

A. S. Tahori
International Centre of Insect Physiology and Ecology (ICIPE)
P. O. Box 30772
Nairobi, Kenya

R. W. Taylor
Tropical Products Institute
Tropical Stored Products Centre
London Road, Slough
Berks SL3 7HL
U.K.

D. L. Whitehead
International Centre of Insect Physiology and Ecology (ICIPE)
P. O. Box 30772
Nairobi, Kenya

PART I

Chemistry and Mode of Action of Natural Products Active Against Pests

Section A

Pesticides: Present Status and New Challenges

CHAPTER 1

Present Insecticides and Approaches to Discovery of Environmentally Acceptable Chemicals for Pest Management

JULIUS J. MENN

Zoecon Corporation, 975 California Avenue, Palo Alto, California 94304, USA

CONTENTS

1.1	Introduction	5
1.2	Pesticide Overview	6
1.3	Agrochemical Industry	7
1.4	Insecticide Development	9
1.5	Approaches to Discovery of Biodegradable Insecticides	11
	1.5.1 Empirical synthesis and screening	11
	1.5.2 Natural products models	13
	1.5.3 Biochemically inspired modifications of known insecticides	17
	1.5.4 Biorational synthesis and screening	20
1.6	Conclusions	22
1.7	Acknowledgements	23
1.8	References	23
1.9	Discussion	26

1.1 INTRODUCTION

Generally insecticidal chemicals will continue to play a major role in insect control and management as an indispensable component in ensuring adequate food and fibre production. One of the key recommendations at the Rockefeller Foundation Conference on the Future of Insecticides in Bellagio, Italy, stated the need to provide means for discovery of new chemicals and development to the user stage for insect control through cooperation with world industries (Metcalf & McKelvey, 1976). This statement is amplified by a massive body of publications and statistical data which documents use patterns, need and future projections involving insect control agents which are still primarily synthetic organic chemicals (Adam, 1977; Braunholtz, 1977; Furtick, 1976; Goring, 1976; Tilak, 1977).

This paper considers the technological resources, strategies, economic and regulatory constraints which impinge on the difficult process of developing a marketable, cost effective, selective, safe, and biodegradable insecticide.

1.2   PESTICIDE OVERVIEW

Most pesticide chemicals are produced in the industrialized countries. Although the rate of growth of pesticide consumption is currently higher in developing countries, the absolute amount in the latter is significantly smaller (Furtick, 1976).

In 1970 the USA consumed 45% of all pesticide chemicals; Western Europe, 23%; Eastern Europe, 13%; Japan, 8%; and the developing countries, 7% (Furtick, 1976). These figures are only approximations as they do not reflect accurate figures for the Soviet Union and People's Republic of China.

Since time immemorial man has used various botanical, mechanical, and mineral means to combat insect pests to protect his food supply. For a brief chronology of forerunners of the synthetic organic insecticides the reader is directed to Boyce (1976). The era of modern insect control with synthetic organic compounds was ushered in with the commercial introduction of the potassium salt of dinitro-ortho-cresol in Germany in 1892 (Haller, 1972). In the early 1930s several organic thiocynanates were marketed in the USA as livestock and household insecticides. However, it was not until Paul Müller, of Geigy in Switzerland, discovered, in 1939, the insecticidal properties of DDT* that the era of modern synthetic organic insecticide chemicals was introduced in full force. Indeed, the science of modern chemical control of insects is a relatively new technological field which had its major inception in synthetic discoveries made primarily after World War II. However, it is of interest to note that as early as the 1930s pesticide chemicals were synthesized with structures approximating a number of compounds in current use. In 1938 scientists at the Geigy Company discovered Mitin FF (Fig. 1.1), a mothproofing agent which is toxic to keratin-eating insects (Spindler, 1978).

In recent years, production and sales of pesticide chemicals have become a multibillion dollar enterprise. Published data (*Chemical Engineering News*, 1977) reported that $7 billion worth of pesticide chemicals were sold

---

* For chemical names, structures and other designations for compounds mentioned in this article, refer to: E. E. Kenaga and R. W. Morgan (1978), *Commerical and Experimental Organic Insecticides* (1978 Rev.), ESA Spec. Public. 1–78, 79pp.

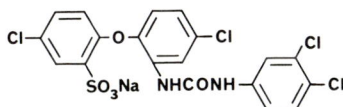

FIG. 1.1 Mothproofing chemical, Mitin FF.

worldwide in 1976 and of these, insecticides accounted for approximately 36% of the total ($2.5 billion).

An examination of the production data for the USA in Table 1.1 (Fowler & Mahan, 1978) indicate that organophosphorus ester insecticides (OPs), including methyl parathion, are currently the predominant class of insecticide chemicals in use. In the chlorinated hydrocarbon insecticide group, toxaphene is the predominant compound, as it still is the major insecticide used for cotton pests in the USA and potato insects in the USSR (M. S. Sokolov, personal communication). The third major group of insecticides in commercial use are the methyl carbamate insecticides which are limited to a small group of insecticides in the USA, notably: aldicarb, bufencarb, carbaryl, carbofuran and methomyl.

Ayers, Ernest & Johnson (1977) projected a slight decrease (1.9%) in the average annual growth rate for insecticde consumption in the USA by the year 1980. This will probably be the result of a combination of factors:

(a) utilization of more efficient insecticides (OPs, pyrethroids and others);
(b) impact of government regulations; and
(c) greater emphasis on utilization of insecticides and other chemical agents such as pheromones which are compatible with integrated pest management programs (IPM).

1.3 AGROCHEMICAL INDUSTRY

The agrochemical industry in the industrialized countries, excluding the Soviet Union and the People's Republic of China, is the major producer of pesticide chemicals. According to Braunholtz (1977), there are currently approximately 100 manufacturers engaged in agrochemicals.

The leading world firms according to dollar volume in agrochemicals are listed in Table 1.2. These twenty companies account for $7.455 billion sales worldwide (Woodburn & Cook, 1979). Further examination of these figures shows that the agrochemical revenues of the European companies exceed those for the USA firms. This is primarily due to greater international marketing networks developed by the European companies, while the American firms have traditionally enjoyed preeminence in the American market.

Julius J. Menn

TABLE 1.1. SYNTHETIC ORGANIC PESTICIDAL CHEMICALS: PRODUCTION OF INSECTICIDES AND FUMIGANTS, UNITED STATES, 1972-77

	1972	1973	1974 (in 1000 pounds)	1975	1976	1977
Aldrin-Toxaphene group[c]	141,858 [a]	145,584 [a]	141,719	[a]	[a]	[a]
Chloropicrin		[a]	4757	5698	6423 [a]	5803
Methoxychlor			3248	5504		
Methyl bromide[d]	24,633 [a]	29,571 [a]	30,452 [a]	36,048 [a]	35,856 [a]	34,684 [a]
Organophosphorus insecticides					189,879	
Methyl parathion	51,076 [a]	48,890 [a]	51.448 [a]	53,668 [a]	[a]	39,695 [a]
Parathion				[a]	[a]	[a]
Other	109,566 [b]	123,715 [b]	135,139 [b]	156,127	[a]	164,350
Toxaphene				59,336	42,164	39,780
All other	236,442	291,409	283,446	349,345	291,762	285,722
TOTAL	563,575	639,169	650,209	665,726	566,084	570,034

[a] Withheld to avoid disclosure. Figures included in "All other".
[b] Separate figure not available.
[c] Includes Aldrin, Chlordane, Dieldrin, Endrin, Heptachlor, Strobane, and Toxaphene.
[d] Fumigant for control of both insects and weeds.
Adapted from: Fowler and Mahan (1978).

TABLE 1.2. AGROCHEMICAL SALES OF WORLD FIRMS, 1977[a]

Company	Agrochemicals ($ millions)	Agrochemicals as % total sales
Bayer	1,300	12.9
Ciba-Geigy	975	19.8
Shell	600	13.1
Monsanto	525	11.4
Rhone-Poulenc	400	7.8
ICI	390	4.4
BASF	365	3.6
Stauffer	325	26.4
Du Pont	320	3.4
American Cyanamid	290	12.0
Dow	280	4.5
Union Carbide	270	3.8
Eli Lilly	260	17.1
Kumiai	194	98.0
FMC	192	8.3
Schering	180	17.9
Hoechst	175	1.6
Rohm & Haas	160	14.3
Diamond Shamrock	130	7.9
Sandoz	124	5.3
TOTAL	7.455	

[a] Adapted from: Woodburn & Cook (1978).

1.4   INSECTICIDE DEVELOPMENT

Much has been written on the logistics and strategy involved in the discovery of new insecticide chemicals (Braunholtz, 1977; Corbett, 1976; Djerassi, Shih-Coleman & Diekman 1974; Goring, 1976; Menn, 1972). In reading these discourses one is invariably struck by the complexity, risk, time and financial constraints which impinge so heavily on the rare discovery which may eventually lead to marketing of a few insecticide chemicals having useful environmental fit and toxicological properties.

The criteria for discovery and successful development of an insecticide are extremely complex as shown in Fig. 1.2 (Braunholtz, 1977). They involve interaction of several key disciplines and functions, which include: synthetic chemistry, biology, agronomy, biochemistry, toxicology, pharmacology, process research, formulations, compliance with regulatory agencies, manufacture and marketing. All these activities are interactive, interrelated, many are interdependent and all, if followed on a critical path, involve time and money constraints as shown by Djerassi *et al.* (1974), who developed critical path maps for insect control agents where time, functions, regulatory constraints and costs are discussed in context of an orderly and sequential development program.

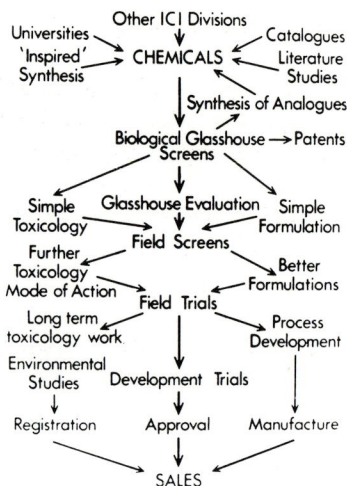

Fɪɢ. 1.2 Flow chart describing discovery and development of a pesticide (from: Braunholtz, 1977).

Traditionally, large resources have been required in the process of discovering and developing successful commercial pesticides. Table 1.3 shows

TABLE 1.3 NUMBER OF COMPOUNDS PASSING THROUGH EACH R&D STAGE PER COMMERCIAL PRODUCT[a]

Activity	1956	1964	1967	1969	1970	1972
Synthesis and initial bio-screen	1800	3600	5500	5040	8000	10,000
Advanced screening	60	36	NA	126	80	NA
Field evaluation	6	4	NA	9	4	NA
Development	2	2	NA	2	2	NA
Sales	1	1	1	1	1	1

[a] From: Braunholtz (1977).

the five stages for selection and identification of one successful candidate based on the success rate of ICI (Braunholtz, 1977). It also illustrates the dramatic rise in numbers of compounds required for screening per one successful commercial product. Undoubtedly in 1979 the industry average is significantly greater than 1/10,000, the figure for 1972. According to the 1978 Ernst and Ernst survey of Industry (Ernst & Ernst, 1978) 92,000 compounds were screened in 1977. In that year only three new products were registered, thus the success ratio is decreasing dramatically. Costs and elapsed time to registration have also risen sharply, averaging $21,600,000 for seven products registered in 1976 and 1977, with an average elapsed time which is currently estimated at 100 months.

While the foregoing appears discouraging, we must bear in mind that there are still major needs for improved, selective and biodegradable new insecticides. It is believed that rational approaches derived from guided empirical synthesis, analog synthesis, natural products, mode of action and biochemistry are vital components in the overall strategy to discover commercially viable, effective and biodegradable pesticides.

1.5   APPROACHES TO DISCOVERY OF BIODEGRADABLE INSECTICIDES

1.5.1   Empirical synthesis and screening

Possibly with the exception of insect growth regulators (IGRs) and a few isolated compounds, most of the insecticides in current use were discovered via conventional, empirical, or guided empirical synthesis, coupled with broad spectrum short-term exposure, screening tests. In this respect, this field approximates the *modus operandi* which led to the discovery of most human drugs (Burger, 1978).

The conventional strategy for pesticides synthesis has been largely based on identification of an active lead followed by analog synthesis programs based on structure activity relations (SAR). This approach has been greatly aided in recent years by the development of computer-based analyses of SAR (Hansch & Fujita, 1964; Topliss, 1972; Wipke, 1974).

Most of the insecticides represented by organophosphorus esters (OPs), *N*-methyl carbamates, the chlorinated hydrocarbons, synthetic pyrethroids, formamidines and benzolphenylureas were derived largely from analog synthesis and SAR optimization. Table 1.4 presents a partial list of synthetic chemical insecticides which were discovered in the past 40 years. A survey of this table quickly reveals that most insecticides discovered since DDT had their origin in empirical and/or guided synthesis. Diflubenzuron represents possibly one of the most important and unexpected recent discoveries involving a new class of selective insecticides (Verloop & Ferrell, 1977). Diflubenzuron and other benzoylphenylureas were synthesized as potential herbicides based on the chemistry of dichlobenil and diuron herbicides, but unexpectedly they showed unique activity in insects. One compound in this class, Du-19111, is a direct derivative of these two herbicides by deletion of two methyl groups from diuron and substituting the bezoyl nitrile nitrogen with a benzoyl carboxy moiety (Fig. 1.3). The resultant compound had no herbicidal activity, but proved to be a larvistatic agent, which caused indirect death of

Julius J. Menn

TABLE 1.4. INSECTICIDES DISCOVERED BY GUIDED EMPIRICISM[a]

1940s	Chlorinated hydrocarbons: DDT, BHC, Aldrin, Chlordane, Toxaphene	Selected, guided synthesis
	OPs: Parathion, Methyl Parathion	Guided synthesis
	Carbamates: Dimetan, Dimetilan, Isolan	Insect-repellent synthesis
1950s	Carbaryl, Malathion, Azinphosmethyl, Phorate	Guided synthesis
	Vinyl phosphates	Selected, guided synthesis
1960s	Fonofos, Trichloronate, Mexacarbate, Bufencarb, Carbofuran, Aldicarb, Methomyl	Guided synthesis
	Synthetic pyrethroids: Resmethrin	From natural product prototype
1970s	Synthetic pyrethroids (2nd generation): Cypermethrin, Permethrin, Decamethrin, Fenvalerate	Guided synthesis
	New OPs: Terbufos, Methamidophos, Acephate, Sulprofos, Profenfos	Guided synthesis
	New carbamates: Bendiocarb, Thiofanox	Guided synthesis
	Benzoylphenyl ureas: Diflubenzuron	Herbicide synthesis

[a] Adapted in part from E. Bergman (Pers. comm., 1979) and J. J. Menn. *Jr. Agric. Food Chem.* (1980).

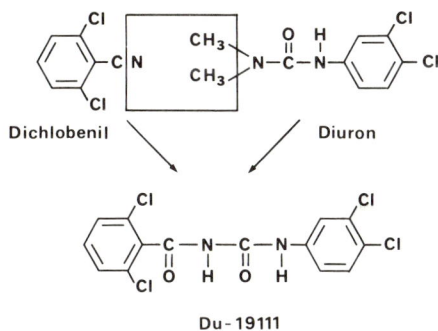

Du-19111

FIG. 1.3. Structures of Dichlobenil, Diuron and Du-19111 (from: Menn, 1980).

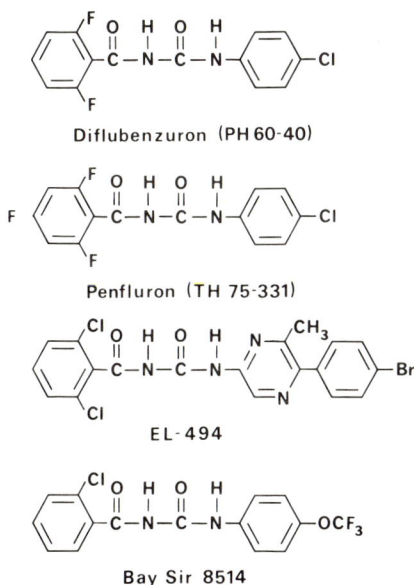

Diflubenzuron (PH 60-40)

Penfluron (TH 75-331)

EL-494

Bay Sir 8514

FIG. 1.4. Structures of promising molt-disrupting compounds (from: Menn, 1980).

Pieris brassicae larvae by interfering with chitin synthesis and deposition (Verloop & Ferrell, 1977). The discovery of the growth regulant activity of Du-19111 was largely due to excellent biological screening; holding the treated insects for a longer period than usually done in conventional insecticidal screens and careful morphological examination of the treated larvae. This finding prompted the synthesis of many analogs in a number of industrial laboratories. Several of the most promising compounds to emerge in this class of insectostatic agents are shown in Fig. 1.4.

Improved bioassay methods and increased diversity in chemical synthesis have led to the discovery, in a guided synthesis program, of a new class of pyrazoline insecticides (Mulder, Wellinga & Van Daalen, 1975). Although PH 60-41 (Fig. 1.5) bears some chemical similarity to the benzoylphenylureas

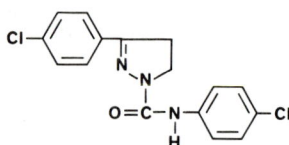

3-(4-chlorophenyl)-1-(4-chlorophenyl-
carbamoyl)-2-pyrazoline

FIG. 1.5 Chemical structure and name of PH 60-41.

and may have been synthesized for herbicidal properties, it demonstrated unique activity, especially on lepidopterous and coleopterous larvae and adults of the latter (Mulder *et al.*, 1975). Intoxicated insects showed severe uncoordinated movement, knockdown, stoppage of feeding and convulsions. Conceivably, compounds of this type have central nervous system and stomach poison effects.

Another serendipitous discovery involved a class of *N*-aryl formamidines which were synthesized initially as herbicides (Hollingworth, 1976). However, they have broad range insecticidal, ovicidal and acaricidal activity. Especially noteworthy are behavioral effects at sublethal doses, thus forestalling high selection pressure on the pest populations, reducing environmental load, and ameliorating undesirable effects on beneficial insect populations (Lund, Hollingworth & Murdock, 1979). The best known member in this series is the insecticide and acaricide chlordimeform (CDM) (Fig. 1.6).

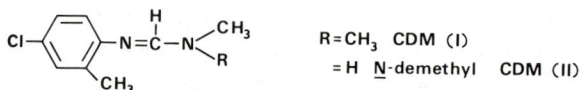

R=CH₃ CDM (I)

=H N-demethyl CDM (II)

FIG. 1.6 Structure of insecticide-acaricide chlordimeform (*N'*-(4-chloro-*o*-tolyl) *N,N*-dimethyl formamidine) (CDM, I) and its *N*-dimethyl derivative (II).

These are but a few selected examples of important insecticides which were discovered in recent years, via the guided synthesis route which followed an empirical screening lead.

1.5.2 Natural products models

Natural products of plant, animal, or microbial origin are a vast source of bioactive substances which have been exploited only to a limited extent as models in synthesis of practical insecticides, acaricides and phagorepellents.

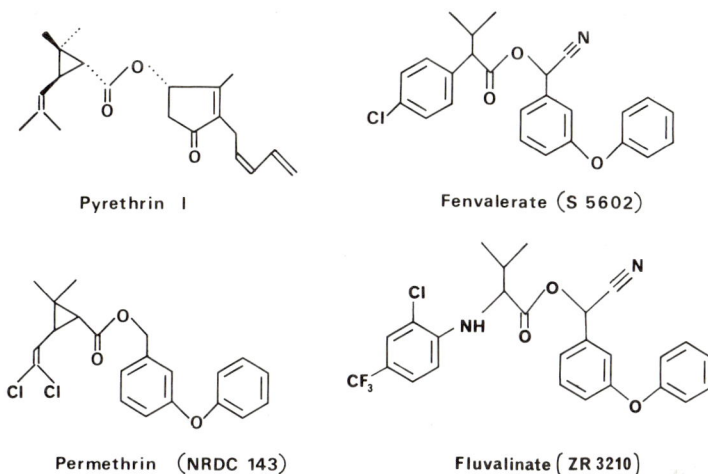

Pyrethrin I

Fenvalerate (S 5602)

Permethrin (NRDC 143)

Fluvalinate (ZR 3210)

Fig. 1.7 Chemical evolution of pyrethroids.

The advent of powerful analytical tools such as ms, nmr, hplc and related techniques have facilitated identification, at an increasing pace, of important bioactive principles isolated from natural product extracts. A number of these model compounds have been described recently by Casida (1976).

Undoubtedly, the most important and significant application of a natural model from botanical origin centres on the insecticidal properties of the extract from pyrethrum flowers, *Chrysanthemum cinerariaefolium*, exploited in Europe by the nineteenth century (Gnadinger, 1936). According to Elliot (1977), the chemical evolution of synthetic pyrethroids can be traced to the prototype, Pyrethrin I (Fig. 1.7), a highly effective insectide, selective toward mammalian species but unstable to sunlight and culminating with decamethrin in optimizing bioactivity and with fenvalerate in the extreme divergence of chemical structure of a synthetic pyrethroid (Fig. 1.7). The chemical evolution of synthetic pyrethroids is elegantly traced and discussed in a recent review by Elliot & Janes (1978). The unique structural features of Pyrethrin I and its related esters offered great opportunity for synthetic variations and modifications, which resulted in the development of a number of highly active synthetic pyrethroids (Fig. 1.7). Several of these have been recently introduced and shown to be extremely potent insecticides (Elliot, 1977; Elliot & Janes, 1978). The relative potency of several selected synthetic pyrethroids, as compared to conventional insecticides in our laboratory on three economic species is shown in Table 1.5.

In contrast to Pyrethrin I and the ensuing synthetic pyrethroids, certain bioactive natural products are too complicated to be synthesized *de novo* or used as prototypes in structural optimization synthetic programs. A pertinent example of the latter involves the remarkable insect phagorepellent properties

TABLE 1.5. INSECTICIDAL ACTIVITY OF TWO SYNTHETIC
PYRETHROIDS AS COMPARED WITH CONVENTIONAL INSECTICIDES[a]

Compound	*Heliothis virescens* L₃, Topical	*Tetranychus urticae* Adult, Spray	*Hypera postica* Adult, Topical	
Permethrin	$0.018 \ \mu g \ larva^{-1}$	15.5 ppm	$0.042 \ \mu g \ adult^{-1}$	
Fenvalerate	0.012	8.4	0.140	
Methyl parathion	0.490	30.0		0.090
Carbaryl	0.580	>1000	66.000	

[a] Data courtesy of G.B. Staal, Zoecon Corporation, Palo Alto, California.

of the seed extract of the China berry tree, *Melia azadirachta* Linn. Crude extracts of parts of the neem tree had been used as an insect antifeedant in India (Tilak, 1977). Gill and Lewis (1971) demonstrated that azadirachtin isolated from the Lilac neem tree was apoplastically transported in plants, thus offering systemic phagorepellency against the desert locust, *Schistocerca gregaria*, Forsch. Subsequently, Zanno, Miura, Nakanishi & Elder (1975) identified the complex structure of azadirachtin from the leaf extract of the Indian Lilac neem tree *Azadirachta indica* (Fig. 1.8).

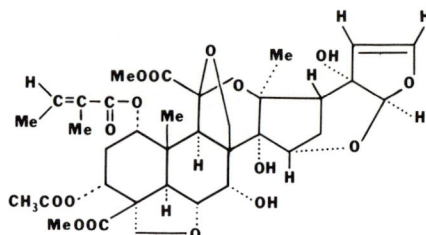

FIG. 1.8 Structure of the insect phagorepellent azadirachtin (from: Zanno *et al.*, 1975).

In the context of this presentation several other natural product models are discussed briefly. Although physostigmine was reputed to be the model for synthesis of stable *N*-methyl carbamate insecticides (Metcalf, 1971), carbamate insecticides in current use such as carbaryl, aldicarb and carbofuran were likely conceived and synthesized empirically (Fukuto, 1976). The alkaloid nicotine was developed as a potent insecticide, but due to its high mammalian toxicity and environmental lability, its development was dropped with the advent of the newer synthetic insecticides (Schmeltz, 1971). Although extensive analog synthesis programs ensued using the nicotine prototype, no useful insecticides emerged. Recently Soloway, Henry, Kollmeyer, Padgett, Powell, Roman, Tieman, Corey & Horne (1979) reported on the synthesis of high active lepidopterous insecticides from nitro-ketenedimethyl mercaptol and substituted diamines. Undoubtedly they were inspired by the nicotine model.

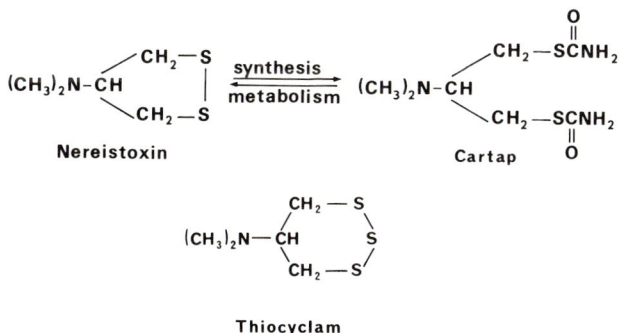

FIG. 1.9 Nereistoxin and its synthetic bisthiocarbamate derivative, cartap.

Nereistoxin, isolated from the marine annelid *Lumbriconereis heteropoda*, proved to be a successful prototype for the synthesis of the novel bis-thiocarbamate rice insecticide, cartap, shown in Fig. 1.9 (Konishi, 1972; Sakai & Sato, 1972). Cartap likely acts as a reversible synaptic blocking agent, possibly in analogous manner to nicotine. It is likely that *in vivo* cartap is biotransformed to nereistoxin or dihydronereistoxin (Casida, 1976).

It is evident that to date, only a very minute proportion of naturally occurring plant or animal derived toxicants have been studied as possible models for synthesis or practical insecticides. The successful advent of synthetic pyrethroids and cartap strongly suggests the untapped modelling potential that exists in nature and awaits further scientific exploitation.

1.5.3 Biochemically inspired modifications of known insecticides

A productive area in designing highly active and useful insecticides has been founded on knowledge gained from metabolic and other related biochemical studies in living organisms and with isolated enzyme systems (Casida, 1979; Corbett, 1974; Hollingworth, 1975). Undoubtedly, this is an area that will further develop as it promises to continue to yield useful insecticides whose structures may not be readily anticipated by the synthetic chemist.

One of the early examples of biochemical activation resulting in the ultimate development of an improved systemic insecticide evolved from the classical studies of Metcalf, Fukuto & March (1957), who demonstrated that thioether oxidation of disulfoton in plants led to improved systemic insecticidal and acaricidal activity. This discovery led to the commercial development of oxydemetonmethyl, a widely used citrus insecticide.

The cyclic OP insecticide salithion was discovered in the course of metabolism studies with the potent neurotoxicant, tri-*o*-tolyl phosphate (TOCP) (Eto, 1974). In the presence of microsomal enzymes (mfo), TOCP is

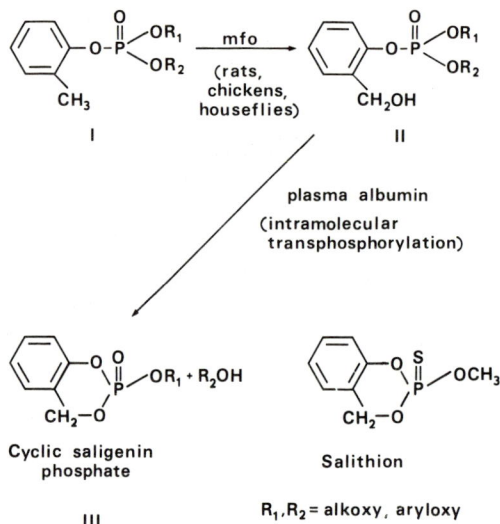

Fig. 1.10 Metabolic rearrangement of trio-*o*-tolyphosphate to cyclic saligenin phosphate.

converted to metabolites I and II (Fig. 1.10). The latter undergoes intra-molecular transphosphorylation, catalyzed by plasma albumin, to form a potent, non-neurotoxic cholinesterase inhibitor, a cyclic phosphate ester of saligenin (III). The thiono analog of compound III was subsequently synthesized and developed as a commercial insecticide in Japan.

While most bioactivation reactions of OP insecticides are oxidative, reaction with endogenous thio groups, such as with glutathion (GSH), may yield active products as was shown with the debromination of the insecticide naled and its conversion to dichlorvos (Fig. 1.11), the putative toxicant (Casida, 1972).

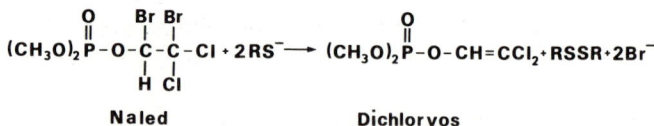

Fig. 1.11 Reaction of naled with thiol groups and conversion to Dichlorvos (from: Eto, 1974).

Organophosphorus ester insecticides containing *N*-alkylamide groups undergo oxidative *N*-desalkylation reactions mediated by oxidative enzymes (mfo). Menzer & Casida (1965) have demonstrated that the insecticide dicrotophos undergoes *N*-desalkylation in animals and plants giving rise to a methylol intermediate and subsequently to the monomethyl analog, the potent insecticide monocroptophos, which is extensively used in cotton culture.

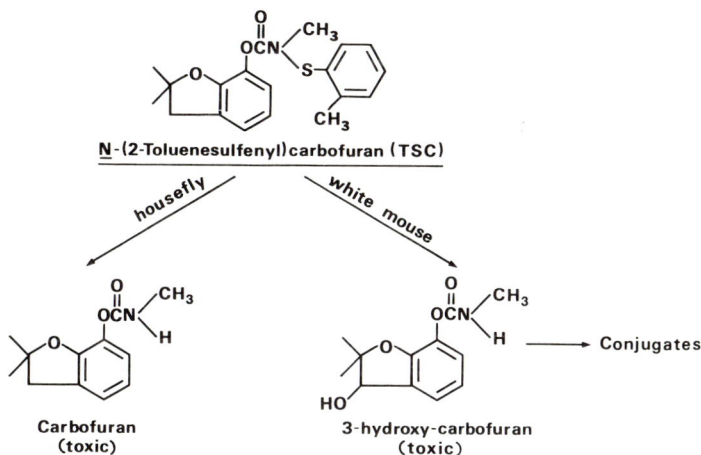

Fig. 1.12 Major differential metabolic steps responsible for selective toxicity of TSC to mammals and insects (adapted from: Fukuto, 1976).

From these studies and others and as a result of added knowledge concerning metabolic pathways of pesticides and their selective toxicity, further useful insights have emerged for biochemically inspired synthesis. Fukuto and his coworkers have effectively explored this concept and successfully applied it through the discovery of several series of selective carbamate insecticides (Fukuto, 1976). These workers turned their attention to derivatizing highly toxic, potent *N*-methyl carbamates and rendering them safer, but yet effective, insecticides. Safety is conferred on these new carbamates by exploiting differential metabolic pathways in insects and mammals brought about by chemical derivatization of the parent toxicant. In the course of these studies the *N*-arylsulfenyl derivative of carbofuran (TSC, Fig. 1.12) retained its insecticidal properties towards the housefly, but advantageously, displayed reduced mammalian toxicity in comparison to carbofuran. Substitution of the nitrogen proton with an arylsulfenyl moiety reduced toxicity to the white mouse 50-fold (Fukuto, 1976; Hollingworth, 1975).

As shown in Fig. 1.12, the major metabolic product of TSC in the housefly is carbofuran, while in the mouse *in vivo* the arylsulfenyl moiety acted as an "opportunity factor", allowing formation of the still toxic 3-hydroxy-carbofuran derivative which, however, was rapidly conjugated and excreted.

Several other instances of discovery of new pesticide prototypes through biochemical studies have been reported recently in the literature. One of the best examples comes from the herbicide field. Sulfoxidation of the *S*-alkyl moiety of thiocarbamate herbicides was considered to be an unlikely chemical reaction, due to instability of the resultant sulfoxide. However, Hubbell and Casida (1977) have demonstrated the formation of thiocarbamate sulfoxides

from corresponding sulfides in rats and corn. Microsomal enzymes catalyze
this oxidation. Casida, Gray & Tilles (1974) have shown that these
thiocarbamate sulfoxides are potent, selective herbicides, and conceivably the
actual herbicidal species generated *in situ* in this class of compounds.

1.5.4   Biorational synthesis and screening

Probably no single avenue to discovery of novel insecticides or insectostatic
agents has elicited as much scientific interest and intellectual appeal than the
so-called biorational approach. Various investigators have coined their own
definitions for what is meant by this approach, and no single definition
appears to be all encompassing. Possibly the brief definition proposed by C.
Henrick (personal communication) is most appropriate in the present context:
"Chemical follow-up to a biological or biochemical breakthrough." Perhaps
the difficulty lies in the fact that very few examples can be cited as fulfilling the
criteria of a biorational discovery. Possibly, the most cogent example revolves
on the discovery of insect growth regulators (IGRs), which mimic the action of
insect juvenile hormones (Menn & Pallos, 1975; Siddall, 1976; Siddall, 1977;
Staal, 1975).

The chemical follow-up on the biological discovery (Roller, Dahm, Sweeley
& Trost, 1967), coupled with innovative biological screening procedures, led
to the discovery of a potent class of farnesane-derived IGRs, most notably
methoprene (Altosid) (Henrick, Anderson, Staal & Ludvik, 1978; Henrick,
Staal & Siddall, 1973). The chemical analogy between methoprene and
juvenile hormone-I (JH-I) and their relative insecticidal potency on *Aedes
aegypti* larvae is shown in Fig. 1.13. We note the remarkable improvement in

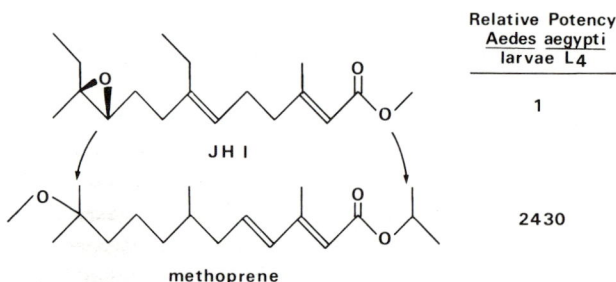

Fɪɢ. 1.13 Chemical structures of JH-I and methoprene and their relative potency versus *Aedes
aegypti* fourth instar larvae (L₄) (adapted from: Siddall, 1977).

activity in methoprene resulting from chemical synthetic optimization
involving replacement of the 10, 11 epoxide by a tertiary methoxyl group and
introduction of a conjugated 2, 4-dienoic ester system (Siddall, 1977).

Out of several thousand IGRs with JH activity synthesized in several industrial laboratories, only methoprene and the closely related analog, kinoprene (Henrick *et al.*, 1978), have found commercial use. Methoprene is registered and sold for:

    (a) control of floodwater and standing-water mosquito larvae,
    (b) horn-fly control in livestock manure,
    (c) control of cigarette beetle and tobacco moth, and
    (d) increased silk production by silkworms.

Further field development of IGRs has been limited due to their high cost per low field stability, the requirement of critical timing of application, and their lack of persistence. This has been disappointing since several of these compounds have shown excellent juvenile hormone activity in the laboratory (Staal, 1975)

Since the early 1970s the concept has been advanced that discovery of an anti-juvenile hormone would have far greater potential utility than the JH-type IGRs, primarily because all of the early moults in the insect's life cycle would be expected to be susceptible, since JH is required throughout most life stages, except the prepupal stage.

A "breakthrough" with very significant ramifications for a biorational lead came with the discovery by Bowers, Ohta, Cleere & Marsella (1976) that two chromenes, Precocene I and II, isolated from the plant *Ageratum houstonianum*, induced precocious metamorphosis and sterilization in several hemipteran species. These effects could be reversed by exogenous administration of natural JHs or IGRs with JH activity. Although intensive chemical follow-up was carried out in academic laboratories (Bowers, 1977; Ohta & Bowers, 1977) and in industrial laboratories, only a few compounds were shown to have limited bioactivity (Staal, 1977). One anti-JH analog, the 6-methoxy-7-ethoxy-2, 2-dimethyl chromene (Precocene III) has shown seven-fold greater activity than that of Precocene-II on the milkweed bug (Ohta & Bowers, 1977). The structure and relative anti-JH activity of the two natural precocenes (I and II) and the synthetic analog, Precocene III, are shown in Fig. 1.14. Although the limited spectrum of activity restricts these compounds to studies of academic interest, nevertheless, they serve as useful prototypes for further chemical synthesis and development of novel bioassay systems. It is likely that more practical models will emerge with higher activity and a broader spectrum of action, but only if the choice of bioassay insect is appropriate, relevant and logical.

The chance discovery of the insecticidal properties of diflubenzuron and other benzoylphenylureas, and the subsequent mode of action studies, led to the biorational development of a number of elegant *in vitro* screening procedures to detect chitin synthesis inhibitors (Misato, Kakiki & Hori, 1979).

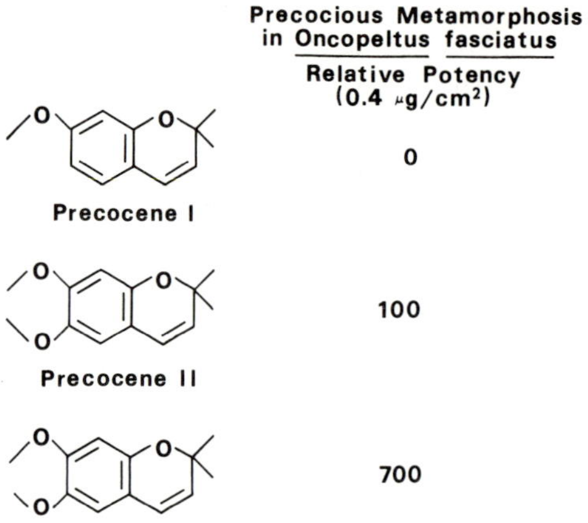

Julius J. Menn

**Precocious Metamorphosis
in Oncopeltus fasciatus**

Relative Potency
(0.4 µg/cm²)

Precocene I

0

Precocene II

100

700

FIG 1.14. Relative potency of Precocene-I, -II, and ethoxy precocene in inducing precocious metamorphosis in second instar milkweed bug nymphs. (Based on data from: Bowers, 1977 and Bowers *et al.*, 1976)

Recent studies by Lund, Hollingworth & Murdock (1979) on the mechanism of action governing the insectostatic properties of CDM (Fig. 1.6), suggest that CDM and/or its demethylated metabolite, *N*-demethyl CDM, elicit a direct excitatory action on non-cholinergic synapses in the central nervous system. These conclusions were deduced from an elegant neurobiochemical assay involving the light organ of the firefly *Photinus pyralis*. These light organs glow in response to treatment with certain biogenic amines, including octopamine. Compounds such as CDM and *N*-demethyl CDM potentiate the light response at the nanogram level per insect. Innovative studies like the foregoing provide the rationale for new biochemical assays inspired by mode of action inputs, which may uncover hitherto latent activity in synthetic compounds previously missed in conventional screening. Furthermore, assays inspired by mode of action studies may provide important leads for synthesis of new chemicals modeled to interfere with a given insect receptor life.

1.6. CONCLUSION

Substantial increases in costs and time to develop a viable insecticide have resulted in drastic decline in the introduction of new insecticides in recent years. Yet the need continues to exist for introduction of selective, biodegradable and safe insecticides. Economic, time and regulatory constraints suggest that new avenues of research need to be explored and utilized in order to maximise returns on research investment for discovery of novel and selective insecticides.

This paper discusses several approaches describing strategies leading to the discovery of selective and biodegradable insecticides and insectostatic agents. While no single approach is the ultimate criterion that may result in a successful discovery, the rational approach is essentially an intelligent program that considers all possible inputs, those discussed here and further elaborated by others (Corbett, 1974). It is evident that we have now accumulated a wealth of knowledge and information to guide us in discovering rationally inspired, useful and selective insecticides. The task ahead is to recognize the knowledge at hand and harness it into productive research and development of future pesticides.

1.7 ACKNOWLEDGEMENTS

I wish to thank the following persons from the Zoecon Research Laboratories: Clive A. Henrick and John B. Siddall for technical comments, Carolyn Erickson for editorial comments, Susan Dillow for the art work, G. B. Staal for photographic material and Sharon H. Strong for diligent typing.

1.8 REFERENCES

Adam, A. V. (1977) Pesticide development and needs in developing countries. *Proc. Int. Congr. Ent., 15th, Washington, DC.*, pp. 741–746.

Ayers J. H., Ernest R. K. & Johnson O. H. (1977) Pesticide industry overview. In *Chemical Economics Handbook* 573, 1000A–573. 1006A. Stanford Research Institute, Menlo Park, California.

Bowers W. S. (1977) Anti-juvenile hormones from plants: chemistry and biology activity. In *Study Week on Natural Products and the Protection of Plants — October 18–23, 1976, Pontificiae Academiae Scientiarum Scripta Varia* (ed. Marini-Bettolo G. B.), pp. 129–142, Vatican City: Elsevier, New York.

Bowers W. S., Ohta T., Cleere J. S. & Marsella P. A. (1976) Discovery of insect anti-juvenile hormones in plant. *Science, NY* **193**, 542–547.

Boyce A. M. (1976) Historical aspects of insecticide development. In *The Future for Insecticides: Needs and Prospects* (eds. Metcalf R. C. and McKelvey J. J., Jr.). pp. 469–488. John Wiley & Sons, New York.

Braunholtz J. T. (1977) Pesticide development and the chemical manufacturer *Proc. Int. Congr. Ent., 15th, Washington, DC.*, Entomol. Soc. Am., College Park, MD, pp. 747–755.

Burger A. (1978) Drug design and development. *J. Med. Chem.* **21**, 1–4.

Casida J. E. (1972) Chemistry and metabolism of terminal residues of organophosphorus compounds and carbamates. In *Fate of Pesticides in Environment, Proceedings of the Second International IUPAC Congress*, Vol. 6 (ed. Tahori A. S.), pp. 295–317. Gordon & Breach, London.

Casida J. E. (1976) Prospects for new types of insecticides. In *The Future of Insecticides: Needs and Prospects* (eds. Metcalf R. L. & McKelvey J. J., Jr.), pp. 349–366. John Wiley & Sons, New York.

Casida J. E. (1979) Pesticide research to maintain and improve plant protection. In *Advances in Pesticide Science*, Vol. 1 (ed. Geissbuhler H.), pp. 45–53. Pergamon Press, Oxford.

Casida J. E., Gray R. A. & Tilles H. (1974) Thiocarbamate sulfoxides: potent, selective and biodegradable herbicides. *Science, NY,* **184**, 573–574.

Chemical & Engineering News (1977) Growth in agrochemical business is slowing. *Chem. & Eng. News* **55**, 10.

Corbett J. R. (1974) *The Biochemical Mode of Action of Pesticides.* 330pp. Academic Press, London & New York.

Corbett J. R. (1976) Developing new pesticides. *Chem Br.***12**, 310–312.

Djerassi C., Shih-Coleman C. & Diekman J. D. (1974) Insect control of the future: operational and policy aspects. *Science, NY,* **186**, 596–607.

Elliot M. (1977) Synthetic pyrethroids. In *Synthetic Pyrethroids,* ACS Symp. Ser. No. 42 (ed. Elliot M.), pp. 1–28. American Chemical Society, Washington, DC.

Elliot M. & Janes N. F. (1978) Synthetic pyrethroids — a new class of insecticides. *Chem. Soc. Rev.* **I**, 473–505.

Ernest & Ernst (1978) *1977 Industry Profile Study.* National Agricultural Chemicals Association, Washington, DC.

Eto, M. (1974) *Organophosphorus Pesticides: Organic and Biological Chemistry.* CRC Press, Cleveland, 387 pp.

Fowler D. L. & Mahan J. N. (1978) *The Pesticide Review.* USDA, ASCS, Washington, DC, 44 pp.

Fukuto T. R. (1976) Carbamate insecticides. In *The Future for Insecticides: Needs and Prospects* (eds. Metcalf R. L. & McKelvey J. J.) pp. 313–346. John Wiley & Sons, New York.

Furtick W. R. (1976) Insecticides in food production. In *The Future for Insecticides: Needs and Prospects* (eds. Metcalf R. L. & McKelvey J. J.), pp. 1–15. John Wiley & Sons, New York.

Gill J. S. & Lewis C. T. (1971) Systemic action of an insect feeding deterrent. *Nature, Lond.* **232**, 402–403.

Goring C. A. I. (1976) Prospects, problems for pesticide manufacturer. *Farm Chemicals* **139**, 18–26.

Gnadinger C. B. (1936) *Pyrethrum Flowers,* 2nd ed. McLaughlin Gormley King Co., Minneapolis.

Haller H. (1972) History of pesticides. In *Industrial Production and Formulation of Pesticides in Developing Countries,* **ID/75**, Vol. 1, pp. 23–24. UNIDO, Vienna: UN, New York.

Hansch C. & Fujita T. (1964). ρ-α-π analysis. A method for the correlation of biological activity and chemical structure. *J. Am. Chem. Soc.* **86**, 1616–1626.

Henrick C. A., Anderson R. J., Staal G. B. & Ludvik G. F. (1978). Insect juvenile hormone activity of optically active alkyl (2E, 4E)-3,7,11-trimethyl-2,4-dodecadienoates and of arylterpenoid analogues. *J. Agric. Food Chem.* **26**, 542–550.

Henrick C. A., Staal G. B. & Siddall J. B. (1973) Alkyl 3, 7, 11-trimethyl-2, 4-dodecadienoates, a new class of potent insect growth regulators with juvenile hormone activity. *J. Agric. Food Chem.* **21**, 354–359.

Hollingworth R. M. (1975) Strategies in the design of selective insect toxicants. In *Pesticide Selectivity,* (ed. Street J. C.), pp. 67–111. Marcel Dekker, Inc., New York.

Hollingworth R. M. (1976) Chemistry, biology activity, and uses of formamidine pesticides. *Environ. Health Pers.* **14**, 57–69.

Hubbell J. P. & Casida J. E. (1977) Metabolic fate of the *N,N*-dialkylcarbamoyl moiety of thiocarbamate herbicides in rats and corn. *J. Agric. Food Chem* **25,** 404–413.

Konishi K. (1972) Nereistoxin and its relatives. In *Insecticides, Proceedings of the Second International IUPAC Congress of Pesticide Chemistry,* Vol. 1 (ed. Tahori A. S.). pp. 179–189. Gordon and Breach Science, New York.

Lund A. E., Hollingworth R. M. & Murdock L. L. (1979) Formamidine pesticides: a novel mechanism of action in lepidopterous larvae. In *Advances in Pesticide Science,* Vol. 3 (ed. Geissbuhler H.) pp. 465–469. Pergamon Press, Oxford.

Menn J. J. (1972) Pest management for the future — an appraisal of industry's role. *Proc. North Central Branch, Entomological Society of America* **27**, 21–29.

Menn J. J. (1980) Contemporary frontiers in chemical pesticide research. *J. Agric. Food Chem.* **28**, 2–8.

Menn J. J. & Pallos F. M. (1975) Development of morphogenetic agents in insect control. *Environ. Letters* **8**, 71–88.

Menzer R. E. & Casida J. E. (1965) Nature of toxic metabolites formed in mammals, insects, and plants from 3-(dimethoxyphosphinyloxy)-N, N-dimethyl-cis-crotonamide and its N-methyl analog. *J. Agric. Food Chem.* **13**, 43–112.

Metcalf R. L. (1971) Structure-activity relationships for insecticidal carbamates. *Bull. Wld. Hlth Org.* **44**, 43–78.

Metcalf R. L., Fukuto T. R. & March R. B. (1957) Plant metabolism of dithio-systox and thimet. *J. Econ. Ent.* **50**, 338–345.

Metcalf R. L. & McKelvey J. J. (Eds) (1976) *The Future of Insecticides: Needs and Prospects* pp. 509–511. John Wiley & Sons, New York.

Misato T., Kakiki K. & Hori M. (1979) Chitin as a target for pesticide action: progress and prospects. In *Advances in Pesticide Science*, Vol. 3 (ed. Geissbuhler H.), pp. 458–464. Pergamon Press, Oxford.

Mulder R., Wellinga K. & Van Daalen J. J. (1975) A new class of insecticides. *Naturwissenschaften* **62**, 531–532.

Nishioka T., Fujita M. & Nakajima M. (1979) Effect of chitin synthesis inhibitors on cuticle formation of the cultured integument of *Chilo suppressalis*. *J. Pestic. Sci.* **4**, 367–374.

Ohta T. & Bowers W. S. (1977) Synthesis in insect anti-juvenile hormones. *Chem. Pharm. Bull.* **25**, 2788–2789.

Roller H., Dahm, K. H., Sweeley C. C. & Trost B. M. (1967) The structure of the juvenile hormone. *Angew. Chem. Int. Ed. Eng.* **6**, 179–180.

Sakai M. & Sato Y. (1972) Destruxins and insecticides of microbial origin. In *Insecticides, Proceedings of the Second International IUPAC Congress of Pesticides Chemistry*, Vol. 1 (ed. Tahori A. S.), pp. 163–177. Gordon and Breach Science, New York.

Schmeltz I. (1971) Nicotine and other tobacco alkaloids. In *Naturally Occurring Insecticides* (eds. Jacobsen M. & Crosby D. G.), pp. 99–136. Marcel Dekker, New York.

Siddall J. B. (1977) Juvenile hormones and their analogs. In *Study Week on Natural Products and the Protection of Plants, Pontificae Academiae Scientiarum Scripta Varia*, Vol. 41 (ed. Marini-Bettolo G. B.), pp. 37–57. Vatican City; Elsevier, New York.

Soloway S. B., Henry A. C., Kolimeyer W. D., Padgett W. M., Powell J. E., Roman S. A., Tieman C. H., Corey R. A. & Horne, C. A. (1979) Nitromethylene insecticides. In *Advances in Pesticide Chemistry*, Vol. 2, pp. 206–217. Pergamon Press, Oxford.

Spindler M. (1978) The role of the Swiss chemical industry in pesticide research. *Chem. Rundschau* **31**, 9–10.

Staal G. B. (1975) Insect growth regulators with juvenile hormone activity. *Ann Rev. Ent.* **20**, 417–460.

Staal G. B. (1977) Insect control with insect growth regulators based on insect hormones. In *Study Week on Natural Products and the Protection of Plants, Pontificae Academiae Scientiarum Scripta Varia*, Vol. 41 (ed. Marini-Betollo, G. B.), pp. 353–383. Vatican City, Elsevier, New York.

Tilak B. D. (1977) Pest control strategy in India. In *Crop Protection Agents — Their Biological Evaluation.* (ed. McFarlane N. R.), pp. 99–109. Academic Press, London, New York & San Francisco.

Topliss J. G. (1972) Utilization of operational schemes for analog synthesis in drug design. *J. Med. Chem.* **15**, 1006A–1011.

Verloop A. & Ferrell, C. D. (1977) Benzolphenylureas — a new group of larvicides interfering with chitin deposition. In *Pesticide Chemistry in the 20th Century*, ACS Symp. Ser. No. 37 (eds. Plimmer J. R., Kearny P. C., Kohn G. K., Menn J. J. & Ries S. K.), pp. 237–270. American Chemical Society, Washington, DC.

Wipke T. (1974) *Computer Representation and Manipulation of Chemical Information* , pp. 147–174. John Wiley & Sons, New York.

Woodburn A. & Cook J. (1979) *Agrochemical Service, Chemicals 60.13*. Wood, Mackenzie & Co., Edinburgh.

Zanno P. R., Miura I., Nakanishi K. & Elder D. L. (1975) Structure of the insect *Phagorepellent azadirachtin*. Application of PRFT/CWD carbon-13 nuclear magnetic resonance. *J. Am. Chem. Soc.* **97**, 1975–1977.

1.9 DISCUSSION

Wandiga: The presentation you have given us was quite interesting to the chemist. Yet at the same time while the chemists are looking at this, we have another problem in the tropical world; and that is we are caught in the dilemma of either to stop using chemicals or to build factories manufacturing these chemicals. This dilemma is a recent one since most of, say, the DDT companies are closing down. These factories, of course, are in the developed countries. In another few years the price of DDT will be 4–5 times its present level. And yet in countries like my own, we cannot afford to stop using them. You talked of manufacturing the newer novel chemicals. What would be the answer in countries like Kenya where we use DDT, not because we want to but because we have to. Some work done here in my department shows that the half-life of DDT may be as short as 120 days in Kenya. Perhaps some of the bad effects are not possible in Kenya.

Menn: Well, I think you have raised a very important point and I am sure that if I could give you the full answer we would be in great shape. I cannot do this but perhaps I can make a few comments. I agree with you that DDT is essential. I was impressed by this while in India. We know that in Ceylon they stopped using DDT and the rise in malaria was catastrophic. Resistance to DDT is also cropping up in India and I think it is rather dangerous situation, because the substitutes are either too expensive or not as long lasting as a control. Now I grant you that the cancellation of DDT registration in the USA, I would say categorically, was precipitous and uncalled for because it is only in a society that is very affluent that one can afford to make such mistakes. Most of the legislators who pass such laws are too far removed from where the real world is. But this is water under the bridge, it was banned. I think it certainly was a mistake. Now as far as DDT plants are concerned, well my colleague here from UNIDO, Dr. Szabo might want to comment. I cannot tell you whether you should build a DDT plant or not. It is still cheaper than most other insecticides, even in the places where it is still being manufactured. There are some other compounds where I think mistakes are being made. I have talked to some Indian industrialists and they are going to build BHC plants. From what they are doing they are going to have fairly high residues and I believe it is the delta isomer, which is carcinogenic. So I wonder why today anyone would want to build a BHC plant. I would say the same for Toxaphene.

On the other hand, in developing your strategies of what you should use, I think planning is the key. Now if we look far enough into the future I can see a different scenario. For example, let us go to the synthetic pyrethroids. Some of these compounds you can use at 10 g per hectare.

So instead of building a plant that will make 10 million pounds per year, you will make one that will produce 100,000 pounds per year. It is going to be a pilot plant and therefore cheaper. Maybe it will be a more versatile plant and perhaps more sophisticated. So if one wants to adopt technologies from the more advanced countries do not start at the beginning, go to the top, look at the more complex molecules, build smaller plants with less environmental impact and in the long run you would be much better off. That is my personal view.

Kumar: I wonder if Dr. Menn would care to comment on the problem of resistance to the natural compounds and their metabolites, because we have already a case where the housefly is becoming resistant to juvenoid compounds, to parasites and rabbits becoming resistant to viruses. How much time would you give these natural products?

Menn: Well, this is a question you can answer on a biochemical and a biological level. I am not a prophet; I can only use examples of compounds that have been in use. There are many compounds like organophosphate esters that have been used for many years without the development of resistance of any consequence.

Gubler: There are obviously differences in their attack by the polyoxidases and there are insecticides which are effective and stay effective against certain insects. We have developed a new phosphate ester against insects in the cotton belt which seems to be highly effective with little signs of the development of resistance.

Menn: There is a compound used in the animal health field known as prolate and this has been used in Australia where they have tremendous problems with *Boophilus microplus*. If you test the McKay strain with prolate the resistance has never exceeded 2- or 3-fold, but to all other compounds there has been enormous resistance.

Bowers: Resistance is a biological fact of life we should all accept as inevitable in almost any situation where an organism is under heavy chemical pressure. There is, however, a wide variation in the rate of development of resistance to compounds. 2,4-D for example, has been used for many years very heavily and I am told that one can measure a certain amount of resistance amongst certain weeds. But the development of that resistance has been very slow and I suspect that as a herbicide, 2,4-D will be with us for many decades without serious problems.

Menn: I agree with Professor Bowers, but I will make another comment. Consider synthetic pyrethroids; we know from the work of the Casida group on *cis–trans* isomers, that the *cis* is more readily attacked by oxidases, the *trans* in permethrin by hydrolases. So one possible solution would be that one should not strive to make pure isomers for insect control, but use a mixture so that you do not have a selection in one direction, they would have to select for both which would be much slower.

Kumar: Dr. Hunter, in a paper in the bulletin of the Entomological Society of America (about 1972), mentioned that screening of about 7500 compounds is required to discover a suitable insecticide. However, you mention that now it takes screening of about 30,000 compounds to market a single insecticide. Would you care to comment on the reasons for the increase in the number of compounds to be screened?

Menn: Well, the 30,000 is a figure which includes all pesticides and all herbicides, so if you cut it in half you may say 15,000. But part of the problem is that when you have compounds like decamethrin or some of the highly active organophosphorus compounds, the natural instinct is to look for compounds that are better, which is difficult. The more complex issue that I try to address myself to is to look at some other attributes of the compounds other than outright kill, and perhaps then the ratio would improve. Another factor is the cost of compounds; there are many excellent compounds that are too expensive to develop.

Bowers: I for one was pleased that you pointed out that the newest, most exciting and most active insecticides were in fact the pyrethroids derived from the natural product which has been in use for thousands of years. Most of the insecticides you discussed came out of a screen, then synthesis, then guided synthesis to the developed product. You made the point that most of these chemicals came that way, and that very few come from natural models. I would submit that the future for the development of insecticides and pesticides in general would be the result of increasing sophistication within the industry and by the consumer. Perhaps natural products will serve as a very important source of new pesticidal molecular models in the future. Histologically, it has been convenient and cheap for industry to undertake random synthesis and to test the intermediates from their other chemical programmes, for pesticidal activities. It has not really been in their interests to spend time and money in fundamental research of plant–insect interactions or on natural products chemistry. I know that industry is now involved in fundamental biochemical studies searching for new niches to attack; is this not right? I am sure we have not exhausted all the new poisons and anti-choline esterases in the world; I imagine we could still make a further 100,000, but perhaps this should not be industry's objective for the future.

Menn: I do not strictly agree with you, and I think that in sorting out the mission that industry has, is where institutions like the ICIPE can play a role. What ICIPE can do is to delve in greater depth into a particular model. Let us say that you have a repellency model. They can establish that you have a model with a biological component and a certain chemical component and then, as I mentioned before, I think that industry can possibly improve on it with the synthetic capabilities that are available. Further biological work can then be done in collaboration with ICIPE. So I think that there may be some avenues for collaboration not duplication. There is no point in an institution like ICIPE doing a lot of synthesis. But the biological and physiological aspect, I think, could be extremely valuable, because in industry, we are very interested in behaviour modification, but I do not know how we can do this on a scale where we look at a lot of substances.

Johnson: I think, however, the key point is how in the first place you choose your model. Which one are you going to choose because you could do an awful lot of work on a compound of moderate activity that really is never going to make the grade. I think, therefore, that the key issue is the initial choice of model.

Menn: This is why I like the neem tree, because this is probably one of thousands of examples that exist and have not been explored. We can also use folk-lore, go to the villages to the people who use folk medicine, etc. I think this is a tremendous resource which is not fully utilized.

Gubler: But it is still highly critical to select the natural model, and not every model will lead to a better compound. We have in industry, as a part of the development phase, a big programme to pick the right model.

Menn: Yes, that is absolutely correct, and I will add to that and say that we also have tremendous advances in microanalytical chemistry. You can have a crude natural product today, and with HPLC and GC coupled to computerized mass spectrometry you can get identifications rapidly. So we can see rapidly what sort of a model we have.

Wandiga: I was very interested in one of your remarks in relation to adopting high technology. If we adopt the attitude of starting from the top we always seem to become dependent on developments elsewhere since there is tremendous development from day to day. I think we just have to start from the lowest base so that the technologies can be absorbed, so that we have a better chance to understand all the mechanisms which go with the development. There is a natural development which the developing countries need to go through.

Menn: I think that is a good point, but you may look at it this way. If you start with an older compound it may be easier to manufacture, but you must also keep in mind that many of these older compounds were registered with very minimal pharmacology and toxicology. It may well be that you will build something and find out later that there is some untoward effect that you could not anticipate, which will eventually neutralize the advances of introducing that particular technology. On the other hand, compounds that have been registered in the last few years, especially in the USA, have the advantage that latent toxicology tests have been exhaustively carried out. You then have a greater confidence level that these will not produce untoward effects in man, his animals and environment.

Otieno: Are the biodegradable products, of the new pyrethroids being developed by Zoecon, known, and if so what are the effects of these compounds on the environment?

Menn: This compound is undergoing extensive field trials and this is the first year of the pilot studies. We have radiolabelled the compound this year and metabolism and environmental studies are underway. So at this time I cannot say much about it but I would say from its molecular structure and its similarities to other synthetic pyrethroids that there will not be too many surprises. In a year from now I can tell you more about it.

Gebreyesus: I was going to ask about this question of choosing a model. It seems fairly evident from the molecular structures that these very diverse structures have similar actions. In most cases we are lacking knowledge of the mode of action at the molecular level. Say, if you talk about antifeeding activity, it is a wide variety of compounds that exhibit antifeeding activity. If you want to choose a model for this activity there is not a certain moiety, a certain molecular function to use as a guideline. So how does one go about choosing a model in such a case?

Menn: I do not know. If I did I would do such research. One contributing factor to the diversity of structures is that the site of activity may vary. One set of compounds may act on the central nervous system while another may bind to sites in the antennae, i.e. have a peripheral effect.

Bowers: I think to begin with you would pick the simplest chemical structure?

Menn: Yes, absolutely.

Johnson: I think the big thing that the ICIPE has got to offer (and this is its trump card) is that they have all these plants on their doorsteps and surveys can be made upcountry to get hold of raw material that would be very difficult for other people to obtain. I think this is a big advantage. After that it is a matter of luck.

Menn: Yes, indeed, and as an extension to this I think that laboratories that have great success in natural products chemistry always tie up with a good botanist or marine botanist. You need a systematic biologist to work with the chemist. Tremendous work has been done at the Harvard herbarium, which is absolutely fantastic, because you can go there, and examine the notes on the specimens, which contain detailed notes on the ecological niche of the plants. So a lot of knowledge is available and such approaches can be utilized.

Nayudama: I was wondering if the approach would be first to identify various practices and products, plants and other materials which are used by the local people. It is not easy, of course, because the villagers keep their secrets. But with polite persuasion it should be possible. The same material can also be collected from other parts of the country. It may be that the plants or practices do not have the effect that they are reported to have so the facts must be sieved from fantasy. Out of 400, four to six might be promising and you stick to that group. The plants should then be identified systematically in relation to other related species. Then conduct the bio-assays and investigate the chemistry of the plants. It may not be necessary to identify the active principles with a view to commercial synthesis. Instead see if the plant material can be used in an extract form. This is important in some Eastern countries where employment potential is an important issue. This available work force can be used to produce the bulk material by farming the plants, etc. If the active principle is not sufficiently active then it can also be used as a model for synthetic analogues which are more active. So the strategy should perhaps be in this direction.

CHAPTER 2

Innovation in Pesticide (Insecticide) Chemistry

KURT GUBLER

Ciba-Geigy Limited, Agricultural Division,
CH-4002, Basel (Switzerland)

CONTENTS

2.1	Introduction	33
2.2	Synthesis Concepts	35
	2.2.1 Random screening	36
	2.2.2 Allround screening	37
	2.2.3 Models of natural compounds	38
	2.2.4 Biochemical design	39
	2.2.5 Chemical modifications	40
	2.2.6 Mathematical models	41
2.3	Summary	43
2.4	References	43
2.5	Discussion	44

2.1 INTRODUCTION

The main purpose of industrial agricultural research is to invent and develop proprietory products, which in turn require a more or less regular output of innovations. Many branches of natural science (chemistry, biology, biochemistry) are involved in this process and only close co-operation between them can lead to successful products. Table 2.1 shows some key activities with strong interdependence.

Although this interdisciplinary collaboration provides the basis for success, it is equally important that industry and academia co-operate intensively. Natural products chemistry, studies on biological agents, pests behaviour and the mode of action of pesticides are typical examples from the domain of universities and agricultural research stations. Industry is readily willing to incorporate its findings into research.

TABLE 2.1. RESEARCH—PLANT PROTECTION

Aim: Invention of Plant Protecting Agents

Chemical Research	Biological Research	Basic Research
Synthesis	Screening Methods	Biochemistry
Structure Elucidation	Screening	Mode of Action
	Biological Agents	Pest Behaviour

Among the various research organizations engaged in agricultural research, industry takes the lion's share. It is estimated that research expenditure amounts to about $400 million annually Table 2.2; this represents 5% of the total world pesticide market which, in 1980, will be around $8,000 million. Chemical companies such as Bayer, Ciba-Geigy, Monsanto, Shell, ICI and

TABLE 2.2. RESEARCH ORGANIZATIONS

Industry

Government

University

International Organisations

World Market 1980	8.000 Mill. $
(Developing Countries 10-20%)	
Research Expenditure	350 - 400 Mill. $

others are strongly involved in agricultural chemical research. The participation of smaller companies has become too risky in view of decreasing success rate and rising development costs.

Table 2.3 shows the rate of new inventions in the field of insecticides and insect regulators over the past 40 years. The golden age began with the discovery of DDT and continued shortly afterwards with such remarkable inventions as the insecticide parathion. Many more products based on this chemistry were developed, making the period between 1961–1970 the most successful one. Completely new inventions, characterized by novel structure and higher selectivity and efficacy, are, however, rare. Chlordimeform (1967), diflubenzuron (1972) and the pyrethroids (1972) are the exceptions. After 1971 it became evident that the innovation curve had exceeded its maximum. The high performance standard of existing products together with increased registration requirements were responsible for this trend. Gilbert tried to

TABLE 2.3. NEW INSECTICIDES

Up to	- 1940	23	Natural Products
1941	- 1950	32	DDT, Dieldrin, Parathion
1951	- 1960	45	Demeton, Diazinon,
			Azinphos-Ethyl, Dichlorvos,
			Carbaryl
1961	- 1970	65	Aldicarb, Chlordimeform
1971	- onwards	17	Diflubenzuron, Permethrin,
			Methoprene

quantify this by plotting the number of screening compounds tested for one commercial product (Fig. 2.1).

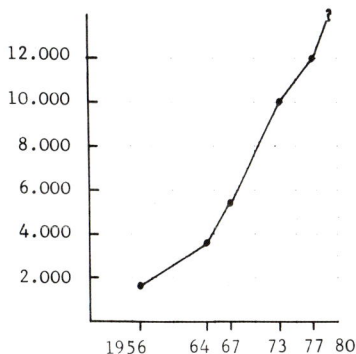

Fɪɢ. 2.1. Number of chemicals screened to produce one commercial compound.

Here we find a not yet interrupted, sharply rising curve, and nobody knows where it will end.

2.2 SYNTHESIS CONCEPTS

There are several methods leading to inventions. Principally we distinguish between two processes: lead finding and lead optimization.

First of all let me try to define the term "lead" or "lead structure". Although each chemist or biologist looks at this in a different way, it is essentially a compound whose chemical structure must not be too complex, which has a specific "minimum activity" and is suitable for a chemical modification programme (lead optimization). A "new lead" is, therefore, a compound with a defined structure whose full biological activity has not yet been reported.

There are several possible ways to tackle *lead finding*. Three important ones, mentioned in Table 2.4, are practised wherever significant research is carried

TABLE 2.4. SYNTHESIS CONCEPTS

Lead Finding ⟶	Lead Optimisation ⟶	Product
- Random Screening Allround Screening	- Chemical Modification	
- Models of Natural Products	- Mathematical Models	
- Biochemical Design	- Other Structure-Activity Relations	

out. The inability to make a reliable prediction of biologically active structures is, however, common to all of them. These methods are in fact purely speculative.

Contrary to lead finding, lead optimization represents a more guided approach to success. Let us look more closely at the main methods of lead finding and lead optimization.

2.2.1   Random screening

Random screening, in the pure sense, is the biological evaluation of chemical compounds with no structural relation to any known pesticide. This was the preferred method at the outset of insecticide research, and DDT was certainly the result of such endeavours (Fig. 2.2).

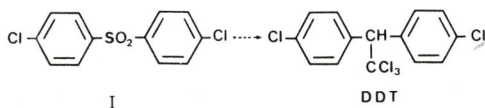

FIG. 2.2 The conception of DDT.

The lead compound was 4,4'-dichlorodiphenyl-sulphone (I) which showed activity against carpet moths. In the chemist's eyes this was a structure with two wings and a middle part which invited deliberate modification. This hypothesis finally led to DDT.

But inventions do not always happen as easily as hind-sight suggests. In particular, selection of the lead compound — among many possibilities — is very delicate and is based on a balanced judgement of the biological efficacy and the chemical potential of the compound.

Random screening is no longer carried out in insecticide (pesticide) research, because the chances of success are too low. It can, however, be useful in conjunction with other methods of lead finding as an unexpected activity might be the beginning of an important development.

2.2.2 Allround screening

As opposed to random screening, allround screening takes in only those compounds which have been synthesized for a specific purpose (guided synthesis). The biological evaluation, however, is done in all major fields and is not restricted to the expected activity. This procedure has its origin in the wide experience of industrial research which has often shown that a compound can surprisingly reveal properties different from those originally anticipated. It is not unusual for such "lead structures" to mark the beginning of a successful era for a new class of agricultural chemicals.

There are two famous examples originating from allround screening in recent years: chlordimeform and diflubenzuron. In both cases the lead structure was synthesized with the intention of creating a herbicide (Figs 2.3 and 2.4).

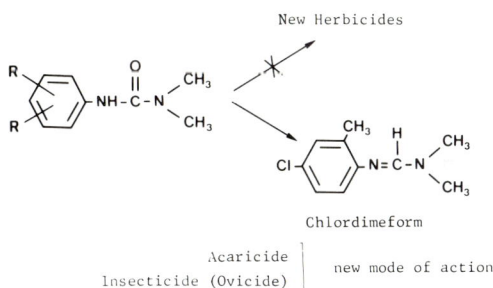

FIG. 2.3. Herbicidal phenyl ureas led to a new ovicide.

Substituted phenylformamidines like phenyl ureas show herbicidal properties. Efforts to improve this activity led to the synthesis of *N*-(2-methyl-4-chlorophenyl)-*N'*-dimethylformamidine. Surprisingly, the compound was found to be active against eggs and immature stages of mites and in addition against eggs and early instars of Lepidoptera.

Although the mode of action is not fully understood, it clearly differs from the mechanism of cholinesterase inhibitors. This aspect is of increasing importance in view of the widespread resistance of many insects to phosphate and carbamate insecticides.

A similar story happened a few years later. Dichlobenil was introduced as a herbicide in 1960. The structure was nevertheless still modified in the hope of improving certain properties. In this connection *N*-(2,6-dichlorobenzoyl)-*N*-'(4-chlorophenyl)-urea was synthesized and also tested as an insecticide. An

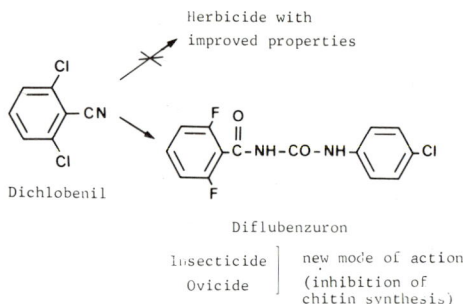

Herbicide with
improved properties

Dichlobenil

Diflubenzuron

Insecticide ⎤   new mode of action
Ovicide ⎦   (inhibition of
                  chitin synthesis)

FIG. 2.4. Dibenzuron evolved from a herbicide.

extensive synthesis programme finally led to diflubenzuron, *N*-(2,6-di-fluorobenzoyl)-*N'*-(4-chlorophenyl)-urea. An interesting feature is again its unique mode of action. The compound inhibits the chitin synthesis specific to invertebrates and is therefore considered to be a selective insecticide.

2.2.3   Models of natural compounds

The search for new chemical models, which originate from natural sources, is attracting increasing attention. Industrial and academic research laboratories are screening extracts from plants, animals (especially marine animals) and microbiological fermentations. The idea is once again to arrive at a new lead structure for which no biological properties have so far been recorded. Countries like Kenya have a vast potential in plant varieties and I believe it would be worthwhile to increase the efforts of ICIPE in identifying new natural compounds with biological activity.

Precocene, an ingredient of an ageratum plant, was recognized as having antijuvenile hormone activity. Unfortunately the insect spectrum was rather narrow and the product was not developed. However, it is not unlikely that modification of the simple precocene structure might lead to a compound with a useful spectrum of activity.

Nereistoxin is a highly toxic compound of the marine annelid *Lumbriconeiris heteropoda*. Its ability to paralyse insects stimulated the isolation and structural characterization of the active component. Skilful variations of this new chemical model led to new synthetic products. They exhibit specific insecticidal properties combined with diminishing mammalian toxicity (Fig. 2.5).

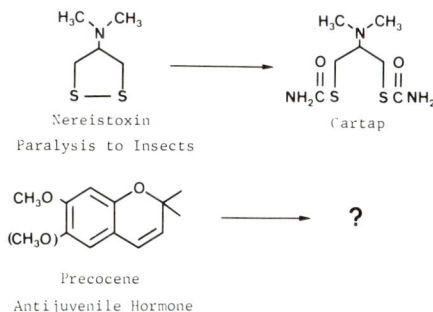

FIG. 2.5. Structures of natural products which have led to novel insecticides.

2.2.4 Biochemical design

The hypothesis of biochemical design is simple and convincing. Thorough knowledge of insects and plants (e.g. specific enzyme systems, metabolites, hormones and their biosynthesis) should eventually lead to the ability to predict a structure or partial structure with definite properties, e.g. enzyme inhibitors, antimetabolites. These high expectations are, however, not confirmed in practice. The action of a pesticide is not only dependent upon its ability to interfere with an essential mechanism in the plant or insect; additional properties such as stability, penetration and translocation ability, and metabolism are necessary.

More realistic expectations can be placed in biochemical knowledge as a useful tool in

(a) the chemical and biological optimization of a lead compound,
(b) the development of new screening methods,
(c) the chemical variation of an enzyme inhibitor according to mode of action knowledge,
(d) the mimicry of hormones and metabolites.

Over the last 20 years papers have been published on the isolation and characterization of insect juvenile hormones, as well as on their biosynthesis and mode of action. The results have considerably stimulated insecticide research. In a worldwide effort research teams modified the juvenile hormone structure and many compounds thereafter have surpassed the characteristics of the original model. But for reasons which cannot be discussed at this juncture commercial success did not follow.

Methoprene is the only compound being commercialized; it is the result of a slight chemical modification. CGA 13353, on the other hand, resembles only slightly the original structure. By replacing two isoprene moieties with two

BIOCHEMICAL DESIGN

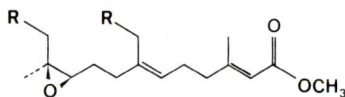

Juvenile Hormones

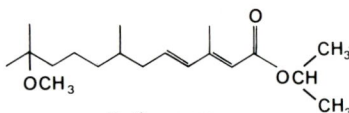

Methoprene

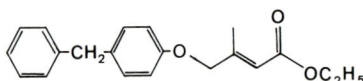

CGA 13353

Fig. 2.6 Mimics of juvenile hormone.

phenyl groups the stability of the original juvenile hormone has been considerably increased at the same time maintaining activity (Fig. 2.6).

2.2.5 Chemical modifications

Chemical modifications probably form the greater part of synthetic research projects. They all start with a given structure whose biological activity is well known. The aim of such projects may be on various levels, but emphasis is placed on the improvement of biological activity, selectivity, mammalian toxicity and manufacturing process. The chemical variability is, of course, limited by the state of art.

The phosphate insecticides form the biggest group of insecticides showing common characteristics. Schrader has proposed several general formulae for phosphates with potential insecticidal activity, e.g. in 1950

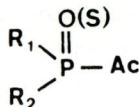

Most of the large numbers of phosphates, being synthesized in many laboratories all over the world, vary the R_1 and R_2 with lower alkoxy, alkylthio or alkylamino groups, R_3 being an organic group bound via O, N or S to phosphorus. The R_3 (Ac) group in particular was modified in various directions. A few examples are given in Fig. 2.7, demonstrating the great variability of a basically simple common formula.

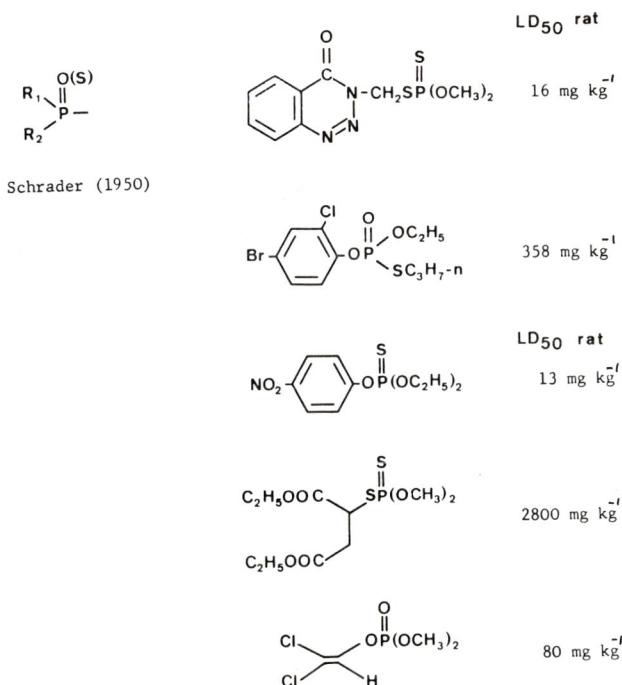

FIG. 2.7. Illustrating the variability of a successful formula.

2.2.6 Mathematical models

Chemical optimization is an expensive activity. Theoretically millions of isomers and analogues can be deduced from a simple general structure. Mathematical equations could perhaps help to rationalize the optimization process and ultimately lead to the prediction of compounds with optimum activity. In the past 20 years some progress has been made on the development of mathematical methods — quantitative structure–activity relationship (QSAR). Hansch is a leader in this field, and his model can help to economize the chemist's work. He tries to correlate changes of the biological response of related compounds with parameters, such as lipophilic, steric and electronic properties of substituents.

Kurt Gubler

$$\triangle F_{BR}^{O} = \triangle F_{E}^{O} + \triangle F_{S}^{O} + \triangle F_{H}^{O}$$

Photosynthesis Inhibitors

$$BR = k \cdot \pi + k' \sigma + k'' E_S + k'''$$

$$BR \; : \; \log \frac{1}{c_{50}} \quad (Hill \; Reaction)$$

FIG. 2.8. Hansch equations.

Since many combinations of such parameters are possible, the statistical significance of each of them is calculated by multiple regression analysis. In this procedure the most suitable equation is evaluated with the help of a computer program. Provided that the resulting correlation coefficient is good ($r \geqslant 0.9$), activities of not yet synthesized compounds can be predicted.

Hansch's classical example in agrochemistry was carried out with mono-substituted ureas (Fig. 2.9). He found a good correlation ($r = 0.944$) for the inihibition of photosynthesis with the Hammett's constant σ and with the partition increments of substituents. If the calculation is done with 3,4-disubstituted ureas, the correlation of the photosynthesis inhibition with only π and σ is very poor. The best fitting equation here is obtained with σ-, E_s- and π-factors. However, the relatively low correlation coefficient and the high standard deviation indicate that some herbicides are located far away from the calculated function.

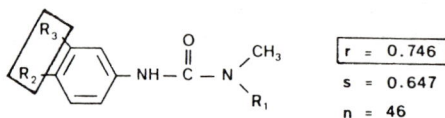

$$r = 0.746$$
$$s = 0.647$$
$$n = 46$$

$$\log \frac{1}{c_{50}} = 4.48 + 0.93\,\pi_3 + 1.28\,\pi_2 + 0.90\,E_{S_2} + 1.43\,\sigma_2$$

FIG. 2.9. Monosubstituted ureas illustrate Hansch's equation.

Other examples of mathematical structure–activity relationships are known, but they do not offer any more substantial information. In spite of such models, however, it remains a valid proposition that small modifications may give rise to major unexpected changes in biological effects.

In summarizing the potential of the various ways leading to new products the question arises as to what progress has been made towards the rational design of new biologically active compounds and here we must admit that a method allowing reliable forecasts does not yet exist.

Certainly, many techniques have improved making the scientist's daily work easier. These include analytical methods, spectroscopy, screening methods, insect-rearing and plant-growing, to name a few. The quality of our current pesticides has reached a high level which new products have to match or even better in many respects (performance, selectivity, toxicity, economy). Real inventions (pioneer discoveries) will obviously become rarer, but, as we have seen, they are still possible. One thing is clear, future research has a genuine need for basic work which may help us to recognize the direction in which we have to go.

2.3 SUMMARY

The search for plant-protecting agents, specifically insecticides, is an old science. For many chemical companies, however, it only became attractive after the discovery of DDT, a forerunner of modern synthetic insecticides. Other insecticides have since been found, each of them signifying advance. The aim of a research organization is to find more efficient, more economic, and safer compounds. Not surprisingly success rates are decreasing.

The backbone of an efficient Research and Development organization is productive chemical synthesis and a versatile biological evaluation system; these necessarily depend on each other. Several approaches to the search for new biologically active compounds (lead generation) as well as to the optimization of a recognized lead structure are known. Some of these methods are illustrated and their importance discussed.

Applied pesticide research is ideally complemented by legal protection of its results, and trends in the future development of patent legislation are discussed with reference to both industrial and developing countries.

2.4 REFERENCES

Anon (1978) *Chem. Ind.* XXX/October, p. 557.
Bowers W. S., Ohta T., Cleare J. S. & Marsella P. A. (1976) Discovery of insect anti-juvenile hormones in plants, *Science* **193**, 542–547.
Gilbert Ch. H. (1978) *Farm Chemicals*, **14** (4), 20–.
Karrer F. E., Scheurer R. & Herzog A. (1974) (unpublished paper given at the International Congress of Pesticide Chemistry III, IUPAC, Helsinki).
Knusli E. (1977) Industrial aspects of the practical use of natural products of derivatives in the protection of crops. Semaine d'Etudes sur le theme "Produits naturels *et la* protection des plantes". *Pontificiae Academiae Scientiarum Scripta Varia* **41**, V, 754–775.
Konishi K. (1972) *Proc. Int. Congress of Pesticide Chemistry*, pp. 179–.
Martin H. & Worthing C. R. (1977) *Pesticide Manual*. British Crop Protection Council (5th edition).
Staal G. B. (1975) Insect growth regulators with juvenile hormone activity. *Ann. Rev. Entomol.* **20**, 417–460.

Verlop A. (1976) Benzoylphenylureas, a new group of larvicides interfering with chitin deposition. *Symposium on Pesticides of the twentieth Century*, ACS Centennial Meeting, New York, April 1976.

Wegler R. (1970) *Chemie der Pflanzenschutz- und Schadlings-bekampfungsmittel*, **1**, p. 248. Springer Verlag, Berlin.

2.5   DISCUSSION

Kumar: I wonder whether industry has seriously considered exploiting some of the compounds found in insect defensive secretion?

Gubler: Yes, some work is going on these compounds, but the outcome of these studies is never published, and knowledge gained in such work is restricted within the industry concerned. May I make it clear that it is not true that we in industry are only interested in highly commercializable projects. We undertake basic studies as well, but over 90% of our work remains unpublished.

Onourah: Is there a Central International Patent Office capable of settling patent claim disputes?

Gubler: No. Most countries have a patent office with a well-stocked search library. These offices are able to advise on the novelty of a patent filed with them. In this way, patent claim disputes are avoided. The arrangement does not, however, preclude a person's right to oppose issued patents on good grounds.

Johnson: It is usually helpful to keep duly signed notebooks. These books help a lot in settling patent disputes.

Gubler: Properly kept and signed notebooks are acceptable as evidence of a claim to a patent right only in the US. Further, the US does not accept professional notebooks from foreign countries in such disputes. However, one can always make a patent quotation in the US. In the event of a dispute, such quotations are acceptable as evidence in US courts. We always do this in Switzerland.

Menn: What percentage of your research is devoted to biochemical design of pesticides?

Gubler: I am unable to tell you the percentage of our work devoted to biochemical design. We have two research groups in this area, one in entomological research and the other in botanical investigations. These groups give us basic data for a better design of partial structures. I hope the foregoing have answered your question.

Patel: Dr. Gubler, you said that over 90% of research findings in industry are not published. Suppose a scientific institution or an individual scientist were interested in obtaining specific information from your unpublished results. Would industry provide the information?

Gubler: Yes, industry would be willing to give the required information after careful consideration of the implications of the request. We will not, however, disclose the whole of the negative results we have accumulated. We normally disclose specific information on specific requests. The purpose for which such information is sought must be fair and must be fully disclosed. We can, for example, give a compound and its properties to a university department, or to an individual. However, we request that results obtained from studies with our specimens, and intended for publication, be examined by ourselves before they are forwarded to the press.

Wandiga: It might be argued that the amount of patentable material or patents produced by a country in a year is a good indication of the scientific output of that nation. What is your view on this and why do you think that developing countries produced an insignificant amount of patents? Is it because of lack of patent laws and offices in these countries which depresses stimulus to produce inventions?

Gubler: I do not think that the amount of inventions made in a year by a country is a measure of that country's scientific output. In some respects, the number of patents coming from the developed countries is decreasing. I am not persuaded that this is an indication of their scientific decline. The chief problem with developing countries is that sometimes they are not aware of scientific practices or progress in other countries. Awareness of what is going on in other countries may provide the necessary stimulus to produce inventions. Provision for patent laws and industrial base to develop patents may also help.

Patel: With regard to Professor Wandiga's question, I would like to add that we have started looking into the question of patent regulation in Kenya. Arrangements are in hand to form a government committee to deal with this issue, and we hope that the matter will be resolved soon in order to provide the necessary stimulus for producing patentable inventions in the country.

Menn: In this regard, ICIPE would benefit if it trained a young technically competent person in patenting procedure.

Bowers: Not all compounds patented are commercialized. Probably only 9% would have a good case for commercial development. I would agree that ICIPE should certainly have knowledge of patent registration procedure, and a capability to ensure that patents are filed on all its inventions with minimum delay.

Kabaara: We have found that some of the agricultural products from the industrially advanced countries arrive at our countries for screening and testing at a time when their use has been discounted in their country of origin. We are sometimes unable to escape the feeling that some multinational corporations making these compounds use our countries not only as dumping ground for undesired products, but as testing ground for unknown effects of their new products, the testing and use of which is objectionable in their own countries. What is Ciba-Geigy doing to improve its image abroad with regards to this practice?

Gubler: We test all our new compounds thoroughly and ensure that they are acceptable before we introduce them to developing countries for use. We have testing stations in Brazil, Japan, South Africa and the Philippines to help us screen our products in all agricultural conditions. I do not think we use developing countries as guinea pigs in our sales promotion efforts.

Waiyaki: Ciba-Geigy recently introduced a new fungicide called "BRAVO" into the Kenyan market for use against coffee berry disease. Could you tell us where the pre-use and testing studies of this product were performed?

Gubler: "BRAVO" has been extensively tested in Kenya by the Coffee Research Station during the last 5 years, before it was recently recommended for application in the country.

Kumar: How do patent laws and patent registration procedure in eastern countries of Europe differ from those in Western Europe?

Gubler: I think that there is no basic difference between patent registration procedure and laws in Eastern European countries and those of Western Europe. There may be minor differences regarding what is patentable.

Menn: For example in the Soviet Union, compositions are not patentable. Further, whereas one can obtain a process patent on a novel procedure for manufacturing a product isolated from natural sources, no patent can be obtained on the product itself. Moreover, the Soviet Union is the most expensive country to file and maintain a patent. China does not have patent laws of her own.

CHAPTER 3

The Search for Fourth-generation Pesticides

WILLIAM S. BOWERS

Cornell University, New York State Agricultural Experiment Station,
Geneva, NY 14456, USA

CONTENTS

3.1   Introduction   47
3.2   History of Insecticide Development   49
3.3   Hormonal Control of Insect Development   50
3.4   Discovery of Anti-juvenile Hormones   52
3.5   Mode of Action of Precocenes   53
3.6   Conclusions   54
3.7   Summary   55
3.8   References   56
3.9   Discussion   56

3.1.   INTRODUCTION

Napolean once observed: "An army marches on it's stomach." Clearly all progess — scientific, industrial and cultural — flourishes when food is abundant. In the history of science, few achievements more directly benefit mankind than those which ensure public health and improve the practice of agriculture. We live with more than one million species of arthropods, most of which are benign or directly beneficial. However, approximately 10,000 species left on their own would prevent us from producing the food and fibre necessary to maintain our present population; of these arthropods, insects are man's greatest competitors for food and fibre; and they are the vectors of the great pestilences such as malaria, yellow fever and plague. In the United States approximately 10% of our crops are destroyed by insects causing 5 to 6 billion dollars in damage every year. This, despite the expenditure of over 650 million dollars for insecticides in the United States alone in 1975. In the developing countries loss of agricultural commodities to insects may range up to 90%, especially where insecticides are unavailable or too expensive.

The excellent public health enjoyed in much of the world is due directly to the successful control of insect disease vectors through the use of insecticides. Nevertheless, 100 million people are afflicted with malaria in sub-saharan Africa and 800,000 die each year. River blindness, carried by a black fly, leaves 700,000 sightless per year in the Volta River Basin alone. Tsetse fly, a vector of sleeping sickness, prevents large areas of the African continent from productive human endeavour. In addition, suffering and mortality is also caused by other insect-transmitted diseases such as onchocerciasis, trypanosomiasis, chagas disease, yellow fever, haemorrhagic dengue. Therefore some appreciation of the importance of insect control must emerge.

During the past decade alone malaria has been eradicated from areas inhabited by 700 million people. It is estimated that the use of DDT in malaria eradication programs has prevented the occurrence of well over 2 billion cases of malaria. Critics who accuse entomologists of aggravating the population explosion recognize that the world population increased by 500 million between 1955 and 1965. This increase is directly due to the development and application of insecticides to control disease vectors and agricultural pests.

It is apparent that our system of agricultural production, combining cultural and chemical methods of pest control, has provided us with a variety and abundance of food never before equalled in the history of mankind, yet there is a widely held myth that somehow more pastoral agricultural methods free from the use of insecticides, fungicides and herbicides could provide tasteful, less expensive food of high nutritional value. There is indeed reason to be concerned about pesticide use in the world. They are used as purposeful environmental contaminants on a very wide scale and it is only relatively recently that we have become aware of the ecological consequences of their ubiquitous presence as micropollutants. Unhappily most pesticides are not only toxic to insects but to other animals and man as well. Concern about our environmental health, wildlife, and the problems of increasing insecticide resistance require new innovative approaches to insect control.

Pesticide use in the United States for the year 1976 shows a consumption of pesticides in agriculture which totals about 661 million pounds. Herbicides account for 394 million pounds, insecticides 162 million pounds, and fungicides 143 million pounds. Non-agricultural usage of pesticides brings the total to about 1 billion pounds active.

There are about 1400 active ingredients. Less than 1% of these are what might be termed "natural" or "biological" such as (a) parasites and predators, (b) natural pheromone and hormone mimics and (c) viruses and bacteria. Of these pesticides only parasites and predators are exempt from Environmental Protection Agency registration requirements. Altosid, à juvenile hormone mimic, was registered in 1972 and gossyplure, a pheromone for the pink bollworm, was registered in 1978. Yet the registration of these materials requires 7 to 10 years of investigations prior to satisfying the requirements of the EPA and an investment of 10 to 15 million dollars. To produce a new pesticide of any kind requires an incredible investment in time and money. Little wonder that many firms formerly active in agricultural chemical research and development have turned to more lucrative, less-regulated enterprises! In order to develop a new strategy for the development of safe pesticides for the future, a look at the past history of pesticides and how they developed is revealing.

3.2 HISTORY OF INSECTICIDE DEVELOPMENT

Insecticides have undergone several distinct stages of evolution. The origins of insecticides are lost in antiquity, but the Chinese used pyrethrum a thousand years ago to control insects. A thriving industry continues in Kenya today producing this natural insecticide from the chrysanthemum flowers. Ancient societies used certain spices to protect their food, such as cinnamon, mustard, nutmeg and pepper, all of which are known to contain insecticides and fungicides.

In the 1880s whitewash was sprayed on grapevines to keep people from stealing the grapes. Then it was discovered that it protected the foliage and fruit from certain insects and perhaps more importantly from certain fungi (Millardet, 1882). The yields of fruit following timed applications was astronomical. The combination of quick lime and copper salts resulted in "Bordeaux mixture" which may have been the first formulated insecticide – fungicide.

In New York, at the Geneva Station about 1902, a botanist, F. C. Stewart, found that the application of Bordeaux mixture increased potato yields tremendously for an investment of literally pennies per acre. New York State farmers since then would never dream of growing potatoes without chemical protection. Soon several inorganic salts including lead arsenate came into use as insecticides and the benefits of chemical plant protection became increasingly evident. Many natural plant-derived toxicants such as nicotine and rotenone also came to be used extensively. We consider these to be the first generation of insecticides. Useful though these compounds were, they failed to provide the plant and public health protection needed by the burgeoning world population.

Near the end of World War II the most significant chemical discovery of all time was made when the insecticidal properties of DDT were discovered. The use of DDT has saved more lives from disease, pestilence and famine than all of the other efforts of man, including all drugs and antibiotics. Its use in public health programs is conservatively estimated to have saved over three billion lives. No other medical or industrial chemical in such broad use has such a demonstrated efficacy and safety record. Yet DDT does accumulate in the environment, and we can measure a few parts per million in our fat; perhaps it is the fat we should worry about, not the tiny DDT residue.

Following the success of DDT in public health and plant protection, numerous related organic pesticides were synthesized and developed by industry. Many of these were much more toxic and persistent than DDT. Many thought the battle against insects had been won, but two seemingly intractable difficulties appeared. Insect pests soon developed resistance to these pesticides and it was found that such persistent compounds remained in the environment for far too long. Effects on wildlife were discovered and it was recognized that the nonselectivity of these organic pesticides destroyed many predators and parasites which, if left to their own devices, would have provided a supplementary degree of insect control.

Prodded by environmental groups, the government began to restrict insecticide use, and even banned several toxicants outright. Seeking less persistent insectides, chemists developed a large number of organophosphorus compounds similar to war gases, which were exceedingly toxic, but nonpersistent. Insects responded rapidly by developing resistance to even these severely poisonous compounds. We think of organochlorine and organophosphorus compounds as the second-generation insecticides.

It is necessary to realize that these insecticides are protoplasmic poisons which act upon biochemical systems fundamental to both vertebrates and invertebrates, and it is clear that the continued use of insectides developed upon this basis will pose a constant hazard to man and domestic animals. Vertebrates and invertebrates share a common evolutionary ancestry and although they have been undergoing divergent evolution for some 600 million years, the common and shared biochemistry leaves little room for the development of selective poisons which discriminate between vertebrates (man) and invertebrates (insects). Although the processes of lipid, carbohydrate, and protein metabolism appear to be little changed and even the genetic code seems quite similar, there are some recognized biochemical innovations which have appeared. The most discernible differences which we recognize are changes within the control mechanisms, i.e. those systems which organize the basic biochemistry including hormones and pheromones which are regulators of insect development, growth, metabolism, diapause and communication. Interference with these control mechanisms offers the greatest promise for the future control of insects.

3.3 HORMONAL CONTROL OF INSECT DEVELOPMENT

Since we believe that development of future chemical methods of insect

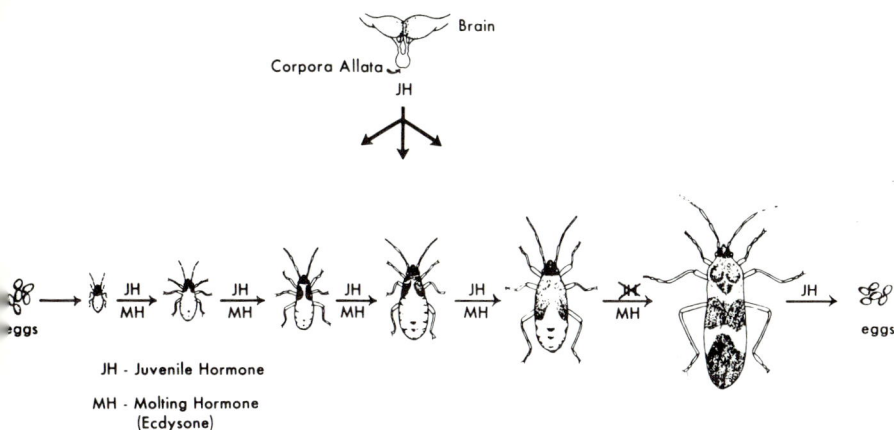

Fig. 3.1. Endocrine regulation of insect moulting, metamorphosis, reproduction.

control must depend upon, and take advantage of the unique aspects of insect metabolism, the insect hormones which control metamorphosis and development stand out as important candidates for research. Wigglesworth (1934) had shown that a differentiation-controlling centre existed in the insect head, and moreover, identified a gland responsible for this action as the corpus allatum. His work paved the way for the discovery by Williams (1956) of a hormone extractable from the cecropia silk worm which would induce juvenilization or continued immature development in insects. This hormone came to be called the juvenile hormone. The participation of the juvenile hormone in insect development is shown in Fig. 3.1. The juvenile hormone acts in concert with the moulting hormones discovered by Becker and Plagge (1939) called ecdysones which are necessary for each of the immature moults which must take place for insects to grow. These hormones were isolated by Butenandt & Karlson (1954) and later characterized by Huber & Hoppe (1965). Since the juvenile hormone could be applied to an immature insect and cause derangement of its development resulting in death of the insect, it became the point of intense investigations by researchers all over the world. Bowers, Thompson & Eubel (1965) synthesized the most common juvenile hormone before it was actually identified from insects by Judy, Schooley, Dunham, Hall, Bergot, & Siddall (1973). A period of intense synthesis of juvenile hormone analogs yielded several worthy of commercial development and one hormonal product is now used in the control of flood water mosquitoes and manure breeding flies (Hendrick, Staal & Siddall, 1973). The juvenile hormones are unique to insects, and therefore represent a third generation of insect-control chemicals completely nontoxic to man and domestic animals. The juvenile hormones, however, are only effective against insects during the latter stages of their immature development. Since the most damaging stages in agriculture are usually the immature stages, juvenile hormone acts too late to prevent most of the plant damage. This is a serious drawback to the further development of juvenile hormone, but as one can see, the juvenile hormone exerts important physiological actions by its presence or its absence during critical stages of insect growth and reproduction. In its presence the insect undergoes immature development continuously. In its absence the insect undergoes metamorphosis into the reproductive adult. If we can treat an insect with juvenile hormone at a time when it should not have it and produce disastrous consequences, then it should also be possible to derange an insect's development if we could interfere with the secretion or production of juvenile hormones at times when it is required.

From classical studies, we know that extirpation of the gland which produces juvenile hormone from an immature insect will induce precocious metamorphosis into a tiny diminutive adult. Since the juvenile hormone is necessary in adult insects for the development of the female ovaries such precociously matured insects are sterile. Clearly a juvenile hormone antagonist would be a two-pronged insecticide. One which would induce precocious metamorphosis of the tiny feeding stages (eliminating them at the damaging state) and/or produce sterile adults which would eliminate future generations.

3.4. DISCOVERY OF ANTI-JUVENILE HORMONES

Insects and plants have undergone constant interaction for millions of years and many plants have developed defensive chemicals to protect them from insect predation. We wondered whether plants might have developed anti-juvenile hormonal chemicals. We began to investigate plants and soon found two chemicals in the common bedding plant, *Ageratum houstonianum* (Bowers, 1976; Bowers, Ohta, Cleere & Marsella, 1976) which would induce precocious metamorphosis and sterilize a variety of pauro-metabolous insects. We called the compounds precocenes (Fig. 3.2). Immature insects contacted with these compounds undergo metamorphosis directly to the adult stage, but the adults are sterile, and they soon die. Normal adult female insects treated with precocenes are permanently sterilized (Fig. 3.3). Mixed developmental stages of insects as they occur in field situations are all adversely affected by the precocenes, i.e. immature insects are forced into premature adulthood and adult insects are permanently sterilized.

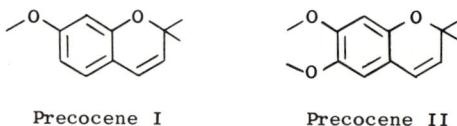

Precocene I   Precocene II

Fig. 3.2. Precocenes I and II. Naturally occurring anti-juvenile hormones from *Ageratum houstonianum.*

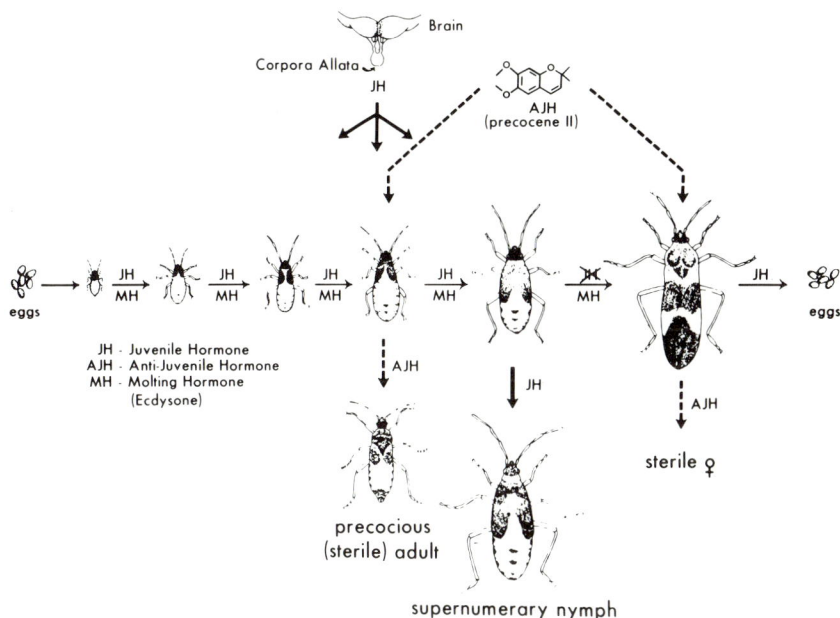

Fig. 3.3. Precocene treatment induces precocious maturation of immature stages and/or sterility in normal adult females.

3.5 MODE OF ACTION OF PRECOCENES

It was discovered by Bowers & Martinez-Pardo (1977) that the precocenes in some way destroy the gland which produces the juvenile hormone in insects. We now realize that the precocenes are activated by metabolism in the insect to produce a compound (the 3,4-epoxide) which reacts with the proteins in the corpus allatum and destroys them (Fig. 3.4). The precocenes, therefore, fulfill the important criteria of highly selective chemicals which attack specific aspects of the insect endocrine system; producing not just a toxic result, but unique derangements of developmental and reproductive processes exclusive to insects.

The foregoing research arose out of my conviction that plants have already developed subtle and effective methods of countering insect predation, and that plants therefore represent a vast natural resource awaiting the efforts of imaginative investigators.

The earliest insecticides were derived from natural sources such as plants, and pyrethrum is one of the outstanding examples. Nicotine and rotenone are others. Following the rise and fall of the indifferently toxic synthetic organic

ACTIVATION — METABOLISM OF PRECOCENE II

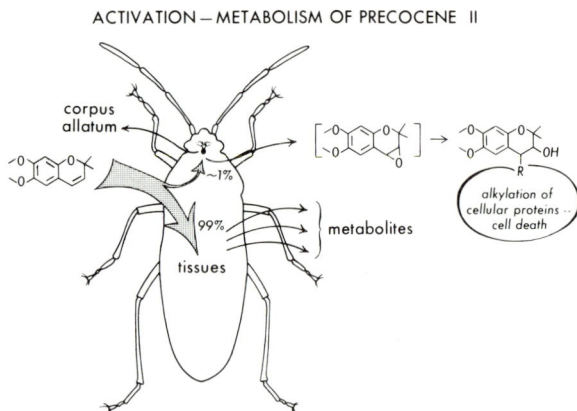

FIG. 3.4. The precocenes undergo metabolic activation to "reactive" epoxides which condense with important cellular elements resulting in destruction of the corpus allatum.

insecticides, we now must turn to new sources for molecular models upon which to pattern our future insecticides; and once again we return to the defensive chemicals developed by plants to protect them from insect predation. In this context, the ICIPE is in a unique situation to seek new insecticidal models from African plants, and in concert with industry they will develop them as effective tools for the control of tropical pests.

3.6 CONCLUSIONS

From the foregoing it should be very clear that our increasing population and the decreasing area of tillable land available for agriculture, combined with the monocultural practices necessary to feed this burgeoning population, necessitate the application of pesticides as protection against the destruction of crops in the field and of stored grain. Protection of the public health from vector-borne diseases continues to place heavy reliance upon the use of insecticides. Thus, our health and welfare depend directly upon the availability of safe and effective insecticides. The history of insecticide development has been especially instructive to us in terms of the great benefits to be derived by the development of effective insecticides as well as the hazards which accompany the indiscriminate use of poisons. Our lessons here have been costly, but educational. Our strategy for the future must therefore rely upon the development of insecticides which, in addition to efficacy, must be safe and selective to the target pest.

While the random screening of industrial by-products remains the principle resource from which new insecticides will be developed; it is the responsibility of fundamental research institutions like the ICIPE to seek to understand those biochemical and physiological differences between vertebrates and invertebrates, and to develop selective means of interfering with insect specific processes. From such studies, new molecular models capable of selective insect control will be discovered and this technology can then be transferred to industry for commercial development. At the ICIPE we have anticipated the need for new strategies of insect control and we have begun the development of anti-hormonal assays on African insects. Soon we will have a battery of insect growth regulator tests ongoing. We are collecting African plants and testing them for effects against various important tropical species. Some of the finest work at the ICIPE has been in natural product chemistry. This chemistry coupled with the excellent endocrinological investigations already in progress will inevitably lead to outstanding discoveries which will doubtless provide many new fourth generation approaches to insect control.

3.7 SUMMARY

The earliest insecticides were natural products derived from plants and included rotenone, cube and pyrethrum powders. To this first generation of poisons one could add the inorganic compounds lead arsenate, copper sulfate and even elemental sulfur.

A quantum leap in chemistry and biological toxicity resulted when the organochlorine insecticides (DDT, Dieldrin, etc.) and organophosphorus and methyl carbamate nerve poisons were generated. Their toxicity and stability combined virtue and vice. The rapid development of insect resistance to the synthetic organic poisons coupled with the alarming accumulation of toxic, teratologic and carcinogenic residues in the environment prompted restrictions on their development and continued use.

A third generation of chemicals effective in insect control arose from basic studies of insect endocrinology and biochemistry. Insect hormones and optimized bioanalogs of the juvenile hormones were developed to interfere with insect metamorphosis. Insects important in the adult stages such as disease vectors could be controlled by hormonal chemicals not even intrinsically toxic to the insect directly. However, immature insect stages which are the most damaging in agriculture were unaffected by the hormone treatment.

Through the study of fundamental insect–plant chemical interactions a new generation of insect growth regulators has been discovered. These regulators interfere with the insect endocrine system in all stages of insect life; and disrupt embryogenesis, metamorphosis, reproduction, diapause and behavior. These simple chemicals from plants were found to act as specific hormonal antagonists by selectively destroying the cells of the endocrine gland responsible for the production of juvenile and gonadotropic hormones.

The success of this effort will require a close relationship to, and interaction with industry. Although insect control for the present and immediate future will continue to rely heavily upon the use of conventional toxicants, our long-range planning must acknowledge the ultimate necessity of developing a variety of chemical, biological, behavioral and cultural options for insect control which depend upon and emphasize the differences between insects and other animals.

These differences can best be discovered and unraveled by organizations devoted to the fundamental studies of insect chemistry, physiology and ecology. Only industry, then, can bring its vast collective resources to insure the optimization, development, registration and marketing of these life-insuring technologies.

3.8   REFERENCES

Becker E. & Plagge E. (1939) Uber das die Puparium-Bildung auslosende Hormon der Fliegen. *Biol. Zbl.* **59**, 326–341.

Bowers W. S. (1976) Discovery of insect antiallatotropins. In *The Juvenile Hormones* (ed. L. I. Gilbert), pp. 394–408. Plenum Press, New York.

Bowers W. S. & Martinez-Pardo R. (1977) Antiallatotropins: Inhibition of corpus allatum development. *Science* **197**, 1369–1371.

Bowers W. S., Ohta T., Cleere J. S. & Marsella P. A. (1976) Discovery of insect anti-juvenile hormones in plants. *Science* **193**, 542–547.

Bowers W. S., Thompson M. & Uebel E. C. (1965) Juvenile and gonadotropic hormone activity of 10,11-epoxyfarnesenic acid methyl ester. *Life Sciences* **4**, 2323–2331.

Butenandt A. & Karlson P. (1954) Uber die Isolierung eines Metamorphosehormons der Insecten in Kristallisierter Form. *Ztschr. Naturf.* **9b**, 389–391.

Hendrick C. A., Stall G. B. & Siddall J. B. (1973) Alkyl 3, 7, 11-trimethyl-2, 4-dodecadienoates, a new class of potent insect growth regulators with juvenile hormone activity. *J. Agric. Food Chem.* **21**, 354–359.

Huber R. & Hoppe W. (1965) Zur Chemie des Ecdysons. *Chem. Ber.* **98**, 2304–2324.

Judy K. J., Schooley D. A., Dunham L. L., Hall M. S., Bergot B. J. & Siddall J. B. (1973) Chemical structure and absolute configuration of a new natural insect juvenile hormone from *Manduca sexta*. *Proc. Natl. Acad. Sci. USA* **70**, 1509–1513.

Millardet P. M. A. (1882) Univesity of Bordeaux.

Ohta T., Kuhr R. J. & Bowers W. S. (1977) Radiosynthesis and metabolism of the insect anti-juvenile hormone, Precocene II. *J. Agric. Food Chem.* **25**, 478–481.

Soderlund D. M., Messeguer A. & Bowers W. S. (1980) Precocene II metabolism in insects: Synthesis of potential metabolites and identification of initial *in vitro* biotransformation products. *J. Agric. Food Chem.* **28**, 724–731.

Stewart F. C. (1802–1911) Potato spraying experiments, NYS Agric. Exp. Stn. Bull. No. 349.

Wigglesworth V. B. (1934) The physiology of ecdysis in *Rhodnius prolixus* (Hemiptera) II. Factors controlling moulting and "metamorphosis". *Quart. J. micr. Sci.* **77**, 191–222.

Williams C. M. (1956) The juvenile hormone of insects. *Nature, Lond.* **178**, 212–213.

3.9   DISCUSSION

Musoke: Have you carried out research to find out why certain insects are more susceptible to precocene than others?

Bowers: We are carrying out metabolic studies which we hope will provide the necessary insight into this phenomenon.

Szabo: (1) How do you assess the future of the fourth generation insecticides from a practical application point of view?

(2) Furthermore, how do you see the effect of resistance to the precocenes and juvenile hormones on their prolonged applicability?

Bowers: (1) I do not think that the precocenes will be used in any widespread agricultural programs to control insect pests and disease vectors at the present time. They might be used in a small way to control some disease vectors like the bedbugs or the triatomine bugs. Nonetheless, I believe that their discovery opens up a wide scope for the new approach to the control of insect pests and disease vectors which involve developing anti-hormonal agents that affect all stages of insect growth, development reproduction and diapause.

(2) I think that more work must be done to obtain more specific anti-juvenile hormones, and to understand the fundamental mechanisms of resistance. Otherwise, the anticipated benefits of anti-juvenile hormones would be eroded by the adverse development of resistance by insects.

Pinchin: We have observed that immature insects treated with precocene take unusually long to metamorphose. Are there any explanations for these observations?

Bowers: There is no clear-cut explanation for those observations. More research is needed to explain them. However, we know that lack of juvenile hormones depresses metabolism in insects, and that precocenes do inflict sub-lethal toxic effects on the insect. We believe that the precocenes are activated by epoxidation which then alkylate important cellular constituents destroying the corpus allatum. These effects probably occur in other cells and tissues in the insect. Taken together these effects might delay ecdysis.

Odero: We have observed that after treatment with anti-juvenile hormones, insects tend to be weak and avoid eating for 2 or 3 days. I think that these observations support Professor Bowers' statement that lack of juvenile hormones depresses metabolic processes in insects.

Odebiyi: Insect resistance to growth regulators is one possible draw back to their widespread development and use to control insect pests and disease vectors. Is there any difference or similarity between the way insects develop resistance to insecticides, and the way they acquire capacity to resist effects of growth regulators?

Bowers: No, I do not think there is a difference. The enzymes responsible for resistance to insecticides play the same role in metabolizing insect growth regulators. Under these circumstances, development of resistance to insect growth regulators would be very similar to the development of resistance to insecticides.

Saxena: How selective are the precocenes against insect pests, parasites and predators?

Bowers: (1) The precocenes are active against a wide range of insect species especially hemimetabolous insects.

  (2) I would not expect the precocenes to be harmful to most insect parasites and predators which are holometabolous species.

Wandiga: Have you made and examined for activity, the analogs for precocenes?

Bowers: Yes. We have made several analogs. Some of them are more specific and active than precocene itself, but none are more advantageous.

Menn: I find it interesting that several analogs of precocene have been made and tested for activity, but none have been patented.

Bowers: Applications for patents have been filed by Cornell University.

CHAPTER 4

New Paths to Novel Environmentally Acceptable Natural Products Chemicals for Pest Management

A. W. JOHNSON

Agricultural Research Council, Unit of Invertebrate Chemistry and Physiology
University of Sussex, Falmer, Brighton, UK

CONTENTS

4.1	Introduction	59
4.2	Designing Insecticides	60
	4.2.1 Pyrethroids	60
	4.2.2 Juvenile hormone analogues	62
	4.2.3 Anti-allatotropins	64
	4.2.4 Diuretic hormones	65
	4.2.5 Pheromones	66
4.3	Conclusion	68
4.4	Summary	69
4.5	Acknowledgements	69
4.6	References	69
4.7	Discussion	70

4.1 INTRODUCTION

The Agricultural Research Council of the UK decided several years ago to establish a number of Research Units, each associated with an individual scientist during his research career. On the retirement of Sir Vincent Wigglesworth the Unit of Insect Research was offered to me as Honorary Director, and as it turned out we have sections on chemistry and biochemistry at the University of Sussex and sections on biochemistry and physiology at the University of Cambridge. Our overall remit is naturally to work with problems of insect control as they affect the UK economy, but we have often used tropical and semi-tropical insects to exemplify particular aspects, particularly when bulk production of insects under laboratory conditions offered special advantages in ease of handling. At the time of setting up the Unit in 1967–1968, many classes of insecticide, especially those containing chlorine, had recently been restricted if not totally banned and alternative types were being actively sought. The competition of insects for available food on a global scale at that time when available food reserves were almost exhausted and now completely exhausted, gave added urgency to the problem. At the same time the increasing difficulty in discovering new synthetic insecticides and the limited patent life of such compounds, coupled with the pressures created by environmental considerations as well as the rapid development of insect resistance in many cases, has made the introduction of new synthetics a lengthy and very costly business. Indeed, quite a number of companies in Europe and the US who had formerly been active in this area have ceased to research and maket insecticides.

4.2 DESIGNING INSECTICIDES

4.2.1 Pyrethroids

In designing useful insecticides, a method which has long been favoured is to model structures on those of active natural products and this approach is best illustrated by the natural pyrethrins. The elucidation of the structures of the natural products, pyrethrins I and II (I, II) and cinerins I and II (III, IV) were established by classical methods more than 20 years ago, and confirmed more recently in every stereochemical detail by modern spectroscopic methods (Fig. 4.1).

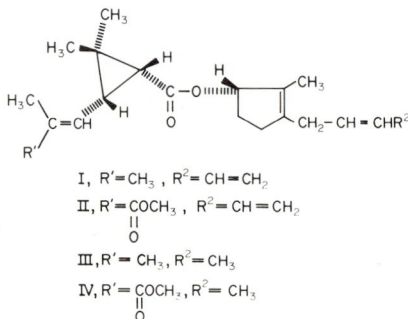

I, $R'=CH_3$, $R^2=CH=CH_2$

II, $R'=COCH_3$, $R^2=CH=CH_2$

III, $R'=CH_3$, $R^2=CH_3$

IV, $R'=COCH_3$, $R^2=CH_3$

FIG. 4.1. The structure of Pyrethrins and Cinerins.

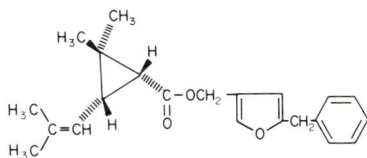

V, (±)−cis−trans isomers

VI, (+)−trans isomer   (a)

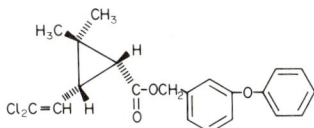

VII

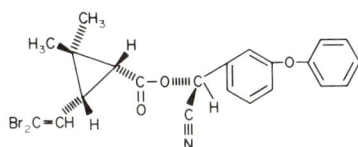

VIII   (b)

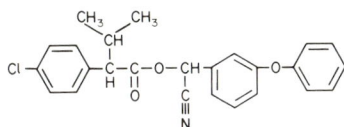

IX

Fig. 4.2. The structure of the Pyrethroids.
a) V   resmethrin   b) VIII   Decamethrin
    VI   bioresmethrin   IX   Fenvalerate
    VII permethrin

The aim of the subsequent synthetic studies was to synthesize non-persistent analogues which possessed low mammalian toxicity and were safe to use near food and on edible market garden and glasshouse crops during the growing period and after harvesting. This was accomplished by Dr. Elliott and his group at Rothamsted, UK (Elliott & Janes, 1979), with the development of resmethrin (V; NRDC 104) and bioresmethrin (VI; NRDC 107) which represented the initial demonstration that synthetic products might prove to be more effective than the naturally occurring insecticides (Fig. 4.2a). The programme was then directed towards the provision of cheaper and more active compounds and especially ones which are sufficiently stable to air and sunlight to be efficient on field crops and yet not too persistent to cause residue problems on edible crops or pollution of the environment. This has proved to be outstandingly successful and among the many products so evaluated, special mention should be made of permethrin (VII; NRDC 143, Fig. 4.2a) selected by ICI for development and the optically active decamethrin (VIII; NRDC 161, Fig. 4.2b) which also is under commercial development; the very low dosage required compensating for the additional structural complexity. An interesting further discovery is the highly active substance fenvalerate (IX) of Sumitomo (Nakayama, Ohno, Aketa, Suzuki, Kato & Yoshioka, 1979) which substitutes an isopropyl group for the gem dimethyl cyclopropane ring and a *p*-chlorophenyl group for the unsaturated cyclopropane substituent (Fig. 4.2b); indeed the ester linkage is almost the sole resemblance to the original pyrethrin structure.

4.2.2 Juvenile hormone analogues

About the time of the first synthetic pyrethrins, a new approach to insect control was suggested depending on the use of compounds essential in the normal metabolism of the insect but applied at an abnormal period in the life cycle — the so-called third-generation insecticides. First to be considered in this context were the juvenile hormones (JH's) which possess linear sesquiterpenoid structures (X, XI, XII — Fig. 4.3).

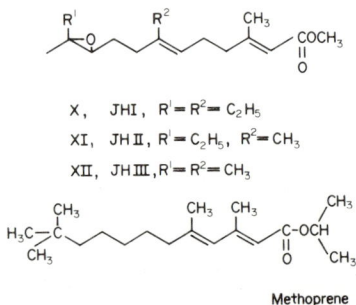

X, JHI, $R^1 = R^2 = C_2H_5$

XI, JH II, $R^1 = C_2H_5$, $R^2 = CH_3$

XII, JH III, $R^1 = R^2 = CH_3$

Methoprene

Fig. 4.3. The formulae of juvenile hormones I, II and III and the synthetic mimic methropreme.

The importance attached to this environmentally acceptable method of insect control offering low persistence and at the same time a very high degree of biological specificity is reflected in the number of syntheses of these molecules —thirty-three at a recent count, though not all of these are stereospecific. Alas, the JH's have failed to fulfil their early promise, largely because of difficulties of application at the precise stage of insect development, although one or two synthetic models are currently marketed — methoprene (Zoecon) being a good example.

Our own studies on JH have centered on the problems raised by its biosynthesis, the transport and metabolism of JH and the nature of the target site receptors. Clearly a rational search for compounds interfering with the natural processes can be organized only after these questions have been answered satisfactorily. A survey of a variety of insects has shown that more than one of the JH's may be involved simultaneously and that the composition of the mixture may well vary between its action as a juvenoid and later in the natural cycle when it is concerned with the reproductive process. In a notable collaboration with Dr. Schooley at Zoecon, my colleague Dr. Richard Jennings established the biosynthetic origin of the ethyl groups of JH I as being derived from propionate via homomevalonate (Schooley, Judy, Bergot, Hall & Jennings, 1976), and confirmed earlier views on other aspects of the biosynthetic pathways to the juvenile hormones (Fig. 4.4).

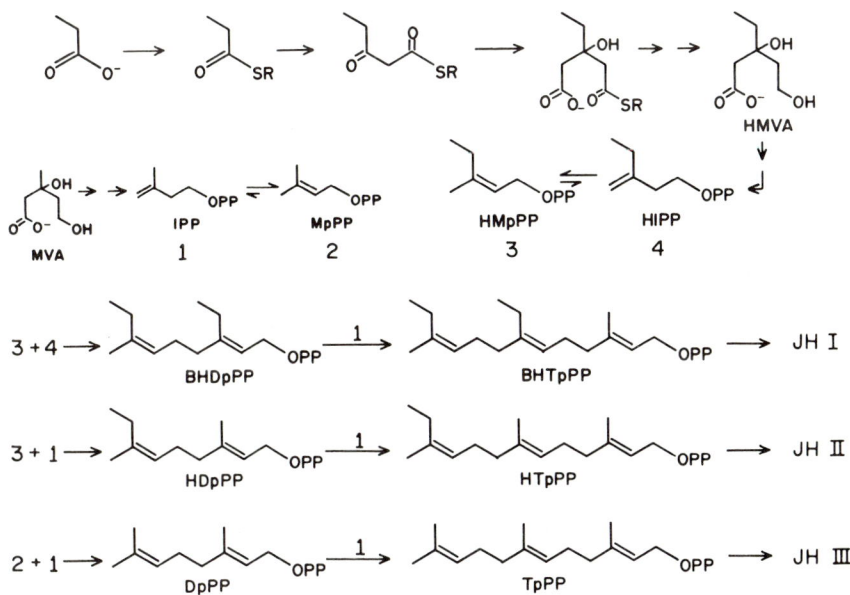

Key:

HMVA = Homomevalonic acid. MVA = Mevalonic acid. IPP = Isopentenyl pyrophosphate. MpPP = Dimethylallyl pyrophosphate or monoprenyl pyrophosphate. HIPP = 3-Ethyl-3-butenyl pyrophosphate or homoisopentenyl pyrophosphate. HMpPP = Z-3-methyl-3-pentenyl pyrophosphate or homomonoprenyl pyrophosphate. BHDpPP = Bishomodiprenyl pyrophosphate. BHTpPP = Bishomotriprenyl pyrophosphate. HDpPP = Homodiprenyl pyrophosphate. HTpPP = Homotriprenyl pyrophosphate. DpPP = Diprenyl pyrophosphate. TpPP = Triprenyl pyrophosphate. PP = pyrophosphate = $-\overset{OH}{\underset{O}{\overset{|}{P}}}-O-\overset{OH}{\underset{O}{\overset{|}{P}}}-OH$

FIG. 4.4. The biosynthetic pathway to the juvenile hormones I, II and III.

More recent studies at Sussex have been aimed at the development of short-term *in vitro* organ culture methods for the quantitative investigation of the biosynthetic performance of corpora allata glands, excised from adult females of mainly the locust *Schistocerca gregaria* and the cockroach *Periplaneta*. However, the rates of JH biosynthesis for individual members of paired corpora allata glands can differ up to 200-fold and in some cases only one of the glands has the ability to manufacture hormone. The chemical work involves the separation and identification of very small amounts of material and requires very precise methods of detection and analysis. For this we have used automated capillary column radio gas chromatography in conjunction with HPLC for the identification of biosynthesized insect hormones, and their immediate precursors by comparison with authentic products. The method affords high resolution plus high sensitivity and isotope discrimination through quantitation by liquid scintillation counting. Picogram quantities of the hormones are produced by corpora allata cultured *in vitro* with medium containing different ^{3}H- and ^{14}C-labelled precursors. The use of high-resolution radio chromatographic systems also enables the latter part of the metabolic pathway leading to the formation of the hormones to be elucidated. Thus late intermediates, especially farnesenic acid which unlike its methyl ester has not yet been identified as a natural precursor, can be supplied exogenously and then is efficiently utilized by the glands which have the ability to *O*-methylate and epoxidize the acid. Other interesting consequences of this work have come from the study of the level of JH synthesis of the corpora allata of *Periplaneta* (adult females) where cycles of synthesis have the same periodicity as the synchronous development of oöcytes. Examination of the utilization of model compounds has shown that the farnesyl binding domain must be appreciably oversize with respect to farnesenic acid as analogues containing additional carbon atoms and even oxygen are acceptable; all of this gives a little more hope in designing inhibitors.

4.2.3. Anti-allatotropins

The natural metabolism of JH at the fifth larval instar naturally led to a search for compounds showing this type of activity and this was rewarded a few years ago by the discovery of the first anti-JHs or antiallatotropins by Bowers in the US (Bowers, Ohta, Cleere & Marsalla, 1976). Two such compounds (XIII, XIV), the so-called precocenes I and II which induce premature maturation, were isolated from extracts of the common bedding plant *Ageratum houstoniamum* (Fig. 4.5).

We have examined these as part of a general examination of JH inhibition and the effect of these compounds on locusts is quite striking. Fortunately these 2,2-dimethylchromenes are relatively easy to synthesize and they are

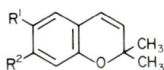

XIII, Precocene I, $R^1 = H$, $R^2 = OCH_3$
XIV, Precocene II, $R^1 = R^2 = OCH_3$

Fig. 4.5. The formulae of the anti-allatotropins precocene I and II.

therefore readily available for physiological study. Metabolism by corpora allata involves, not unexpectedly, epoxidation of the non-aromatic double bond and hydration of the oxide to the 3,4-diol, although with Malpighian tubules further conjugation also occurs before excretion as the major pathway although it is speculative that the oxide plays an alternative role in fulfilling its anti-JH properties.

4.2.4 Diuretic hormones

The juvenoids are, of course, only one class of insect hormone and other classes could be selected for evaluation as insect-control agents. We ourselves have put in a good deal of effort into the isolation of the diuretic hormone of *Rhodnius prolixus*, a tropical blood-sucking insect particularly prevalent in Brazil where it acts as a vector of Chaga's disease. Interference with diuresis in other insects may be the fundamental mode of action of certain insecticides. The diuretic hormone of *Rhodnius* is synthesized by the mesothoracic ganglia from where it is released into the blood stream to exert its powerful effect on the Malpighian tubules and thus promote diuresis. The biological testing requires both skill and patience but is effective and has formed the basis of the isolation work. However, our attempts in this area have shown that the problem is complex. It appears that the hormone occurs in two forms (Hughes, 1979), of large and small molecular weight, probably bound and unbound to protein. It is possible to isolate the large molecular weight form from mesothoracic ganglia by sonification and homogenization followed by fractionation by treatment with aqueous ammonium sulphate of increasing strengths and centrifugation. Delipidation of the ganglionic masses before extraction (chloroform: methanol) followed by gel filtration gives evidence for the low molecular weight form (m.w. 1500–5000). A curious observation is that 5-hydroxytryptamine shows physiological effects similar to that of the diuretic hormone and indeed has been isolated by us from mesothoracic ganglia and identified as the bispentafluoropropionyl derivative. However, the amounts of 5-hydroxytryptamine present are insufficient to account for the observed diuretic effects.

4.2.5   Pheromones

Another area where the study of naturally occurring chemical substances affecting insect behaviour has led to promising leads for insect control has been the pheromones which are used to transmit messages to other members of the same species. Pheromones are subdivided according to the physiological response they evoke in the insect; thus there are sex pheromones attracting the sexes to each other, trail pheromones and alarm pheromones. Although the biological utilization of these substances is seen to the highest degree in the social Hymenoptera, it is now clear that pheromones are utilized in all the families of insects so far examined. One of the earliest studies on pheromones was that of Butenandt and his colleagues (Butenandt, Beckmann & Hecker, 1961), on the sex attractant of the silkworm moth, *Bombyx mori*, where it was shown eventually that the compound involved was *trans*-10, *cis* -12--hexadeca-dienol, $CH_3(CH_2)_2CH^c{:}CH.CH^t{:}CH(CH_2)_9OH$, the structure of which was confirmed by total synthesis. Work on other insect species followed rapidly, for example the identification of the queen substance of bees as *trans*-9-oxo-2-decenoic acid, $CH_3CO(CH_2)_5CH{:}CH. COOH$ by Butler (Butler, Callow & Johnston, 1962) at Rothamsted, UK. These researches led to a very intensive effort directed at numerous species of lepidoptera and particularly toward the identification of the female sex pheromones. These proved to be mainly medium-chain (C_{12}–C_{18}) unsaturated alcohols or the corresponding acetates and they form the largest group of identified pheromones. Furthermore, these structures can be synthesized by relatively straightforward methods so that they are easily available for further biological testing and screening.

The detection of these pheromones by the insect is a function of the antennae, which in response to a positive chemical signal pass an electric signal to the brain. This observation led to the development of the electroantennogram in which excised antennae are exposed to the chemical under test and the electrical response is measured (Schneider, 1962). Such an apparatus represents a very sensitive method for the *in vitro* assessment of synthetic chemicals and natural concentrates. It has proved to be invaluable in this work although it is fair to add that incorrect conclusions have been drawn in the past from such studies and formulae which later need revision were claimed for certain pheromones. The refinement of analytical methods, particularly mass spectrometry coupled with gas or high-pressure liquid chromatography, has now shown that the natural insect pheromones rarely comprise only one individual compound and that usually there are minor components present which exert a synergistic biological effect on the main component. No convincing explanation for this synergism has been advanced to date. Furthermore, there is no clear picture as yet concerning the direct relationship between insect taxonomy and the chemical structure of the pheromones; clarification must await the results of studies on new species.

Because the chemical structures of the components of the lepidoptera, sex pheromones are relatively simple and because they exert their influence over a considerable area, they appear to offer a promising method of insect control. Being volatile, as well as natural products, they do not present environmental problems and, moreover, the insect does not develop resistance to them. However, the cost of these synthetic chemicals is relatively high and field application is therefore restricted to high value crops, especially certain fruit crops. Insect traps based on this principle are available and have been used, for example, for control of the Gypsy moth and codling moth, although on a limited scale.

Use of chemicals which affect insect behaviour, allomones and kairomones as well as pheromones, has had more limited applications in other insect families, although with the Diptera a product, muscalure, which is a C_{23}-olefine, has been formulated in a preparation for application in the home. No convincing control of the mosquito has been achieved by use of pheromones as yet and this may well be a case where the use of conventional insecticides such as DDT will be more effective for many years still. In the Hymenoptera, reference has been made already to the historically important queen substance of bees. Much of the later work in this area with bees, wasps, ants and termites has involved studies of trail pheromones, normally used by the insect in its quest for food, and alarm pheromones, used to alert the colony to any outside disturbance. Once again these pheromones have usually proved to be mixtures of compounds, often several, and the original idea of the pheromone as a single substance is largely illusory. Much of the social behaviour of this family of insects is under pheromonal control and a great deal of further work is needed in this area.

It is arguable that the greatest need for insect control is in the Coleoptera, which present special problems of application for any chemical, because of their type of habitat and life cycle. Problems concerning control of wood-boring beetles, for example, still await a practical solution. Nevertheless considerable progress has been made in elucidating the nature of the pheromones utilized by these beetles for purposes of aggregation, responses which in turn lead to courtship and mating. Again, there are few, if any, species which appear to use single compounds for pheromone messages and in most of the cases examined, especially wood borers, complex mixtures of chemicals are involved. The situation is further complicated by reactions of the insect to the stimulus of chemical constituents of the host plant. In this connection negative stimuli or repulsion by chemicals is also well recognized. Special attention has been paid to studies of Scolytidae because of their connection with Dutch Elm Disease, and although the nature of several compounds possessing pheromone-like action is now known, problems of application still remain formidable. Of the three types of chemical known to be concerned in the attraction of beetles to elm trees (Blight, Wadhams &

wenham, 1978) one, α-cubabene (XV) is a metabolite of the tree and is concerned in the initial attraction (Fig. 4.6). The other two, 4-methyl-3-heptanol (XVI) and multistriatin (XVII), are concerned with the attraction between male and female beetles and act as an aggregation pheromone. The stereochemistry of the isomers involved is very precise and has been determined by preparation of each of the isomers and careful evaluation of their attractant properties (Fig. 4.6).

$$CH_3(CH_2)_2CH(CH_3)CH(OH)CH_2CH_3$$

XVI

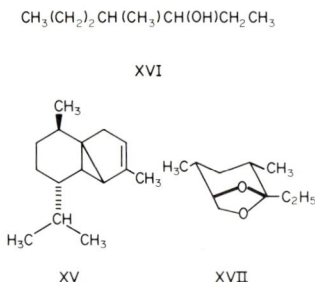

XV   XVII

FIG. 4.6. L-α-cubabene (XV) 4-methyl-3-heptanol (XVI) and multistriatin (XVII) formulae.

4.3 CONCLUSION

Thus, like the insect hormones the commercial utilization of pheromones has had only limited success to date, but with several notable exceptions, which include the control of a variety of fruit flies and moths. Reference has been made to the codling moth and useful control of light infestations of the pink bollworm and the boll weevil in cotton has been reported. Much research is in progress but the application problems presented to departments of forestry are immense. A useful application of pheromones in the field has been for the monitoring of population densities of insects, which may vary considerably over relatively small areas. Such information may be used to suggest the optimum time for insecticide application and large-scale evaluations of this sort have been reported from the US, Canada and Holland. Another possible application of these compounds on the large scale is the disruption of the insects' normal mating habits and here, too, some success has been reported in the protection of certain special crops.

Thus, both the hormones and the pheromones of insects have a role to play in the overall control of insects by chemicals. They present obvious advantages because of lack of resistance problems and low persistence, but on the other hand, they are often costly to prepare and they present special problems with regard to application. At the molecular level, current studies on biosynthesis and mode of action, e.g. the nature of receptor sites, may well suggest the design of synthetic molecules which are even more effective in their action. I have omitted reference to anti-feedant substances deliberately, as so much of the work was done here at ICIPE in Nairobi by Professors Nakanishi and Meinwald, two very distinguished Directors of Research of the Centre, but clearly this is another area which will require very careful evaluation in the future.

4.4 SUMMARY

A brief review of the current status of pyrethroid insecticides is presented followed by an account of insect hormones and pheromones and their value for insect control in the field.

4.5 ACKNOWLEDGEMENTS

The original work referred to in this paper from the ARC Unit of Invertebrate Chemistry and Physiology at the University of Sussex has been carried out by Drs. R. Aston, G. T. Brooks, J. F. Grove, L. Hughes, R. Jennings, F. Mellon, G. E. Pratt, A. Ottridge and R. Weaver, Mr. A. Hamnett and Miss A. Blight and it is a pleasure to acknowledge their scientific ability and experimental skills.

4.6 REFERENCES

Blight M. M., Wadhams L. J. & Wenham M. J. (1978) Volatiles associated with unmated *Scolytus scolytus* beetles on English elm: differential production of α-multistriatin and 4- methyl-3-heptanol, and their activities in laboratory assay. *Insect Biochem.* **8**, 135–142.
Bowers W. S., Ohta T., Clere J. S. & Marsalla P. A. (1976) Discovery of insect anti-juvenile hormones in plants. *Science* **193**, 542–547.
Butenandt, A., Veckmann R. & Hecker E. (1961) I. Biological testing and isolation of the pure sexual attractant, bombykol. *Z. Physiol. Chem.* **324**, 71–83. II. Constitution and configuration of bombykol. *Ibid.*, pp. 84–87.
Butler C. G., Callow R. K. & Johnston N. C. (1962) The isolation and synthesis of queen substance, 9-oxodec-*trans*-2-ecoic acid. *Proc. Roy. Soc.*, Ser B, **155**, 417–432.
Elliott M. & Janes N. F. (1979) Recent structure–activity correlations in synthetic pyrethroids. In *Advances in Pesticide Science*, Part 2 (ed. Geissbuhler H.), pp. 166–173. Pergamon Press, Oxford.
Hughes L. (1979) Further investigations of the isolation of diuretic hormone from *Rhodnius prolixus*. *Insect Biochem.* **9**, 247–255.
Nakayama I., Ohno N., Aketa K. I., Suzuki, Y., Kato T. & Yoshioha H. (1979) Chemistry, absolute structures and biological aspect of the most active isomers of fenvalerate and other recent pyrethroids. In *Advances in Pesticide Science*, Part 2 (ed. Geissbuhler H.), pp. 174–181. Pergamon Press, Oxford.
Schnider D. (1962) Electrophysiological investigation on the olfactory specificity of sexual attracting substances in different species of moths. *J. Insect Physiol.* **8**, 15–30.

Schooley D. A., Judy K. J., Bergot B. J., Hall M. S. & Jennings R. C. (1976) Determination of the physiological levels of juvenile hormone in several insects and biosynthesis of the carbon skeletons of the juvenile hormones. In *The Juvenile Hormones* (ed. Gilbert L. I.), pp. 101–117. Plenum Publishing Corporation, New York.

4.7 DISCUSSION

Menn: I have a question on the use of pheromones as control agents. To my knowledge, they have only been used as monitoring agents. I think there has been no successful cases of actual control on a large scale?

Johnson: I would disagree, but only qualitatively in that I am sure they have been used in a minor way for the control of codling moth, and certainly in the UK they have been used for the Pea moth. Also the boll weevil has been trapped in this way.

Bowers: I have worked with Professor Roelofs, who is probably the foremost pheromone researcher in the world, and to my knowledge he has been able to perform one or two control experiments with pheromones. One was with the red-banded leaf roller, another with the codling moth, and another with the grape berry moth. They were able to use mass trapping techniques to depress the population sufficiently to provide an acceptable method of control. The amount of pheromones, the number of traps, the isolation of the crop from other infestations were optimized, but under no circumstances was this economical or would be attempted again. There have been experiments in mass confusion which have been successful against the pink bollworm. Again this was an herculeian measure taken to control the moth in a very idealized situation. In the case of the pink bollworm this was performed by Professor Shorey at Riverside in California and the experiment was a success in that the moths were confused sufficiently, that an economic population was never achieved. However, I understand that there was no commercial crop produced from the experiment because other pests were quite able to destroy the crop. Much has been said about the use of pheromones, sex pheromones in particular, as monitoring tools and they have been used mainly for this purpose but even here with only moderate success. Unless or until we get some better ideas as to the use of pheromones I think they are not going to provide any real degree of control in the immediate future. This does not mean that the investment in pheromone research is not a good one, because the knowledge of how insects communicate with each other and with plants, etc., is very important because some day this will, I am sure, lead to control measures. Our techniques and our ideas today are still very primitive, but this work must continue.

Johnson: I attended a demonstration of these things recently in Rothamsted in the UK, where they were monitoring the pea moth. They were able to give the farmers 2 weeks notice of when the maximum concentration of pea moth was likely to be. This, of course, would allow the farmer to take adequate control measures in good time to reduce crop losses.

Kumar: I would say that, as yet, there are indeed no spectacular applications of pheromones on a large scale, with similar success to, say, DDT. Pheromones are being used in some cases on African pests, for example the African armyworm and the maize stem borer, *Busseola fusca*. Work is also being carried out at ICIPE on termite pheromones. In Ghana, progress is being made in the study of the pheromones of *Eldana saccharina* in collaboration with the scientists at the Tropical Products Institute in the UK. Here I would like to pay tribute to our British colleagues who have always been ready to assist our studies, especially in the characterization and synthesis of these pheromones. I hope that this collaboration will continue.

Menn: I think that confusion techniques still leave much to be desired. Another problem with pheromones, of course, is the presence of mixtures of isomers, both geometric and optical. In the case of the gypsy moth pheromone, dispalure, the minor isomer in fact inhibits the effect of the major isomer. In other cases, of course, the minor isomers are synergists and are very necessary for the action of the pheromone. This sort of thing makes the situation very complex.

Wandiga: Could you comment on the philosophical and ethical framework scientists should adopt on the basis of which we advise the use of insect-control agents? If you consider the situation which arose from the introduction of DDT where the residue problems, etc., occurred, what position should the scientist, the chemist, take?

Johnson: Well, I think this is where developments in the fields of pheromones, juvenile hormones, etc., will play an increasingly important role in providing environmentally safer methods of insect control. We have learnt from experience that we must look for environmentally acceptable control agents. But the chemists can only provide the chemicals and their job is done.

Bowers: I agree that the chemists did their job and that it was the biologists who abdicated their responsibilities. Today, however, we are going towards a more integrated approach in which insect scientists form various disciplines which are playing a joint role.

Kumar: I would like to disagree with the chairman that the blame for lack of progress in areas such as forecasting, economic threshold, etc., be laid on the shoulders of the biologists. I think that the agricultural community was so intoxicated by the success of the early insecticide applications that research in vital areas of applied entomology were discouraged and were termed "economic niceties". It was only after the advent of such problems as resistance and residues that funds became available for basic research in applied entomology.

Bowers: I disagree emphatically! That might have been your problem, but in the USA agricultural scientists, including biologists, were beguiled by the efficacy of DDT, and discontinued their research for alternative measures of insect control of their own volition. The biologists must accept the blame, and not simply claim a lack of support.

Pinchin: As a practical application of Professor Johnsons diuretic hormone research in *Rhodnius*, I would like to suggest that inhibition of this hormone (thus causing delay in urination and subsequent reduction of the risk of *Trypanosoma* transmission) may be more important than promoting death by dehydration through stimulating the hormone.

Johnson: I agree; however we are still a long way from practical applications since we do not yet know enough about these hormone systems.

Saxena: I would like to make two comments with reference to rice pests. The application of synthetic pyrethroids in rice fields seem to lead to a resurgence of resistance in brown plant hopper. The second point is that a lot of work is now being done on antifeedants and these seem to offer considerable promise.

CHAPTER 5

The Search for New Agrochemical Leads from Natural Sources

R. J. PRYCE

Shell Research Limited, Shell Biosciences Laboratory,
Sittingbourne Research Centre, Sittingbourne, Kent ME9 8AG, UK

5.1	Introduction	73
5.2	Biologically Active Natural Products	74
	5.2.1 The boletic acids	74
	5.2.2 The viniferins — phytoalexins from grapevines	77
	5.2.3 Chemical activation of the antifungal defence mechanism in rice	80
	5.2.4 Novel amino acids from seed of *Ateleia herbert smithii*	84
5.3	Natural Products Research at Shell Biosciences Laboratory	85
5.4	The Role of Industry in Natural Products Research	86
5.5	Summary	87
5.6	Acknowledgement	88
5.7	References	88
5.8	Discussion	88

5.1 INTRODUCTION

Natural products which have biological activities in insects, plants and plant pathogenic microorganisms provide a continual inspiration to the agrochemical business in its search for new products to control pests and improve crop yields. In principle, the natural products could be used in agriculture if they had the right biological and physical properties. However, in practice, quantities required for field use would normally be too great unless the product was microbially produced and could be obtained in quantity, by fermentation methods. For the most part, therefore, biologically active natural products are looked upon as leads for chemical synthesis of structurally or topographically related mimics. The mimics might also have more favourable biological and physical properties than the original natural products.

At the Shell Sittingbourne Research Centre we have had an interest in biologically active natural products spanning many years, and our activity in this area has recently increased considerably. The approach of our natural products activity to date can be illustrated with some recent examples.

(a) Our search for novel antifungal antibiotics against plant pathogens led to the discovery of the boletic acids from the edible toadstool, *Boletus elegans*.

(b) Plants' natural defence mechanisms against fungal attack have been examined in the case of grapevines and rice. The chemicals concerned, the viniferins and momilactones respectively, have been characterized. In the case of the viniferins some chemical mimicry, by what could be a biosynthetically related synthesis, has been possible. On the other hand, in rice, we have shown that a synthetic chemical, a fungicide, is capable of helping the plant to make its own defensive compounds, the momilactones.

(c) Amino acids with novel structures often fall into the classification, biologically active natural products. Two such compounds, a cyclobutane amino acid and an aza-bicyclo (2,2,1) hexane amino acid from the insect-resistant seed of *Ateleia herbert smithii* will be described.

The work to be presented in these areas has been a collaborative effort by a number of chemists and biologists; acknowledgement to them is made by reference.

Finally, I would like to describe the way in which we are now, with increased vigour, looking for new biologically active natural products which could lead to new agrochemical products. At the same time I would like to make some comments on the role of industry in the finding and development of biologically active natural products for pest and disease control.

5.2   BIOLOGICALLY ACTIVE NATURAL PRODUCTS

5.2.1   The boletic acids

A few years ago we had an extensive programme looking for new, microbially produced antibiotics, for use against plant fungal pathogens. The approach was entirely classical. Soil samples from all round the world were screened for bacteria and fungi which were capable of producing growth inhibitors of the plant pathogenic fungi, *Pythium debaryanum*, *Botrytis cinerea* and *Pyricularia sasakii*. Active isolates were grown in liquid media and both culture filtrates and mycelial extracts were tested for antibiotic activity *in vitro*. Then, active extracts were tested against a range of fungal diseases on plants. In addition to the screening of soil samples, other microbial sources were examined in a similar manner. One such source that we examined was locally collected Basidiomycetes from which mycelial cultures were produced in the laboratory. The edible toadstool *Boletus elegans*, the Larch *Boletus*, produced culture filtrates when grown in Sabouraud's dextrose broth which were highly anti-fungal *in vitro* and looked very interesting *in vivo* against a number of plant pathogens (Table 5.1).

TABLE 5.1   EFFECTS OF METHYLENE CHLORIDE EXTRACTS OF *BOLETUS ELEGANS* ON CROP PLANTS INNOCULATED WITH PATHOGENIC FUNGI (UK PATENT NO. 1 503 492)

Crop	Disease	Pathogen	Disease rating[a]		% Control
			Treated	Untreated	
Barley	powdery mildew	*Erysiphe graminis*	0	9.3	100
Potato	late blight	*Phytophthora infestans*	0	10.0	100
Rice	leaf blast	*Pyricularia oryzae*	0	9.0	100
Vine	downy mildew	*Plasmopara viticola*	0.14	6.7	98
Wheat	brown rust	*Puccinia recondita*	0.7	10.0	93

[a]Mean of three replicates.
    Methylene chloride extracts of a 3 litre culture of *Boletus elegans*, dried and made up to 60 ml with distilled water to which a drop of Triton X-155 was added then sprayed to just before run off onto the plants. Plants were innoculated 24 h later with spore suspensions of the pathogens. Disease development was assessed after 7–10 days on a 0–10 scale where 0 represents complete control of the disease.

By quite conventional extraction procedures and chromatography the antibiotic from *B. elegans* was isolated. It was a liquid which has never shown any sign of crystallization and was produced in the culture filtrates at 0.2–2.0 g litre^{-1}. Spectroscopy alone was not particularly revealing and a little degradative chemistry rapidly showed that this antibiotic was a mixture of homologous glycolipids which we have called the boletic acids. Their structures are shown in Fig. 5.1(I) and the position of the acetate and malonate groups on the β-D-mannose moiety remains unknown. The work leading to these structures for the boletic acids is summarized in Scheme 5.1 and the NMR spectra of the boletic and alloboletic acids and their methyl esters are shown in Table 5.2. In this investigation, methylene chloride extracts of a 3-litre culture of *Boletus elegans*, dried and made up to 60 ml with distilled water to which a drop of Triton X-155 was added. Plants sprayed with this preparation were then innoculated 24 h later with spore suspensions of the pathogens. Disease development was assessed after 7–10 days on a 0–10 scale where 0 represents complete control of the disease.

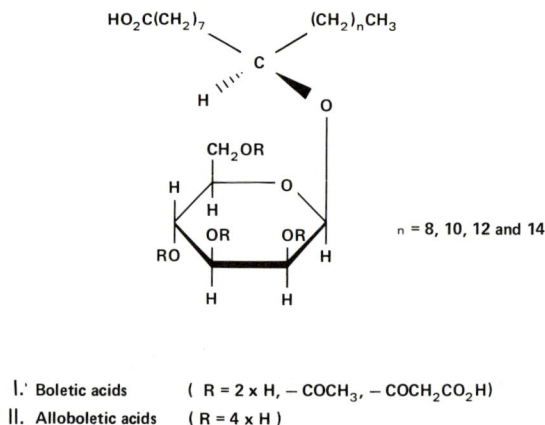

I.' Boletic acids   (R = 2 x H, − COCH$_3$, − COCH$_2$CO$_2$H)
II. Alloboletic acids (R = 4 x H)

Fig. 5.1. Structure of boletic and alloboletic acids.

The structure of the boletic acids (Fig. 5.1, I) allows for the possibility of equilibria between lactone and hydroxy acid forms and malonate half and diesters. These forms are shown diagrammatically in Fig. 5.2 and may account for (i) the four interchangeable components seen on chromatography, (ii) observed variation in the *in vivo* fungicidal activity of different samples, and (iii) the discrepancy of the mean molecular weight as determined by equivalent weight titrations. This type of equilibrium between lactone and hydroxy acid forms of more complex glycosides of other hydroxy fatty acids from fungal and plant sources has been noted by other workers (Tulloch Hill & Spencer, 1968). Antibiotic properties of some of these naturally occurring glycosides of hydroxy fatty acids have also been observed (Valette, 1938; Khanna & Gupta, 1967; Okabe & Kawasaki, 1970; Wagner & Kazamaeir, 1971).

The alloboletic acids (Fig. 5.1, II) (the boletic acids minus the acetate and malonate on the mannose) are about four times more active *in vitro* than the boletic acids (Fig. 5.1, I) against *Pythium debaryanum*. In these antibiotic disc assays the alloboletic acids show a minimum detectable activity at 4 µg-disc compared with the boletic acids at 13 µg disc^{-1}. Against other plant pathogenic fungi *in vitro* the alloboletic acids appear generally to have a higher specific activity than the boletic acids.

The specific activities, however, were not good enough and the isolated boletic acids and alloboletic acids were too feeble for use *in vivo* as anti-fungal agents. The boletic acids were present in very large quantities (up to 50 g litre^{-1}) in crude extracts of cultures which gave the promising *in vivo* results shown in Table 5.1. Unfortunately this problem, of not knowing the specific biological activities of biologically active natural products until they have been isolated, is unavoidable.

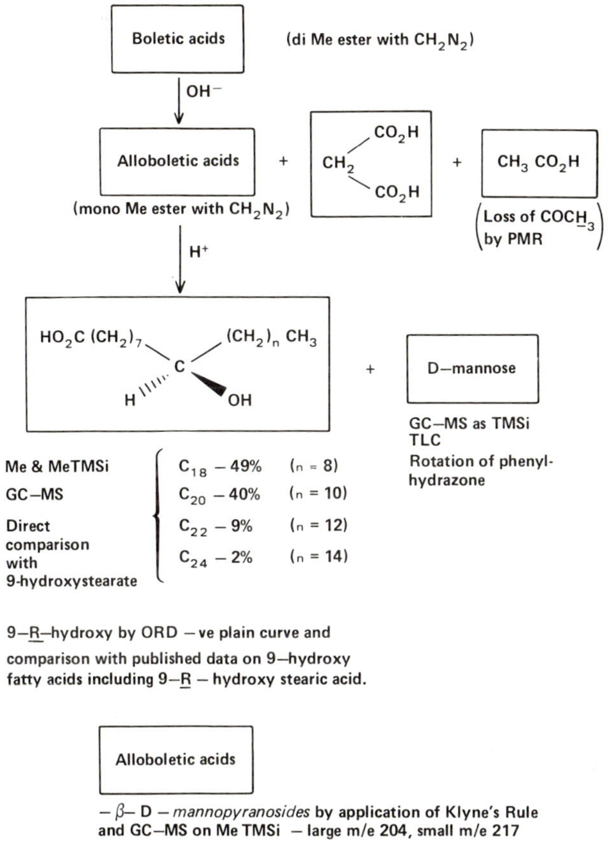

Boletic acids (di Me ester with CH_2N_2)

$\downarrow OH^-$

Alloboletic acids + $CH_2 \diagdown^{CO_2H}_{CO_2H}$ + $CH_3 CO_2H$

(mono Me ester with CH_2N_2)

$\left(\begin{array}{l}\text{Loss of } CO\underline{C}H_3 \\ \text{by PMR}\end{array}\right)$

$\downarrow H^+$

$HO_2C (CH_2)_7 \diagdown \diagup (CH_2)_n CH_3$
$H \diagdown^{C}\diagup OH$
+ D—mannose

GC—MS as TMSi
TLC
Rotation of phenyl-
hydrazone

Me & MeTMSi
GC—MS

Direct
comparison
with
9-hydroxystearate

$\left\{\begin{array}{ll} C_{18} - 49\% & (n = 8) \\ C_{20} - 40\% & (n = 10) \\ C_{22} - 9\% & (n = 12) \\ C_{24} - 2\% & (n = 14) \end{array}\right.$

9—R—hydroxy by ORD — ve plain curve and
comparison with published data on 9—hydroxy
fatty acids including 9—R — hydroxy stearic acid.

Alloboletic acids

— β— D — *mannopyranosides* by application of Klyne's Rule
and GC—MS on Me TMSi — large m/e 204, small m/e 217

SCHEME 5.1. Steps leading to elucidation of the structures for the boletic acids.

5.2.2 The viniferins — phytoalexins from grapevines

Grapevines are an extremely valuable crop which, in Europe, suffers from
two important diseases, vine downy mildew (*Plasmopara viticola*) and soft rot of
the berries (*Botrytis cinerea*). We have for some time been investigating the
grape vine's own defence mechanisms against fungal attack with a view to
enhancing or exploiting it. Here is a summary of some of the work that we have
done (Langcake & Pryce, 1976, 1977a,b,c; Pryce & Langcake, 1977) on the
phytoalexins, anti-fungal compounds which are produced in grapevine leaves
in response to infection and injury.

It was possible to isolate three pure phytoalexins from infected or UV
irradiated grapevine leaves and these have been named α-, γ- and ε-viniferin
(Fig. 5.3). Structures (III) and (IV) respectively have been proposed for α-and

TABLE 5.2 100 MHz ^{1}H NMR DATA AND ASSIGNMENTS FOR BOLETIC ACIDS AND DERIVATIVES (τ VALUES; TETRAMETHYLSILANE INTERNAL STD; J AND $W_{1/2}$ IN Hz

Compound (solvent)	—CH_2—CH_3 ca. 15 × —CH_2—	—CO—CH_3	—O_2C—CH_2—CH_2	>CH—O— and —CO—CH_2—CO—	—OCH_3	—OCH_3	Anomeric —O—C—O— —$CHOH$ H	—OH
Boletic acids (I) (CDCl₃)	9.15 (t) J6 / 8.76 (br)	7.94 (s)	7.68 (t) J7	(9H)5.00–6.80 (br m)	—	—	4.62 (br s) $W_{1/2}$7	4 × OH 3.80 (br s) with py HOD
Boletic acids (I) (D₅-py)	9.15 (t) J6 / 8.75 (br)	7.99 (m)	7.51 (t) J7	(9H)4.80–6.50 (br m)	—	—	(0.5H)3.93 (br s)$W_{1/2}$5 and (0.5H)4.42(br m)	—
Methyl boletates (CDCl₃)	9.14 (t) J6 / 8.75 (br)	7.92 (s)	7.71 (t) J7	(9H)4.90–6.90 (br m)	6.34 (s)	6.24 (s)	4.60 (br d) J 2.4	2 × OH ca. 6.7 (br)
Alloboletic acid (II) (CDCl₃)	9.16 (t) J6 / 8.77 (br)	—	7.71 (t) J7	(7H)5.80–7.00 (br m)	—	—	5.50 (br s) $W_{1/2}$ 7	5 × OH 4.30 (br s) with py HOD
Alloboletic acid (II) (D₅-py)	9.15 (t) J6 / 8.75 (br)	—	7.53 (t) J7	(7H)5.30–6.80 (br m)	—	—	5.03 (s) $W_{1/2}$ 3	—
Methyl allobole-tates (CDCl₃)	9.14 (t) J6 / 8.75 (br)	—	7.71 (t) J7	(7H)5.90–7.00 (br m)	6.35 (s)	—	5.50 (s) $W_{1/2}$ 4	4 × OH ca. 6.7 (br s)

Possible equilibrating forms of the boletic acids

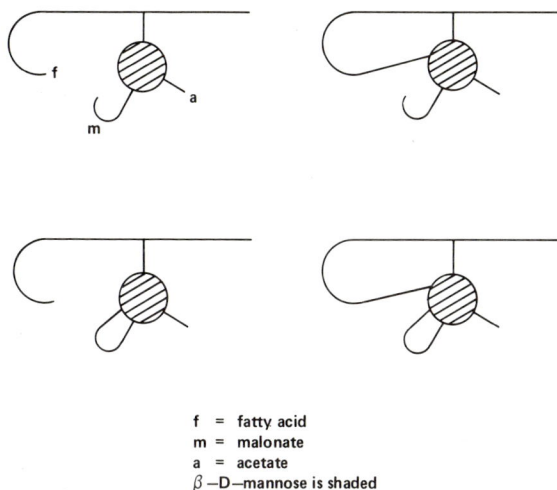

f = fatty acid
m = malonate
a = acetate
β —D—mannose is shaded

Fɪɢ. 5.2. Possible equilibrating forms of the boletic acids in diagrammatic form.

ε-viniferin based chiefly on spectroscopic evidence. None of the viniferins could be detected in healthy vine leaves. Similarly, the trihydroxy-*trans*-stilbene, resveratrol (V) could only be detected in infected or injured vine leaves (Fig. 5.3).

Measurements of the kinetics of the production of resveratrol and the viniferins supported the hypothesis that the viniferins could be produced in vine leaves by oxidative oligomerization of resveratrol as shown in Scheme 5.2. γ-viniferin has not been characterised but our evidence suggested that it is a higher oligomer of resveratrol. The viniferins are moderately anti-fungal *in vitro* against all the phytopathogenic fungi tested (Table 5.3).

In an attempt to mimic the oxidative oligomerization of resveratrol in the laboratory, *trans*-resveratrol was treated with horseradish peroxidase — hydrogen peroxide. The major product, isolated in 41% yield, was shown to be the resveratrol dimer (VI) chiefly on spectroscopic evidence; no ε-viniferin could be detected in the product. In this dimer, which was, unlike the viniferins, racemic coupling had occurred through the *p*-hydroxy substituted ring. In the viniferins coupling appears to have proceeded through the 3,4-dihydroxy substituted ring. Despite the structural differences between (VI) and ε-viniferin sufficient similarity remains to make this resveratrol dimer moderately anti-fungal against the same pathogens as ε-viniferin. In fact (VI) appears to have a slightly higher activity than ε-viniferin (Table 5.3).

Some other 4-hydroxystilbenes have been oxidatively dimerized to ana-

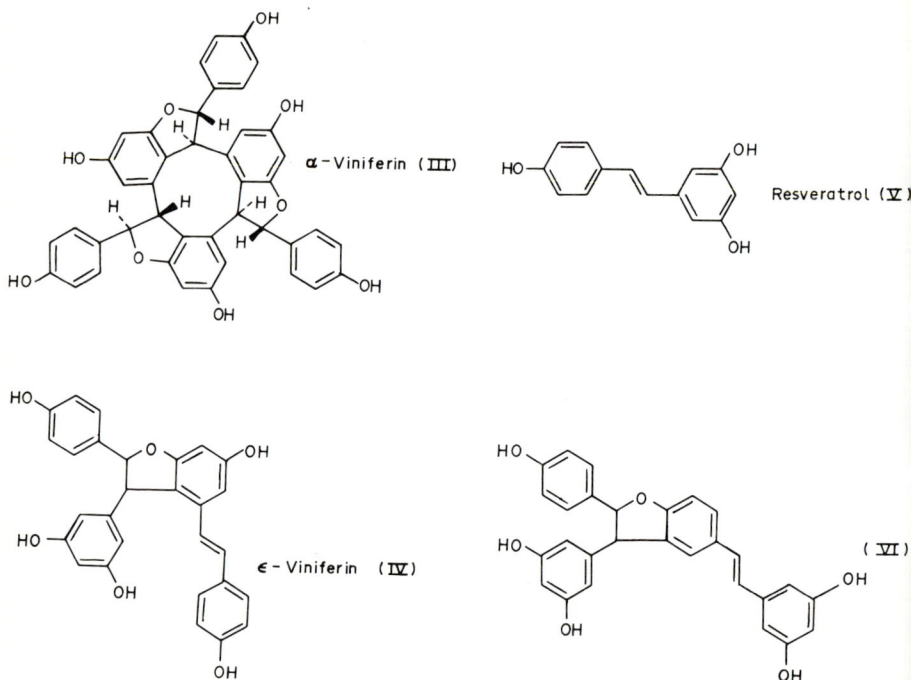

Fig. 5.3. Structures of α- and ε-viniferins (III and IV) resveratrol monomer (V) and dimer (VI).

logues of (VI) and these are shown in Scheme 5.3. Unfortunately, neither ε-viniferin, the resveratrol dimer (VI), nor the synthetic dimers shown in Scheme 5.3 showed any more than protectant activity against *P. viticola* infections of grapevine leaves (80 to 100% control at 1 kg ha^{-1} spray).

5.2.3   Chemical activation of the anti-fungal defence mechanism in rice

The possibility that a chemical to combat plant disease might function by activating the natural resistance mechanisms of the host had been suggested previously (Day, 1974), but there were few well-documented examples. We were able to show that 2,2-dichloro-3,3-dimethyl cyclopropane carboxylic acid (WL 28325) (VII) could assist the rice plant in its defence against the rice blast fungus, *Pyricularia oryzae*.

Drop diffusates from infected rice leaves, some of which had been treated with WL28325 prior to innoculation with *P. oryzae*, were assessed for their ability to inhibit germ tube growth from spores of *P. oryzae*. The result, which is shown in

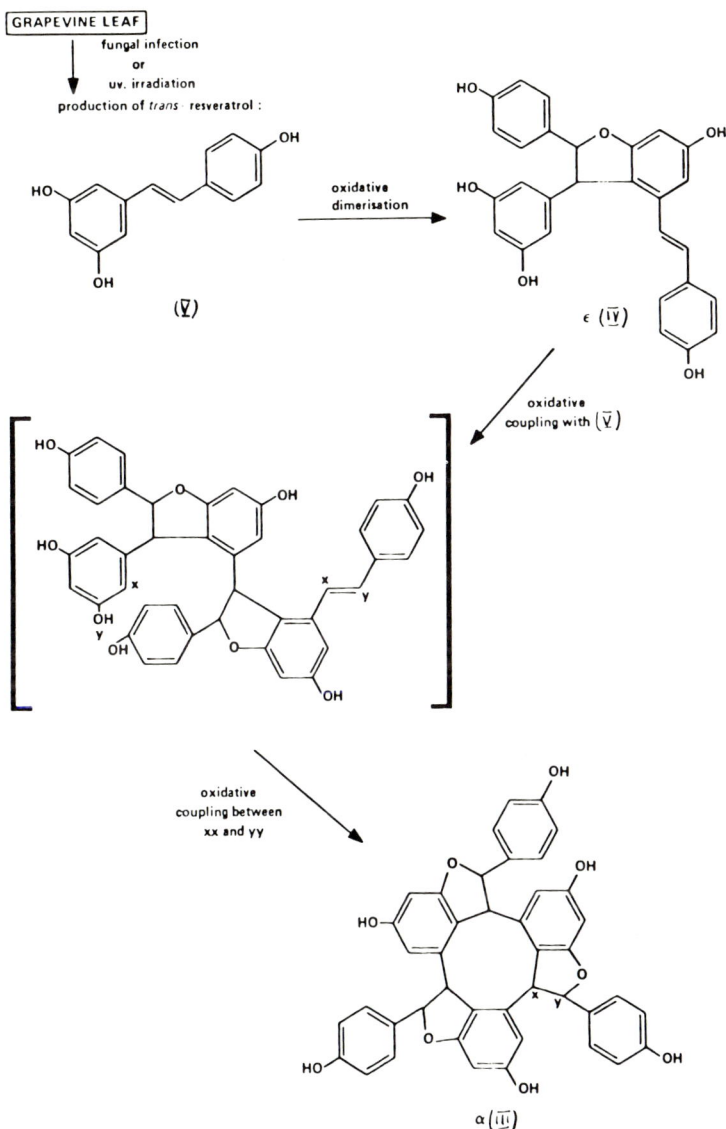

SCHEME 5.2. Oxidative oligomerization of resveratrol to viniferins.

Fig. 5.5, clearly shows that pretreatment with the synthetic fungicide assists the rice leaf to produce anti-fungal compound(s) in response to the infection. Anti-fungal compounds are produced in the non-WL 28325 treated leaves but more slowly.

TABLE 5.3 ANTI-FUNGAL ACTIVITY OF RESVERATROL AND DERIVATIVES

		Spore germination tests			Plasmopara viticola	
	Tlc plate assay (a)	*Botrytis cinerea*	*Cladosporium cucumerinum*	*Pyricularia oryzae*	Zoospore release	Zoospore motility
Resveratrol	>60	>200	>200	>200	>200	>200
α-viniferin	1	97	47	28	35	11
γ-viniferin	9	>200	150	57	>100	>100
ε-viniferin	2.5	100	37	n.t.	19	12.5
Compound VI	2	125	30	30	16	5

Minimum amount detectable (μg) by *Cladosporium cucumerinum* tlc plate assay.
Concentrations (μg ml^{-1}) causing 50% inhibition of spore germination.
Concentrations (μg ml^{-1}) causing 50% inhibition of release of zoospores from sporangia of *Plasmopara viticola*, or the motility of the zoospores after their release.

Bioassays were carried out in duplicate and the results represent the mean values of at least two independent determinations.

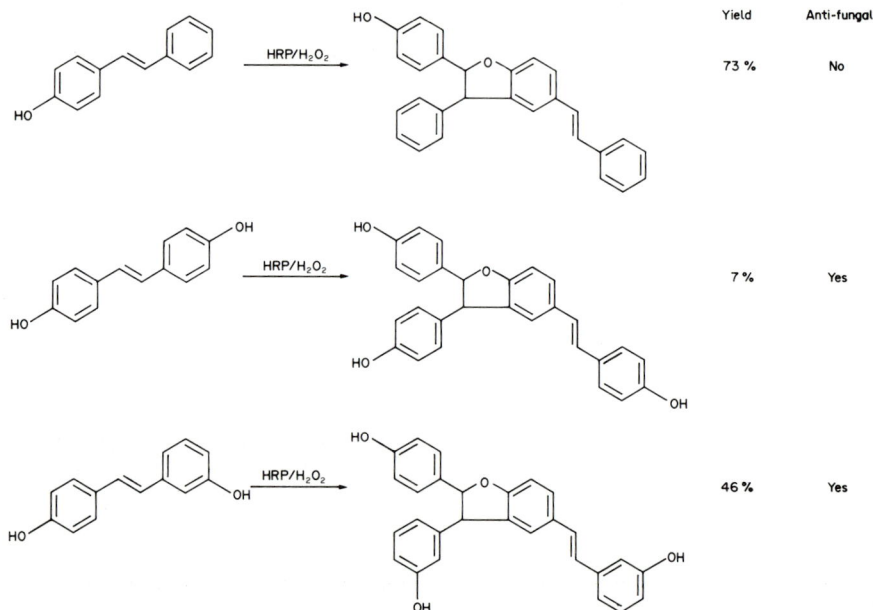

SCHEME 5.3. Oxidative dimerization of 4-hydroxy stilbenes to resveratrol dimer.

Subsequent extraction and isolation work showed that two anti-fungal compounds were produced in response to both normal and WL 28325-treated infections in rice leaves. The compounds were not detectable in extracts of healthy rice leaves and treatment with WL 28325 did not alone cause their

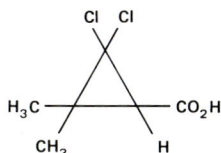

WL 28325 (VII)

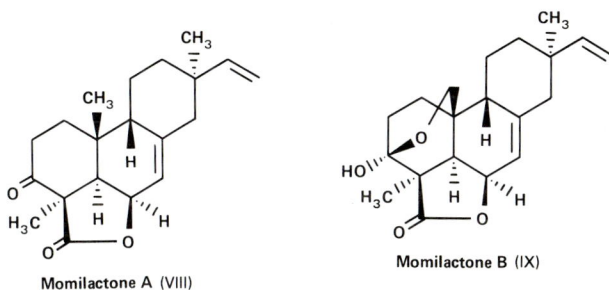

Momilactone A (VIII)

Momilactone B (IX)

FIG. 5.4. The structure of chemicals which activate the antifungal defence mechanism in rice.

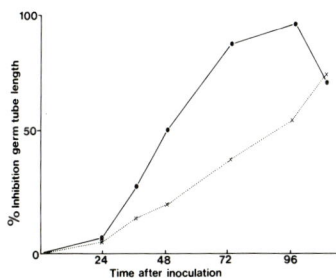

FIG. 5.5. Effect of WL 28325 treatment and *P. oryzae* infection on the inhibitory activity of drop diffusates against germ tube growth. O—O WL 28325 treated and infected, □—□ untreated and infected. Each point is the mean of three replicate analyses.

production. Production of the two phytoalexins in rice leaves could also be caused by UV irradiation but in this case pretreatment with WL 28325 had no influence.

The phytoalexins were isolated and, by MS, PMR and optical rotation comparison with published data (Kato, 1973, 1977) they were shown to be the 9-β-pimaradiene diterpenes, momilactones A (VIII) and B (IX). Momilactones A and B had previously been isolated as plant growth and germination inhibitors from rice husks (Fig. 5.4) (Kato *et al.*, 1973, 1977).

Concentrations of the phytoalexins giving 50% inhibition of germ tube growth of *P. oryzae* were 5 and 1 µg ml^{-1} respectively for momilactones A and B. GC-MS analyses of drop diffusates from infected leaves showed that the amounts of the momilactones that were present could therefore account for all the anti-fungal activity detected.

The unique activity of WL 28325 in increasing the capacity of rice to synthesize phytoalexins in response to infection is worthy of further study. Such a study might provide new insights into ways in which host defence mechanisms can be manipulated to control plant pests and diseases.

5.2.4 Novel amino acids from seed of *Ateleia herbert smithii*

In collaboration with Professor E. A. Bell, of King's College, London, we have been examining a number of novel amino acids from legume seeds. Unusual amino acids often show some biological activity resulting from their ability to interact with the metabolite of the protein amino acids (Bell *et al.*, 1980).

The legume *Ateleia herbert smithii* which grows in Costa Rica produces small, hard seeds at 2- to 3-year intervals. These seeds are preyed upon by the larvae of the weevil *Apion ateleia* (Curculionidae). This weevil kills 50–99% of the seed crop of each tree. On the other hand, the seeds are ignored by at least 100 other species of seed predators in the same habitat. This observation suggests that the seeds are protected against most seed predators, but that one species of weevil is able to circumvent that protection.

An analysis of free, non-protein amino acids (potentially protective compounds) in the seed of this species revealed the presence of major concentrations of one acidic and one neutral amino acid which could not be identified chromatographically and electrophoretically with any previously known amino acids.

After a considerable amount of MS, PMR and ^{13}C-NMR work on these two amino acids and the assistance of CONGEN (Cahart, Smith, Brown & Djerassi, 1975) very unusual cyclobutance containing structures could be written for them both. Their structures were conclusively shown to be 2,4-methano-glutamic acid (1-amino-1,3-*trans*-dicarboxycyclobutane) (X) and 2,4-methanoproline (2-carboxy-2,4-methano-pyrrolidine) (XI) by X-ray crystallography (Fig. 5.6). The latter amino acid contains the previously unknown 2-aza-bicyclo (2,1,1) hexane ring system. Unlike most amino and imino acids, these compounds did not show optical activity.

Neither of these new amino acids has yet shown any activity against whole insects. However, in a locust neuromuscular preparation (the isolated retractor unguis) (Clements & May, 1974) the glutamic acid analogue (X) did

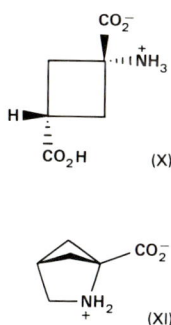

Fig. 5.6. The structures of 2,4-methano-glutamic acid (X) and 2,4-methanoproline (XI) derived from seeds of a legume; *Ateleia herbert.*

show some weak glutamate-like activity at concentrations 10 to 100 times more than required for glutamic acid. More compound would be required to establish this glutamate activity fully. On the other hand, the amino acid (XI) showed no glutamate antagonism or agonism in this test.

The two novel amino acids were also tested for their effects on seedling growth (Table 5.4). Whilst the proline analogue showed little or no activity, the glutamate analogue did show significant inhibition of shoot and root growth. Our interest in novel amino acids will continue.

5.3 NATURAL PRODUCTS RESEARCH AT SHELL BIOSCIENCES LABOR-
ATORY

As can be seen from the few selected examples of our work above, we have had a considerable interest in biologically active natural products spanning many years. For the most part our interests were concerned with natural products related to particular crop/disease situations. Recently, we have directed more effort into the study of biologically active natural products, where we are looking at natural sources simultaneously for insect, plant and fungal active compounds. This is a multidisciplinary approach employing the skills of chemists, biologists, biochemists and microbiologists all working towards the same goal.

Natural materials for examination are selected quite randomly or because we are aware of some clues about the types of biological activity that they may contain. In any event, all of the extracts of these natural materials are screened

TABLE 5.4 SEEDLING GROWTH AFTER APPLICATION OF THE TWO
NOVEL AMINO ACIDS FROM LEGUME SEEDS

Compound	Dose (ppm)	Shoot/ Root	Barnyard[a] grass	Cress[b]	Lettuce[c]
(structure: cyclobutane with CO_2^-, $\cdots\text{NH}_3^+$, H, CO_2H)	0	Shoot	100±17.8	100±16.6	100±12.3
		Root	100±19.1	100±25.1	100±14.1
	1	Shoot	103.1	103.2	100
		Root	90.7	97.8	90.7
	10	Shoot	87.5	80.3[d]	100
		Root	101.1	104.3	84.2[d]
	100	Shoot	90.9	87.2	96.8
		Root	71.3[d]	78.9	35.7[d]
(structure: cyclopentane with CO_2^- and NH_2^+)	0	Shoot	100±17.3	100±23.0	100±13.3
		Root	100±25.1	100±31.0	100±25.7
	1	Shoot	88.8	103.3	90.3
		Root	100.0	100.0	112.8
	10	Shoot	106.8	97.8	93.5
		Root	89.4	119.1	113.4
	100	Shoot	82.5[d]	89.5	96.8
		Root	86.3	121.7	105.9

[a]After 6 days growth; [b]after 7 days growth; [c]after 10 days growth; [d]significantly different from control at $P = 0.05$.
NB: Growth is expressed as a percentage of the weight of the control plants.

in special small-scale screens against insects, plants (including observations of herbicidal and growth regulatory effects) and plant pathogenic fungi.

The objective of this programme is to find new leads for chemical synthesis or, where the natural source is microbiological, to investigate the possibility of producing the potential new agrochemicals by fermentations. It represents a conscious attempt to widen our horizons in our search for new pest- and disease-control agents and we feel confident that the investigation of biologically active natural products is an exciting, complementary approach to the more conventional one of organic synthesis and screening alone. This approach to natural products also complements, and contributes to, our continuing fundamental biochemical and physiological studies which are directed towards finding new agrochemicals.

5.4 THE ROLE OF INDUSTRY IN NATURAL PRODUCTS RESEARCH

The title of this session was "The role of industry in developing naturally occurring pest control agents". Clearly, industry can have its own separate role and is usually well equipped in terms of skills and equipment to fulfil its own function. On the other hand, the contribution that non-industrial scientists make in this area is substantial and it is clear that our industry-based research could benefit from outside collaboration. We at Shell Biosciences Laboratory recognize this, and already collaborate with a few natural products researchers in academia. The advantages that industry can offer the non-industrial researcher are its collective biological and chemical skills and facilities. An industrial research laboratory like our own often has a wide selection of biological testing facilities which would be unlikely to be available to the non-industrial scientist. On the chemical side also, industry can often offer specialist equipment and expertise. There are also many areas where the academic or government scientists can contribute specialist skills and facilities to industry.

Another very important contribution that industry can make to any mutually satisfactory collaboration with non-industry-based researchers is industry's ability to develop a discovery and put it into the field for practical agricultural use. There must be many interesting ideas and observations which could be put to good use, but are lost because of the somewhat separate lives led by industry and non-industrial researchers. To avoid this wastage, it would be better if researchers in government and academic research institutes were able to collaborate with industry more freely from a very early stage in research programmes. Collaboration and pooling of skills, resources and ideas is what is required to achieve the objective of researches into biologically active natural products for pest and disease control.

5.5 SUMMARY

Natural products that have biological activities in insects, plants and plant pathogenic microorganisms provide a continual inspiration to the agrochemicals business in its search for new products to control pests and improve crop yields. The final products can be either the natural products themselves or synthetic chemicals which are structurally or topographically related to them. At Shell Biosciences Laboratory, Sittingbourne, we have had an interest in biologically active natural products spanning many years and our activity in this area has recently increased considerably. We are now, more than ever, searching for new agrochemical leads from natural sources.

The approach of our natural products activity to date is illustrated by some recent examples. Our search for novel anti-fungal antibiotics led to the discovery of a novel class of glycolipid antibiotics from *Boletus elegans*. Plants' natural defence mechanism against fungal attack have been examined in the case of grapevines and rice, and the defensive chemicals concerned, the viniferins and momilactones respectively, have been characterized. In the former case some chemical mimicry has been possible while in the latter, synthetic chemicals can assist the plant's own defence mechanism.

Amino acids with novel structures often fall into the classification, biologically active natural products. Two such compounds, a cyclobutane amino acid and an aza-bicyclo (2,1,1) hexane imino acid from the insect-resistant seed of *Ateleia herbert smithii*, will be described.

5.6 ACKNOWLEDGEMENTS

I wish to thank my colleagues Drs. R. A.–S. A.–Y. Ah Sum, P. Langcake and B. N. Herbert of Shell Research Ltd., Sittingbourne, who collaborated with me in the research reviewed in this chapter.

5.7 REFERENCES

Applewhite T. H., Binder R. G. & Graffield W. (1967) *J. Org. Chem.* **32**, 1173–1178.
Baker C. D. & Gunstone F. D. (1963) *J. Chem. Soc. Perkin*, pp. 759–.
Bell E. A., Qureshi M. Y., Pryce R. J., Janzen D. H., Lemke P. & Clardy J. (1980) 2,4-Methanoproline (2-carboxy-2,4-methano-pyrrolidine) and 2,4-methanoglutamic acid (1-amino-1,3-dicarboxycyclobutane) in seeds of *Ateleia herbert smithii* Pittier (Leguminosae). *J. Am. Chem. Soc.* **102**, 1409–1412.
Cahart R. E., Smith D. H., Brown H. & Djerassi C. (1975) *J. Am. Chem. Soc.* **97**, 5755–.
Cartwright D., Langcake P., Pryce R. J., Leworthy D. P. & Ride J. P. (1977) *Nature, Lond.* **267**, 511–.
Clements A. N. & May T. E. (1974) *J. exp. Biol.* **61**, 421–.
Day P. R. (1974) *Genetics of Host-Parasite Interactions*, p. 169, Freeman W. H., San Francisco.
Dejongh D. C., Radford T., Hribar J. D., Hanessian S., Bieber M., Dawson G. & Sweely C. C. (1969) *J. Am. Chem. Soc.* **19**, 1728–.
Kato T. *et al.* (1973) *Tetrahedron Letters*, 3861–.
Kato T. *et al.* (1977) *J. Chem Soc.* 250–.
Khana S. N. & Gupta P. C. (1967) *Phytochem.* **6**, 735–.
Langcake P. & Pryce R. J. (1976) *Physiol. Plant Path.* **9**, 77–.
Langcake P. & Pryce R. J. (1977a) *Experientia*, **33**, 151–.
Langcake P. & Pryce R. J. (1977b) *Pytochem.* **16**, 1193–.
Langcake P. & Pryce R. J. (1977c) *Chem. Comm.* 208–.
Okabe H. & Kawasaki T. (1970) *Tetrahedron Letters* 3123–.
Pryce R. J. & Langcake P. (1977) *Phytochem.* **16**, 1452–.
Schroepfer G. J. & Block K. (1965) *J. Biol. Chem.* **240**, 54.
Tulloch A. (1968) *Can. J. Chem.* **46**, 3727–.
Tulloch A. P., Hill A. & Spencer J. F. T. (1968) *Can. J. Chem.* 3337–.
Valette G. & Liber A. (1938) *Compt. Rend. Soc. Biol.* 362–.
Wagner H. & Kaxamaier P. (1971) *Tetrahedron* 3233–.

5.8 DISCUSSION

Bowers: I find the first set of your compounds interesting because they are related to the vinylstilbenes found in tropical vines. Some of these stilbenes are known to display potent fungicidal activity.

Marini-Bettolo: I should also like to add that longicidines also display potent fungicidal activity which have some potential for exploitation.

Jacobson: I would like to comment on the need for close co-operation between government, industry and university agencies regarding the question of research. While I fully support your feelings about this concept, I wish to point out that there should be more closer collaboration/co-operation between government agricultural research stations, academia and industry than is hitherto the case.

Graham-Bryce: I find Dr. Pryce's view in this matter rather dismissive of the role of public sector. He seems to imply that co-operation is an activity which industry can do without, and goes on to confuse the issue further by suggesting that it is a bonus that industry could take advantage of, as if it was theirs. In my view, the public sector has three important roles which industry cannot fulfil under the commercial constraints pertinent to industry. The first is a watchdog function over matters relating to the environment, etc., the second is to explore and expose principles which are really not economical for industry to take on, but which in the long run, can benefit industry. The third role is to provide a solution to factor-oriented problems which are viable but not likely to be economic in the eyes of industry.

Djerassi: This sort of statement has been made quite frequently, but I do not see much being done by the public sector in solving factor-oriented problems at all. Emphasis on special products with minor uses is not attractive to large industry.

Graham-Bryce: Both the public sector and industry need one another in promoting new products. Full development and registration of a new product for a minor use is likely to be costly to both. Therefore both must share the responsibilities involved, i.e. collaborate effectively.

Szabo: Because of individual approach and the desire to go it alone, many novel but quite useless inventions have been produced. This underlies the worthlessness of the one-sided approach and therefore begs for a collective and beneficial approach involving all interested institutions.

Djerassi: I fully agree that collaborative approach stands a better chance of success than lack of co-operation.

Pryce: Industry would prefer to be involved in part of the work right from start. An approach to industry to partake in the development of a biological activity of a compound or to promote a product at any other time would be unlikely to lead to the desired results.

Feeny: Commenting on the subject of co-operation in research amongst organizations, I would like to add that industry is quite keen on following leads from work produced by phytochemical research institutions. This interest, I believe, would form a good basis for collaboration. The public sector is producing a vast amount of work on really all frontiers of research. Therefore, I think it is in the future interests of industry that they should at least collaborate with institutions producing such vast and unrestricted research material.

Djerassi: Industry is keen on their economy and therefore unlikely to be interested in public sector developed products like phermones with limited potential for development. On the other hand, industry has to be secretive because of the stiff competition with other concerns. The secrecy with which industry does research is not beneficial to the public sector especially in the area of information exchange beneficial to the promotion of science in general.

Pryce: I think that as long as industry remains secretive about its research, the confidence necessary for its co-operation with the public sector is likely to be undermined to an extent that will totally exclude any meaningful collaboration in research. A carefully worked-out balance is desirable for the mutual existence of these sectors.

Ellis-Pratt: We have had a very nasty experience with industry on collaborative research. In our particular experience we requested industry (and they agreed) to establish to what extent precocene-like activity is a measure of epoxidase activity. Despite several attempts, we have not been able to obtain any useful result from the firms which did the experiments for us. It sounds like an isolated case, but it can frustrate effort to uncover novel products for commercial development. It is also very damaging for frank exchange of scientific views between academia and industry.

Djerassi: I fully agree with these feelings. For the past few years development of plant-protection chemicals has been the domain of a few large, highly diversified companies. And the public sector has sort of only come into play during the stage of introducing the compounds to the market. We have cases where the public sector has asked for further studies on the safety of a new product simply because the government was not kept informed of industry's intentions at the early stage. Less than 30% of the plant-protection work done by industry should involve government agencies, the academia and specialized laboratories at an early stage of development of such products.

Section B

Natural Pyrethrin and Synthetic Pyrethroids

CHAPTER 6

Natural Pyrethrin as an Insecticide: Problems of Chemical Activity, Industrialization and Use

DISMAS A. OTIENO

Chemistry and Biochemistry Research Unit, The International Centre of Insect Physiology and Ecology (ICIPE), P.O. Box 30772, Nairobi, Kenya

CONTENTS

|---|---|---|
6.1	Introduction	93
6.2	Photochemical Rearrangements	94
	6.2.1 The chrysanthemic acids	94
	6.2.2 Rethrolones and rethrins	94
6.3	Thermal and Acid-catalysed Rearrangements	96
	6.3.1 The chrysanthemic acids	96
	6.3.2 Rethrolones	98
	6.3.3 Rethrins	100
6.4	Problems with Development of the Pyrethrins as All-purposes Crop Protection and Disease Vector Control Chemicals	101
	6.4.1 Problems with production of pyrethrum flowers	101
	6.4.2 Approaches to the problems associated with pyrethrum production and the instability of the pyrethrins	102
6.5	Conclusion	103
6.6	Summary	103
6.7	References	103
6.8	Discussion	104

6.1 INTRODUCTION

The natural pyrethrin insecticides have the desirable environmental properties of being both non-toxic to mammals and non-persistent. These features, combined with their broad spectrum of insecticidal activity and rapid biodegradibility to non-toxic residues, have made the pyrethrins ideal agents for the control of insect pests and disease vectors. In spite of their superior environmental qualities, the general instability of the natural pyrethrins has restricted their development considerably as all-purpose crop-protection and disease vector-control agents. During the past decade, a number of studies have been made of the fundamental chemical processes involved during the degradation of the natural pyrethrins under photochemical, thermal and acid or base conditions. In addition, more studies have been done on the agronomic methods useful for pyrethrum production. This paper summarizes the progress made in these areas of research.

6.2 PHOTOCHEMICAL REARRANGEMENTS

The natural pyrethrins (1) exhibit many chemically sensitive structural features, and for convenience, it has been found appropriate to examine the independent chemical behaviour of the various chrysanthemic acid (2) and rethrolone (3) (Fig. 6.1) of which molecules the esters are constructed.

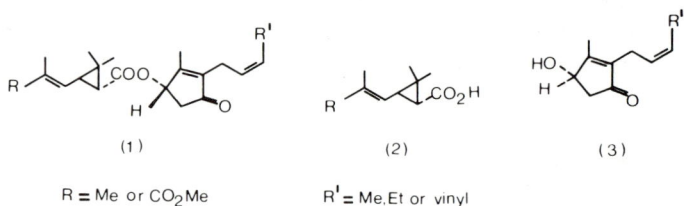

R = Me or CO$_2$Me   R' = Me,Et or vinyl

Fig. 6.1. The natural pyrethrins (1) and the various chrysanthemic acid (2) and rethrolone (3) molecules of which they are constructed.

6.2.1   The chrysanthemic acids

The solution photochemistry of naturally derived (1R)-*trans*-chrysanthemic acid (4) (Fig. 6.2) has been summarized previously by Bullivant & Pattenden (1976). The fundamental photochemical transformation corresponds to a *trans*-to-*cis* isomerization about the cyclopropane ring, leading to a photostationary 2:1 equilibrium mixture of recemic modifications of the *cis*- and *trans*-acids (5 and 4, respectively). Two minor photorearrangements are cleavage of the 1,3-bond and radical recombination leading to the butanolide (7), and fragmentation to 3-methylbut-2-enoic acid (6) (Fig. 6.2). Separate studies with (1R)-*trans*-chrysanthemic acid (4) have established that the photochemical isomerization to (5) proceeds with almost complete racemization. By contrast, irradiation of thin films of rethrin esters derived

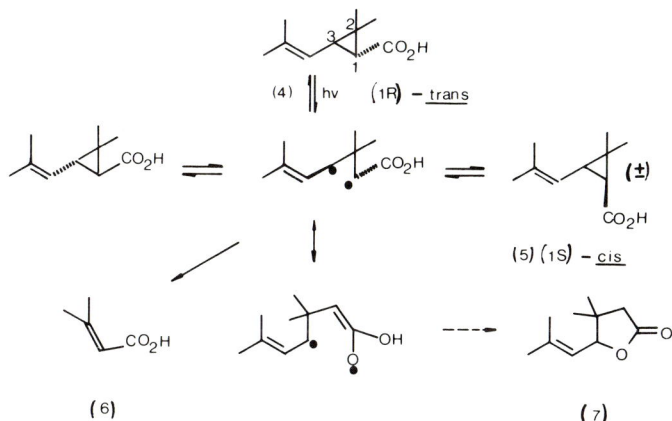

FIG. 6.2. Photochemical rearrangement of (1R)-*trans*-chrysanthemic acid (4).

from chrysanthemic acid, in the presence of molecular oxygen, leads to a variety of products (8–10) (Fig 6.3) resulting from specific ixodation of the double bonds in the chrysanthemic acid portions of the molecules (Chen & Casida, 1969); whether or not these products result from involvement of single or triplet oxygen is not clear.

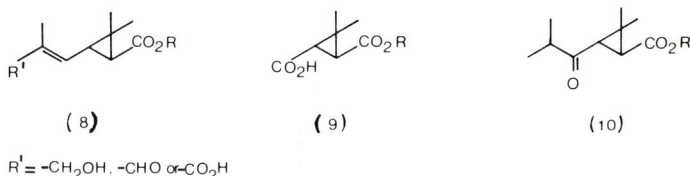

$R' = -CH_2OH, -CHO \text{ or } -CO_2H$

FIG. 6.3. Irradiation products of thin films of rethrin esters derived from chrysanthemic acid, in the presence of oxygen.

Ueda & Matsui (1971) have shown that irradiation of solutions of the di-*tert*-butyl ester (11) of "chrysanthemum dicarboxylic acid" 3-(2-carboxy-prop-1-enyl)-2,2-dimethylcyclopropanecarboxylic acid results in simultaneous isomerization about the double bond and the cyclopropane ring, leading to a mixture of isomers (12), (13) and (14) (see Fig. 6.4)

6.2.2 Rethrolones and the rethrins

The rethrolones (3) feature a side chain which is sensitive towards photochemical (Z)-(E) isomerization. Thus photolysis of either the methyl ether or the acetate derivatives of cinerolone (3) (R^1 = methyl) and

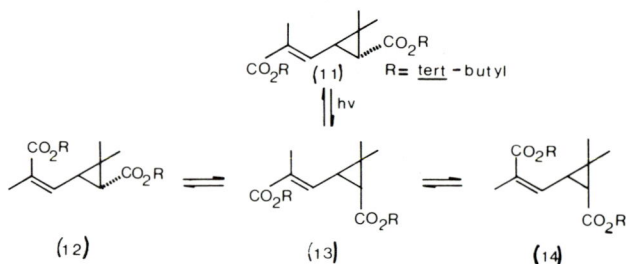

Fig. 6.4. Photochemical products of the *tert*-butyl ester of "chrysanthemum dicarboxylic acid"(11).

jasmololone (3) (R^1 = ethyl) leads (Bullivant & Pattenden, 1976) to photostationary 9:1 equilibrium mixtures of the (E)- and (Z)- isomers.

6.3   THERMAL AND ACID-CATALYSED REARRANGEMENTS

6.3.1   The chrysanthemic acids

The cyclopropane ring system in the chrysanthemic acids (2) (R = methyl or —COOH) and their derivatives is well known to be susceptible to cleavage by thermal processes, and a range of rearrangement and breakdown products has been observed (Crombie, 1971). Recent work (Otieno, Pattenden & Popplestone, 1977) has shown that acid catalysed thermal rearrangement of chrysanthemic acid (4) indeed follows a very complex pathway leading to isomeric products formally resulting from cleavage of all three cyclopropane ring bonds. Thus, treatment of the acid with pyridinium chloride at 210°C led to the dienoic acid (17) (20%), and to a mixture of neutral products (55%) containing the diene (18), the butanolides (7), (20) and (21) and the pentonolide (19). These conversions are rationalized as shown in Figs. 6.6 and 6.7. Thus the conversion of chrysanthemic acid to the dienoic acid (17) and the diene (18) proceeds with cleavage of the 2,3-bond, via a thermal homodienyl 1,5-H shift leading to the intermediate (22), followed by decarboxylation to (18) and isomerization to (17). The butanolide (20) is then derived from the dienoic acid (17). Both the pentenolide (19) and the butanolide (21) result from cleavage of the cyclopropane 1,2-bond in (4) (Fig. 6.7), and cleavage of the remaining 1,3-bond in (4) accounts for the formation of the butanolide (7).

It has been shown (Crombie, Doherty, Pattenden & Woods, 1971; Otieno, 1977) that thermal rearrangement of chrysanthemum dicarboxylic acid (2, R=COOH) at 260°C results in the formation of a complex mixture of acidic and neutral products (Fig. 6.8). The principal components of the acidic

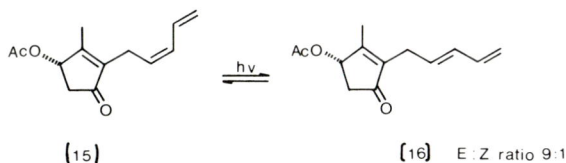

(15)   (16)   E : Z ratio 9 : 1

Fig. 6.5. Photochemical isomerization of (Z)-pyrethrolone acetate (15) to the (E)-isomer (16).

and neutral fractions have been separated by a combination of preparative thin-layer and gas–liquid chromatography and identified as the dienoic acids (26), (27) and (28), the cyclopropanecarboxylic acid (23), 2,6-xylenol (30) the cyclohexadienone (29), the butanolide (24) and the pentenolide (25). The majority of these products result from an initial thermal homodienyl 1,5-H shift in the dicarboxylic acid (2) (R=COOH) followed by decarboxylation to (26) and (28), cyclodehydration to (29) and 1,3-sigmatropic shift leading to (30) (see Fig. 6.8).

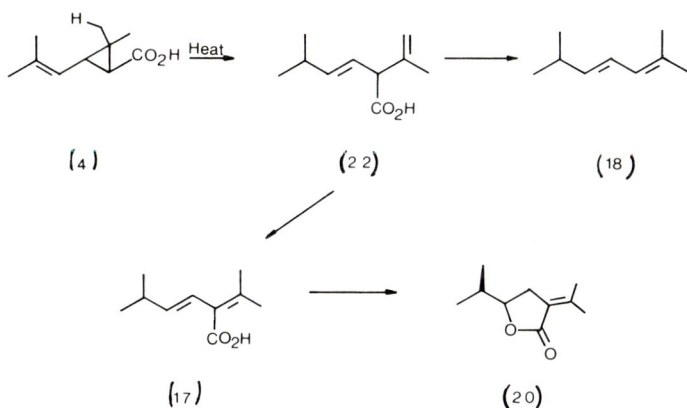

(4)   (22)   (18)

(17)   (20)

Fig. 6.6. Rearrangement products obtained by thermal-acid-catalysis of chrysanthemic acid (4).

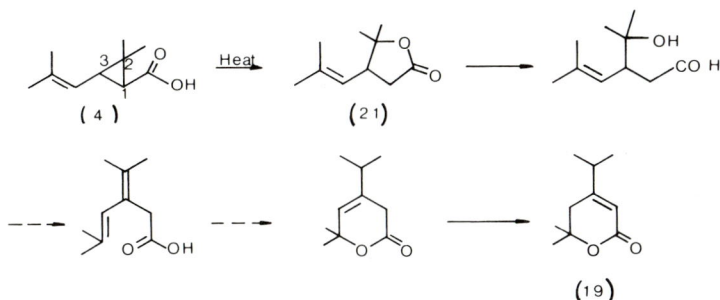

(4)   (21)

(19)

Fig. 6.7. Further products obtained by thermal acid-catalysis of chrysanthemic acid (4).

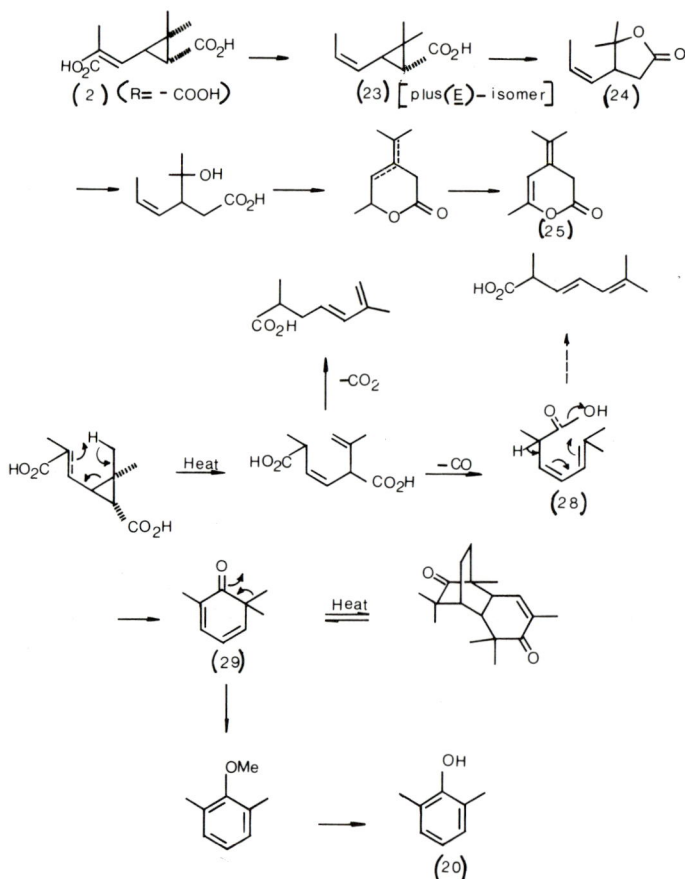

FIG. 6.8. Products derived from chrysanthemum decarboxylic acid $(2, R = \text{—COOH})$ following thermal rearrangement.

6.3.2   Rethrolones

Early work by Elliott (1964) has shown that thermal rearrangement of pyrethrolone (3) (R^1 = vinyl) at 200°C leads to the iso-compound (31) by way of a concerted 1,5-H shift process (Fig. 6.9). By contrast, Pattenden & Storer (1974) have shown that treatment of the rethrolones (3) (R = H, methyl or ethyl) with pyridinium chloride at 210°C produces the corresponding cyclopentenediones (32) (R = H, methyl or ethyl respectively), whereas the pyrethrolone gives largely the dione (33) accompanied by the isomers (34) and (35) (Fig. 6.9).

Detailed investigations have established that these interesting rearrangements probably proceed via intermediate enols (36) (Fig. 6.10). The

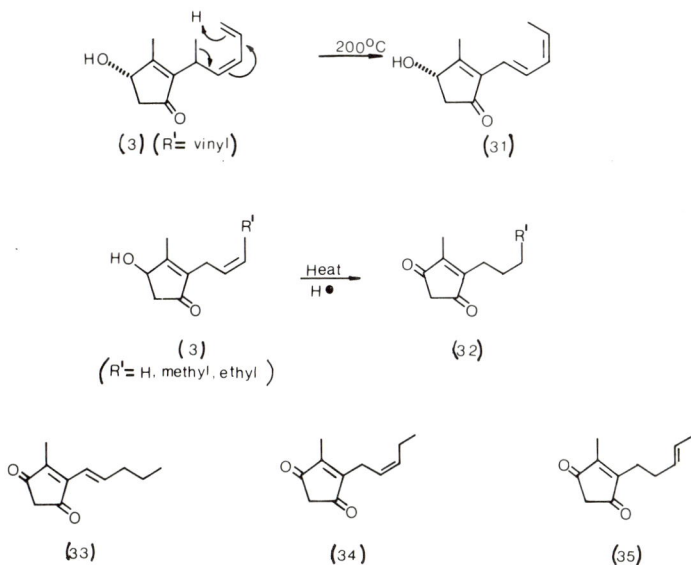

FIG. 6.9. Thermal rearrangements of pyrethrolone 3 (R = vinyl) and rethrolones 3 (R = H, methyl or ethyl); 33, 34 are isomeric cyclopent-4-3n-1,3-diones given by pyrethrolone upon acid catalysis.

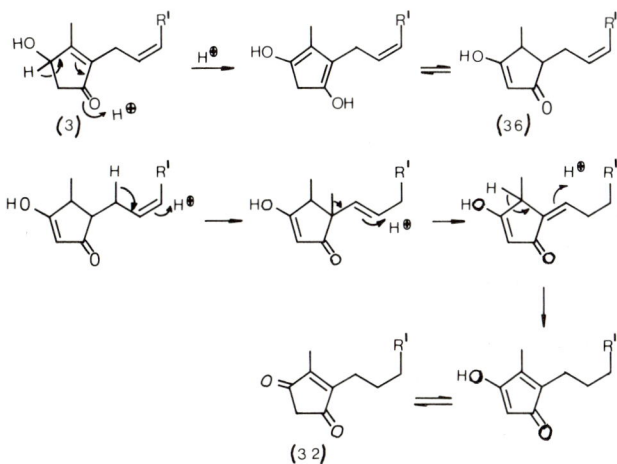

FIG. 6.10. Thermal rearrangements and isomerizations of rethrolones.

conversions (3) to (36) are envisaged as acid-catalysed enonedienol type rearrangements and conversions (36) to (32) as double-bond migrations proceeding by a series of prototropic shifts (see Fig. 6.10). A mixture of enolic forms of cyclopentanedione (36) (R = H) is also obtained (Elliott, 1965) by treatment of allethrolone with a solution of sodium methoxide in methanol. In an analogous treatment of pyrethrolone, Pattenden & Storer (1974) found that the enol (37) was produced (Fig. 6.11).

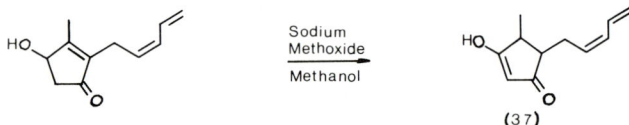

Fɪɢ. 6.11. Thermal base-catalysis of pyrethrolone leading to the enol 37.

6.3.3   Rethrins

Thermal rearrangement of pyrethrin 1(3) (R = methyl, R^1 = vinyl) at 400°C leads to a complex mixture of products from which chrysanthemic acid (4), the diene (18) and the interesting bicyclic ketone (39) have been isolated (Nakada, Yura & Murayama, 1972). The bicyclic ketone (39) results from an electrocyclic reaction in the isocompound (38) obtained from concomitant 1,5-H shift in the pentadienyl side-chain, and elimination of the chrysanthemic acid residue in the pyrethrin I (Fig. 6.12).

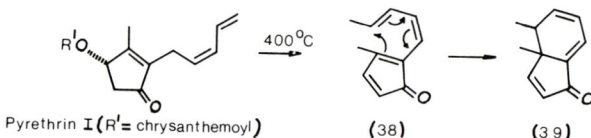

Fɪɢ. 6.12. Thermal rearrangement of pyrethrin I.

In a similar manner, thermal rearrangement of allethrin (1) (R = methyl, R^1 =H) at 400°C produces a mixture of (4) and (18) accompanied by the butanolide (21) and also the isomeric indanones (24) and (43) (Nakada, Yura & Murayama, 1971). The indanones result from decarbonylation of the corresponding Diels–Alder dimers, for example, (4) produced from the cyclopentadienone intermediate (40) (Fig. 6.13).

The acid-catalysed rearrangement of allethrin in pyridinium chloride at 210°C has also been investigated by Otieno (1977) and shown to lead to a mixture of products (7) and (17–21) resulting from rearrangement of the chrysanthemic acid portion of the ester, and to the cyclopentenedione (32) (R^1 =H) from the isomerization of the allethrolone portion of the ester; the enol

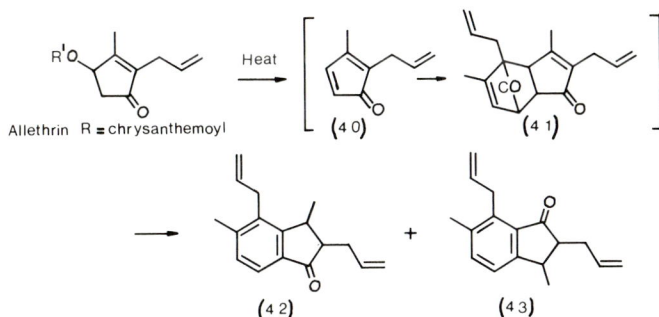

FIG. 6.13. Pyrolysis products of allethrin.

ester (44) is implicated in this scheme (see Fig. 6.14).

The corresponding behaviour of "chrysanthemum dicarboxylic acid" (2) (R=COOH) in acid or base environments has not been reported, and only inconclusive results have been obtained from attempted studies of the stability of the pyrethrins in soil and under microbial activity.

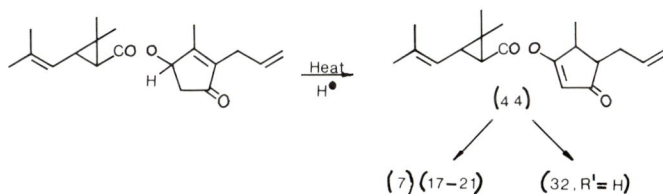

FIG. 6.14. Acid-catalysed thermal rearrangement of allethrin.

6.4 PROBLEMS WITH DEVELOPMENT OF THE PYRETHRINS AS ALL-PURPOSES CROP PROTECTION AND DISEASE VECTOR-CONTROL CHEMICALS

The chemistry of the pyrethrins described in sections 6.2 and 6.3 of this paper shows that the pyrethrins are highly unstable in the field and can quickly be converted to non-toxic products. This is one disadvantage of the pyrethrins which frustrates their commercial development as crop protection and disease vector-control chemicals. In addition to this setback, the pyrethrins are expensive and their supply in the world market is inadequate and very unpredictable.

6.4.1 Problems with production of pyrethrum flowers

Several problems contribute to the inconsistent supply of pyrethrum products in the world market. In the first place, pyrethrum farming requires agricultural conditions suitable for the production of lucrative cash crops like tea and coffee, and many farmers in pyrethrum-growing areas tend to favour the latter cash crops because of favourable marginal returns which they can use to buy food and other necessities. In the second place, pyrethrum farming is a labour-intensive exercise, and a large-scale venture invariably proves cost prohibitive. Thirdly, cheap technology for large-scale weeding and picking of pyrethrum flowers is not yet available. Thus, small-scale farming of the plant is a standard practice in nearly all pyrethrum-producing countries. But experience with small-scale pyrethrum farming in Kenya shows that this scheme too has serious problems with (a) the organization of the small farms into co-operative societies, (b) the establishment of dependable mechanisms for handling and distributing the sale proceeds arising from delivery of the crop to pyrethrum-processing and marketing organizations and (c) competition with food crops which must be grown in the same plots.

6.4.2 Approaches to the problems associated with pyrethrum production and the instability of the pyrethrins

There are several approaches to overcome the problems associated with pyrethrum production. The principal pyrethrum-growing countries choose those approaches suitable to their own situations. In Kenya, for example, the handicaps of small-scale pyrethrum farming are being arrested by (a) increasing the base-rate payments to the farmer, (b) adjusting the role of co-operative societies in pyrethrum production and in delivery of the flowers to the processing factory, in a manner profitable to the farmer, and (c) by applying the recently discovered tissue culture principles and methods to propagate clonally, young pathogen-free pyrethrum plants which grow to uniform heights and which are known to have uniform pyrethrin content upon maturity. These measures have led to the increase in acreage under pyrethrum cultivation in Kenya, and are therefore regarded as a move in the right direction.

The advances made in the study of the instability of the pyrethrin insecticides, together with achievements from research in pyrethrin formulation and application techniques, have been usefully adapted to prolong the effective life of the pyrethrins in field tests. For example, a combination of an oxidant (2,6-dioctadecyl-*p*-cresol) and an ultraviolet screening agent (benzyl cinnamate) in an oil formulation has been found (Miskus & Andrews, 1972) to stabilize significantly the pyrethrins for a period up to 4 h. Such stabilized formulations have been used to control forest and agricultural insect pests with quite encouraging results. In addition to this approach, advances in slow-release methods like microencapsulation are also slowly being adapted to extend the effective life of the pyrethrin molecules in field applications.

6.5 CONCLUSION

Therefore, taken together with the moves to improve the production of pyrethrum flowers, the measures discussed enhance prospects for the expanded use of the pyrethrins to improve food production and forestall insect threat to livestock and human survival. Pyrethrum has a good future.

6.6 SUMMARY

The problems which restrict the development of the natural pyrethrin insecticides, as all-purpose crop-protection and disease vector-control agents, are discussed and the present knowledge about the degradation of the pyrethrins under photochemical, thermal and acid or base conditions is summarized.

6.7 REFERENCES

Bullivant M. J. & Pattendent G. (1976) Photodecomposition of natural pyrethrins and related compounds. *Pesticide Science* **7**, 231–235.

Chen Y. L. & Casida J. E. (1969) Photodecomposition of pyrethrin I, allethrin, phthalthrin and dimethrin. *J. Agricl Food Chem.* **17**, 208–215.

Crombie L., Doherty C. F., Pattenden G. & Woods D. K. (1971) The acid thermal decomposition products of natural chrysanthemum dicarboxylic acid. *J. Chem. Soc.* (C), pp. 2739–2743.

Elliott M. (1964) The pyrethrins and related compounds. III. Thermal isomerisation of *cis*-pyrethrolone and its derivatives. *J. Chem. Soc., Lond.*, pp. 888–892.

Elliot M. (1965) The pyrethrins and related compounds. VI. The structures of the "enols" of pyrethrolone. *J. Chem. Soc. Lond.*, pp. 3097–3101.

Miskus R. P. & Andrews T. L. (1972) Stabilization of thin films of pyrethrins and allethrin. *J. Agric. Food Chem.* **20**, No. 2, 313–315.

Nakada Y., Yura Y. & Murayama K. (1971) Pyrolysis of allethrin. *Bull. Chem. Soc. Japan* **44**, 1724.

Nakada Y., Yura Y. & Murayama K. (1972) Studies on chrysanthemate derivatives. II. Pyrolysis of pyrethrin I. *Bull. Chem. Soc. Japan* **45**, 2243.

Otieno D. A. (1977) Thermal and photochemical studies of pyrethrins. Ph.D. Thesis, University of Nottingham, UK, pp. 90–92.

Otieno D. A., Pattenden G. & Popplestone C. R. (1977) Thermal acid-catalysed rearrangements of natural chrysanthemic acids. *J. Chem. Soc. Perkin Trans. I*, pp. 196–201.

Pattenden G. & Storer R. (1974) Acid-catalysed transformations of substituted 4-hydroxy-2-(prop-2-enyl) cyclopent-2-enones. *J. Chem. Soc. Perkin Trans. I*, pp. 1606–1611.

Ueda K. & Matsui M. (1971) Studies on chrysanthemic acid. XXI. Photochemical isomerisation of chrysanthemic acid and its derivatives. *Tetrahedron* **27**, 2771–2774.

6.8 DISCUSSION

Menn: Thank you Dr. Otieno. Can the pyrethrins compete with the synthetic pyrethroids on a cost-effectiveness basis?

Otieno: The natural pyrethrins possess qualities which I believe would enable them to compete with the synthetic pyrethroids on a cost/effectiveness basis.

Odhiambo: Commenting on Dr. Otieno's answer to Dr. Menn's question I want to bring to your attention that one of the important features of the pyrethrins is that they have a naturally-occurring repellency factor (or factors). These have greatly enhanced the value of pyrethrins, adding to their knockdown and kill qualities. It is true that we do not yet fully understand the chemistry and biological behaviour of the repellents in the natural pyrethrum plant or its extract, but there is no doubt that the repellency factors are a vital aspect of the pyrethrins.

Arunga: I have two questions on synergism of the pyrethrins. We know that the knockdown and kill properties of the pyrethrins are synergized by piperonyl butoxide. Yet to date, no compound has been found which can enhance the repellency effects of these compounds. My questions are as follows:

(1) Can we really synergize the repellency activity of the natural pyrethrins?

(2) I am aware of work which was initiated sometime ago by the Pyrethrum Board of Kenya, to search for synergists from local plant sources in Kenya. Can you tell us the progress which has been made on this work?

Otieno: (1) I do not know for sure whether we can synergize repellency effects of the pyrethrins. We will first need to have a full understanding of the mode of action of repellent factors in pyrethrum in order to see a possibility for enhancing their repellent effects. To date, as Professor Odhiambo said, we do not have this knowledge.

(2) No outstanding progress has been made in these efforts.

Johnson: Can you please give us an overview of how the pyrethrum industry in Kenya operates?

Otieno: In Kenya, the pyrethrum industry is controlled by the Pyrethrum Board of Kenya — a statutory body established by an Act of Parliament with responsibility to regulate the quantity of pyrethrum flowers produced each year, and to undertake processing, sales and marketing of pyrethrum and its products on behalf of the farmers growing the crop.

Johnson: What sort of pyrethrum products are marketed by the Pyrethrum Board of Kenya? Can you briefly describe the composition of these products in terms of pyrethrin content?

I have one more question. I wonder how insects flying from a pyrethrum plant to another survive the lethal effects of the pyrethrins?

Otieno: (1) The Pyrethrum Board of Kenya markets a high-quality pale extract containing 25% or 50% pyrethrins for incorporation into aerosol formulations, and a fine pyrethrum powder standardized to contain 1.3 or 0.6% pyrethrins for the manufacture of mosquito coils. Further, some crude extract (ole-resin) used in grain protection and agricultural insecticides, is also sold to MGK Company in the US.

    (2) The natural pyrethrins are contact insecticides with high boiling points. Insects flying from immature pyrethrum plant to another do not risk contacting these compounds or their vapour, and will therefore survive their lethal action.

Abe: In connection with Professor Johnson's question, are there any insect pests of pyrethrum?

Otieno: Yes, the athropod pests *Thrips tabaci* and *Thrips nigropilosus* are known to cause appreciable damage to the pyrethrum plant.

Jondiko: You have said that the pyrethrins are unstable in heat, light and in acidic or basic conditions. Can you explain the photostability of the pyrethrins in the flowers? Is there any chemical basis for this, and if so, is the compound responsible known and has it been used to make a pyrethrin formulation with long effective life?

Otieno: Pyrethrum flowers kept in the sun or under hot conditions for unduly long periods lose their pyrethrin content most likely via thermal and photodegradation route. There is no known chemical basis for the stability of the pyrethrins in pyrethrum flowers. It is possible that the natural products which occur together with the pyrethrins in pyrethrum flowers, confer protection to the pyrethrins against light, but the mechanisms fundamental to this process have not been elucidated.

Bowers: I would like to comment on this point. I believe that the possibility that a naturally occurring chemical compound (or compounds) is responsible for protection of the pyrethrins in the pyrethrum plant is quite attractive, and basic research into this phenomenon should be undertaken. I think that new knowledge gained here might be beneficial in formulating a pyrethrin preparation with long effective life and capable of application as an agricultural insecticide.

Kahumbura: Dr. Otieno, can you explain how a burning mosquito coil repels insects? It seems to me that under the burning conditions, the pyrethrins would thermally decompose and we would be losing their efficacy?

Otieno: There are several schools of thought regarding how the pyrethrins are liberated from a mosquito coil to repel insects. One of the most widely believed concepts is that a short distance from the burning tip of the coil, the pyrethrins are vaporized to the gaseous form in which they display repellent activity. It is thought that under these temperature conditions, the expected pyrolysis of the pyrethrins is insignificant. Several questions raised by this theory have not, however, been resolved unequivocally.

Onuorah: In view of the instability feature of the pyrethrins and the reported declining levels of production in Kenya and stiff competition from synthetic pesticides, how do you rate the future of the pyrethrin insecticides?

Otieno: At present, in spite of the progress made in the manufacture of synthetic pyrethroids, the pyrethrins will continue to compete successfully in the insecticide market. Further, the safety of the pyrethrins and their low persistence make them a strong contender as environmentally safe insecticide. I therefore believe that the market for the natural pyrethrins is strong. I also think that to consolidate this position, the production of pyrethrum flowers must be improved, and so must formulation and application techniques for the pyrethrins.

Young: Commenting on the question concerning the future outlook for pyrethrum, I would like to point out that unless a stable supply of pyrethrum can be assured, pesticide formulators will continue the present trend toward synthetic pyrethrins thereby making the future of the pyrethrins look bleak. I hope that the matter is considered a most urgent priority by the Pyrethrum Board of Kenya.

Taylor: Dr. Otieno mentioned the amount of pyrethrum flowers produced in Kenya in 1978. Can he tell us how this compares with production in the other years?

Otieno: The annual production of pyrethrum flowers in Kenya has been steadily falling since the 1974/75 period. The table below illustrates this trend.

Production period	Amount of flowers produced (in metric tonnes)
1974/75	15,046
1975/76	14,267
1976/77	11,482
1977/78	8437

Up to 22 August 1979 since the beginning of the 1978/79 production period, 7095 tonnes of flowers had been received by the pyrethrum processing factory in Nakuru. There is a strong possibility that this trend will be discontinued in the 1979/80 production period and thereafter production will begin to rise steadily again.

Kahumbura: Would you like to comment on breeding experiments done with pyrethrum flowers and how these experiments have altered the pyrethrin content of the flowers?

Otieno: Yes. Considerable breeding experiments have been done with the pyrethrum plant. As a result of these studies, we now have pyrethrum clones with pyrethrin content quite in excess of 2%. However, this achievement has had no widespread use because we have found that flowers with more than 2% pyrethrins present unresolved operational difficulties in extraction and refining.

CHAPTER 7

Development of Synthetic Insecticides from Natural Products: Case History of Pyrethroids from Pyrethrins

JOHN E. CASIDA

Pesticide Chemistry and Toxicology Laboratory, Department of Entomological Sciences, University of California, Berkeley, California 94720, USA

CONTENTS

7.1	Introduction	109
7.2	Need for Pyrethrum Extract and Pyrethroids	110
7.3	Composition of Pyrethrum Extract and Pyrethroids	110
7.4	Pyrethrin Chemistry and Biology	112
	7.4.1 Instability of pyrethrins	112
	7.4.2 Stabilizing pyrethroids	114
	7.4.3 Persistence of pyrethroids	115
	7.4.4 Action of pyrethroids on insects and insect nerves	116
	7.4.5 Pest resistance to pyrethroids	117
	7.4.6 Pyrethroid action in mammals	118
7.5	Expanding Concept of Pyrethroid Structure and Action	119
7.6	Natural to Synthetic: Pyrethrins to Pyrethroids	120
7.7	Acknowledgements	120
7.8	Summary	120
7.9	References	121
7.10	Discussion	123

7.1 INTRODUCTION

Natural products can be used directly in pest control and as model compounds for research and development programmes to produce analogues with favourable economics and enhanced potency and utility. This approach involves identification of the active component(s) followed by synthesis and structure optimization. It has been most successful in obtaining pyrethroids

NPIPM-F

suitable for crop protection from the natural but unstable pyrethrins. A review of this case history may provide insights into needs and prospects for development of synthetic insecticides from other types of natural products.

7.2   NEED FOR PYRETHRUM EXTRACT AND PYRETHROIDS

Pest insect control over the past 35 years has been achieved largely with chlorinated hydrocarbons, organophosphorus compounds and methylcarbamates. These synthetic compounds are generally contact and broad spectrum insecticides acting on the nervous system. Pyrethrum extract is the best example of a natural product which acts as an insect nerve poison. The pyrethrins have one advantage over all other insecticides, natural or synthetic, which is their low toxicity to mammals, verified by both recent toxicological tests and a long history of safe use.

The supply of pyrethrum flowers is often far below the level required on a worldwide basis. Early shortages resulted from limited production areas and distribution networks. As these became organized the culture and marketing were sometimes disrupted by world wars, by adverse weather and by economic, political and social changes in the countries of pyrethrum production. Each period of limited pyrethrum supply accelerated research to find appropriate synthetic substitutes.

Until recently, there appeared to be a seemingly endless array of new and relatively inexpensive synthetic contact insecticides. During this period only pyrethroid devotees continued research with the goal of understanding the unique properties of the pyrethrins and modifying them for crop protection. The demise of chlorinated hydrocarbons in the past decade has left a gap most easily filled by other contact insecticides acting quickly on the nervous system. The required compounds had to have a residual activity from a few days to a few weeks and offer little or no problems of undue persistence, food-chain accumulation and adverse toxicology. Pyrethroids of enhanced photostability were developed just in time to help fill this gap in crop-protection capabilities. Structures of pyrethroids considered are given in Fig. 7.1.

7.3   COMPOSITION OF PYRETHRUM EXTRACT AND PYRETHROIDS

Different factors govern the purity of the pyrethrins and pyrethroids as normally used. Natural products are difficult to economically separate from a myriad of co-extractives but importantly, the desired compound is usually stereochemically pure. Compounds from directed synthesis are easier to purify but are often obtained as isomer mixtures. It is therefore to be expected

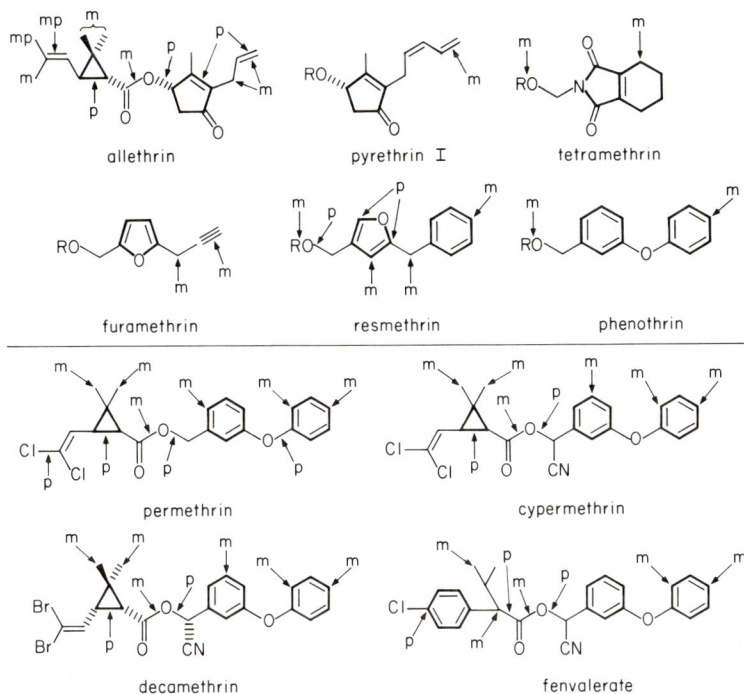

FIG. 7.1. Structures of six chrysanthemates and four pyrethroids with halogen-containing acid moieties. The chrysanthemates are shown as 1R, *trans* isomers although they are used as various isomer mixtures. Sites are shown for metabolic (m) and photochemical (p) attack (Casida and Ruzo, 1980; Ruzo, 1980; and references cited therein).

that the purity and composition will differ greatly for the pyrethrins in pyrethrum extract and the insecticidal components in pyrethroids. Pyrethrum extract is readily analysed by GLC or HPLC allowing precise standardization of composition, relative to pyrethrins I and II and the less active cinerins and jasmolins (Head, 1973; Mourot, Boisseau & Gayot, 1978). The impurities vary with the extraction and purification procedures. Refined pyrethrum extract contains approximately 40% pyrethrins I and II, 15% cinerins and jasmolins, 20% fatty acids and 25% of other mostly unidentified ingredients (Head, 1973). Pyrethroids vary from about 12.5% content of the most active ingredient (e.g. cypermethrin, based on the 1R,*cis*,αS isomer in a 50:50 *cis,trans*,αRS mixture) to almost 100% active ingredient for the compounds of high stereochemical purity (e.g. S-bioallethrin, [1R,*trans*]resmethrin and decamethrin). Production of stereochemically pure pyrethroids is becoming more common with improvements in isomer-specific syntheses, resolution of optical antipodes, isomerization and recycling of inactive byproducts.

The toxicological properties of impurities are evaluated in part in the normal testing of technical and commercial grade insecticides. Problems have not been encountered with impurities in pyrethrins and pyrethroids except earlier for an allergenic factor(s) causing dermatitis on exposure to crude pyrethrum extract; this problem is resolved on preparing refined pyrethrum extract (Barthel, 1973; Rickett & Tyszkiewicz, 1973; Williams, 1973).

7.4   PYRETHRIN CHEMISTRY AND BIOLOGY

7.4.1   Instability of pyrethrins

Most insecticidal natural products are limited in their potency and application by instability to light and rapid metabolism. The pyrethrins are useful only under conditions in which they contact the pest without extensive exposure to light and air. Attempts to improve their stability by adding antioxidants or UV screens or by formation of inclusion compounds have been only marginally successful (Miskus & Andrews, 1972; Yamatomo, Ohsawa & Plapp, 1977). The photoproducts of rethrins (Fig. 7.2) are best understood for S-bioallethrin under normal use conditions (Ruzo, Gaughan & Casida, 1980). The instability of the pyrethrins is due in part to photo-oxidation of the isobutenyl group and, probably even more important, of the pentadienyl group (Chen & Casida, 1969). The solution photochemistry of jasmolin I is partially defined under anaerobic conditions (Bullivant & Pattenden, 1976). The pyrethrate side chain undergoes very facile E/Z photoisomerization (Ohsawa & Casida, 1979).

Insects treated with pyrethrum extract or allethrin generally recover following knockdown unless a synergist such as piperonyl butoxide is also used. This enhances potency by inhibiting oxidative detoxification which occurs at the isobutenyl, pentadienyl and allyl moieties (Fig. 7.3) (Casida & Maddrell, 1971; Elliott, Jones, Kimmel & Casida, 1972; Yamamoto & Casida, 1966; Yamamoto, Kimmel & Casida, 1969). The former site of attack is defined in houseflies, rats and their microsomal oxidase systems. The latter two sites are firmly identified only in rats. The carbomethoxy group of pyrethrin II is hydrolysed in all of these systems.

Rapid metabolism of the pyrethrins has important toxicological implications beyond the necessity to add a synergist for enhanced insecticidal activity. It is a critical safety feature for mammals evident by the very low toxicity of pyrethrins I and II following oral administration relative to that

PHOTOREACTIONS OF RETHRINS

reactions of alcohol side chain (R)

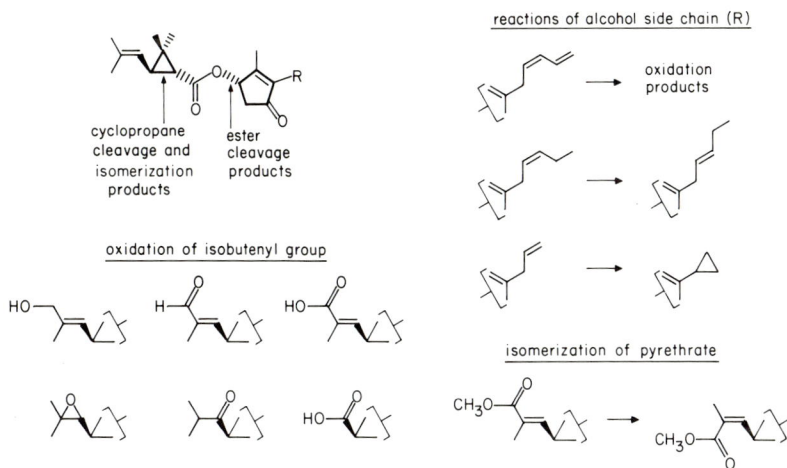

cyclopropane cleavage and isomerization products

ester cleavage products

oxidation products

oxidation of isobutenyl group

isomerization of pyrethrate

FIG. 7.2. Photoreactions of rethrins examined with pyrethrin I, Jasmolin I, S-bioallenthrin and methyl pyrethrate.

METABOLISM OF RETHRINS

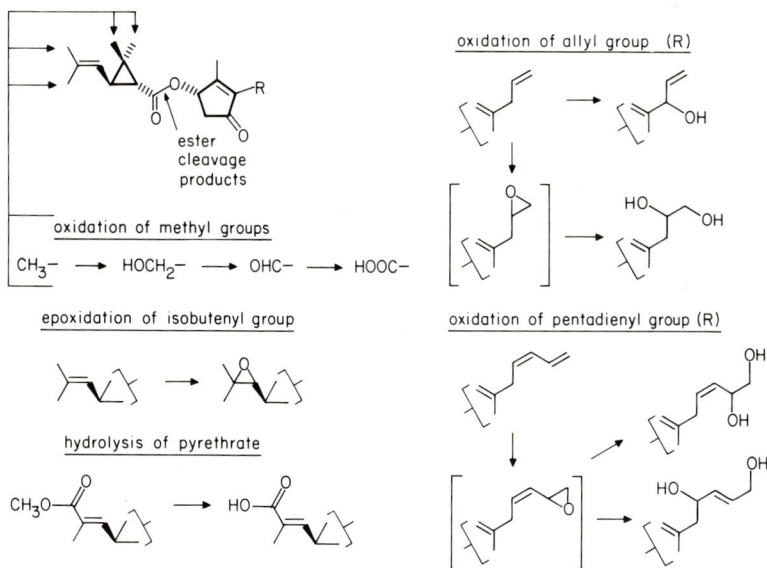

ester cleavage products

oxidation of allyl group (R)

oxidation of methyl groups

$CH_3- \longrightarrow HOCH_2- \longrightarrow OHC- \longrightarrow HOOC-$

epoxidation of isobutenyl group

oxidation of pentadienyl group (R)

hydrolysis of pyrethrate

FIG. 7.3. Metabolism of rethrins examined with pyrethrins I and II and S-bioallethrin.

TABLE 7.1 ORAL AND INTRAVENOUS TOXICITIES OF PYRETHRINS
AND TWO PYRETHROIDS TO FEMALE RATS

	Toxic dose, mg kg^{-1}*	
Pyrethroid	Oral LD_{50}	IV ~ LD
Pyrethrin I	260–420[a]	5[b]
Pyrethrin II	> 600[a]	1[b]
[1R,*trans*]Resmethrin	> 8000[b]	340[b,c]
Decamethrin	85[d]	3[d]

[a]Male rats with dimethylsulphoxide as carrier vehicle (Kimmel, Elliott & Janes, 1971).
[b]Female rats with glycerol formal as carrier vehicle (Verschoyle & Barnes, 1972).
[c]This toxicity is probably attributable to the impurity, [1R,*cis*]resmethrin present in the [1R,*trans*]resmethrin sample (Gray & Conneors, 1980, personal communication).
[d]Female rats with polyethyleneglycol 200 as carrier vehicle (Roussel Uclaf/Procida, 1977).

following intravenous treatment (Table 7.1).

7.4.2 Stabilizing pyrethroids

An unstable compound can be stabilized only to a limited extent by formulations and additives to minimize photodegradation. There is also a definite limitation in the effectiveness of synergists to retard metabolism. Additional stabilization requires a change to related chemicals, i.e. replacing the pyrethrins with pyrethroids. Many of the structural modifications that enhance potency (Elliott & Janes, 1978; Yoshioka, 1978) are also those that retard photodecomposition and metabolic detoxification (Fig. 7.1).

Improved photostability is achieved by replacing the pyrethrolone moiety of pyrethrin I by other alcohol moieties giving allethrin, tetramethrin, resmethrin and phenothrin where oxidation of the isobutenyl group has now become a factor or the major factor limiting stability. The acid moiety is photostabilized by replacing the isobutenyl group by the more stable and isosteric dihalovinyl group (i.e. substitution of methyl by chloro or bromo substituents) or by removing the cyclopropane ring on shifting from dihalovinyldimethylcyclopropanecarboxylates to chlorophenylisovalerates. The most stable and effective commercial esters are 3-phenoxybenzyl and α-cyano-3-phenoxybenzyl derivatives, i.e. permethrin, cypermethrin, decamethrin and fenvalerate. The stabilized compounds still photodecompose by a variety of reactions including ester cleavage, cyclopropane ring opening and closing with concomitant isomerization, and decarboxylation of the chlorophenyl isovalerates (Fig. 7.1) (Holmstead, Casida & Ruzo, 1977; Ruzo, 1980).

The relative stability of different pyrethroids in insects is sometimes interpretable from the effectiveness of synergists, i.e. high synergistic factors are associated with rapid pyrethroid metabolism. Metabolism data for insects are at present insufficient to directly intercompare structure–insecticidal activity and structure–biodegradability relationships. Only one systematic study (Soderlund & Casida, 1977a) with mouse liver microsomes provides sufficient data for comparisons (Table 7.2). Esterase but not oxidase metabolism is severely retarded on shifting from esters of *trans*- to *cis*-cyclopropanecarboxylates and from esters of primary to secondary alcohols. The α-cyano-3-phenoxybenzyl and chlorophenyl-isovalerate substituents retard both esterase and oxidase metabolism.

TABLE 7.2 METABOLISM RATES OF VARIOUS PYRETHROIDS
INCUBATED WITH MOUSE LIVER MICROSOMAL ENZYMES (BASED ON
SODERLUND AND CASIDA, 1977 a,b)

| | Metabolism rate relative to pyrethrin I oxidation, % | | | |
| | Esterase | | Oxidase | |
Pyrethroid	1R, *trans*	1R, *cis*	1R, *trans*	1R, *cis*
	Primary alcohol esters			
Tetramethrin	60	7	148	179
Furamethrin	476	10	48	83
Resmethrin	188	< 7	48	69
Phenomethrin	140	< 10	64	88
Permethrin	183	< 5	71	62
	Secondary alcohol esters			
Pyrethrin I	< 7[a]		100[b]	
Allethrin (αS)	< 5		105	
Cypermethrin (αRS)	40	< 5	10	12
Decamethrin (αS)		< 7		< 10
Fenvalerate (αRS)[c]	10	< 5	14	26

[a]Pyrethrin II and cinerin II are hydrolysed faster than pyrethrin I and cinerin I.
[b]Pyrethrin II and cinerins I and II are oxidized at rates similar to pyrethrin I.
[c]The R acid is arbitrarily positioned under *trans* and the S acid under *cis*.

7.4.3 Persistence of pyrethroids

The trend in developing pyrethroids for crop protection has been to select compounds for potency and environmental stability. It was therefore necessary to evaluate whether or not these structural changes yielded compounds of acceptable persistence in mammals. The most potent commercial pyrethroid insecticide, decamenthrin, provides a suitable example in this respect. Decamethrin is metabolized in mammals and *in vitro* in enzyme systems by ester hydrolysis and five sites of hydroxylation (Fig. 7.1) (Ruzo, Unai & Casida, 1978; Ruzo, Engel & Casida, 1979; Shono, Ohsawa &

Casida, 1979). Decamethrin and its halogen-containing and aryl moieties are quickly and almost completely eliminated from orally-treated rats and mice, except for a small amount of the parent compound which is moderately persistent in fat. Remarkable insecticidal potency is achieved without sacrificing mammalian biodegradability. The pyrethroids in general are quite biodegradable as long as they are not localized at sites inaccessible to esterase and oxidase attack. Compounds which are *cis*-cyclopropane-carboxylates generally have greater persistence in biological systems than the corresponding *trans*-isomers (Abernathy, Ueda, Engel, Gaughan & Casida, 1973; Casida, Gaughan & Ruzo, 1979). Environmental persistence is dominated by their photochemical stability and low volatility and water solubility. Undue persistence of the parent pyrethroid can be avoided with suitable use restrictions.

The pyrethrins and pyrethroids are biodegraded with relative ease but their complex structure and multiple functional groups inevitably lead to an unusually large number and variety of metabolites and photoproducts. The chemistry and activity of these persisting or terminal products are of considerable interest. The mixtures of photoproducts of the pyrethrins are less toxic than the pyrethrins when injected into mice or applied topically to houseflies. This is also usually the case with the large number of pyrethroids examined (Chen & Casida, 1969; Ueda, Gaughan & Casida, 1974). The photoproducts and metabolites of chrysanthemates include epoxides and other reactive but transitory intermediates. Resmethrin photolysis gives an unidentified toxicant(s) (Ueda *et al.*, 1974). Allethrin is converted to unidentified mutagen(s) formed both as metabolite(s) (Matsuoka, Hayashi & Ishidate, 1979) and photoproduct(s) (Kimmel, Ruzo & Casida, 1980, unpublished results). Fortunately, many of the environmental metabolites and photoproducts are the same as compounds formed in mammals; under these circumstances, the long-term effects of the environmental products are evident in part from the normal chronic feeding studies with rats.

7.4.4 Actions of pyrethroids on insects and insect nerves

Pyrethrins and pyrethroids are exceptionally potent as knockdown, repellent and antifeeding agents. Potent synthetic compounds used extensively as knockdown agents are allethrin, tetramethrin and furamenthrin (Fig. 7.1). Knockdown activity is associated with polar substituents conferring rapid entry into the insect, i.e. with pyrethrin II of the pyrethrins (Briggs, Elliott, Farnham & Janes, 1974). Polarity may also be associated with rapid metabolism; so, this property conferring knockdown often also permits later recovery. Pyrethrins are especially important in flushing cockroaches from their hiding places to contact lethal doses of insecticide, in space sprays for rapid knockdown, and in mosquito coils or other delivery systems to disorient or kill mosquitoes in or near human habitats. The agricultural pyrethroids have considerable insect antifeeding activity, which is a valuable property in minimizing crop damage even though it may impede contact with or ingestion of a lethal dose of pyrethroid. Pyrethroids differ greatly in their potency for repelling insects and mites, perhaps in relation to their volatility and excitant action on sensory nerves.

Pyrethrins and pyrethroids act in insects and other arthropods on both sensory and motor nerves and on the central nervous system, sensory nerves being generally most sensitive (Gammon, 1980b; Lowenstein, 1942; Narahashi, 1971, 1976; Wouters & Van Den Bercken, 1978). They interact with sodium, potassium and perhaps calcium channels in axonal membranes. The pyrethrins, pyrethroids and DDT are more toxic at lower than at higher temperature, a definite limitation in some areas of pest control. This negative temperature coefficient may result simply from the dual effect of both the low temperature and the pyrethroid in prolongation of the sodium conductance increase. At the organismal level, the negative temperature coefficient could also be due to changes in the significance of action on different nerves within an insect. Nerve excitation, the primary mode of action of pyrethrins and pyrethroids, leads to release of neurosecretory hormones (e.g. diuretic hormone; Casida & Maddrell, 1971) which in turn cause changes that may ultimately result in death.

Pyrethroids appear to fall into two different types based on the symptoms they produce in poisoned insects and their actions on invertebrate nerves (Gammon, 1980a,b; Gammon & Casida, 1980, unpublished results, Narahashi, 1980, personal communication). Insects poisoned with pyrethrins and several non-cyano pyrethroids become uncoordinated and hyperactive, followed by paralysis and death; these are referred to as Type I symptoms. Decamethrin, cypermethrin and fenvalerate containing the α-cyano group gives a prominent convulsive phase in the poisoning, designated as Type II symptoms. The Type I but not the Type II effects are associated with repetitive firing in the cercal sensory nerves of the American cockroach (Gammon & Casida, 1980). Thus, structural modifications of the pyrethrins have led to pyrethroids with differences in their site(s) of action or with a new mode of action.

7.4.5   Pest resistance to pyrethroids

Resistance to pyrethrins was recognized quite early to involve DDT cross-resistance from the extensive use of this chlorinated hydrocarbon (Busvine, 1951; Fine, 1963). It is established in a few cases and assumed more generally that a nerve-insensitivity factor (known in Diptera as the knockdown resistance or kdr factor) confers this type of resistance to DDT, pyrethrins and pyrethroids (Farnham, 1977; Gammon, 1980 a,b; Miller, Kennedy & Collins, 1979; Osborne & Hart, 1979; Tsukamoto, Narahashi & Yamasaki, 1965). Genetic studies with houseflies to isolate individual pyrethroid-resistance factors also establish the importance in some strains of one or more of metabolic detoxification, reduced penetration rate and modifier characters (Farnham, 1973, 1977; Plapp & Casida, 1969). In an extreme case from laboratory selection for these multiple resistance factors, houseflies are highly resistant to every pyrethroid tested (Sawicki, 1978, 1980, personal communication).

The use of more stable pyrethroids has selected resistant strains of several species which, combined with the DDT-pyrethrins resistance cases, account for a total of twenty-six resistant pest species (Sawicki, 1980, personal communication). In some cases resistance develops rapidly with intensive pressure. The extent and rate of spread of resistant populations will strongly influence the use of pyrethroids and their favourable economics. Resistance to pyrethroids will also ultimately affect the use pattern of pyrethrum extract. Careful consideration is currently being given to alternative strategies to delay the selection of resistant strains, e.g. use chrysanthemates first and stabilized pyrethroids later, use synergists to revert resistant strains to susceptibility, and use pyrethroids acting in different ways or varying in stability. It is now clear that resistance is an inevitable limitation in the use of pyrethroids on many dipterans, lepidopterans and other pests. Of interest, but of unknown practical importance, are observations of negatively correlated cross-resistance between fenvalerate and organophosphorus pesticides, i.e. malathion against several leaf-hoppers (Ozaki, 1979) and azinphos-methyl against two-spotted mites (Chapman & Penman, 1979).

7.4.6 Pyrethroid action in mammals

Two types of pyrethroid action are evident from acute toxicity studies with rats (Aldridge, 1980; Barnes & Verschoyle, 1974; Ray & Cremer, 1979; Verschoyle & Barnes, 1972; White, Verschoyle, Moradian & Barnes, 1976). The pyrethrins, allethrin (1R, *cis*), resmethrin and most of the other non-cyano compounds produce aggressive sparring behaviour, tremors and increased body temperature (Type I effects). Decamethrin (Table 7.1) and most cyanophenoxybenzyl pyrethroids give cholinergic salivation, pawing, choreoathetosis and tonic seizures (Type II effects) possibly associated with an action mediated via the basal ganglia. Based on the limited evidence available, the two classes of compounds giving the different types of actions in mammals may be similar to or even identical with those giving the different effects in insects. On considering compounds of similar insecticidal activity, the target site in the mammalian brain may be highly sensitive to one isomer of a pyrethroid (e.g. (1R, *cis*) resmethrin) but not to another isomer (e.g. (1R, *trans*) resmethrin, Table 7.1) (Gray, Cannors, Hoellinger & Nam, 1980, (personal communication); White *et al.*, 1976) Target site sensitivity is only one factor in animal susceptibility. Metabolic detoxification is also important since esterases and oxidase inhibitors greatly increase the toxicity of some pyrethroids (Gaughan, Engel & Casida, 1980, unpublished results; Casida, 1977b).

Lifetime feeding studies on a variety of pyrethroids are currently in progress or have been recently completed by the relevant industrial laboratories. Although the results of these investigations are not published at present, the subacute toxicity studies with rats give generally favourable findings for the pyrethrins and pyrethroids. Effects on the nervous system are normally transient with no permanent lesion. However, changes in liver pathology or function are detected in rats on high dietary levels of pyrethrins, several chrysanthemates and permethrin in feeding studies of 3 to 18 months duration (Bond, Mauger & De Feo, 1973; Miyamoto, 1976). Hepatic microsomal mixed-function oxidases can also be induced (Carlson & Schoenig, 1980; Glickman, Elcombe & Lech, 1980; Springfield, Carlson & De Feo, 1971). Some pyrethroids are considerably more potent than pyrethrum extract in producing cellular and enzyme changes in liver and other organs. Each pyrethroid must be fully evaluated on its own, since small changes in toxicant structure can importantly influence long-term effects. On the other hand, difficulties encountered with one compound cannot be interpreted as carrying over to related compounds, particularly with the diversity of functional groups present in the pyrethroids.

7.5 EXPANDING CONCEPT OF PYRETHROID STRUCTURE AND ACTION

Pyrethrin I served as a prototype for the pyrethroids. Conversely, the pyrethroids are pyrethrin analogues derived by isosteric replacements retaining critical stereochemical features (Elliott & Janes, 1978). The replacements can be so extensive that no functional group of the pyrethrins remains in the analogue. The ester group can be replaced with an equivalent amide, ketone or oxime ether, albeit with significant loss of activity (Berteau & Casida, 1969; Berteau, Casida & Narahashi, 1968; Davies, 1979). These findings establish that pyrethroids differ from organophosphorus and methylcarbamate insecticides in not forming a covalent derivative at the nerve target or receptor site. The cyclopropane ring is not required for pyrethroid-like activity (Yoshioka, 1978). Pyrethroids are already subdivided into two types of action, based on the symptoms they produce and their effects on nerves in insects and mammals. No single nerve assay adequately differentiates pyrethroids from non-pyrethroids. Some recent insecticides were prepared as hybrids between pyrethroids and DDT-type molecules (Holan, O'Keefe, Virgona & Walser, 1978). It is no longer possible to easily recognize by their chemistry or action whether or not a few of the newer compounds should be designated as pyrethroids. However, these are exceptional cases and most potent neurotoxins maintaining the essential stereochemical features of pyrethrin I are clearly products from a long and systematic series of structural modifications from the natural product.

7.6 NATURAL TO SYNTHETIC: PYRETHRINS TO PYRETHROIDS

The pyrethrins were uniquely favourable prototypes because of their high potency and rapid contact action as insect nerve poisons and their established effectiveness and safety over a long period of use. Extensive modifications produced and continued to yield pyrethroids improved in ease of synthesis, cost, potency, stability and utility. The term pyrethroid may convey the notion of safety but this point must be fully investigated with each new compound. Even with their large differences in persistence and sometimes even in mode of action the synthetic compounds to date retain many of the favourable properties of the pyrethrins. The case history of pyrethroids from concept to goal to reality is a resounding success but, due to a paucity or lack of other equally favourable natural products, the precedent will not be easy to repeat.

7.7 ACKNOWLEDGEMENT

The author is indebted to the International Centre of Insect Physiology and Ecology and the United Nations Development Programme for the opportunity to participate in the Scientific Working Group on the Use of Naturally Occurring Plant Products in Pest and Disease Control. Drs. Luis Ruzo, Derek Gammon and Isaac Ishaaya of the Pesticide Chemistry and Toxicology Laboratory at Berkeley are acknowledged for helpful comments. A portion of the Berkeley studies were supported by the National Institutes of Health (Grant PO1 ES00049), the Environmental Protection Agency (Grant R805999), Mobil Foundation Incorporated (Edison, New Jersey), MGK Corportion (Minneapolis, Minnesota) and Roussel-Uclaf (Paris, France).

7.8 SUMMARY

Insecticidal natural products are useful models for structure modification to develop new synthetic insecticides. The pyrethrins are broad spectrum, contact insecticides acting as nerve poisons. They also have low mammalian toxicity and a long history of safe use. These unique characteristics and the inconsistent supply of pyrethrum flowers, prompted synthesis of several chrysanthemate insecticides which readily found acceptance as substitutes for pyrethrum extract. Further optimization, guided by potency changes on sequential structural modifications and by expanding knowledge of pyrethroid metabolism and photochemistry, led to exquisitely potent compounds of sufficient persistence for use in crop protection. Although acting in insects and mammals as at least two different types of neurotoxins, the currently-important pyrethroids are readily biodegraded and retain many of the favorable characteristics of the pyrethrins. The case history of pyrethrins to pyrethroids illustrates the advantages and limitations of using natural products as prototypes. It sets a precedent which will be difficult to repeat since it started from a botanical insecticide with uniquely favourable properties.

7.9 REFERENCES

Abernathy C. O., Ueda K., Engel J. L., Gaughan L. C. & Casida J. E. (1973) Substrate-specificity and toxicological significance of pyrethroid-hydrolyzing esterases of mouse liver microsomes. *Pestic. Biochem. Physiol.* **3**, 300–311.

Aldridge W. N. (1980) Mode of action of pyrethroids in mammals: summary for toxicity and histological, neurophysiological and biochemical studies at Carshalton. Table Ronde Roussel Uclaf.

Barnes J. M. & Verschoyle R. D. (1974) Toxicity of new pyrethroid insecticide. *Nature, Lond.* **248**, 711.

Barthel W. F. (1973) Toxicity of pyrethrum and its constituents to mammals. In *Pyrethrum, The Natural Insecticide* (ed. Casida J. E.) pp. 123–142. New York: Academic Press.

Berteau P. E. & Casida J. E. (1969) Synthesis and insecticidal activity of some pyrethroid-like compounds including ones lacking cyclopropane or ester groupings. *J. agric. Food. Chem.* **17**, 931–938.

Berteau P. E., Casida J. E. & Narahashi, T. (1968) Pyrethroid-like biological activity of compounds lacking cyclopropane and ester groups. *Science* **161**, 1151–1153.

Bond H., Mauger K. & De Feo J. J. (1973) Interactions in the toxicity of pyrethrum, synergists, and other chemicals to mammals. In *Pyrethrum, The Natural Insecticide* (ed. Casida J. E.) pp. 177–194. New York: Academic Press.

Briggs G. G., Elliott M., Franham A. W. & Janes N. F. (1974) Structural aspects of the knockdown of pyrethroids. *Pestic. Sci.* **5**, 643–649.

Bullivant M. J. & Pattenden G. (1976) Photochemistry of 2-(prop-2-enyl) cyclopent-2-enones. *J. Chem. Soc. Perkin* **1**, 249–256.

Busvine J. R. (1951) Mechanism of resistance to insecticide in houseflies. *Nature, Lond.* **168**,193–195.

Carlson G. P. & Schoenig G. P. (1980) Induction of liver microsomal NADPH cytochrome c reductase and cytochrome P-450 by some new synthetic pyrethroids. *Toxicol. Appl. Pharmacol.* **52**, 507–512.

Casida J. E., Gaughan, L. C. & Ruzo L. O. (1979) Comparative metabolism of pyrethroids derived from 3-phenoxybenzyl and α-cyano-3-phenoxybenzyl alcohols. In *Advances in Pesticide Science*, Part 2 (ed. Geissbuhler H.), pp. 182–189. New York: Pergamon.

Casida J. E., Kimmel E. C., Elliott M. & Janes N. F. (1971) Oxidative metabolism of pyrethrins in mammals. *Nature, Lond.* **230**, 326–327.

Casida J. E. & Maddrell S. H. P. (1971) Diuretic hormone release on poisoning *Rhodnius* with insecticide chemicals. *Pestic. Biochem. Physiol.* **1**, 71–83.

Chapman R. B. & Penman D. R. (1979) Negatively correlated cross-resistance to a synthetic pyrethroid in organophosphorus-resistant *Tetranychus urticate. Nature, Lond.* **281**, 298–299.

Chen Y-L. & Casida J. E. (1969) Photodecomposition of pyrethrin I, allethrin, phthalthrin, and dimethrin. Modifications in the acid moiety. *J. agric. Food Chem.* **17**, 208–215.

Davies J. H. (1979) Benzyl ethers of aryl alkyl ketoximes: a novel pyrethroid-like group of alkyl insecticides. *Int. Symp. Chemistry of the Pyrethroids, Oxford, England,* 18 July.

Elliott M. & Janes N. F. (1978) Synthetic pyrethroids — a new class of insecticide. *Chem. Soc. Reviews (London)* **7**, 473–505.

Elliott M., Janes N. F., Kimmel E. C. & Casida J. E. (1972) Metabolic fate of pyrethrin I, pyrethrin II, and allenthrin administered orally to rats. *J. agric. Food Chem.* **20**, 300–313.

Farnham A. W. (1973) Genetics of resistance of pyrethroid-selected houseflies, *Musca domestica* L. *Pestic. Sci.* **4**, 513–520.

Farnham A. W. (1977) Genetics of resistance of houseflies (*Musca domestica* L.) to pyrethroids I. Knock-down resistance. *Pestic. Sci.* 8, 631–636.

Fine B. C. (1963) The resent status of resistance to pyrethroid insecticides. *Pyrethrum Post* **7** (2) 18–21 and 27.

Gammon D. W. (1980a) Pyrethroid resistance in a strain of *Spodoptera littoralis* is correlated with decreased sensitivity of the CNS *in vitro. Pestic. Biochem. Physiol.* **13**, 53–62.

Gammon D. W. (1980b) The mode of action of pyrethroids in insects — a review. *Table Ronde Roussel Uclaf.*

Glickman A. H., Elcombe C. R. & Lech J. J. (1980) Induction of hepatic microsomal enzymes by permethrin. *Toxicology* (in press).

Head S. W. (1973) Composition of pyrethrum extract and analysis of pyrethrins. In *Pyrethrum, The Natural Insecticide* (ed. Casida J. E.), pp. 25–53. New York: Academic Press.

Holan G., O'Keefe D. F., Virgona C. & Walser R. (1978) Structural and biological link between pyrethroids and DDT in new insecticides. *Nature Lond.* **272**, 734–736.

Holmstead R. L., Casida J. E. & Ruzo L. O. (1977) Photochemical reactions of pyrethroid insecticides. *Am. Chem. Soc. Symp. Ser.* **42**, 137–146.

Lowenstein O. (1942) A method of physiological assay of pyrethrum extracts. *Nature, Lond.* **150**, 760–762.

Matsuoka A., Hayashi M. & Ishidate M., Jr. (1979) Chromosomal aberration tests on 29 chemicals combined with S9 mix *in vitro. Mutation Res.* **66**, 277–290.

Miller T. A., Kennedy J. M. & Collins C. (1979) CNS insensitivity to pyrethroids in the resistant kdr strain of house flies. *Pestic. Biochem. Physiol.* **12**, 224–230.

Miskus R. P. & Andrews T. L. (1972) Stabilization of thin films of pyrethrins and allethrin. *J. agric. Food. Chem.* **20**, 313–315.

Miyamoto J. (1976) Degradation. metabolism and toxicity of synthetic pyrethroids. *Environ. Hlth. Perspectives* **11**, 15–28.

Mourot D., Boisseau J. & Gayot G. (1978) Separation of pyrethrins by high-pressure liquid chromatography. *Anal. Chim. Acta* **97**, 191–193.

Narahashi T. (1971) Mode of action of pyrethroids. *Bull. Wld. Hlth. Org.* **44**, 337–345.

Narahashi T. (1976) Effects of insecticides on nervous conduction and synaptic transmission. In *Insecticide Biochemistry and Physiology* (Ed. Wilkinson C. F.), pp. 327–352. New York: Plenum Press.

Ohsawa K. & Casida J. E. (1979) Photochemistry of the potent knockdown pyrethroid kadethrin. *J. agric. Food Chem.* **27**, 1112–1120.

Osborne M. P. & Hart R. J. (1979) Neurophysiological studies of the effects of permethrin upon pyrethroid resistant (kdr) and susceptible strains of dipteran larvae. *Pestic. Sci.* **10**, 407–413.

Ozaki K. (1979) Suppression of resistance in the plant — and leafhopper through synergistic combinations. *U.S.-Japan Cooperative Science Program Seminar* (Organizers Georghiou, G. P. and Saito T.), Palm Springs, Calif., Dec. 3–7.

Plapp F. W., Jr. & Casida J. E. (1969) Genetic control of house fly NADPH-dependent oxidases: relation to insecticide chemical metabolism and resistance. *J. econ. Ent.* **62**, 1174–1179.

Ray D. E. & Cremer J. E. (1979) The action of decamethrin (a synthetic pyrethroid) on the rat. *Pestic. Biochem. Physiol.* **10**, 333–340.

Rickett F. E. & Tyszkiewicz K. (1973) Pyrethrum dermatitis. II. The allergenicity of pyrethrum oleoresin and its cross-reactions with the saline extract of pyrethrum flowers. *Pestic. Sci.* **4**, 801–810.

Roussel Uclaf/Procida (1977) K-Othrin (NRDC 161). Technical Bulletin 43 pp.

Ruzo L. O. (1980) Photodecomposition of pyrethroid insecticides. *Table Ronde Roussel Uclaf.*

Ruzo L. O., Engel J. L. & Casida J. E. (1979) Decamethrin metabolites from oxidative, hydrolytic, and conjugative reactions in mide. *J. agric. Food Chem.* **27**, 725–731.

Ruzo L. O., Gaughan L. C. & Casida J. E. (1980) Pyrethroid photochemistry: *S*-bioallethrin. *J. agric. Food Chem.* **28**, 246–249.

Ruzo L. O., Unai T. & Casida J. E. (1978) Decamethrin metabolism in rats. *J. agr. Food Chem.* **26**, 918–925.

Sawicki R. M. (1978) Unusual response of DDT. *Nature, Lond.* **275**, 443–444.

Shono T., Ohsawa K. & Casida J. E. (1979) Metabolism of *trans-* and *cis-*permethrin, *trans-* and *cis-*cypermethrin, and decamethrin by microsomal enzymes. *J. agric. Food Chem.* **27**, 316–325.

Soderlund D. M. & Casida J. E. (1977a) Effects of pyrethroid structure on rates of hydrolysis and oxidation by mouse liver microsomal enzymes. *Pestic. Biochem. Physiol.* **7**, 391–401.

Soderlund D. M. & Casida J. E. (1977b) Substrate specificity of mouse-liver microsomal enzymes in pyrethroid metabolism.. *Am. Chem. Soc. Symp. Ser.* **42**, 162–172.

Springfield A. C., Carlson G. P. & De Feo J. J. (1971) Increased microsomal enzyme activity due to pyrethrum administration. *Toxicol. Appl. Pharmacol.* **19**, 394.

Tsukamoto M., Narahashi T. & Yamasaki T. (1965) Genetic control of low nerve sensitivity to DDT in insecticide-resistant houseflies. *Botyu-Kagaku* **30**, 128–132.

White I. N. H., Verschoyle R. D., Moradian M. H. & Barnes J. M. (1976) The relationship between brain levels of cismethrin and bioresmethrin in female rats and neurotic effects. *Pestic. Biochem. Physiol.* **6**, 4191–500.

Williams C. H. (1973) Tests for possible teratogenic, carcinogenic, mutagenic, and allergenic effects of pyrethrum. In *Pyrethrum, The Natural Insecticide* (ed. Casida J. E.), pp. 167–176. New York: Academic Press.

Wouters W. & Van Den Bercken J. (1978) Action of pyrethroids. *Gen. Pharmacol.* **9**, 387–398.

Ueda K., Gaughan L. C. & Casida J. E. (1974) Photodecomposition of resmethrin and related pyrethroids. *J. agric. Food Chem.* **22**, 212–220.

Verschoyle R. D. & Barnes J. M. (1972) Toxicity of natural or synthetic pyrethrins to rats. *Pestic. Biochem. Physiol.* **2**, 308–311.

Yamamoto I. & Casida J. E. (1966) *O*-Demethyl pyrethrin II analogues from oxidation of pyrethrin I, allethrin, dimethrin, and phthalthrin by a housefly enzyme system. *J. econ. Ent.* **59**, 1542–1543.

Yamamoto I., Kimmel E. C. & Casida J. E. (1969) Oxidative metabolism of pyrethroids in houseflies. *J. agric. Food Chem.* **17**, 1227–1236.

Yamamoto I., Ohsawa K. & Plapp F. W., Jr. (1977) Effect of the inclusion compounds of pyrethroids and methyl parathion on certain cotton insects. *J. Pestic. Sci.* **2**, 41–49.

Yoshioka H. (1978) Development of fenvalerate, a new and unique synthetic pyrethroid containing the phenylisovaleric acid moiety. *Rev. Plant Protection Res.* **11**, 39–52.

7.10 DISCUSSION

Graham-Bryce: May I make a comment on the environmental persistence of the pyrethroids. You referred to the low volatility and the low solubility, but, in considering environmental persistence, should we not place at least equal weight on the lipophilicity or partition properties, and just as you mentioned the importance of adsorption by fat depots, adsorption onto soil organic matter for example, which is also related to the water partition coefficient, can be very important. I think that with most of the pyrethroids that you have considered the degradation in the soil solution would be relatively rapid, as you've indicated, but you could nevertheless get quite long persistence in soil if you just measured the total decline in concentration, because a portion would be held principally by the soil organic matter. I suppose one might relate that to two other examples. DDT, which would also be strongly adsorbed by soil organic matter, would in addition be relatively stable in the soil solution and would therefore have a high persistence. On the other hand, paraquat, the bipyridinium herbicide, is even more strongly adsorbed than either of the above, and in this case by the mineral soil constituents, and is very inactive because of this. Nevertheless, it's very persistent if you imagine the total residue, since the material adsorbed is not degraded. What I'm really asking for is a much more detailed consideration of what environmental persistence means in relation to the potential for biological effect.

Casida: I fully agree with all of your comments. If the compounds were not as readily tied up in the soil, they might be more effective as soil insecticides which is an area where we also have great needs. But as far as I am aware that has not been a favourable property so far.

Jotwani: I wanted to know about this repellent action which is associated with the pyrethroids. Would this not be detrimental in the case of crops where there is cross pollination by insects.

Casida: I cannot give a very good answer. I can only talk around the subject. If the compound stimulates the insects to move and alight continuously, it could cause problems in pollination, disease transmission, etc. Anything that keeps them moving can cause problems. If it keeps them from alighting at all it becomes quite a different situation and in the field I am not sure what would happen. But in the lab, frequently they will not land and feed.

Saxena: Resurgence of the brown plant hopper on the rice plant following the application of decamethrin or decis is well documented. What are your comments on this observation. Is this an effect on the insect or on the rice plant itself. In fact, we use decamethrin in the field to induce BPH populations used in our variety screening programme. We do not use it for insect control.

Casida: I have no idea of the mechanism. The reports that I'm aware of on the resistance of rice leaf hopper to pyrethroids are based on Japanese observations and in some of those cases there is a negative correlation of cross resistence with malathion. As far as I know there has been no specific study of the mechanism of resistance.

Saxena: We are trying to study the mechanism of resurgence. We are carrying out experiments with brown plant hopper populations which were not previously exposed to insecticides; these are insectary-reared populations. The reproductive potential of the insect increases significantly, the growth period of the larvae is reduced significantly, and the pre-oviposition period of females becomes very short, which leads to highly significant buildup in the population.

Casida: In the context of the type of discussion we are having here, I would ask you, does the phenomenon follow the same structure/activity relationships as the insecticidal activity. Of the eight isomers of decamethrin is it only the active isomer (NRDC 161) which gives the type of increased productivity of which you are talking?

Saxena: We are using two insecticides. Methyl parathion was used very extensively for BPH control, but after some time we found the phenomenon of hopper burn, because of resurgence. Methyl parathion was widely used and resistance to it had been well documented in BPH. But decis, because of its novelty, was used extensively for control of vegetable insect pests. They started trying it in the field and without prior exposure to the hopper, resistance was immediately found. What I am suggesting is that there might be some effect of the decis itself on the rice plant that perhaps alters the chemistry of the rice plant.

Casida: Based on studies with other plants it is unlikely that very much decis would enter the plant itself. Based on the doses that have been used, corrected for entry, it must be extremely potent if it is altering the plant in any way. Which is another way of saying I doubt if that is the explanation.

Bowers: I have a philosophical question. Professor Djerassi pointed out that the development time of potentially useful compounds is something like 10 years. Assuming, and I think its a safe assumption, that the pyrethroids will become one of the major insecticides to fill the gap left by the chlorinated hydrocarbons and also the phosphates and carbamates in the future, it is likely that one or two companies, one or two products or structures will dominate the field. It would be reasonable to assume that under heavy pressure there will be rapid development of resistance in major crop pests and disease vectors, and yet you point out that there may be differences in the mode of action in some of these pyrethroids which could be used to advantage to diminish the onset of resistance. If this is so, how does one get this information on the toxicological properties and proper methods of application of these products to the farmer so that the compounds may be used in a rational manner?

Djerassi: This is certainly an extraordinarily important question, and rather than just answer you superficially, let's wait for the answer till after Dr. Elliot's talk because, really, this is pertinent to his talk also. Then we can ask both of them this same question, if that's okay with you.

Bowers: Fine!

CHAPTER 8

Developments in the Chemistry and Action of Pyrethroids

MICHAEL ELLIOTT

Department of Insecticides and Fungicides, Rothamsted Experimental Station
Harpenden, Hertfordshire, AL5 2JQ, UK

CONTENTS

8.1 Introduction 127
8.2 The Natural Pyrethrins 128
8.3 Synthetic Compounds 129
    8.3.1 Allethrin and related synthetic pyrethroids; biological activities 129
    8.3.2 Benzyl chrysanthemates 133
    8.3.3 Resmethrin and bioresmethrin 134
    8.3.4 Development of acidic components 136
8.4 Photostable Pyrethroids 138
    8.4.1 α-Cyanobenzyl esters 139
    8.4.2 (R)- and (S)-Isomers of α-cyanoesters: decamethrin 140
8.5 Other Photostable Pyrethroids 141
8.6 Activities of Various Combinations of Pyrethroid Acids and Alcohols 142
8.7 Present Situation of Pyrethroids 143
8.8 Characteristics of Pyrethroid Insecticides 145
8.9 Summary 145
8.10 Acknowledgement 146
8.11 References 146
8.12 Discussion 149

8.1 INTRODUCTION

The great diversity of structures of natural products is a valuable source of guidance from which to develop related synthetic compounds with modified properties. Natural compounds, if examined thoroughly in the broad surveys used by many agrochemical companies, probably compare favourably in this respect with randomly selected synthetic compounds (Pryce, 1982).

The natural pyrethrins are particularly appropriate as a subject for study because useful general principles can be adduced from the process by which extremely powerful insecticides were derived from the structures of the natural compounds. After discovering compounds with biological activity, and determining their structures, the choice of related compounds to be synthesized and the biological assays used to evaluate them are crucial to the success. The following survey in historical sequence of the development of the present range of synthetic pyrethroids will show this.

8.2   THE NATURAL PYRETHRINS

The natural pyrethrins were suitable foundations for further exploration because they are very effective insecticides with low mammalian toxicity (Gnadinger, 1936, 1945; Crombie & Elliott, 1961; Elliott & Janes, 1978). However, their range of applications is severely limited by their instability in light and because they are expensive. The insecticides are extracted from the flowers of a daisylike plant *Chrysanthemum cinerariaefolium* (Trev.) Sch. Bip. which grows best above 1500 metres, for example in Kenya, Tanzania or Ecuador. Pyrethrum is a relatively expensive insecticide because the crop must still be harvested by hand and supplies depend on good weather.

Early investigations had indicated that the insecticidal compounds were esters, but knowledge of their structures is based mainly on the work, from 1910–1916, of Staudinger & Ruzicka (1924). These chemists examined the structures of the natural products hoping that simpler compounds with similar properties could be synthesized. The structures (Fig. 8.1) are those now accepted from the original work of Staudinger and Ruzicka and subsequent degradative, synthetic and spectroscopic studies by a succession of investigators, principally La Forge and co-workers in America and Harper and Crombie and their collaborators in the UK.

The six esters are derived from two acids and three alcohols. The dimethylcyclopropane acids have R configurations at C-1 and unsaturated side chains *trans* to the carboxyl group at C-3. Pyrethric acid (from which the II series esters are derived) has a methoxycarbonyl group in place of the methyl group of chrysanthemic acid (which gives the I series esters). The alcoholic components are now known to be methylcyclopentenolones with Z-pentadienyl, Z-butenyl or Z-pentenyl side chains at position 2; Staudinger and Ruzicka did not detect that the alcoholic component was heterogeneous, nor that there was a double bond in the C_5 ring. Pyrethrin I and pyrethrin II

PYRETHRIN I CINERIN I JASMOLIN I
(35%) (10%) (5%)

PYRETHRIN II CINERIN II JASMOLIN II
(32%) (14%) (4%)

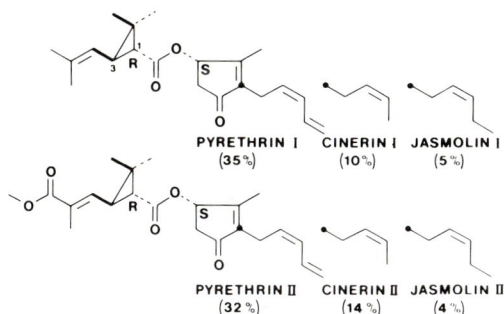

Fig. 8.1. The natural pyrethrins.

are the more important constituents, present in larger quantities in a typical extract.

8.3 SYNTHETIC COMPOUNDS

Staudinger and Ruzicka not only established the main features of the structures of the natural esters, but also laid firm foundations for our knowledge of the relationship between their chemical structures and insecticidal activities. Using a simple but effective assay with cockroaches collected from the crevices of their laboratory, they showed that unsaturation in the side chain of both acidic and alcoholic components was important. From results with thirty-two esters of the natural alcohol mixture and eighty-two (+)-*trans*-chrysanthemates, they concluded that only esters with structures very close to those of the natural compounds were highly active and that the acids and alcohols themselves and their simple derivatives were inactive. Their observations of activity in the chrysanthemates of substituted benzyl alcohols such as piperonyl and cuminyl alcohols and the inactivity of benzyl chrysanthemate itself were especially significant in the light of subsequent developments. They suspected that the optical form at the chiral centres in acid and alcohol was important; this was confirmed later.

8.3.1 Allethrin and related synthetic pyrethroids; biological activities

No other significant advances with synthetic compounds were made until the work, from 1935 onwards, of Haller, La Forge and their collaborators in the United States. They revised the previously suggested structures of the alcohol (Fig. 8.2) and discovered the second component, cinerolone (La Forge & Barthel, 1944, 1945 a,b,c). This was potentially simpler to synthesize than

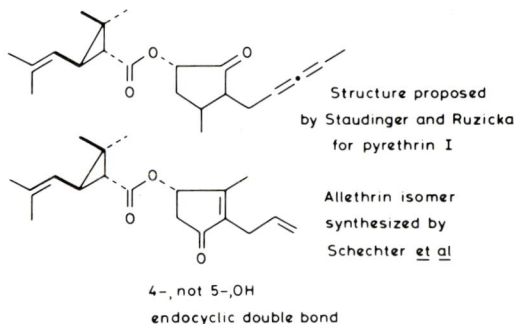

Structure proposed
by Staudinger and Ruzicka
for pyrethrin I

Allethrin isomer
synthesized by
Schechter *et al*

4-, not 5-,OH
endocyclic double bond

FIG. 8.2. Development of allethrin.

pyrethrolone and Schechter *et al.* developed a general route to cyclopentenolones with mono-unsaturated side chains (Schechter, Green & La Forge,
1949). In 1949 they first made allethrin with the simple allyl side chain
starting from the accessible intermediate allyl chloride and with (+)-*cis
trans*-chrysanthemic acid (Campbell & Harper, 1945), as acid component.

Some of the first tests with the new compound in America were against
houseflies by methods which emphasized assessment of knockdown (Laforge,
Gersdorff, Green & Schechter, 1952). In contrast, those insects which might
indicate potential for controlling agricultural, horticultural and public health
pests were examined at Rothamsted. Two synthetic esters, (+)-allethrolone
esterified with synthetic (+)-*cis,trans*-chrysanthemic acid or with naturally
derived (+)-*trans*-chrysanthemic acid (now called respectively, allethrin and
bioallethrin) were compared with the mixture of natural esters (Elliott,
Needham & Potter, 1950) (Table 8.1). In *Musca domestica* (housefly) tests,
where rapidity of knockdown was emphasised, the two synthetic compounds
were respectively three and six times as effective as the natural esters, but for
kill of houseflies, allethrin was no more effective than the natural esters and it
was considerably less active than the natural pyrethrins against three other
species.

These results, summarized in Table 8.1, did not provide a clear basis for
relating activity with structure and to make progress, simplification by
comparing each natural constituent separately with the synthetic compounds
was necessary. Examination of small samples of the natural esters available by
displacement chromatography (Table 8.2) (Lord, Ward, Cornelius & Jarvis,
1952; Ward, 1953) had indicated that pyrethrin I was generally the most
active constituent, but for extended tests larger samples of the esters,
manipulation of which was complicated by instability, were necessary. In
re-examining procedures for isolating the components of the natural esters
(+)-pyrethrolone was separated from the other alcohols as a crystalline
monohydrate, and thence obtained pure for the first time (Elliott, 1964).

TABLE 8.1 POTENCIES OF ALLETHRIN AND BIOALLETHRIN RELATIVE
TO NATURAL PYRETHRINS (= 1)

		Allethrin[a]	Bioallethrin[b]
Musca domestica (knockdown)	USA	3	6
Musca domestica (kill)	USA	1	
Periplaneta americana	USA	0.4	
Blatella germanica	USA	0.2	
Oncopeltus fasciatus	USA	0.2	
Plutella maculipennis	RES	1.8	4
Macrosiphon solarnifolii	RES	0.06	0.1
Oryzaephilus surinamensis	RES	0.4	0.5
Phaedon cochleariae	RES	0.2	0.4

[a]($\pm$)-Allethronyl ($\pm$)-*cis,trans*-chrysanthemate.   RES = Rothamsted
[b]($\pm$)-Allethronyl ($\pm$)-*trans*-chrysanthemate.

Esterification by means of the readily accessible chlorides of the natural acids, then gave supplies of pyrethrins I and II (Fig. 8.1) for chemical and biological examination, limited only by the availability of natural extract to process (Elliott & Janes, 1969).

TABLE 8.2 RELATIVE POTENCY OF THE NATURAL PYRETHRINS

	Pyrethrin I[a]	Cinerin I	Pyrethrin II	Cinerin II
MB, toxicity (acetone, topical)	10	4	3	2
HF, toxicity (acetone, topical)	10	4	18	7
HF, toxicity (kerosene spray)	10	7	3	2
HF, susceptibility to synergism	10	17	4	9
HF, knockdown (dodecane, topical)	10	8	26	10

[a]Used as standard and given potency 10.   MB = mustard beetle   HF = housefly.

Figure 8.3 shows the results of comparing pyrethrin I with S-bioallethrin, the most active, fully resolved isomer of allethrin (Laforge, Green & Schechter, 1954) and with the mixture of allethrin isomers against houseflies and mustard beetles (*Phaedon cochleariae*, Fab.).

Relative potencies are shown in mg kg^{-1} of body weight, to facilitate comparisons between insect species of differing size and weight. Unexpectedly, pyrethrin I was 100 times as effective against mustard beetles as against houseflies on this basis. Further, S-bioallethrin was only 1/40th and allethrin less than 1/200th as active as pyrethrin I against mustard beetles. In contrast S-bioallethrin was 5 times more active than pyrethrin I to houseflies, but this relative activity was only impressive because pyrethrin I is so relatively ineffective against this insect. The mixed isomers of allethrin, often considered as a standard pyrethroid, are even less active against both insects. Subsequent tests with other species showed pyrethrin I to be as active to them as to mustard beetles, and the relative inactivity against houseflies to be atypical.

	MUSCA (HF) DOMESTICA mg. kg⁻¹	PHAEDON (MB) COCHLEARIAE mg. kg⁻¹ (Relative potencies)

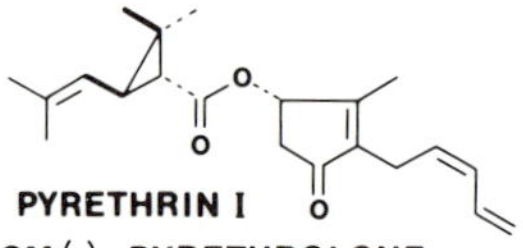

PYRETHRIN I
(FROM (+)- PYRETHROLONE PURIFIED AS MONOHYDRATE)

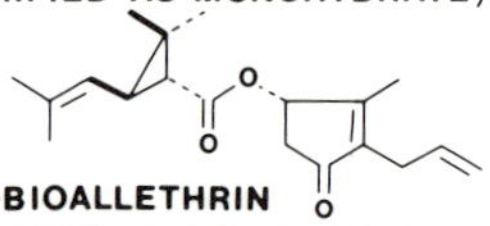

S-BIOALLETHRIN
(LAFORGE, BY SYNTHESIS AND RESOLUTION)
ALLETHRIN (MIXED ISOMERS)

Compound	MUSCA (HF) DOMESTICA mg. kg⁻¹	PHAEDON (MB) COCHLEARIAE mg. kg⁻¹ (Relative potencies)
PYRETHRIN I	30	0·3 (100)
S-BIOALLETHRIN	6	12 (~3)
ALLETHRIN (MIXED ISOMERS)	20	80 (0·4)

FIG. 8.3 Lethal doses of pyrethrin I and allethrin.

Therefore the terminal double bond on the side chain of pyrethrin I greatly modified insecticidal activity, increasing it more than 30-fold in some instances in comparison with the related isomer of allethrin. This finding emphasizes the importance of using pure compounds, particularly optical and geometrical isomers, in structure-activity correlations, so that important and subtle effects are detected and can be exploited. The comparison of pure pyrethrin I and the allethrin isomers influenced the standards by which the status of successive synthetic compounds were judged in the Rothamsted work. Results were subsequently assessed in the light of these findings by criteria which differed from those where the level of activity of allethrin, even the mixed isomers, was considered typical of that of pyrethroids in general, not only against houseflies, but also against other species.

The work with the natural constituents and the allethrin isomers also emphasized the two types of response to these compounds by various insect species. These are a rapid knockdown action particularly shown with flying insects such as houseflies and a very powerful killing action demonstrated with other insects. Kill is associated particularly with the I series esters and knockdown, which is reversible, with the more polar II compounds. Therefore a significant conclusion from this work is that appropriate insect species and test procedures must be selected so that the influence of structural changes on the particular response important in the study can be assessed. For example, in the Rothamsted work, killing power was more significant than rapid knockdown. Therefore potencies to mustard beetles were considered as typical of a range of important species, and were emphasized. Pyrethrin I (and not allethrin, as in some other studies) was the most appropriate reference compound for this and in following discussions it is assigned a potency of 100, so that, for example, the values for S-bioallethrin and allethrin (shown in parentheses on Fig. 8.3) are about 3 and 0.4, respectively.

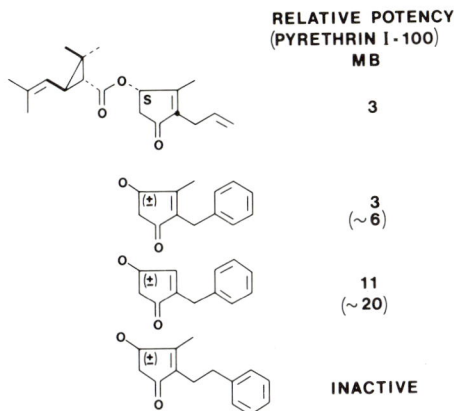

FIG. 8.4. Aromatic analogues of allethrin.

Many compounds related to allethrin were synthesized in the period 1950–1960. None was significantly better than allethrin itself against any species and all were therefore much less active than pyrethrin I. The discovery that the conjugated diene side chain in pyrethrin I was important, suggested that an aromatic system might also be effective so the benzyl analogue (Gersdorff & Mitlin, 1954) was evaluated (Fig. 8.4). Allowing for the racemic alcohol this modification brought about doubled activity. The "non-rethrin" (no methyl group on the ring) (Elliott, Janes & Payne, 1971), was even more active. When the phenyl group was separated from the ring by an additional methylene link, activity was lost (Elliott, Farnham, Janes, Petersen & Needham, unpublished results). These examples indicated that esters with modified planar rings to which an aromatic side chain was attached by one methylene group, were active.

8.3.2 Benzyl chrysanthemates

A significant advance in synthetic pyrethroids was made in 1958 by Barthel, who had worked with Laforge (see Barthel, 1961). Barthel examined further the activity of benzyl chrysanthemates, a class first synthesized by Staudinger & Ruzicka (1924). Figure 8.5 shows his and other related compounds (Elliott, Janes, Jeffs, Needham & Sawicki, 1965) arranged in columns of three levels of activity. Barthel found that 2,4-dimethylbenzylchrysanthemate was an active insecticide, but the related 2,5-compound was not. This suggested an analogy between the moderately active 2,3-dimethylcyclopentonone ester (Gersdoff & Mitlin, 1953), which Barthel and Chen had synthesized, and the inactive 3-methyl compound (Elliot, Harper & Kazi, 1967). The 4-position on the benzene ring might therefore correspond to the 2-position of the cyclopente-

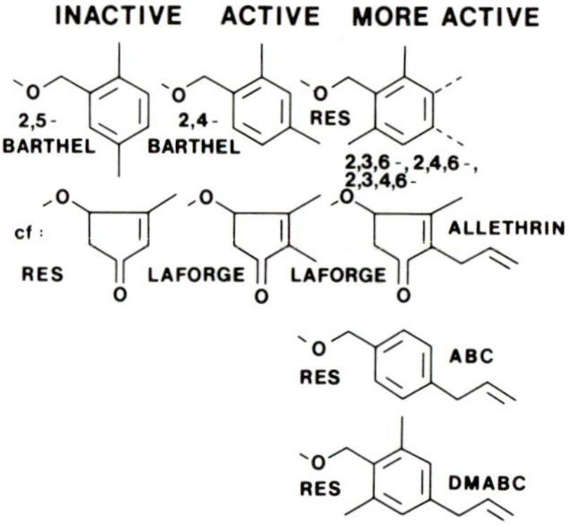

Fig. 8.5. The biological activity of the benzyl chrysanthemates.

none, where it was known that unsaturated groups such as allyl produced activity. 4-Allylbenzyl chrysanthemate was therefore synthesized; it was sufficiently active to encourage examination of other related esters. 4-Allyl 2,6-dimethylbenzyl chrysanthemate (Elliott *et al.*, 1965; Elliott & Janes, 1967; Elliott, Janes & Pearson, 1967) was the most effective of a number of variations, and was the first compound in the series significantly more active against mustard beetles. Various polymethyl benzyl chrysanthemates were also moderately active insecticides (Elliott, Farnham, Ford, Janes & Needham, 1972).

8.3.3   Resmethrin and bioresmethrin

Benzyl chrysanthemates with unsaturated side chains were established in these studies as a new group of synthetic pyrethroids and examined further. 4-Benzylbenzyl chrysanthemate was more potent than the allyl compound, as in the cyclopentenone series, and the furfuryl compound, also active, showed that an alternative type of aromatic substitution was also effective (Fig. 8.6). The readily accessible 5-benzylfurfuryl ester was investigated because the furan ring was similar in size to cyclopentenone; again activity was sufficient to encourage investigation of other variations of this combination. A dramatic 20-fold increase in activity was obtained when the point of attachment of the alcohol was shifted from the 2- to the 3-position (Elliott, Farnham, Janes, Needham & Pearson, 1967; Elliott, Janes & Pearson, 1971; Elliott, Farnham, Janes & Needham, 1974). The product, resmethrin, was the first synthetic pyrethroid with activity against a range of insect species of the same order as pyrethrin I; it was about 40 times as active as allethrin. On an appropriate furan nucleus, therefore, aromatic unsaturation was as effective as the *cis*-dienyl system on the C_5 ring of pyrethrin I, but as in allethrin the simpler

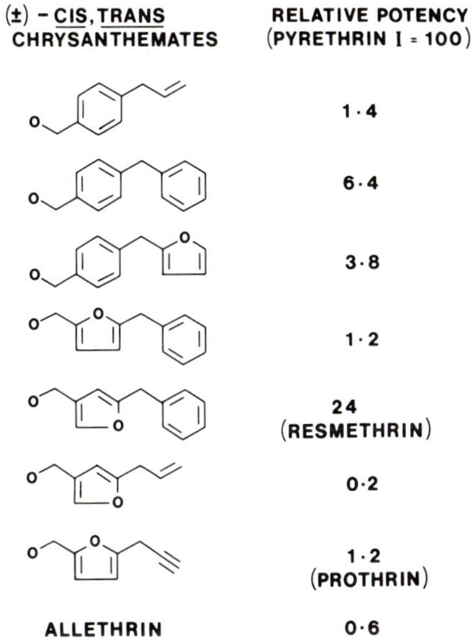

(±) – CIS,TRANS CHRYSANTHEMATES

RELATIVE POTENCY (PYRETHRIN I = 100)

1·4

6·4

3·8

1·2

24
(RESMETHRIN)

0·2

1·2
(PROTHRIN)

ALLETHRIN 0·6

Fig. 8.6. Development of resmethrin.

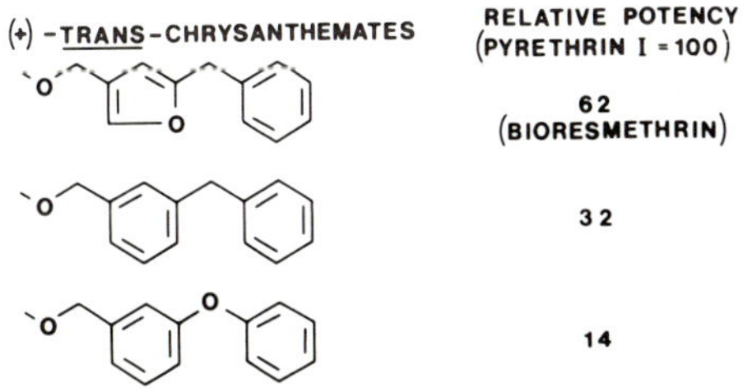

(+) –TRANS–CHRYSANTHEMATES

RELATIVE POTENCY (PYRETHRIN I = 100)

62
(BIORESMETHRIN)

32

14

Fig. 8.7. Development of 3-phenoxybenzyl esters.

allyl side chain was not adequate. Acetylenic unsaturation, as in the compound prothrin, produced as a volatile knockdown agent for flying insects, is barely more effective than allyl.

Resmethrin and bioresmethrin, the corresponding (+)-*trans* chrysanthemate, were developed commercially (Anon., 1971). Bioresmethrin and pyrethrin I had similar levels of activity to many insect species, and therefore provided a common basis for further development of active structures. More compounds related to both were synthesized, among them 3-benzylbenzyl chrysanthemate (Fig. 8.7) selected because the angle between the bonds at the *m*-position was very similar to that between the 3- and 5-position of the furan (Elliott, 1971; Elliott & Janes, 1979). The activity of this compound, greater than that of the 4-benzylbenzyl ester synthesized earlier, suggested that this was an important feature.

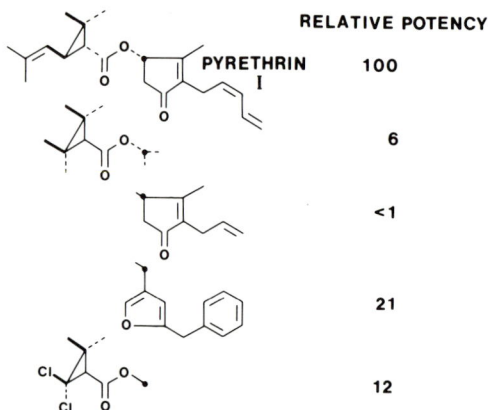

FIG. 8.8. The relative insecticidal potency of the pyrethrin acidic components.

This active 3-benzylbenzyl ester made possible assessment of the importance of the methylene group itself as a link between the rings, and therefore 3-phenoxybenzyl esters were examined. Although the phenoxy chrysanthemate was less active than the corresponding benzyl ester, 3-phenoxybenzyl alcohol gave very active and important esters with other acids subsequently developed following the progression now described.

8.3.4   Development of acidic components

Many variations of the acidic components of pyrethroids were also investigated (see review, Elliott & Janes, 1978), but few of the new esters synthesized approached the activity of (+)-*trans* chrysanthemates, and the significance of the conclusions was limited because allethrolone was the esterifying alcohol. The symmetrical tetramethyl cyclopropane carboxylic acid (Fig. 8.8) (Matsui & Kitahara, 1967) was an interesting modification but the pyrethrolone ester was only 6% as active as pyrethrin I and that from allethrolone, relatively active compared with allethrin against houseflies, was ineffective against mustard beetles. Even the ester of the furan alcohol was less active than the chrysanthemate and the dichlorodimethyl isostere (Barlow, Elliot, Farnham, Hadaway, Janes, Needham & Wickham, 1971) was still less potent.

Against several species examined in the earlier phases of development the (+)-*cis*-chrysanthemate was less active than bioresmethrin, the (+)-*trans* chrysanthemate (Fig. 8.9). The only modification which produced a significant increase in activity was discovered (Velluz, Martel & Nomine, 1969), in the ethanochrysanthemate, which was more active than pyrethrin I. This result, among others, stimulated synthesis of a number of compounds related

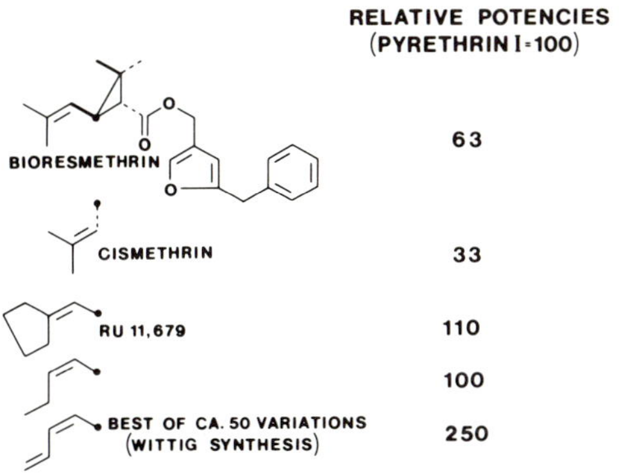

RELATIVE POTENCIES
(PYRETHRIN I = 100)

BIORESMETHRIN   63

CISMETHRIN   33

RU 11,679   110

100

BEST OF CA. 50 VARIATIONS
(WITTIG SYNTHESIS)   250

FIG. 8.9. The potency of acidic components II.

to (+)-*trans* chrysanthemates with various unsaturated side chains. Subtle modifications such as changing the *iso*butenyl substituent to *cis*-butenyl, produced significant increases in activity. Esters of the *cis*-butadienyl, substituted cyclopropane acid, discovered in the course of this phase of the work were some of the most active insecticides known (Elliott, Farnham, Janes, Needham & Pulman, 1973, 1976; Elliott, Janes & Pulman, 1974).

Such increases in activity produced by changing the nature of the unsaturation at this site in the molecule indicated that an acetylene-substituted compound should be investigated. The monochlorovinyl compound (Fig. 8.10) was synthesized as intermediate to it; the benzyl furan ester of this was itself as active as the monomethyl isostere and stimulated preparation of the dichloro compound, the isostere of the chrysanthemate, which was nearly 3 times as active (Elliott *et al.*, 1973; Elliott, Farnham, Janes, Needham & Pulman, 1975).

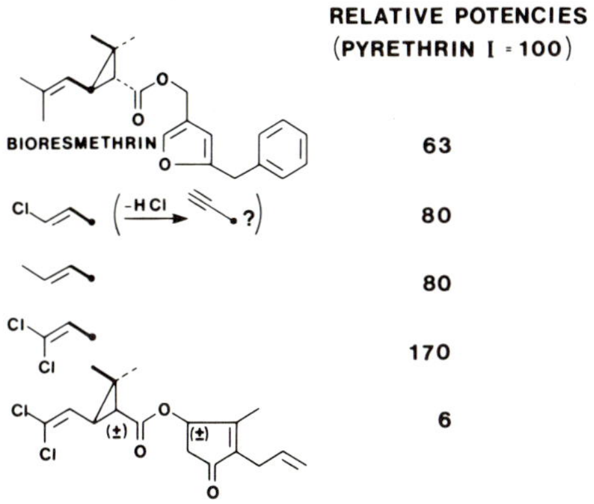

RELATIVE POTENCIES
(PYRETHRIN I = 100)

BIORESMETHRIN   63

Cl (−HCl →, ?)   80

80

170

6

FIG. 8.10. The development of dihalovinyl cyclopropane carboxylates.

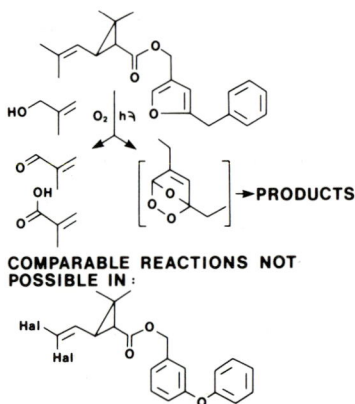

FIG. 8.11. Photodecomposition of resmethrin.

Farkas, Kourin & Sorm (1959) had earlier prepared allethrolone esters of the racemic *trans* and *cis* forms of this acid, but dismissed them as without interest because they had diminished knockdown of houseflies. Without an approach aimed at establishing the optimum structure for kill, in the course of which 5-benzyl-3-furylmethyl alcohol, essential for detecting the value of the dichlorovinyl acid originated, it is doubtful if the importance of the dihalovinyl esters would have been discovered.

8.4   PHOTOSTABLE PYRETHROIDS

The much enhanced activity of the dichlorovinyl furan ester suggested that esters of this acid with other alcohols might also have advantages, such as photostability. Casida's studies (Fig. 8.11) of the photodecomposition of resmethrin and other chrysanthemates (Chen & Casida, 1969; Ueda *et al.*, 1974) had identified two main centres of instability; the *trans*-methyl group of the isobutenyl side chain in the acid and the furan ring in the alcohol. No comparably reactive sites should be present in dihalovinylcyclopropane esters of 3-benzyl and 3-phenoxybenzyl alcohols.

3-Benzyl and 3-phenoxybenzyl dichlorovinyl esters were therefore synthesized (Elliott, 1971; Elliott, Farnham, Janes, Needham, Pulman & Stevenson, 1973; Elliott & Janes, 1979). Unlike the chrysanthemates, the 3-phenoxy compounds in this series were more active than the 3-benzyl; so *cis*-and *trans*-difluoro, dichloro and dibromo esters of the furan and 3-phenoxy-benzyl alcohols (Elliott *et al.*, 1975) compared in Fig. 8.12 with the *cis* and *trans*-chrysanthemates, were prepared. The 3-phenoxy compounds were all, as expected, much more stable in light than the furans. Moreover, the 3-phenoxybenzyl dichlorovinyl esters were more active than would have been

NOT PHOTOSTABLE

PHOTOSTABLE

	X			X	
1R,trans	Me	63		Me	31
	Cl	200		Cl	390
	Br	170		Br	180
	F	110		F	26
1R,cis	X			X	
	Me	25		Me	13
	Cl	180		Cl	200
	Br	160		Br	230
	F	130		F	110

PERMETHRIN (NRDC 143) ISOMERS

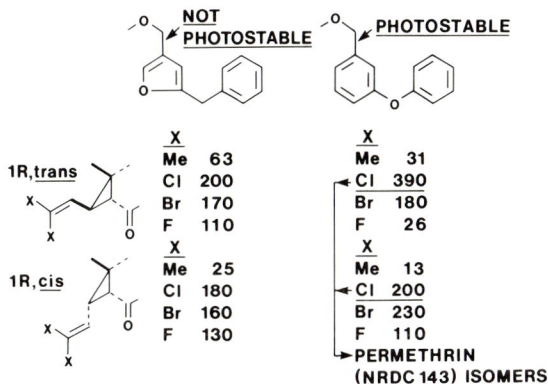

FIG. 8.12. Chrysanthemates and dihalovinyl esters compared.

predicted, 3-phenoxybenzyl chrysanthemates having been only half as active as furan chrysanthemates.

The isomers indicated are components of the photostable pyrethroid insecticide permethrin, now developed commercially (Ruscoe, 1979) and discussed further below.

8.4.1 α-Cyanobenzyl esters

After slow progress for many years, the demonstration in permethrin of the possibility of highly active, photostable pyrethroids rapidly stimulated further developments.

One important advance arose from chrysanthemates of some α-acetylenic alcohols synthesized earlier (BASF, 1968) (Fig. 8.13). Although novel structures, none of the esters was sufficiently active to excite much interest. The Sumitomo chemists investigated esters of a range of pyrethroid acids with α-acetylenic and then with the structurally analogous α-cyano-3-phenoxy-benzyl alcohols and obtained much more active compounds (Matsuo, Itaya, Okuna, Mizutani & Ohno, 1973; Matsuo, Itaya, Mizutani, Ohno, Fujimoto, Okuno & Yoshioka, 1976). As these reports appeared, the great activity of various esters of the dihalovinyl acids had just been discovered, so combinations with α-cyano-3-benzyl and 3-phenoxybenzyl alcohols were examined. Although 3-substituted benzyl esters of the dihalo acids were themselves very active insecticides by all previous standards, their α-cyano analogues were 2–3 times more active still, and so reached unprecedented levels of insecticidal potency.

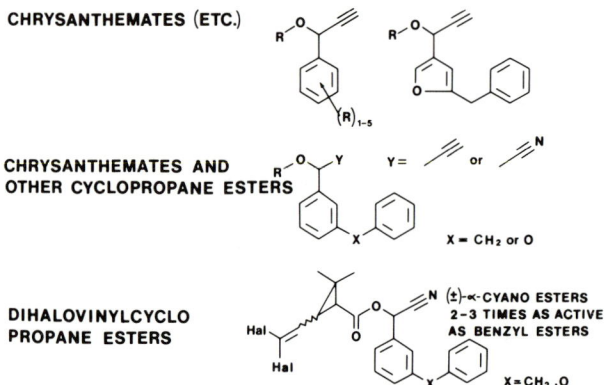

FIG. 8.13. Structures of substituted benzyl esters.

8.4.2 (R)- and (S)-Isomers of α-cyanoesters: decamethrin*

By introducing an α-cyano substituent, a new chiral centre was generated, and with it the challenge to separate, and to determine the relative activities of esters of the two isomers.

α-Cyano alcohols are not themselves easy to resolve but the required separation was achieved (Fig. 8.14) as their *cis*-dibromo vinyl esters which were selected on the basis that *cis*-substituted cyclopropane acids more frequently give crystalline derivatives than *trans*-isomers.

Optically active (1R,*cis*)-dibromovinyl acid was obtained from tertiary butyl (+)-*cis*-chrysanthemic acid by the route indicated and esterified with racemic α-cyano-3--phenoxybenzyl alcohol.

One of the pair of diastereoisomers obtained crystallized and was purified by recrystallization. The crystalline isomer, m.p. 100°, was by far the most powerful insecticide of any class tested at Rothamsted; it was more active than the non-crystalline isomer from the mother liquors. As this isomer already contained two heavy atoms, Owen (1975) of the Molecular Structures Department at Rothamsted was able to perform the X-ray analysis on it directly and establish the new chiral centre to be S. The liquid R-cyanohydrin ester obtained from 3-phenoxybenzaldehyde and hydrogen cyanide in the presence of D-oxynitrilase was, as anticipated, less active (Elliott, Farnham, Janes, Needham & Pulman, 1974; Elliott, Farnham, Janes & Soderlund, 1978; Elliott, Janes, Pulman & Soderlund, 1978).

---

*The name "decamethrin" has been widely used to describe NRDC 161, (α-S)-cyano-3-phenoxybenzyl (1R,*cis*)-3(2,2-dibromovinyl)-2.2-dimethyl-cyclopropanecarboxylate, but it has not been approved as a common name. The alternative common name, recently submitted for approval, is "deltamethrin".

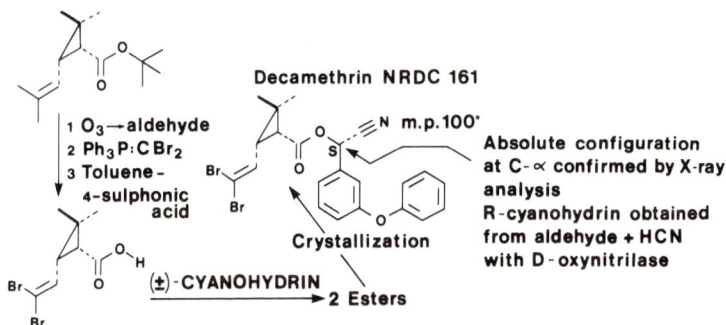

FIG. 8.14. Optical resolution of α-cyano esters: decamethrin.

Related dihalovinyl esters were examined. To purify and then determine the relative potencies of α-S and α-R isomers, some of which, unlike the *cis*-bromo compounds did not crystallize from mixtures, isomeric pairs were separated by high-pressure liquid chromatography and their purities were confirmed by nuclear magnetic resonance spectroscopy. The results here (Fig. 8.15), typical of others (Elliott *et al.*, 1978), show the remarkably low activities of the α-R isomers compared with the outstanding potencies of the (1R, *cis*)-dichloro-and dibromo-α-S isomers. The crystalline isomer decamethrin, with three resolved optical centres, is already manufactured in a process developed by Martel (1978). The final stage in this procedure converts α-R isomers to α-S with a base such as triethylamine in a solvent from which the solid α-S isomer crystallizes.

8.5 OTHER PHOTOSTABLE PYRETHROIDS

Another very significant phase in the rapid recent development of synthetic pyrethroids was initiated by the discovery of the insecticidal activity of arylisovalerates (Ohno *et al.*, 1974, 1976). In testing a range of esters of 5-benzyl-3-furylmethyl alcohol these workers found that the α-ethylphenylacetate was active (Fig. 8.16). Of many related compounds examined, α-isopropyl-4-chlorophenylacetates were amongst the most active and readily accessible. A direct relation with the cyclopropyl pyrethroids was established by showing that only esters of pyrethroid alcohols were active and further, that esters of the S-form of the acid (corresponding to 1R dimethylcyclopropane carboxylic acids) were more potent than those from the R isomer. Esters of the new acids with α-cyano-3-phenoxybenzyl alcohol are especially effective insecticides. However, the activity against mustard

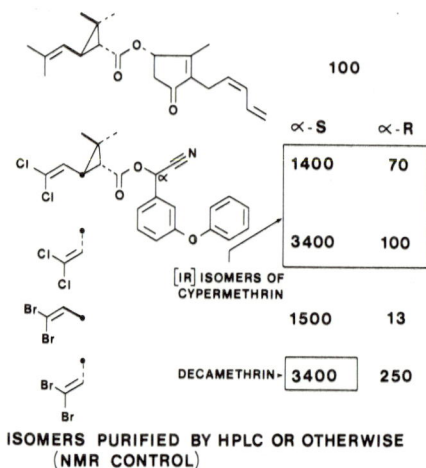

FIG. 8.15. Relative potencies of pyrethrin I and dihalovinyl esters.

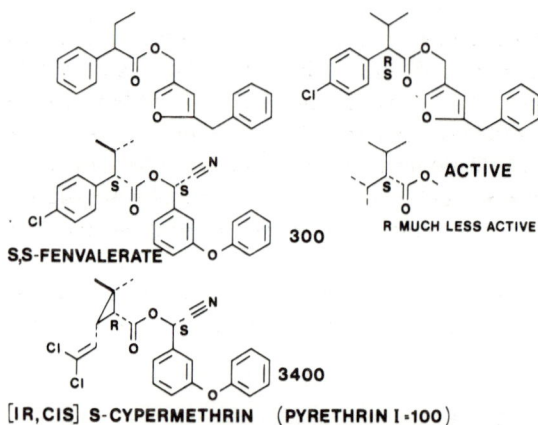

FIG. 8.16. Development of fenvalerate.

beetles of the most potent isomers of the newer simpler structures is less than that of the fully resolved dihalovinyl acid esters discovered earlier.

8.6 ACTIVITIES OF VARIOUS COMBINATIONS OF PYRETHROID ACIDS AND ALCOHOLS

This survey has indicated that, although some measurable activity is attained in many esters from pyrethroid components, only relatively few

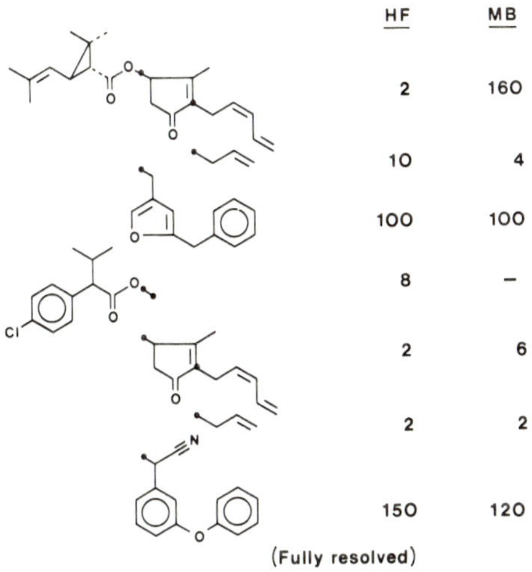

Fig. 8.17. Relative potency of various combinations of pyrethrins.

combinations are extremely potent. The full potential of the arylisovalerates is realized only with α-cyano-3-phenoxybenzyl alcohols, as the results in Fig. 8.17 against houseflies and mustard beetles illustrate. Although the structures of pyrethrolone and chrysanthemic acid, the prototype alcoholic and acidic components of pyrethroid insecticides, contain all the features now known to be necessary for insecticidal activity, the most effective combinations have been evolved only by successive modifications of acids and alcohols and by testing promising combinations of them. There are as yet no guiding principles to predict the most effective compounds of a series against any particular insect pest: practical investigations are still essential.

8.7   PRESENT SITUATION OF PYRETHROIDS

Figure 8.18 illustrates pyrethrin I and some of the pyrethroids with characteristics suitable for commercial development or with significant structural features; despite many modifications in all these compounds a relation with pyrethrin I remains. Greater increases in activity and valuable modification of other properties (for example, the outstanding knockdown activity of Kadethrin) have probably been achieved in the pyrethroid series than from any previous insecticidal lead compound; indeed there can be few natural products with biological activity of whatever category from which comparable developments have been possible. Decamethrin (NRDC 161) (Elliott *et al.*, 1974) and related compounds such as the dichloro- (Elliott *et al.*, 1978a) and chlorotrifluoromethyl (Bentley, Cheetham, Huff & Sayle, 1980)

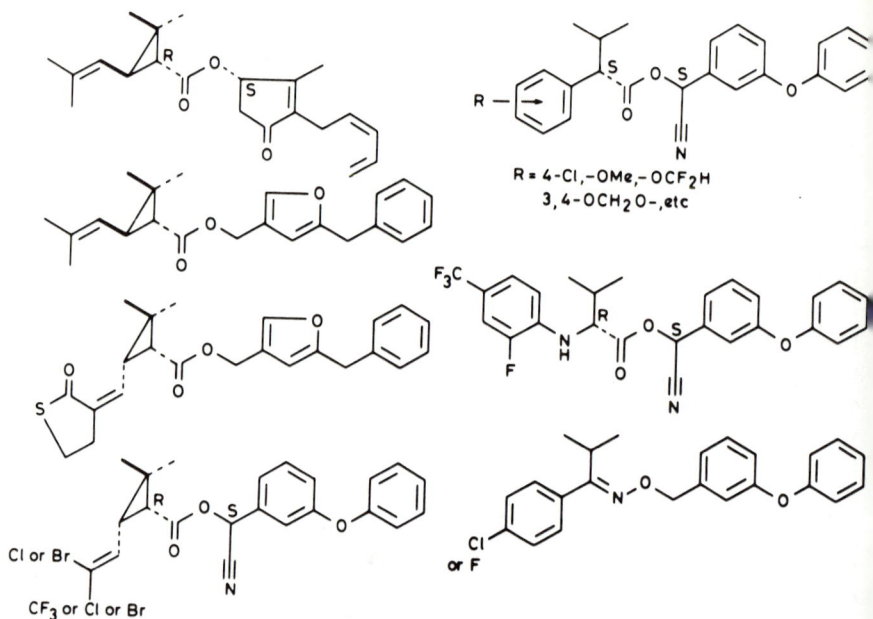

FIG. 8.18.

analogues (NRDC 182 and cyhalothrin, respectively) are very active to various insect species (median lethal doses in the range 0.004–0.06 mg kg^{-1} by topical application) so that field rates as low as 1.0 g ha^{-1} in particular applications are effective (Elliott & Janes, 1978; Elliott *et al.*, 1978b). An especially significant feature of the development of these potent compounds has been that the isosteric replacements involved have also removed the centres of photochemical attack, and the newer members of the series are as stable, or even more stable in the field than many established organophosphate and carbamate insecticides. A further important advance was the discovery of the insecticidal activity of iso-propylphenyl acetates; various para substituents on the aromatic ring produce a powerful new series of insecticides, as indicated in Fig. 8.18. The acyclic series has recently been extended in the aniline derivatives (Henrick, Garcia, Staal, Cerf, Anderson, Gill, Chinn, Labovitz, Leippe, Woo, Carney, Gordon & Kohn, 1980), where greatest activity requires specific substituents, for example, trifluoromethyl and fluorine. Finally some isopropyl and cyclopropyl oximes (Henry, 1978; Nanjyo, Katsuyama, Kariya, Yamamura, Hyeon, Suzuki, & Tamura, 1980) have insecticidal activity of a similar type to established pyrethroids and extend in a remarkable way the range of active structures. Industrial applications of chiral intermediates has made it possible to produce commercial quantities of some of the compounds resolved at two (Kadethrin) or three (decamethrin) chiral centres (Martel, 1978).

8.8 CHARACTERISTICS OF PYRETHROID INSECTICIDES

Synthetic pyrethroids resemble the chlorinated hydrocarbon insecticides in their small solubilities in water and lipophilic characteristics (Briggs, Elliott, Farnham, Janes, Needham, Pulman & Young, 1976). Lipoid solubility is probably essential for the insecticidal action of this group of compounds; metabolic detoxification of pyrethroids involves generating more polar derivatives. When applied to plants they are adsorbed onto the waxy layers where they are available to kill insects, but in contact with soil microorganisms they are rapidly detoxified, so that unlike the chlorinated hydrocarbons they do not persist unnecessarily. Similarly pyrethroids are rapidly detoxified in mammals (Hutson, 1979), as they encounter powerful esterases and mono-oxygenases before reaching sensitive nervous centres but penetrate rapidly to sensitive sites in insects, which are therefore quickly affected.

Some of the synthetic pyrethroids such as bioresmethrin and phenothrin have sufficiently low mammalian toxicity to be suitable for applications near food and for disinfestation of edible crops immediately before harvest. The synthetic pyrethroids, developed from unique natural compounds, are therefore a group of lipophilic insecticides with a generally more favourable combination of properties than other established groups. They illustrate well one way in which natural products can be used, in this instance indirectly, in pest and disease control.

8.9 SUMMARY

The natural pyrethrins, six closely related esters isolated from the flowers of *Chrysanthemum cinerariaefolium*, are valuable insecticides which kill a range of insect species rapidly with little risk to other organisms. They are too unstable to control pests of agricultural crops, but their structures have served as excellent prototypes for developing a range of compounds with modified properties. Some synthetic pyrethroids developed within the past few years are exceptionally active insecticides and have a more favourable combination of properties than many established organochlorine, organophosphate and carbamate insecticides.

Features in the structures of the natural and earlier synthetic esters associated with insecticidal activity, low mammalian toxicity and instability in light were identified by systematic synthesis and testing of related compounds. These studies led to the discovery of the most active insecticides known; the newer compounds are more stable in sunlight but rapidly degraded by metabolizing systems such as those of mammals and soil microorganisms.

Numerous structural variations at several sites in the molecules give active compounds (not all esters), but properties optimum for any particular application are attained in relatively few of the many combinations possible.

8.10 ACKNOWLEDGEMENT

I thank my colleague Norman Janes for much help in preparation of this contribution.

8.11 REFERENCES

Anon. (1971) Plant to make synthetic pyrethrin. *Chem. Engng News* **49** (2), 32.

Barlow F., Elliott M., Farnham A. W., Hadaway A. B., Janes, N. F., Needham P. H. & Wickham J. C. (1971) Insecticidal activity of the pyrethrins and related compounds. IV. Essential features for insecticidal activity in chrysanthemates and related cyclopropane esters. *Pestic. Sci.* **2**, 115–118.

Barthel W. F. (1961) Synthetic pyrethroids. *Adv. Pest Control Res.* **4**, 33–74.

BASF (1968) Esters chrysanthemique substitues et produits parasiticides contenant ces esters. Belgian Patent 738112, 14 pp.

Bentley P. D., Cheetham R., Huff R. K. & Sayle J. D. (1980) Highly halogenated chrysanthemic acid analogues and their esters. *Pestic. Sci.* **11**.

Briggs G. G., Elliott M., Farnham A. W., Janes N. F., Needham P. H., Pulman D. A. & Young S. R. (1976) Insecticidal activity of the pyrethrins and related compounds. VIII. Relation of polarity with activity in pyrethroids. *Pestic. Sci.* **7**, 236–240.

Campbell I. G. M. & Harper S. H. (1945) Experiments on the synthesis of the pyrethrins Part 1. Synthesis of chrysanthemum monocarboxylic acid. *J. Chem. Soc.* **74**, 283–6.

Chen Y. L. & Casida J. (1969) Photodecomposition of pyrethrin I, allethrin, phthalthrin and dimethrin. Modifications in the acid moiety. *J. agric. Food Chem.* **17**, 208–215.

Crombie L. & Elliott M. (1961) Chemistry of the natural pyrethrins. *Fortschr. Chem. Org. Natstoffe* **19**, 120–164.

Elliott M. (1964) The pyrethrins and related compounds. V. Purification of (+)-pyrethrolone as the monohydrate, and the nature of "pyrethrolone-C". *J. Chem. Soc.* Part V, No. 999, 5225–5228.

Elliott M. (1971) The relationship between the structure and activity of pyrethroids. *Bull. Wld Hlth Org.* **44**, 315–324.

Elliott M., Farnham A. W., Ford M. G., Janes N. F. & Needham, P. H. (1972) Insecticidal activity of the pyrethrins and related compounds. V. Toxicity of the methylbenzyl chrysanthemates to houseflies (*Musca domestica* L.) and mustard beetles (*Phaedon cochleariae* Fab.). *Pestic. Sci.* **3**, 25–28.

Elliott M., Farnham A. W., Janes N. F. & Needham P. H. (1974) Insecticidal activity of the pyrethrins and related compounds VI. Methyl-, alkenyl- and benzyl-furfuryl and -3-furylmethyl chrysanthemates. *Pestic. Sci.* **5**, 491–496.

Elliott M., Farnham A. W., Janes N. F., Needham P. H. & Pearson B. C. (1967) 5-Benzyl-3-furylmethyl chrysanthemate: a new potent insecticide. *Nature, Lond.* **213**, 493–494.

Elliott M., Farnham A. W., Janes N. F., Needham P. H. & Pulman D. A (1973) Potent pyrethroid insecticides from modified cyclopropane acids. *Nature, Lond.* **244**, 456–457.

Elliott M., Farnham A. W., Janes N. F., Needham P. H. & Pulman D. A. (1974) Synthetic insecticide with a new order of activity. *Nature, Lond.* **248**, 710–711.

Elliott M., Farnham A. W., Janes N. F., Needham P. H. & Pulman D. A. (1975) Insecticidal activity of the pyrethrins and related compounds VII. Insecticidal dihalovinyl analogues of *cis* and *trans* chrysanthemates. *Pestic. Sci.* **6**, 537–542.

Elliott M., Farnham A. W., Janes N. F., Needham P. H. & Pulman D. A. (1976) Insecticidal activity of the pyrethrins and related compounds X. 5-Benzyl-3-furylmethyl 2,2-dimethylcyclopropanecarboxylates with ethylenic substituents at position 3 on the cyclopropane ring. *Pestic. Sci.* **7**, 499–502.

Elliott M., Farnham A. W., Janes, N. F., Needham P. H., Pulman D. A. & Stevenson J. H. (1973) A photostable pyrethroid. *Nature, Lond.* **246**, 169–170.

Elliott M., Farnham A. W., Janes, N. F. & Soderlund D. M. (1978) Insecticidal activity of the pyrethrins and related compounds. Part XI. Relative potencies of isomeric cyano-substituted 3-phenoxybenzyl esters. *Pestic. Sci.* **8**, 112–116.

Elliott M., Harper S. H. & Kazi M. A. (1967) Experiments on the synthesis of the rethrins XIV. Rethrins and the cyclopentadienone related to 3-methylcyclopent-2-enone. *J. Sci. Food Agric.* **18**, 167–171.

Elliott M. & Janes N. F. (1967) Synthesis of 4-allyl- and 4-benzyl-2,6-dimethylbromobenzene from 2,6-xylidine. *J. Chem. Soc. C*, 1780–1782.

Elliott M. & Janes N. F. (1969) Pyrethrin II and related esters obtained by reconstitution. *Chem. Ind.* No. 9, 270–271.

Elliott M. & Janes N. F. (1978) Synthetic pyrethroids—a new class of insecticide. *Chem. Soc. Rev.* **7**, 473–505.

Elliott M. & Janes N. F. (1979) Recent structure–activity correlations in synthetic pyrethroids. In *Advances in Pesticide Science* (ed. Geissbuhler H.), pp. 166–173. Pergamon Press, Oxford.

Elliott M., Janes N. F., Jeffs, K. A., Needham, P. H. & Sawicki, R. M. (1965) New pyrethrin-like esters with high insecticidal activity. *Nature, Lond.* **207**, 938–940.

Elliott M., Janes N. F. & Payne M. C. (1971) The pyrethrins and related compounds. Part XI. Synthesis of insecticidal esters of 4-hydroxy-cyclopent-2-enones (nor-rethrins). *J. Chem. Soc.*, C, 2548–2551.

Elliott M., Janes N. F. & Pearson B. C. (1967) Pyrethrins and related compounds IX. Alkenylbenzyl and benzylbenzyl chrysanthemates. *J. Sci. Food Agric.* **18**, 325–331.

Elliott M., Janes N. F. & Pearson B. C. (1971) The pyrethrins and related compounds XIII. Insecticidal methyl-, alkenyl- and benzyl-substituted furfuryl and furylmethyl chrysanthemates. *Pestic. Sci.* **2**, 243–248.

Elliott M., Janes N. F. & Pulman D. A. (1974) The pyrethrins and related compounds. Part XVIII. Insecticidal 2,2-dimethylcyclopropanecarboxylates with new unsaturated 3-substituents. *J. Chem. Soc. Perkin* **1**, 2470–2474.

Elliott M., Janes N. F., Pulman D. A. & Soderlund D. M. (1978) The pyrethrins and related compounds. Part XXII. Preparation of isomeric cyano-substituted 3-phenoxybenzyl esters. *Pestic. Sci.* **9**, 105–111.

Elliott M., Needham P. H. & Potter C. (1950) The insecticidal activity of substances related to the pyrethrins. I. Toxicities of two synthetic pyrethrin-like esters relative to that of the natural pyrethrins and the significance of the results in the bioassay of closely related compounds. *Ann. appl. Biol.* **37**, 490–507.

Farkas J., Kourim P. & Sorm F. (1959) Relation between chemical structure and insecticidal activity in pyrethroid compounds. I. An analog of chrysanthemic acid containing chlorine in the side chain. *Colln Czech. Chem. Commun.* **24**, 2230–2236.

Gersdorff W. A. & Mitlin N. (1953) The relative toxicity to house flies of the methyl and ethyl analogues of allethrin. *J. econ. Ent.* **46**, 945–948.

Gersdorff W. A. & Mitlin N. (1954) The relative toxicity of some aryl analogs of allethrin to house flies. *J. econ. Ent.* **47**, 888–890.

Gnadinger C. B. (1936) *Pyrethrum Flowers*, 2nd edition, 380 pp. McLaughlin Gormley King Co., Minneapolis.

Gnadinger C. B. (1945) *Pyrethrum Flowers—Supplement to Second Edition*, pp. 381–690. McLaughlin Gormley King Co., Minneapolis.

Henrick C. A., Garcia B. A., Staal G. B., Cerf D. C., Anderson R. J., Gill K., Chinn H. R., Labovitz J. N., Leippe M. M., Woo S. L., Carney R. L., Gordon D. C. & Kohn G. K (1980) 3-Methyl-2-phenylaminobutanoates and 2-(2-isoindolinyl)-3-methyl-butanoates, two novel groups of synthetic pyrethroid esters not containing a cyclopropane ring. *Pestic. Sci.* II.

Henry A. C. (1978) Benzyl oxime ethers U.S. Patent 4,079,149, 6 pp.

Hutson D. H. (1979) The metabolic fate of synthetic pyrethroid insecticides in mammals. In *Progress in Drug Research*, Vol. 23 (ed. Jucker E.), pp. 215–252. Birkhauser, Basel.

Laforge F. B. & Barthel W. F. (1944) Constituents of pyrethrum flowers. XVI. Heterogeneous nature of pyrethrolone. *J. org. Chem.* **9**, 242–249.

Laforge F. B. & Barthel W. F. (1945a) Constituents of pyrethrum flowers. XVII. The isolation of five pyrethrolone semicarbazones, *J. org. Chem.* **10**, 106–113.

Laforge F. B. & Barthel W. F. (1945b) Constituents of pyrethrum flowers. XVIII. The structure and isomerism of pyrethrolone and cinerolone. *J. org. Chem.* **10**, 114–120.

Laforge F. B. & Barthel W. F. (1945c) Constituents of pyrethrum flowers. XIX. The structure of cinerolone. *J. org. Chem.* **10**, 222–227.

Laforge F. B., Gersdorff W. A., Green N. & Schechter M.S. (1952) Allethrin-type esters of cyclopropanecarboxylic acids and their relative toxicities to house flies. *J. org. Chem.* **17**, 381–389.

Laforge F. B., Green N. & Schechter M. S. (1954) Allethrin. Resolution of dl-allethrolone and synthesis of the four optical isomers of *trans*- allethrin. *J. org. Chem.* **19**, 457–462.

Lord K. A., Ward J., Cornelius J. A. & Jarvis M. W. (1952) Chromatographic separation of the pyrethrins. *J. Sci. Food Agric.* **3**, 419–426.

Martel J. (1978) Definitions chimiques des pyrethrinoides photostables. *Phytiat.-Phytopharm.* **27**, 5–14.

Matsui M. & Kitahara T. (1967) Studies on chrysanthemic acid. Part XVIII. A new biologically active acid component related to chrysanthemic acid. *Agric. Biol. Chem.* **31**, 1143–1150.

Matsuo T., Itaya N., Mizutani T., Ohno N., Fujimoto K., Okuno Y. & Yoshioka H. (1976). 3-Phenoxy-α-cyano-benzyl esters, the most potent synthetic pyrethroids. *Agric. Biol. Chem.* **40**, 247–249.

Matsuo T., Itaya N., Okuno Y., Mizutani T. & Ohno N. (1973) *Ger. Offen.* **2**, 231–312, 59 pp.

Nanjyo K., Katsuyama N., Kariya A., Yamamura T., Hyeon S., Suzuki A. & Tamura S. (1980) New insecticidal pyrethroid-like oximes. *Agric. Biol. Chem.* **44**, 217–218.

Ohno N., Fujimoto K., Okuno Y., Honda T. & Yoshioka H. (1974) A new class of pyrethroidal insecticides; -substituted phenylacetic acid esters. *Agric. Biol. Chem.* **38**, 881–883.

Ohno N., Fujimoto K., Okuno Y., Mizutani T., Hirano M., Itaya N., Honda T. & Yoshioka H. (1976) 2-Arylalkanoates, a new group of synthetic pyrethroid esters not containing cyclopropanecarboxylates. *Pestic. Sci.* **7**, 241–246.

Owen J. D. (1975) Absolute configuration of the most potent isomer of the pyrethroid insecticide α-cyano-3-phenoxybenzyl *cis*-3-(2,2-dibromovinyl)-2,2-dimethylcyclopropanecarboxylate by crystal analysis. *J. Chem. Soc. Perkin Trans.* **1**, 1865–1868.

Pryce R. J. (1982) The search for new agrochemical leads from natural sources. In this book Chapter 5 (ed. Whitehead D. L. & Bowers, W. S.). Pergamon Press, Oxford & New York.

Ruscoe C. N. E. (1979) The impact of the photostable pyrethroids as agricultural insecticides. *Proc. 1979 British Crop Protection Conf.—Pests and Disease*, pp. 803–814.

Schechter M. S., Green N. & Laforge F. B. (1949) Constituents of pyrethrum flowers. XXIII. Cinerolone and the synthesis of related cyclopentenolones. *J. Am. Chem. Soc.* **71**, 3165–3173.

Staudinger H. & Ruzicka L. (1924) Insektentotende Stoffe. 1–10 Mitteilung. Uber Isolierung und Konstitution des wirksamen Teils des dalmatinischen Insektenpulvers. *Helv. chim. Acta* **7**, 177–201, 201–211, 212–235, 236–244, 245–259, 377–390, 390–406, 406–441, 442–448, 448–458.

Ueda K., Gaughan L. C. & Casida J. E. (1974) Photodecomposition of resmethrin and related pyrethroids. *J. agric. Food Chem.* **22**, 212–220.

Velluz L., Martel J. & Nomine G. (1969) Synthese d'analogues de l'acide *trans*-chrysanthemique. *C.r. hebd. Sean. Acad. Sci. Ser. C, Paris* **268**, 2199–2203.

Ward J. (1953) Separation of the pyrethrins by displacement chromatography. *Chem Ind.*, pp. 586–587.

8.12 DISCUSSION

Bowers: Commenting on the issue of resistance, I think that if we had plant protection chemicals whose mode of action was clear to us, we could issue instructions for their use which would limit development of resistance to the compounds by the insects to manageable levels.

Elliott: Commenting on Professor Bowers' suggestion regarding insect resistance to the pyrethroids, I would say that insect resistance to the pyrethroids has not reached a stage where it constitutes a practical problem. In my view, the prospects for introducing selectively active compounds which can be made by large firms on a large scale is a much more daunting problem than insect resistance to pyrethroids because the cost of such an exercise is prohibitive at present rates. On the other hand, structural modifications on the parent pyrethrin structure have led to products with novel activities, e.g. decamethrin. A stage has now been reached when possible changes on the pyrethrin structure can not lead to products with predictable activity as has hitherto been the case. Therefore, it seems to me that we now have to look for model compounds with potent insecticidal activity from plants. However, I do not know how the resources necessary for discovering and developing biologically active compounds from plant sources can best be used. Nonetheless, I believe that some involvement by government agencies in financing phytochemical work on a national or international basis is necessary for success.

Casida: Further to Dr. Elliott's views on insect resistance to pesticides, I think there are a number of issues to consider. In the first place, the mechanism of metabolic detoxification of pesticides by insects must be thoroughly understood. Secondly, modifications on the structures with established activity must not lead to products that are environmentally undesirable. In this respect, I think the pyrethrins have been exceptionally lucky in that modifications have led to only few objectionable products.

Elliott: From studies limited to laboratory bioassays only, all our synthetic pyrethroids deal with resistant and non-resistant insects with equal effectiveness.

Graham-Bryce: I would like to go back to Professor Bowers' question on resistance to insecticides. It seems to me that there are two parts to his question. The first part is how do you select compounds for use and the second part is how do you actually implement strategies in practice. On the selection of compounds, differences in modes of action are quite useful, but in terms of resistance, that is not really the key question. The question is that there are differences in capacity to select resistance mechanisms. This is not quite the same thing at all. The problem of cross resistance is surely an extreme and unpredictable one in that there are cases of cross resistance between organochlorides and pyrethroids, between organophosphates and pyrethroids and there are cases of differences among organophosphates. Some organophosphates are still active against populations that are resistant to others. It seems to me that until one can get useful predictive information, it is impossible to devise a rational strategy. You have to make the best guess.

On the question of how do you manage the use of pesticides, surely it is a political one. It either requires some kind of regulation or it could be done by some kind of agreement among manufacturers. In the UK, for example, there are some movements in the latter direction in that some manufacturers are now suggesting on labels that some of their compounds must not be used exclusively but should be used in rotation. This I think is a right move.

On Professor Djerassi's thought that the pyrethroids were going to be the last conventional powerful insecticides, I do not quite understand why he actually made that statement. I think that the alacrity with which the pyrethroids were taken up demonstrates that whenever you have something potentially good in conventional insecticides, it it likely to be very successful and there are still a number of gaps like systemic activity or soil activity which need to be filled by knowledge.

Djerassi: I think the prospects for having another group of molecules to serve the purpose which the pyrethrins did in the development of the pyrethroids are poor. Further, I think that it is most unlikely that anybody in Agrochemical industry is going to work on the minor uses of insecticides and this is purely for economic reasons. The industry will, however, continue to work for major leads in chemical modifications with a potentially good large-scale use, especially in crop protection.

Bernays: Certain pyrethroids are quite effective against tsetse flies, even at very low concentrations. Do you know why?

Elliott: No. We have not really examined that, and I do not really know for sure whether your statement is disputable or not.

Graham-Bryce: I think this relates to the mode of action in laboratory tests on a weight by weight basis. Not on the susceptibility of the flies.

Whitehead: Tsetse flies are also particularly susceptible to endosulphan. Perhaps a biochemical reason can be found to link this with the susceptibility to decamethrin.

CHAPTER 9

Microencapsulated Natural Pyrethrum—an Improved Insect Repellent

J. SIMKIN

Technical Division, Pennwalt Corporation, King of Prussia, PA19406, USA

and

R. GALUN

Department of Zoology, The Hebrew University of Jerusalem, Jerusalem, Israel

CONTENTS

9.1	Introduction	151
9.2	Rationale for Microencapsulation	152
9.3	Advantages of using Microcapsules	154
	9.3.1 Field tests	155
9.4	Encapsulation of Natural Pyrethrum	156
	9.4.1 Materials and methods	156
	9.4.1.1 Preparation of microencapsulated formulations	156
	9.4.1.2 Repellency tests	156
	9.4.2 Results and discussion	157
9.5	Conclusion	160
9.6	Summary	160
9.7	Acknowledgements	161
9.8	References	161
9.9	Discussion	161

9.1 INTRODUCTION

Tsetse, vectors of trypanosomiasis, and ticks, vectors of several serious livestock diseases, continue to plague the health and economy of great parts of

Africa. Different approaches to controlling these vectors have been attempted over the years, including some new chemical control techniques. However, vast areas with favourable climatic and edaphic conditions for livestock production are still infested and cannot be used for raising cattle.

Previous efforts to protect people and domesticated animals by repelling the tsetse have been disappointing. To be practical, protection of livestock against tsetse flies and ticks by means of repellents requires a repellency of at least 1 week. Attempts were made in the forties (Hornby & French, 1943; Findlay, Hardwicke & Phlep, 1946; Holden & Findlay, 1944), and then discontinued for almost 30 years. Recently, the usefulness of some new repellents against the tsetse has been evaluated. Of nine repellents, including Deet, tested on human skin, none gave 100% protection for more than 2 h (Schmidt, 1977). To date, the only compound conferring up to 2 days' protection against ticks and tsetse is natural pyrethrum (Glynne Jones & Sylvester, 1966). However, pyrethrum decomposes very rapidly due to UV radiation from sunlight, and it also oxidizes readily.

9.2   RATIONALE FOR MICROENCAPSULATION

Galun (1975) proposed attempting to increase the residual repellency of pyrethrum using microencapsulation as a means of reducing evaporation rate, minimizing photodegradation, avoiding oxidation and slowing down absorption of the material into the skin (Galun, 1975).

The technique, of microencapsulation of pesticides, was pioneered by Pennwalt Corporation. The rationale for using this system is based upon the following advantages of microencapsulation over conventional pesticide formulations, such as an emulsifiable concentrate (EC): increased residual activity; greater effectiveness — less pesticide required; reduced mammalian toxicity; reduced volatility.

Pennwalt's process is based upon a patented interfacial polymerization procedure (Vandegaer, 1971). Simply stated, it involves a mechanical dispersion of a hydrophobic pesticide in an aqueous medium. The pesticide contains a soluble monomer such as a diacid chloride. When the droplets have been reduced to the desired size, say 35 to 50 microns, a water-soluble monomer (e.g. diamine) is added. An immediate polymerization occurs at the droplet interface producing a thin plastic film around the surface (Scheme 9.1).

Sometimes it is desirable to modify the polymer wall by introducing crosslinking. For example, a polyfunctional isocyanate can provide three reactive sites. This yields a crosslinked capsule wall which can modify the release rate of pesticides (Scheme 9.2).

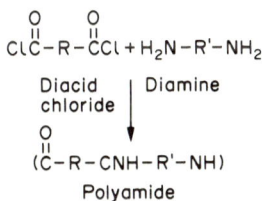

Scheme 9.1.

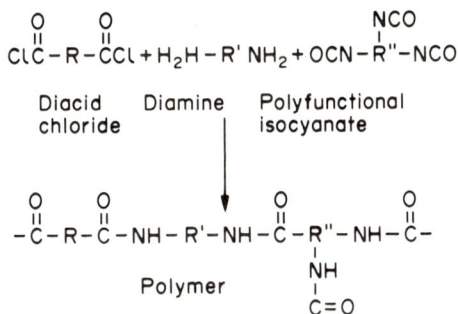

Scheme 9.2.

Several types of polymeric capsule walls are possible. The selection depends upon empirical evaluation of encapsulant compatibility and the nature of the release produced. Examples of the different polymeric walls possible by this process are shown in Table 9.1.

TABLE 9.1. COMPOSITION OF CAPSULE WALL POLYMER

Polyamide	$(\overset{O}{\overset{\|}{C}}\sim\overset{O}{\overset{\|}{C}}NH\sim NH)_n$
Polyester	$(\overset{O}{\overset{\|}{C}}\sim\overset{O}{\overset{\|}{C}}O\sim O)_n$
Polyurea	$(\overset{O}{\overset{\|}{C}}NH\sim NH\overset{O}{\overset{\|}{C}}NH\sim NH)_n$
Polyurethane	$(\overset{O}{\overset{\|}{C}}NH\sim NH\overset{O}{\overset{\|}{C}}O\sim O)_n$
Polycarbonate	$(\overset{O}{\overset{\|}{C}}O\sim O\overset{O}{\overset{\|}{C}}O\sim O)_n$

$\sim$ denotes (CH_2) 1 or >1

Pennwalt have been utilizing a polamide polyurea copolymer wall for many of our agricultural applications.

9.3 ADVANTAGES OF USING MICROCAPSULES

A unique feature of these microcapsules is the flexibility in approaches to modify the release rate of a pesticide. Thus the release rate can be varied by the nature of the monomer, the degree of crosslinking, the thickness of the capsule wall and by the capsule size. An increase in the degree of crosslinking or wall thickness generally results in a reduced release of the encapsulant.

An important advantage of microencapsulation is reduction of oral and dermal toxicity of pesticides. The acute oral LD_{50} for the very toxic technical methyl parathion is 10 mg kg^{-1} for the mouse versus a typical 120 mg kg^{-1} found for methyl parathion in encapsulated form. Similar differences exist in dermal toxicity (Table 9.2).

TABLE 9.2 TOXICITY OF ENCAPSULATED METHYL PARATHION

	Acute LD_{50}	mg kg^{-1}
	Oral — Mouse	Dermal — Rabbit
Encapsulated	120	2250
Technical	10	178

A most important feature of microencapsulation is the sustained release of encapsulant. This is most clearly demonstrated in bioassay experiments.

When microencapsulated methyl parathion (Penncap-M[R] — registered trademark of Pennwalt Corporation) is sprayed onto plants a prolonged insecticidal activity is produced in comparison to treatment with an EC formulation. Table 9.3, which is based upon an experiment in which bean plants were infested with Japanese beetles, illustrates this effect.

TABLE 9.3 COMPARATIVE ACTIVITY (JAPANESE BEETLE ON BEAN PLANT)

		% Mortality after			
Treatment	AI (mg ft^{-2})	2 days	3 days	5 days	8 days
Penncap-M[R]	5	100	100	100	80
Methyl parathion EC	5	100	100	100	0
Control		0	0	0	0

Upon infesting the plants 5 days after spraying, the EC treatment no longer shows toxicity whereas the microencapsulated formulation is active even after 8 days.

The persistency effect can also be demonstrated by plotting percent mortality versus the time interval between spraying and infestation. The graph (Fig. 9.1) is typical of data obtained from greenhouse bioassay tests. In

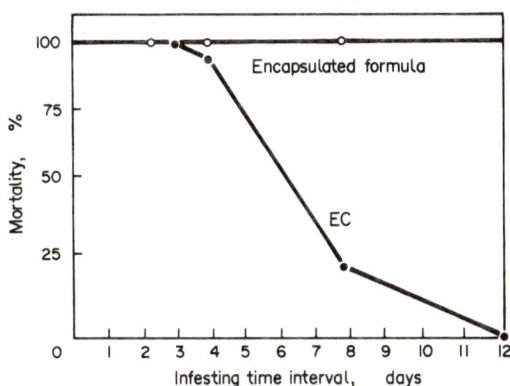

FIG. 9.1. Encapsulated organophosphate insecticide is superior to emulsified against cockroaches.

this instance, a microencapsulated organophosphorus insecticide was tested against cockroaches and compared with an EC of equivalent active ingredient. The microencapsulated formulation continued to give 100% control long after the EC formulation was no longer active.

9.3.1 Field tests

Field tests have confirmed the significantly increased residual activity of microencapsulated products even though these increases are usually not as dramatic as in indoor tests. Typically for Penncap-M^R, a two- to three-fold increase in persistence over the emulsifiable concentrate is observed.

In view of the interest in increasing the persistence of natural pyrethrum, we have conducted bioassay tests using microencapsulated synergized natural pyrethrum. Cardboard cartons were treated indoors and then infested with houseflies at various intervals. The graph Fig. 9.2 demonstrates a significant difference between the persistence of the microencapulated formulation and an EC applied at equivalent amount of AI.

Because of the flexibility in tailoring the release rate of pesticide, we have been able to demonstrate utility for a variety of pesticides in a number of different applications: foliar, stored product, soil pesticides; pesticides for insects affecting man, livestock and household pests; insect attractants for use in traps and poison baits; disruption of mating with sex pheromones; and herbicides.

The application of microencapsulation techniques in the formulation of persistent insect repellents is a logical extension of the above.

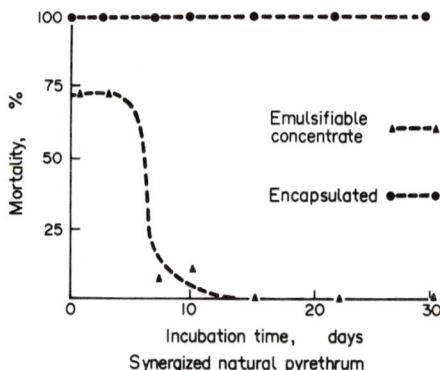

FIG. 9.2. Encapsulated and emulsified synergized natural pyrethrum tested against *Musca domestica* at 10 mg ft^{-2}.

9.4 ENCAPSULATION OF NATURAL PYRETHRUM

The present paper reports on the repellency and toxicity of twenty-seven formulations of microencapsulated natural pyrethrum, with different capsule walls exhibiting dissimilar release rates of the repellent/pesticide. For comparison, non-encapsulated natural pyrethrum and microencapsulated synthetic pyrethroids were also tested.

9.4.1 Materials and methods

9.4.1.1 Preparation of microencapsulated formulations This was carried out at the Pennwalt Corporation Technological Centre, Pennsylvania, USA. In some formulations, piperonyl butoxide synergist was added, in ten-fold concentrations relative to the pyrethrum, before encapsulation. Varying degrees of release were effected by using formulations having variation in wall thickness and degree of crosslinking. Each of these characteristics has an independent effect on release rate. Samples were chosen for testing on the basis of estimated release characteristics derived from experience with other encapsulants. No direct determination of release, by chemical analysis, was performed. All test samples contained microcapsules ranging in size from 30 to 50 microns.

9.4.1.2 Repellency tests Repellency and toxicity tests were performed by the Department of Zoology, The Hebrew University of Jerusalem, Jerusalem, Israel (Galun *et al.*, 1980).

(a) *Tsetse*. The tsetse used were *Glossina morsitans*. Tests were carried out with either hungry teneral flies, 48 h after emergence, or with non-teneral flies, 48 h after the last blood meal. Five ml of the test formulation was applied to semi-shaved bellies of guinea pigs and spread evenly over an area of 9 × 16 cm. The same treatment was carried out at 28° C and 70% relative humidity. The treated guinea pig was immobilized on its back, by using a specially designed restrainer table. Fly cages were then placed on the treated bellies thus exposing the flies to the repellent for 5 min. Immediately upon removal of the cage, the number of engorged flies was recorded and all the flies transferred to a clean cage. Mortality was recorded 24 h following testing. Having observed that unencapsulated pyrethrum protected the guinea pigs for 5 to 6 days, the test was standardized by infesting with flies on the sixth day following treatment and testing daily until feeding was observed.

(b) *Ticks*. The ticks used were starved adult female *Ornithodoros tholozani*. Testing was carried out under the conditions above, using the same restraining table for the guinea pigs. A special glass apparatus for testing tick repellency and toxicity was harnessed to the guinea pig according to the method of Bar Zeev & Gothilf (1973). *Ornithodoros* adults do not climb on glass and are therefore in contact with the treated belly area for the whole period of exposure, 10 min for repellency testing, 6 hr for toxicity testing. Repellency testing was begun 2 days following application, and carried out daily using different batches of ticks, ten ticks per batch, and continued until feeding was observed. The number of attached and of engorged ticks was recorded. Mortality was recorded 1 week after exposure since delayed mortality seems to occur. Only formulations conferring 7 or more days of protection against tsetse were tested for toxicity to ticks.

9.4.2 Results and discussion

Table 9.4 shows repellency data on tsetse with non-synergized natural pyrethrum. Formulations of non-encapsulated natural pyrethrum, at a concentration of 0.52 mg cm^{-2} gave guinea pigs complete protection from tsetse bites for 6 days. Ten microencapsulated formulations were tested; all but two showed increased repellency over the non-encapsulated standard. Table 9.4 gives data on four of the best formulations. The maximum persistency obtained was 11 to 15 days. The figures in parentheses are maximum days of complete protection.

The determination of minimum effective dose (MED) is another approach to evaluation. This measures the minimum concentration giving 90% protection from biting when tested 48 h after application to the test animal. Note that the best formulation is almost four times as effective as the standard. Treatments were repeated after 8 months sample storage and the same duration of protection was obtained. This indicates the long shelf life of the microencapsulated materials.

TABLE 9.4. COMPARATIVE NATURAL PYRETHRUM REPELLENT
STUDY — TSETSE, *GLOSSINA MORSITANS* (NON-SYNERGIZED
PYRETHRUM ON GUINEA PIG)

Formulation[b]	Test concentration (mg cm^{-2})	Repellency, days	Minimal[a] effective dose
Not encapsulated	1.05	6	0.052
Not encapsulated	0.52	6	
Not encapsulated	0.26	4	
FI	0.52	11	0.026
MI	0.6	10(15)	0.014
F2	0.52	9	0.052
SI	0.8	7(9)	0.027

[a]Minimum concentration mg cm^{-2} giving 90% protection from biting, tested 48 h after application.
[b]See Table 9.7.

Table 9.5 shows repellency data for synergized natural pyrethrum formulations. Note that microencapsulation results in an improvement in efficacy in comparison with non-encapsulated formulations. However, there is a reduction in the duration of complete repellency when we compare the data against that for the non-synergized formulation noted in the previous table.

TABLE 9.5. MICROENCAPSULATED NATURAL PYRETHRUM
REPELLENT/MORTALITY STUDY — TICK, *ORNITHODOROS THOLOZANI*
(NON-SYNERGIZED PYRETHRUM ON GUINEA PIG)

Formulation	Test concentration (mg cm^{-2})	Repellency, days	Mortality% A[a]	B[b]
Not encapsulated	0.52	2	100	—
M4	0.90	5	30	90
S3	0.90	3	30	—
F5	0.52	3	60	60
FI	0.52	2	100	80
Synergized pyrethrum (1:10 piperonyl butoxide) on guinea pig				
Not encapsulated	0.52	2	100	—
F3	0.55	3	100	—
S2	0.38	2	100	100

[a]Ten min exposure 2 days after application — recorded 7 days later.
[b]Six h exposure 7 days after applic

Although the piperonyl butoxide synergist greatly improves the toxicity of natural pyrethrum, its dilution effect appears to limit the repellent efficacy.

Table 9.6 lists the results of tests comparing some synthetic pyrethroids with natural pyrethrum. Despite their high toxicity, these synthetics did not confer protection from tsetse bites even when tested only 2 days following treatment. Microencapsulation apparently imparts no improvement in repellence of these pyrethroids.

TABLE 9.6 COMPARATIVE SYNTHETIC PESTICIDES REPELLENT
STUDY — TSETSE, *GLOSSINA MORSITANS* (ON GUINEA PIG)

Pesticide	Formulation	Test concentration (mg cm^{-2})	Repellency days
Bioresmethrin	Not encapsulated	0.52	2
Bioresmethrin	Microencapsulated	0.52	2
Resmethrin	Microencapsulated	1.0	2
Permethrin	Microencapsulated	0.9	2
Natural pyrethrum	Not encapsulated	0.52	6
Natural pyrethrum	Microencapsulated	0.52	11

Table 9.7 depicts results of a repellent mortality study on the tick
Ornithodoros tholozani. Starved adult females are repelled for shorter periods
than the tsetse flies. This table demonstrates the enhanced duration of
repellency that microencapsulation produces, as much as 5 days of complete
repellency versus less than 2 days with non-synergized natural pyrethrum.
Piperonyl butoxide as synergist reduces the efficacy of natural pyrethrum with
respect to repellency against ticks. Some interesting observations are evident
when considering toxicity effects. This is important since ticks killed within
24 h of their attachment to the host are less likely to transmit pathogens which
usually require several days of activation to be virulent. Hard ticks attach and
feed for days, while toxicants can act over a longer period. In one test, ticks
were exposed for 10 min to guinea pigs which had been treated 2 days prior to
testing. In a companion test, the ticks were given 6 h exposure to an area
treated 7 days prior to testing. In both cases toxicity readings were made
7 days after exposure. While non-encapsulated natural pyrethrum is expected

TABLE 9.7. COMPARATIVE NATURAL PYRETHRUM
REPELLENT/MORTALITY STUDY — TICK, *ORNITHODOROS THOLOZANI*
(NON-SYNERGIZED PYRETHRUM ON GUINEA PIG)

Formulation	Test concentration (mg cm^{-2})	Repellency days	Mortality% A[a]	B[b]
Not encapsulated	0.52	2	100	—
M4	0.90	5	30	90
S3	0.90	3	30	—
F5	0.52	3	60	60
FI	0.52	2	100	80
Synergized pyrethrum (1:10 piperonyl butoxide) on guinea pig				
Not encapsulated	0.52	2	100	—
F3	0.55	3	100	—
S2	0.38	2	100	100

[a] 10 min exposure 2 days after application — recorded 7 days later.
[b] 6 h exposure 7 days after application.
[c] F = fast, M = medium, S = slow (permeation rate from capsule).

to be ineffective after a brief period of exposure to the environment, it was observed that persistency of toxicity was achieved by microencapsulation.

9.5 CONCLUSION

Tsetse and ticks are vectors of serious diseases which plague man and livestock. Research objectives are to protect the animals from pathogens which the parasites carry. Natural pyrethrum is effective both as a repellent and a toxicant against tsetse and ticks, but its lability in the environment is a limiting factor. This research has demonstrated that controlled release techniques can produce an extension of full protection relative to conventional natural pyrethrum formulations. Even after complete protection was no longer observed, a high protection level, sometimes greater than 90%, continued for a significant period. Thus, it appears that microencapsulation of natural pyrethrum could be a viable, economically feasible approach. The evaluations discussed were performed in the laboratory, it now appears to be of sufficient interest to warrant its evaluation in the field.

9.6 SUMMARY

Tsetse and ticks, vectors of several serious diseases of man and livestock, continue to plague the health and economy of vast areas of Africa. Because of its known insect-repellent property, natural pyrethrum has been used to control these insects. Although highly effective when applied to livestock, the duration of complete protection in the field is only 2 days. This characterizes the major problem with utilizing natural pyrethrum — the rapid decomposition by the environment. To be practical as a repellent for livestock, the desired duration of repellency is at least 1 week.

Microencapsulation of natural pyrethrum has been proposed for improving the persistency of its repellent/pesticide properties toward tsetse and ticks.

This paper reports on an evaluation of twenty-seven formulations of microencapsulated natural pyrethrum. The major variation consists of differences in the release rate of the encapsulant by modification of the capsule wall composition. Comparison is also made with non-encapsulated pyrethrum and with some synthetic pyrethroids. Results of tests show that microencapsulation makes possible a significant increase in the persistence of the repellency and toxicity of natural pyrethrum toward tsetse and prolongs the repellency and toxicity toward the tick, *Ornithodoros tholozani*. The synthetic pyrethroids tested were comparatively ineffective even after microencapsulation.

9.7 ACKNOWLEDGEMENTS

The research described was supported in part by the United States Agency for International Development. We also wish to acknowledge the valuable suggestions and assistance of Dr. M. N. Ben-Eliahu and Dr. D. Ben-Tamar of the Department of Zoology, Hebrew University, who performed the bioassay tests.
The data in section 9.4.1.2(a) is reproduced with kind permission of IAEA, Vienna (Galun *et al.*, 1980).

9.8 REFERENCES

Bar Zeev M. & Gothilf S. (1973) Laboratory evaluation of tick repellents, *J. Med. Entomol.* **110**, 71–.
Galun R. (1975) Protection of livestock from tsetse bites by means of repellents. *Pyrethrum Post* **13**, 2.
Galun R., Ben-Eliahu M. N., Ben-Tamar D. & Simkin J. (1980) Long-term protection of animals from tsetse bites through controlled-release repellents. In *Isotope and Radiation Research on Animal Diseases and Their Vectors* (Proceedings of a symposium, 7–11 May 1979) published by Int. Atomic Energy Agency, Vienna. IAEA-SM-240/23, pp. 207–218.
Glynne Jones G. D. & Sylvester N. K. (19-6) Pyrethrum as an insect repellent. Part I: Literature reviews. *Pyrethrum Post* **8**, 28–.
Hornby H. E. & French M. H. (1943) Introduction to the study of tsetse repellents in the field of veterinary medicine. *Trans. R. Soc. Trop. Med. Hyg.* **32**, 41–.
Holden J. R. & Findlay G. M. (1944) Pyrethrum as a tsetse repellent; human experiments. *Trans. R. Soc. Trop. Med. Hyg.* **38**, 199–.
Findlay G. M., Hardwick J. & Phlep A. I. (1946) Tsetse fly repellents. *Trans. R. Soc. Trop. Med. Hyg.* **40**, 341–.
Schmidt M. L. (1977) Relative effectiveness of repellents against *Simulium damnosum* and *Glossina morsitan* in Ethiopia. *J. Med. Entomol.* **14**, 276–.
Vandegaer J. E. (1971) US Patent 3,577,515, May.

9.9 DISCUSSION

Whitehead: I would like to ask you a question about your repellency tests. Is the repellant still effective 48 h after the last meal in tsetse flies? You didn't quite make it clear to me how long the flies had starved before they were used in your tests.
Simkin: You are talking about how long the flies had been starved?
Whitehead: Yes.
Simkin: I do not know how long they had been starved. This work was done by Professor Galun. And what I know is that the flies were starved for a while before the test.
Otieno: Can you give a little bit of detail about how you quantified the degree of protection in your repellency tests.

Simkin: The major approach to determine the repellency was that there had to be truly 100% repellency. In other words, none of the flies indicated that they were engorged or that they were feeding. We went on diluting our test solutions, and for each dilution, we counted the number of flies that were engorged. As soon as we saw any, that was the end of repellency as far as we were concerned.

Casida: In your introduction, you indicated certain aspects of microencapsulation, e.g. to increase the persistence, you have to reduce the acute oral toxicity. So if you change the persistence and you change toxicity, are you therefore required to do new toxicology and residue studies on the compound in the microencapsulated form with microencapsulation ingredients: otherwise don't you have to re-register the insecticide as a single compound?

Simkin: Yes, we have to go through all that tussle. So far two compounds have been registered with microencapsulation as a factor. One of them is microencapsulated natural parathion. Then we had to go through the same procedure with the second one, i.e. the microencapsulated diazamide which is used mainly as a household insecticide. We knew right at the beginning that we would have difficulty with the excessive requirements for proof of acceptability.

Casida: Are you in a position to say whether there is a change in the "no-effect" level when diazinon or methlyl parathion is microencapsulated? In other others words, can you tell us whether the level at which there is no effect in feeding studies changes from 5 parts per million?

Simkin: With methyl parathion, it is active as a technical material and in microencapsulated form.

Bowers: Piperonyl butoxide is usually used as a synergist with pyrethrins. Did you use it for this purpose as well to see if it enhances repellency of the encapsulated products?

Simkin: Yes, we tried it to see if it could increase repellency. Unfortunately, it did not.

Macharia: The natural pyrethrin insecticide is a contact insecticide. It has low volatility. Is it possible that the actual repellency is due to pyrethrin I and II series?

Simkin: We took the natural pyrethrum extract as a whole for our tests. Whether there is a different factor there, other than the pyrethrins, which is responsible for repellency, I do not know for sure, but I believe that there must be some contact between the insect and the repellent, even though the pyrethrins have low volatility.

Marini Bettolo: I just wanted to mention that there is an excellent paper from Brazil where the author has shown that using rubber with decamethrin completely deterred the vectors of chaga's disease from biting their hosts.

Simkin: We did not assess the whole spectrum of the synthetic pyrethroids. I think it is worth trying rubber instead of the polymer we have used in our microencapsulation studies.

Djerrasi: Is it expensive to use these microencapsulated insecticides?

Simkin: It is not very expensive. We are using US$100 per pound of insecticide and this as you can see is not expensive. However, if you are dealing with an expensive insecticide, then the whole exercise can be costly.

Saxena: I would like to point out that our experience with microencapsulated materials, e.g. the pheromones, in the tropical climates did not give encouraging results. We are therefore not enthusiastic or optimistic about the success of this mode of application of pesticides in the tropical areas of the world.

Djerassi: That question should have been addressed to Graham-Bryce, who put up a good case for an empirical formulation which can be used across various climates.

Bowers: Are you talking about man or animal as a target of protection using encapsulated repellents?

Simkin; Both. But I am concerned chiefly with increasing the level of protection for cattle.

Section C

Plant Host/Pest Relations: The Bases of Plant Resistance

CHAPTER 10

Coevolution of Plants and Insects

PAUL FEENY

Department of Entomology, Cornell University, Ithaca, New York, USA

CONTENTS

10.1	Introduction	167
10.2	Chemical Defenses	169
	10.2.1 Cruciferous plants	169
	10.2.2 Umbelliferous plants	171
	10.2.3 Ephemeral plants	173
10.3	Insect Resistance in Foliage of Trees	175
10.4	Relationship Between Plant Defense and Plant Apparency	175
10.5	Implications for Agriculture	177
10.6	Conclusion	180
10.7	Summary	180
10.8	Acknowledgements	181
10.9	References	181
10.10	Discussion	184

10.1 INTRODUCTION

Coevolution is the series of evolutionary changes that occurs in two or more populations (usually of different species) as a result of the reciprocal selective action of the populations on one another (Arms & Camp, 1979; Janzen, 1980). When the selective effects of two populations on each other are so strong that some of the adaptations in one population can be attributed uniquely to the selective action of the other and vice versa, the coevolution is said to be "tight". Such coevolution may occur, for example, between a plant population that is pollinated only by a single species of bee and a population of the bees that relies on these plants as their sole source of nectar. Coevolution that occurs between arrays of populations, each generating selective pressures as a group, is referred to as "diffuse" (Janzen, 1980).

The term coevolution was introduced by Ehrlich & Raven (1964) to describe the diffuse coevolution that is believed to occur between plants and phytophagous insects. The phenomenon had been referred to earlier as "parallel evolution" by Brues (1920) and as "reciprocal adaptive evolution" by Fraenkel (1959). According to the hypothesis, the selective pressures exerted by phytophagous insects have been responsible in large measure for the evolution and elaboration of toxic natural products (secondary compounds) in plants. This in turn has selected for the evolution by insects of mechanisms for detoxifying toxins in their food and for using characteristic plant compounds, including the original toxins, as behavioral cues for finding and recognizing their food plants. The picture is then one of a coevolutionary "arms race", in which plants are constantly under selective pressure to improve and diversify their defenses while insects are under corresponding pressure to evolve new counter-measures.

The coevolution hypothesis has not gone unchallenged. One argument against it is based on the observation that insects in natural habitats tend to be rare relative to the available plant biomass and hence are unlikely to be able to influence plant evolution appreciably (Jermy, 1976). This argument can be countered by arguing that the effects of herbivores do not have to be catastrophic in order to influence the outcome of competition between plant genotypes and hence to have a selective impact on a plant population. It is countered also by a number of recent studies demonstrating that insects can indeed have a major impact on plant survivorship, growth and reproduction in natural communities (e.g. Varley & Gradwell, 1962; Breedlove & Ehrlich, 1968, 1972; Janzen, 1970, 1971; Harris, 1973; Morrow & Lamarche, 1978; Rausher & Feeny, 1980). The criticism is weakened yet further by studies showing that plant defenses attributable to a particular insect species are lacking in populations of the same plant species that have probably never been exposed to attack by that insect (e.g. Janzen, 1975).

A second criticism of the coevolution hypothesis is that closely related insect species are frequently found to feed on botanically very distant plant species. This situation, it is argued, cannot have arisen by coevolution between plants and their associated insect herbivores (Jermy, 1976). The problem here is that insects adapted to the chemistry of one plant species are preadapted to feeding on other plants, however distant botanically, that share similar chemistry. The resulting radiations of insect taxa onto plants of similar chemistry (e.g. Dethier, 1941, 1954) tend to obscure underlying patterns of coevolution. It is not simply possible, by casual inspection, to deduce whether or not a particular insect–plant association is the result of coevolution or whether the insect is a relatively recent arrival, preadapted to exploit the plant as a result of its evolutionary association with other plant species (Janzen, 1980). Thus present-day patterns of host–plant use by insects cannot by themselves be used as evidence either for or against the hypothesis of coevolution between plants and insects. They are likely to result partly from coevolution and partly from "sequential evolution", the process postulated by Jermy (1976).

A third and much more convincing criticism of the coevolution hypothesis, as formulated by Fraenkel (1959) and by Ehrlich & Raven (1964), is that it places undue emphasis on insects as the prime selective agents favoring the evolution of secondary compounds in plants (Jermy, 1976). There is no question that a great many bacteria, viruses, fungi and animals, as well as competing plants and even the physical environment, have influenced the evolution and elaboration of secondary compounds in plants (Whittaker & Feeny, 1971; Rosenthal & Janzen, 1979). Coevolution between insects and plants must therefore be viewed as one part of a broader evolutionary scene that includes coevolution between plants and a diverse range of enemies (Feeny, 1975, Janzen, 1980). The relative contribution of insects as selective agents shaping the secondary chemistry of any particular plant population will, unfortunately, be hard to assess with confidence.

With these caveats in mind, I will illustrate chemical aspects of presumed coevolution between plants and their enemies by reviewing some recent studies of interactions between plants of the families Cruciferae and Umbelliferae and some of their associated insects. This will lead to more general speculation about the evolution of chemical defenses in plants, and to some possible implications for the role of natural products in pest control.

10.2 CHEMICAL DEFENSES

10.2.1 Cruciferous plants

The family Cruciferae consists of about 3000 species, distributed throughout the world but concentrated in temperate regions with a Mediterranean climate (Hedge, 1976). Many species have become weeds associated with man-made disturbance throughout the world, and a few, most notably the cole crops of the genus *Brassica*, have become important economically. The class of natural products that most typifies the family is that of the glucosinolates (mustard oil glucosides). More than 300 species have been tested for these compounds so far and all have been shown to contain at least one and usually several of the eighty or so known glucosinolates (Kjaer, 1960, 1976; Ettlinger & Kjaer, 1968; Van Etten & Tookey, 1979). Hydrolysis of glucosinolates, typically to form the corresponding volatile isothiocyanates (Fig. 10.1), is catalyzed by enzymes (myrosinases) that are stored separately from the

$$CH_2 = CH - CH_2 - C \underset{S - Glucose}{\overset{N - O - SO_3^-}{\diagdown}}$$

Allylglucosinolate

Brassica
nigra

Thlaspi
arvense

$$CH_2 = CH - CH_2 - NCS$$

Allylisothiocyanate

$$CH_2 = CH - CH_2 - SCN$$

Allylthiocyanate

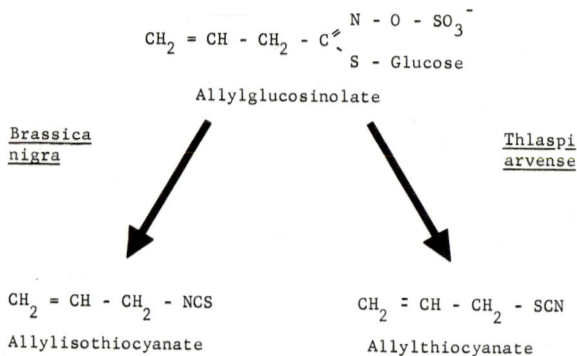

FIG. 10.1. Allylglucosinolate is Hydrolyzed to allylisothiocyanate in *Brassica nigra* and most other plants in which it occurs. In a few species, including *Thlaspi arvense*, it is hydrolyzed instead of allylthiocyanate.

glucosides in the plant tissues. These enzymes come into contact with their substrates when the tissues are bruised or otherwise damaged (Kjaer, 1976; Bjorkman, 1976).

Isothiocyanates, including several of those produced by crucifers, have been known for some time to be powerful antibiotics (Virtanen, 1965), fungicides (Walker, Morell & Foster, 1937), inhibitors of seed germination (Bell & Muller, 1973), and insecticides for non-phytophagous insects (Brown, 1951; Lichtenstein, Morgan & Mueller, 1964) (Fig. 10.1). More recently, they have been shown to inhibit the growth of some phytophagous insects that occur in the same environments as crucifers and pose potential threat to these plants. The concentration of allylglucosinolate in the foliage of wild *Brassica nigra* (black mustard) plants in New York State is about 0.4% of fresh weight (Feeny & Rosenberry, 1981). By feeding larvae of the black swallowtail butterfly, *Papilio polyxenes*, on celery leaves that had been cultured in aqueous solutions of allylglucosinolate, Erickson & Feeny (1974) showed that this compound, at naturally occurring concentrations, inhibited larval growth and caused substantial mortality. Subsequently, Blau, Feeny, Contardo & Robson (1978) were able to demonstrate that the effect is due not only to inhibition of feeding by larvae but also to actual toxicity of allylglucosinolate; the larvae gained weight more slowly than would be expected from their (reduced) feeding rate.

In spite of their content of glucosinolates, cruciferous plants are attacked by many insect species, some of which are serious pests of cruciferous crops. One such pest is the European cabbage butterfly, *Pieris rapae*, the larvae of which feed only on glucosinolate-containing plants. Slansky & Feeny (1977) found that variation in growth by the larvae of this species on a range of different crucifer species and crop varieties could be explained rather well in terms of plant nitrogen content. They concluded that glucosinolates, though known to serve as feeding stimulants for *Pieris* larvae (Verschaffelt, 1911; Schoonhoven, 1972), play at best a minor role in comparison with plant nutrients as determinants of larval growth rate. This conclusion has been substantiated by Blau *et al.* (1978), who showed that allylglucosinolate was not toxic to *P. rapae* larvae, even when these were reared on crucifer (collard) leaves in which the allylglucosinolate content had been boosted artificially to levels many times higher than those occurring naturally. Glucosinolates thus seem to be "qualitative" barriers to attack by insects (Feeny, 1975). Against insects, such as *P. polyxenes*, that do not normally attack crucifers or other glucosinolate-containing plants, they are effective toxins at quite low concentrations; against those species that have evolved mechanisms to tolerate the compounds, however, they are ineffective over even quite large ranges of concentration.

There are several genera of crucifers which are known to contain, in addition to glucosinolates, compounds belonging to other classes of natural products. The genera *Erysimum* and *Cheiranthus*, for example, contain cardenolides (cardiac glycosides), plants of the genus *Iberis* contain cucurbitacins, and those of the genera *Lunaria* and *Capsella* contain alkaloids (Gheorghiu, Constantinescu & Ionescu-Matiu, 1959; Hegnauer, 1964; Nielsen, 1978). These compounds, or close relatives of them, are known or suspected to be potential toxins (Rosenthal & Janzen, 1979). Since plants in these genera are often rejected by or support poor growth of insects (including *P. rapae*) that feed on other crucifers, it is tempting to conclude that the novel compounds represent secondary lines of defense, perhaps evolved in response to attack by insects and other enemies that have evolved counter-measures against glucosinolates (Feeny, 1976, 1977; Slansky & Feeny, 1977; Nielsen, 1978). In an explicit test of this hypothesis (Usher, 1979) larvae of *P. rapae* were found to grow slightly better on artificial diets containing cucurbitacins or k-strophanthin, a mixture of cardenolide (strophanthidin) glycosides, than on control diets. Although these results run counter to the hypothesis outlined above, Nielsen, Larsen & Sorensen (1977) and Nielsen (1978) report experimental evidence in its favor. They found that cucurbitacins, at naturally occurring concentrations, are potent feeding inhibitors to *Phyllotreta nemorum*, a crucifer-feeding beetle that does not eat *Iberis* species. Furthermore, cardenolides (including strophanthidin) are potent feeding inhibitors to *P. undulata*, *P. tetrastigma* and *Phaedon cochleariae*, crucifer-feeding beetles that do not eat *Cheiranthus* and *Erysimum* species.

10.2.2 Umbelliferous plants

The family Umbelliferae comprises about 2800 species, mostly perennial herbs that are comparable in distribution and ecology to crucifers; many weedy umbellifer species occur together with crucifers in disturbed habitats associated with human activities. Among the unusual natural products found in umbellifers are the furanocoumarins, a group of more than sixty compounds that are formed biosynthetically by prenylation of 7-hydroxycoumarins and cyclization of the prenyl group to form a furan ring (Nielsen, 1970). Most common are the linear furanocoumarins, such as xanthotoxin, in which prenylation occurs at the 6-position, leading to a 6,7-attachment of the furan ring (Fig. 10.2). Linear furanocoumarins have been reported from almost all species of the subfamily Apioideae so far examined and are found also in plants of the Rutaceae and a few other families (Hegnauer, 1964-1973; Nielsen, 1970; Towers, 1980). Much less common are the angular furanocoumarins, which are restricted in distribution to only a few genera of the subfamily Apioideae and the genus *Psoralea* in the family Leguminosae. In these compounds, prenylation occurs at the 8-position of the hydroxycoumarin nucleus, leading to a 7,8-attachment of the furan ring (Fig. 10.2) (Nielsen, 1970; Towers, 1980).

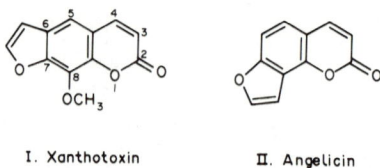

I. Xanthotoxin   II. Angelicin

FIG. 10.2. Xanthotoxin (8-methoxypsoralen) is a linear furanocoumarin. Angelicin is an example of an angular furanocoumarin.

Linear furanocoumarins (psoralens) form covalently bonded complexes with pyrimidine bases, cross-linking the strands of DNA molecules, when exposed to ultraviolet light (Rodighiero, Chandra & Wacker, 1970). This is the probable basis of their established toxicity to viruses, fungi, mammals and other groups of organisms (Towers, 1980). These compounds have also been shown to act as feeding inhibitors to insects, including *Spodoptera exempta* (T. Gebreyesus) and *S. litura* (Yajima, Kato, & Munakata, 1977). Berenbaum (1978) has shown that xanthotoxin, incorporated into artificial diets at concentrations found naturally in umbellifer foliage, caused 100% mortality of larvae of the southern armyworm, *S. eridania*. She showed that this effect was due not only to feeding inhibition but also to toxicity of the compound to the larvae. When larvae were fed similar diets beneath a filter that specifically excluded ultraviolet light, survival was enhanced greatly, suggesting that the mode of toxicity is that outlined above. Armyworm larvae are notoriously polyphagous and possess unusual levels of mixed-function oxidase activity in their midguts (Krieger, Feeny & Wilkinson, 1971). That xanthotoxins are so toxic to them suggests that linear furanocoumarins would probably be toxic to a very broad array of insect herbivores and that insects are likely to have played a significant role as selective agents favoring the evolution and diversification of these compounds (Berenbaum, 1978).

That the linear furanocoumarins are effective barriers against the colonization of umbellifers by insects is suggested by the relatively impoverished insect fauna of umbellifers (Jermy, 1976) and by the unusually high proportion of species that live, as larvae, in rolled leaves. Berenbaum (1978) showed that the leaves fail to transmit ultraviolet light and she suggests that the leaf-rolling habit may be a preadaptation for feeding on plants containing linear furanocoumarins. Of the insects that feed on apioid umbellifers without the shielding afforded by leaf rolls, larvae of several butterflies of the genus *Papilio* are particularly conspicuous. Using the method of Blau *et al.* (1978), Berenbaum (1981) showed that xanthotoxin, ingested at ecologically realistic levels, is not toxic to the larvae of *P. polyxenes* and even has a slight stimulatory effect on larval feeding rate. In another experiment, the larvae reached normal pupal weights when reared on an artificial diet containing 1% xanthotoxin, a concentration higher than that found in umbellifer foliage (Berenbaum, 1981). It appears, then, that the linear furanocoumarins represent another example of a "qualitative" barrier to insect evolution.

Angular furanocoumarins probably evolved more recently than their linear relatives; they are present, in addition to linear furanocoumarins, in only a few genera of umbellifers. They do not crosslink DNA molecules in the presence of ultraviolet light, nor are they as toxic to viruses, fungi and other organisms as are the linear compounds (Scott, Pathak & Mohn, 1976; Coppey, Averbeck & Moreno, 1979). They are, however, toxic to insects, including at least some of those that can tolerate linear furanocoumarins. Angelicin was found to inhibit growth of *P. polyxenes* larvae when fed at concentrations found naturally in the foliage of *Angelica atropurpurea* (M. Berenbaum & P. Feeny, 1981). The inability of *P. polyxenes* larvae to tolerate angelicin is likely to be genetically based since it was manifested even by larvae reared from eggs on *A. atropurpurea* and thus exposed to angelicin in their early instars. The effect cannot therefore be attributed merely to absence of larval conditioning (Jermy, Hanson & Dethier, 1968) or enzyme induction (Brattsten, Wilkinson & Eisner, 1977). Angular furanocoumarins may thus represent a further step in the coevolutionary interaction between plants and insects.

10.2.3 Ephemeral plants

Crucifers and umbellifers seem to share similar modes of chemical resistance to insects. Each family contains a relatively unique class of secondary compounds of demonstrable toxicity to several insects that do not normally attack the plants but that occur in the same habitats. In each family, there are associated insects that have overcome the toxicity of the respective compound classes and that can ingest them, at ecological concentrations, with evident impunity. Finally, each plant family contains species that seem to have elaborated novel compounds, atypical of the family as a whole, that are probably toxic to at least some of the insect species adapted to the compounds more typical of the family. The phytochemical literature suggests that such modes of chemical defense are typical of herbaceous plants generally, each family typified by a major class of compounds but also containing species with compounds novel for the family (e.g. Hegnauer, 1964–1973; Rosenthal & Janzen, 1979).

If the defensive compounds of crucifers, umbellifers and perhaps other herbaceous families are susceptible to counter-adaptation by insects, how do these plants continue to survive in nature? The answer appears to be that they rely to a great extent on escaping discovery from those insects that are capable of feeding on them. This is indicated by many experiments in which cultivated crucifers, planted in monocultures, have been shown to be attacked by greater numbers of crucifer-feeding insects than plants grown nearby in diverse meadow vegetation (Pimentel; 1961, Root, 1973; Smith, 1976). The relatively low densities of insects on the isolated plants are due, in part, to interference with the host-finding abilities of crucifer insects by odors emanating from nearby plants of other species. This phenomenon, long known to organic gardeners, has been called "associational resistance" (Tahvanainen & Root, 1972).

The difficulty of finding appropriate hostplants in space and time is presumably the reason that specialist insects have evolved sophisticated visual and chemosensory mechanisms for host finding and recognition (e.g. Schoonhoven, 1973; Rausher, 1978; Dethier, 1980). Often the compounds used as recognition cues are the very same compounds that are believed to have been evolved originally as mechanisms of defence. At least twenty species of crucifer-feeding insects, for example, are known to make use of glucosinolates or isothiocyanates for host location and recognition (Schoonhoven, 1972; Nielsen, 1978).

In view of the exploitation of defensive compounds as behavioral cues by adapted insects, there is a second reason why one might expect selection to favor plant genotypes that have modified their secondary chemistry. Not only might this provide a new line of resistance against previously adapted insects but it could also alter the chemical profile of the plant in such a way as to render it less susceptible to discovery by these insects. The cruciferous herb *Thlaspi arvense* seems to be an example of such "biochemical escape". Like *Brassica nigra*, this plant produces only one glucosinolate, allylglucosinolate, in any quantity in its foliage. Unlike *B. nigra*, however, this compound releases not allylisothiocyanate but its geometric isomer, allylthiocyanate, by enzymic hydrolysis when the tissues are damaged (Fig. 10.1) (Gmelin & Virtanen, 1959; P. Feeny & L. Rosenberry, unpublished results). The allylthiocyanate is toxic to insects (Brown, 1951) but, unlike the isothiocyanate, it is not used as a host-finding cue by the crucifer-feeding flea beetles *Phyllotreta cruciferae* and *P. striolata*. Traps baited with allylthiocyanate caught almost no beetles whereas traps baited with the isothiocyanate in the same experiment caught many. Moreover, colonization of three-plant "islands" of *T. arvense* was found to be accelerated greatly by addition of vials containing solutions of allylisothiocyanate (P. Feeny, J. Gaasch & L. Contardo, in preparation). Though it will be hard and perhaps impossible ever to prove that insects were responsible for this modification in the chemistry of *T. arvense*, such a change confers a demonstrable advantage to the plants by reducing their rate of discovery by insects.

10.3   INSECT RESISTANCE IN THE FOLIAGE OF TREES

The young foliage of many trees can escape, to some extent, the potential onslaught of phytophagous insects because of the unpredictability of leaf-flushing date each season (Feeny, 1976). Not surprisingly, therefore, there are parallels between the defensive chemistry of herbaceous plants and that which seems to typify the defensive chemistry of the immature foliage of trees (Rhoades & Cates, 1976; Feeny, 1976). The mature foliage of trees, by contrast, is eminently predictable in space and time and one might expect, therefore, to find that it is protected by defences that are less susceptible to counter-adaptation by herbivorous insects than are those of herbaceous plants. This appears to be the case. In a survey of the larval growth rates of twenty species of Lepidoptera reared on their natural food plants, Scriber & Feeny (1979) found a general relationship between larval-growth and plant-growth form: tree-feeding species grow only half as fast as do herb-feeding species, even when the species compared belong to the same genus. Presumably all the insect species tested are reasonably well adapted to the qualitative chemical barriers in their food-plants. The variation in growth, then, must be attributed to "quantitative" barriers that seem to set absolute limits to the growth rates of insects of a given growth form — barriers that are "dosage-dependent" in their effects and that are resistant to counter-adaptation. These mechanisms of resistance in the mature foliage of trees probably include leaf toughness, low nutrient and water content, and in some cases, perhaps, the presence of digestibility-reducing compounds such as tannins and resins (Rhoades & Cates, 1976; Feeny, 1976).

10.4   RELATIONSHIP BETWEEN PLANT DEFENSE AND PLANT APPARENCY

A likely relationship between plant resistance and plant escape, proposed

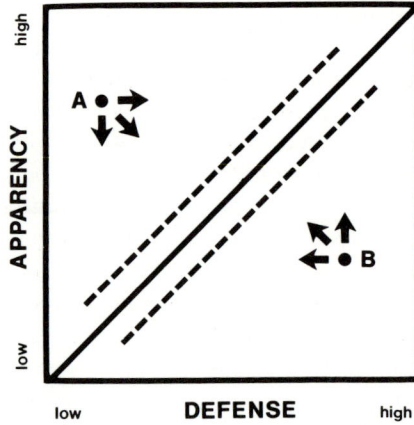

Fɪɢ. 10.3. A graphical model illustrating the likely relationship between plant apparency and levels of plant defense. Plants at ecological and evolutionary equilibrium with their enemies are considered to lie along the solid diagonal line. Herbaceous plants will tend to be found towards the lower end of the line and trees towards the upper end. The broken lines enclose a zone in which the defenses of the different tissues of a plant may vary, for a given apparency, as a function of tissue value. Plants (A) that find themselves above the equilibrium line will tend to return to it, as a result of predation (reducing apparency), for example, or as a result of selection for increased defense. Plants (B) that find themselves below the line will also tend to return towards it, perhaps as a result of becoming more abundant or by selection for reallocation of resources away from defense and towards other adaptations that maximize reproductive success.

simultaneously by Rhoades & Cates (1976) and Feeny (1976), can be represented by a simple graphical model (Fig. 10.3). For convenience, the susceptibility of a plant or plant tissue to discovery by insects or other enemies is referred to as its "apparency" (Feeny, 1976). The mature foliage of trees, at one extreme, is presumed to be highly "apparent"; herbaceous plants, by contrast, are thought to be relatively "unapparent". The main feature of the model is the diagonal line representing the hypothesized set of possible equilibrium values between apparency and defense. Plants that find themselves above this line, perhaps because of an increase in density, the appearance of a new herbivore or pathogen, or the evolution of improved host finding by an existing enemy, have inadequate defense for their new level of apparency; natural selection will then be expected to move the population gradually back to the line by reducing apparency and/or by increasing defense. Plants that find themselves below the line, perhaps because some major enemy has disappeared, or because they are at an unusually low density, will also be expected to shift back to the line. Selection will act against individuals that "waste" metabolic resources for defense that is not justified by the level of apparency and will favor plant genotypes in which resources are reallocated so as to maximize reproductive success under the new conditions. Plants could also return to the equilibrium line as a consequence of becoming more abundant, perhaps due to the loss of a major herbivore.

The different types of tissue comprising a plant are likely to differ in "value" or potential contribution to plant fitness; the loss of seeds or even young foliage to herbivores, for example, is likely to be more catastropic to a plant than is the loss of mature leaves. McKey (1974) has suggested that levels of defense in plants are likely to be related to the different values of different tissues. This relationship can be incorporated into the present model by thinking of the relationship between apparency and defense as a zone (indicated by broken lines in Fig. 10.3) rather than as a line. For a given level of apparency, one would expect seeds to be better defended than mature leaves.

This model is undoubtedly simplistic, even if the underlying hypothesis relating apparency to defense is upheld by future research. For the sake of simplicity, the relationship has been portrayed as a straight line, yet there is no particular reason for believing that such a relationship would be linear. Levels of defense in plants will probably be found to depend on other things than apparency and tissue value; these will have to be included in the model, if they do not result in rejection of the major hypothesis. The model was conceived originally in terms of "static" plant defenses yet already there are many well-established examples of chemical defenses that are only synthesized or elaborated when tissues are actually damaged by animals or pathogens (e.g. Janzen, 1979). The presence of such inducible defenses would presumably allow undamaged plant individuals to maintain lower levels of "static" defense than the model would predict for a given level of apparency. Finally, the model makes no attempt to disentangle the relative contributions of morphological and chemical characteristics to plant defense (e.g. Norris & Kogan, 1980). In spite of its simplistic nature, however, the model seems to be a useful initial framework within which to think of ecological influences on plant resistance to insects.

10.5 IMPLICATIONS FOR AGRICULTURE

Our understanding of the chemistry of insect–plant interactions remains scant in spite of accelerated interest over the past couple of decades. Nevertheless, the work on crucifers and umbellifers outlined here, along with comparable studies on other plant families, so far lends support to the hypothesis that many natural products serve a defensive role in plants, as suggested by Fraenkel (1959). Research results are also consistent, in general, with hypothesis of Ehrlich & Raven (1964) of diffuse coevolution between plants and their enemies, including phytophagous insects (Rosenthal & Janzen, 1979). Probable or possible implications of current ideas for agriculture are best considered under two headings. In the first place we can group those phenomena that occur in the immediate present (in "ecological time") and that can be discussed without reference to evolutionary change. In the second category are included implications of effects that involve evolutionary change and thus take place over a longer period ("evolutionary time").

A major ecological effect of agriculture is to make naturally unapparent plants more apparent, not only to the farmer but also to many of the plants' natural enemies (Feeny, 1977). The growing of plants in monocultures, in large fields, and in synchrony over wide areas tends to make the plants easier for insects to track in space and time and hence to render each individual plant more apparent. For the plants to survive, let alone provide economic yields, in the face of this increased apparency, corresponding increases in defense must be supplied (Fig. 10.3). This can be achieved either by the application of defensive compounds (pesticides) or by planting genotypes that have been bred for increased resistance to pests (Maxwell & Jennings, 1980). Both methods have their practical drawbacks yet both have proved enormously successful and remain the twin bulwarks of agricultural pest control. It is interesting to note that much of the progress in both of these areas has been achieved empirically, without the aid of a clear understanding of the underlying biological interactions.

An important implication of research on insect–plant coevolution is that levels of defense, whether bred into plants or applied as pesticides, could be reduced considerably if the apparency of crop plants were to be held down. Intercropping of different crop species or varieties, for example, would be expected to increase associational resistance. In spite of its application in traditional subsistence agriculture, this technique has only recently begun to attract interest as a procedure that might be applied to larger scale agriculture. Other techniques for reducing plant apparency include planting varieties that lack specific compounds used as host-finding cues (cf. the *Thlaspi arvense* example discussed earlier), and to restore some form of crop rotation to provide less predictability to the pattern of crops in space and time. It is interesting to note that biological control techniques, now an increasing component of Integrated Pest Management, have the effect of reducing plant apparency simply by reducing the densities of pest insects.

Perhaps the most important insight at the evolutionary level to be gained from coevolutionary studies is that almost any insecticidal chemical, whether natural or man-made, has only a certain lifetime of effectiveness in any given ecosystem and that this lifetime is profoundly affected by the apparency of the compound or the plants containing it (Beck & Schoonhoven, 1980). This is particularly true of compounds that act as "qualitative" barriers, which correspond approximately to "vertical" resistance (Horber, 1980). The greater the rate at which insects in a population encounter such a compound, the sooner the population is likely to evolve resistance to it (Southwood, 1973). This is probably the major reason why natural defensive compounds, deployed in plants of low apparency, may have lifetimes of thousands or millions of years whereas the defensive compounds present in or applied to crop plants, highly apparent to insects, may remain effective for only a few seasons. Thus, in addition to the ecological benefits outlined above, there are compelling evolutionary reasons for reducing the apparency of crop plants or, indeed, of individual pesticide compounds. One might predict, for example, that alternating the use of two different compounds on a crop, or applying the compounds to different fields in the same area, would prolong the lifetime of effectiveness of each compound.

In devising ways to increase the effective lifetimes of insecticidal compounds, whether naturally occurring or man-made, it will become ever more important to understand the genetic bases of insect resistance that operates at the behavioral level ("non-preference", etc.) and insect resistance that operates at the metabolic level ("antibiosis"). Much is made of the potential for repellents; yet if a repellent is not backed up by the presence of a toxic (metabolic) effect, there may be nothing to prevent the rapid evolution of insect genotypes that lack the repellent response and hence can attack the crop in question. If, as seems likely, insect adaptations for host recognition and host tolerance are controlled by separate, unlinked genes or polygenes, then at least two independent mutations must occur for an adapted specialist insect to be able to colonize a chemically distinct crop or variety, unless the insect population is preadapted at either the behavioral or metabolic levels (see Bush, 1974. 1975). Thus the rate at which insects adapt to a novel plant variety or pesticide compound would probably be reduced greatly by the presence of toxicity in addition to repellency, and the lifetimes of the active compounds increased correspondingly. The stability of resistance is likely to be influenced substantially by the host–plant specificities of the insects in question; resistance based chiefly on repellents might have more chance of stability against polyphagous insects than against oligophagous species (Beck & Schoonhoven, 1980).

Much of the discussion in this paper may give the impression that wild plants rely on only one or two lines of chemical resistance — glucosinolates in crucifers, furanocoumarins in umbellifers, etc. Reality is likely to prove more complex. One would certainy expect plants to retain a variety of different defense mechanisms simultaneously since the addition of each is likely to lessen substantially the rate of counter-adaptation by insects. Plants with a mixture of defense compounds are thus likely to survive longer in the flora of an area than those genotypes with fewer lines of defense. Should further research substantiate this kind of reasoning, it would also provide a cogent argument for applying mixtures of pesticide compounds, rather than individual compounds, and for combining compounds that are known to be detoxified in other insects by mechanisms controlled by different genes.

What, finally, do coevolutionary studies have to suggest about the future role of natural products in the control of pests and diseases? First, there is now overwhelming evidence that natural products have played a major role as defensive adaptations in the survival and evolution of plants. The staggering diversity of these compounds throughout the plant kingdom must surely provide a continuing source of clues for novel insecticides for decades to come. For particularly effective compounds that are unusually expensive to synthesize, it may prove economically feasible to harvest plants commercially, as is done for pyrethrins. Second, there is the realization that plants have faced many and perhaps all of the problems we have faced, albeit at generally lower levels of apparency and without having to provide economic yields, and have clearly evolved solutions. In trying to decide what mixtures of compounds to apply to crops in order to minimize the evolution of insect resistance, for example, it might be wise to discover what mixtures of compounds are deployed by plants. In searching for compounds that are effective against herbivorous insects but that will not affect natural predators or parasites, it might prove profitable to compare the kinds of chemical defenses evolved by plants with those evolved by insects (e.g. Rodriguez & Levin, 1976).

10.6   CONCLUSION

I conclude on a somewhat different note with an appeal for the conservation of substantial areas of different types of natural plant communities as sources for future research. Lowland tropical rain forests, in particular, are being cut down and destroyed at an alarming rate throughout the world. These regions, such as the remaining sections of the Kakamega Forest in Western Kenya, are the richest communities on the planet. They are natural laboratories in which natural selection, the ultimate screening agent, has been at work for millions of years. Coevolutionary interactions in lowland rainforests are more elaborate than anywhere else (Gilbert & Raven, 1975; Gilbert, 1977) and the diversity of natural products they contain must be prodigious. If these communities were to disappear, the loss of future research on natural products would be immeasurable and permanent.

10.7   SUMMARY

Chemical aspects of coevolution between insects and plants are illustrated by a discussion of the role of glucosinalates in the Cruciferae and of furanocoumarins in the Umbelliferae. These compounds are effective forms of resistance, at both the "behavioral" and "metabolic" levels, against non-adapted insects. Once overcome by adaptation, however, they seem to have little effect on adapted insect populations and may even be exploited as behavioral cues for host–plant recognition. Crucifers, umbelliferas and other herbaceous plants survive the depredations of adapted insects in part by their ephemeral life histories, which reduce their risk or rate of discovery by enemies. Evolutionary responses to adapted insects (and pathogens) seem to have included diversification of chemical defense compounds and elimination of compounds used by insects for host recognition.

Research on coevolution suggests that agricultural strategies should emphasize (1) a diversity of compounds that confer resistance at the metabolic level (toxicity), rather than merely at the behavioral level (repellency), and (2) techniques for reducing the apparency of crop varieties to insects. These strategies promise both an immediate increase in associational resistance and also a reduction in the rate of evolution of counter adaptations by insect pests.

10.8 ACKNOWLEDGEMENTS

I thank May Berenbaum for helpful comments on the manuscript. The research was supported by research grants DEB76-20114 and DEB-7922130 from the National Science Foundation.

10.9 REFERENCES

Arms K. & Camp. P. S. (1979) *Biology*, 970 pp. Holt, Rinehart & Winston, New York.
Beck S. D. & Schoonhoven L. M. (1980) Insect behaviour and plant resistance. In *Breeding Plants Resistant to Insects* (eds. Maxwell F. G. & Jennings P. R.), pp. 115–135. John Wiley & Sons, New York.
Bell D. T. & Muller C. H. (1973) Dominance of California annual grasslands by *Brassica nigra*. *Amer. Midl. Naturalist* **90**, 277–299.
Berebaum M. (1978) Toxicity of a furanocoumarin to armyworms: a case of biosynthetic escape from insect herbivores. *Science* **201**, 532–534.
Berenbaum, M. (1981) Effects of linear furanocoumarines on an adapted specialist insect (*Papillio polyxenes*) *Ecol. Entomol.* (in press.)
Berenbaum, M. & Feeny, P. (1981) Toxicity of angular furanocoumarines to swallowtails: Escalation in a coevolutionary armsrace? *Science*, **212**, 927–929.
Bjorkman R. (1976) Properties and function of plant myrosinases. In *The Biology and Chemistry of the Cruciferae* (eds. Vaughan J. G., Macleod A. J. & Jones B. M. G.), pp. 191–205. Academic Press, New York.
Blau P. A., Feeny P., Contardo L. & Robson D. S. (1978) Allylglucosinolate and herbivorous caterpillars: a contrast in toxicity and tolerance. *Science* **200**, 1296–1298.
Brattsten L. B., Wilkinson C. F. & Eisner T. (1977) Herbivore–plant interactions: mixed-function oxidases and secondary plant substances. *Science* **196**, 1349–1352.
Breedlove D. E. & Ehrlich P. R. (1968) Selective pressure of grazing butterfly larvae on flowering time of *Lupinus amplus* through control of seed production. *Science* **162**, 671–672.
Breedlove D. E. & Ehrlich P. R. (1972) Coevolution: Patterns of legume predation by a lycaenid butterfly. *Oecologia* **10**, 99–104.

Brown A. W. A. (1951) Insect Control by Chemicals, 817 pp. Wiley, New York.
Brues C. T. (1920) The selection of food-plants by insects with special reference to Lepidopterous larvae. *Am. Natur.* **54**, 313–332.
Bush G. L. (1974) The mechanism of sympatric host race formation in the true fruit flies (Tephritidae). In *Genetic Mechanisms of Speciation in Insects* (ed. White M. J. D.), pp. 3–23. Australian and New Zealand Book Co., Sydney.
Bush G., L. (1975) Sympatric speciation in phytophagous parasitic insects. In *Evolutionary Strategies of Parasitic Insects and Mites* (ed. Price P. W.), pp. 187–206. Plenum Publishing Corp., New York.
Coppey J., Averbeck D. & Moreno G. (1979) Herpes virus production in monkey kidney and human skin cells treated with angelicin or 8-methoxypsoralen plus 365 nm light. *Photochem. Photobiol.* **29**, 797–801.
Dethier V. G. (1941) Chemical factors determining the choice of food plants by *Papilio* larvae. *Am. Natur.* **75**, 61–73.
Dethier V. G. (1954) Evolution of feeding preferences in phytophagous insects. *Evolution* **8**, 33–54.
Dethier V. G. (1980) Evolution of receptor sensitivity to secondary plant substances with special reference to deterrents. *Am. Natur.* **115**, 45–66.
Ehrlich P. R. & Raven P. H. (1964) Butterflies and plants: a study in coevolution. *Evolution* **18**, 586–608.
Erickson J. M. & Feeny P. (1974) Sinigrin: a chemical barrier to the black swallowtail butterfly, *Papilio polyxenes. Ecology* **55**, 103–111.
Ettlinger M. G. & Kjaer A. (1968) Sulfur compounds in plants *Recent Advances in Phytochemistry* **1**, 59–144.
Feeny P. (1975) Biochemical coevolution between plants and their insect herbivores. In *Coevolution of Animals and Plants* (eds. Gilbert L. E. & Raven P. H.), pp. 3–19. Univ. Texas Press, Austin & London.
Feeny P. (1976) Plant apparency and chemical defense. *Rec. Adv. in Phytochem.* **10**, 1–40.
Feeny P. (1977) Defensive ecology of the Cruciferae. *Ann. Missouri Bot. Gdn.* **64**, 221–234.
Feeny, P. & Rosenberry, L. (1981) Seasonal variation in the glucosinolate content of *Brassica nigra* and three *Dentaria* sp. in central New York State. *Biochem. Syst & Ecol.* (in press.)
Fraenkel G. S. (1959) The raison d'etre of secondary plant sustances. *Science* **129**, 1466–1470.
Cheorghiu A., Constantinescu A. & Ionescu-Matiu E. (1959) Extragera si separarea alcaloizilor fluorescenti din *Capsella bursa pastoris* L. *Stud. Cercet. Biochim.* **2**, 403–406.
Gilbert L. E. (1977) The role of insect–plant coevolution in the organization of ecosystems. In *Le Comportement des Insectes et Les Signaux Issus du Milieu Trophique* (ed. Labeyrie V.), pp. 399–413. CNRS, Paris.
Gilbert L. E. & Raven P. H. (1975) *Coevolution of Animals and Plants*, 246 pp. University of Texas Press, Austin and London.
Gmelin R. & Virtanen A. I. (1959) A new type of enzymatic cleavage of mustard oil glucosides. Formation of allylthiocyanate in *Thlaspi arvense* L. and benzylthiocyanate in *Lepidium ruderale* L. and *Lepidium sativum* L. *Acta Chem. Scand.* **13**, 1474–1475.
Harris P. (1973) Insects in the population dynamics of plants. In *Insect–Plant Relationships* (ed. Van Emden H. F.), pp. 201–210. Blackwell Scientific Publ., Oxford.
Hedge I. C. (1976) A systematic and geographical survey of the Old World Cruciferae. In *The Biology and Chemistry of the Cruciferae* (eds. Vaughan J. G., Macleod A. J. & Jones B. M. G.), pp. 1–45. Academic Press, London, New York and San Francisco.
Hegnauer R. (1964–1973) *Chemotaxonomie der Pflanzen*, Vols. 3–6. Birkhauser Verlag, Basel and Stuttgart.
Horber E. (1980) Types and classification of resistance. In *Breeding Plants Resistance to Insects* (eds. Maxwell F. G. & Jennings P. R.), pp. 15–21. John Wiley & Sons, New York.
Janzen D. H. (1970) Herbivores and the number of tree species in tropical forests. *Am. Natur.* **104**, 501–528.
Janzen D. H. (1971) Seed predation by animals. *Ann. Rev. of Ecology and Systematics* **2**, 465–492.
Janzen D. H. (1975) Behaviour or *Hymenaea courbaril* when its predispersal seed predator is absent. *Science* **189**, 145–147.

Janzen D. H. (1979) New horizons in the biology of plant defenses. In *Herbivores: Their Interaction with Secondary Plant Metabolites* (eds. Rosenthal G. A. & Janzen D. H.), pp. 331–350. Academic Press, New York.

Janzen D. H. (1980) When is it coevolution? *Evolution* **34**, 611–612.

Jermy T. (1976) Insect–host–plant relationship: Co-evolution or sequential evolution? *Symp. Biol. Hung.* **16**, 109–113.

Jermy T., Hanson F. E. & Dethier V. G. (1968) Induction of specific food preference in lepidopterous larvae. *Ent. Exp. Appl.* **11**, 22–231.

Kjaer A. (1960) Naturally derived isothiocyanates (mustard oils) and their parent glucosides. *Fortschr. Chem. Org. Naturstoffe* **18**, 122–176.

Kjaer A. (1976) Glucosinolates in the Cruciferae. In *The Biology and Chemistry of the Cruciferae* (eds. Vaughan J. G., Macleod A. J. & Jones B. M. G.), pp. 207–219. Academic Press, New York.

Krieger R. I., Feeny P. P. & Wilkinson C. F. (1971) Detoxication enzymes in the guts of caterpillars: an evolutionary answer to plant defenses? *Science* **172**, 579–581.

Lichtenstein E. P., Morgan D. G. & Mueller C. H. (1964) Naturally occurring insecticides in cruciferous crops. *J. Agric. Food Chem.* **12**, 158–161.

Maxwell F. G. & Jennings P. R. (1980) *Breeding Plants Resistant to Insects.* John Wiley & Sons, New York.

Mckey D. (1974) Adaptive patterns in alkaloid physiology. *Am. Natur.* **108**, 305–320.

Morrow P. A. & Lamarche V. C., Jr. (1978) Tree ring evidence for chronic insect suppression of productivity in subalpine *Eucalyptus. Science* **201**, 1244–1246.

Nielsen B. E. (1970) Coumarins of umbelliferous plants. *Dansk Tidsskr. Farm.* **44**, 111–286.

Nielsen J. K. (1978) Host plant discrimination within Cruciferae: Feeding responses of four leaf beetles (Coleoptera: Chrysomelidae) to glucosinolates, cucurbitacins and cardenolides. *Ent. Exp. Appl.* **24**, 41–54.

Nielsen J. K., Larsen L. M. & Sorensen H. (1977) Cucurbitacin E. and I in *Iberis amara*: Feeding inhibitors for *Phyllotreta nemorum. Phytochemistry* **16**, 1519–1522.

Norris D. M. & Kogan M. (1980) Biochemical and morphological base of resistance. In *Breeding Plants Resistant to Insects* (eds. Maxwell F. G. & Jennings P. R.), pp. 23–61. John Wiley & Sons, New York.

Pimentel D. (1961) Species diversity and insect population outbreaks. *Ann. Ent. Soc. Am.* **54**, 76–86.

Rausher M. D. (1978) Search image for leaf shape in a butterfly. *Science* **200**, 1071–1073.

Rausher M. D. & Feeny P. (1980) Herbivory, plant density, and plant reproductive success: The effect of *Battus philenor* on *Aristolochia reticulata. Ecology* **61**, 905–917.

Rhoades D. F. & Cates R. G. (1976) Toward a general theory of plant antiherbivore chemistry. *Recent Adv. Phytochem.* **10**, 168–213.

Rodighiero G., Chandra P. & Wacker A. (1970) Structural specificity for the photoinactivation of nucleic acids by furanocoumarins. *FEBS Letters* **10**, 29–32.

Rodriguez E. & Levin D. A. (1976) Biochemical parallelisms of repellents and attractants in higher plants and arthropods. *Recent Adv. Phytochem.* **10**, 214–270.

Root R. B. (1973) Organization of a plant-arthropod association in simple and diverse habitats: the fauna of collards (*Brassica oleracea*). *Ecol. Monogr.* **43**, 95–124.

Rosenthal G. A. & Janzen D. H. (1979) *Herbivores: Their Interaction with Secondary Plant Metabolites.* Academic Press, New York.

Schoonhoven L. M. (1972) Secondary plant substances and insects. *Recent Adv. Phytochem.* **5**, 197–224.

Schoonhoven L. M. (1973) Plant recognition by lepidopterous larvae. In *Insect/Plant Relationships* (ed. Van Emden H. F.), pp. 87–99. Blackwell Scientific Publ., Oxford.

Scott B. R., Pathak M. A. & Mohn G. R. (1976) Molecular and genetic basis of furanocoumarin reactions. *Mutation Research* **39**, 29–74.

Scriber J. M & Feeny P. (1979) Growth of herbivorous caterpillars in relation to feeding specialization and to the growth form of their food plants. *Ecology* **60**, 829–850.

Slansky F., Jr. & Feeny P. (1977) Stabilization of the rate of nitrogen accumulation by larvae of the cabbage butterfly on wild and cultivated food-plants. *Ecol. Monogr.* **47**, 209–228.

Smith J. G. (1976) Influence of crop background on aphids and other phytophagous insects on Brussels sprouts. *Ann. Appl. Biol.* **83**, 1–13.

Southwood T. R. E. (1973) The insect/plant relationship — an evolutionary perspective. In

Insect/Plant Relationships (ed. Van Emden H. F.), pp. 3–30. Blackwell Scientific Publ., Oxford.

Tahvanainen J. O. & Root R. B. (1972) The influence of vegetational diversity on the population ecology of a specialized herbivore, *Phyllotreta cruciferae* (Coleoptera: Chrysomelidae). *Oecologia* **10**, 321–346.

Towers G. H. N. (1980) Photosensitizers from plants and their photodynamic action. *Progress in Phytochemistry* **6**, 183–202.

Usher B. F. (1979) A typical secondary compound in the family Cruciferae: Tests for toxicity to *Pieris rapae*, an adapted crucifer-feeding insect. M.S Thesis, Cornell University, Ithaca, N.Y.

Van Etten C. H. & Tookey H. L. (1979) Chemistry and biological effects of glucosinolates. In *Herbivores: Their Interaction with Secondary Plant Metabolites* (eds. Rosenthal G. A. & Janzen D. H.), pp. 471–500. Academic Press, New York.

Varley G. C. & Gradwell G. R. (1962) The effect of partial defoliation by caterpillars on the timber production of oak trees in England. *Proc. 11th Int. Congr. Ent.* **2**, 211–214.

Vershaffelt E. (1911) The cause determining the selection of food in some herbivorous insects. *Proc. Acad. Sci. Amsterdam* **13**, 536–542.

Virtanen, A. I. (1965) Studies on organic sulphur compounds and other labile substances in plants. *Phytochem.* **4**, 207–228.

Walker J. C., Morell S. & Foster ll. H. (1937) Toxicity of mustard oils and related sulfur compounds to certain fungi. *Am. J. Bot.* **24**, 536–541.

Whittaker R. H. & Feeny P. P. (1971) Allelochemics: Chemical interactions between species. *Science* **171**, 757–770.

Yajima T., Kato N. & Munakata K. (1977) Isolation of insect anti-feeding principles in *Orixa Japonlea* Thunb. *Agric. Biol. Chem.* **41**, 1263–1268.

10.10 DISCUSSION

Callow: Further to your nice and elegant model of plant deterrency against insects, I would like to point out that plants do not only evolve defence chemicals against insects but also against bacteria and fungi.

Feeny: Evolution of plant resistance to insect attack is a complicated phenomenon. There are cases where a plant produced a germination stimulant which is a potent antibiotic. So, presumably, in such cases, the same evolution process has gone on to the same extent.

Callow: I agree, but in those cases, the arrival of a pathogen is purely a random process. It has no relation to the apparency of resistance. However, I think this is a relationship which I do not think is as good as you propose.

Feeny: I am sufficiently enchanted with the model I have just described in the paper, but I would not bet my life on it. I do not think that it is better to put fungi on a plant and observe how the plant develops resistance to the fungi. It works better if the plant evolves resistance in the natural way.

Bowers: I would like to make a comment about this discussion at this stage. Indeed, the spores of fungi do not fall randomly over a population of plants. There must be a source of infection to produce the spores. These must be distributed to uninfected plants. Random disposal of spores is more likely if plants are widely distributed as occurs in Nature than in monoculture. But if you take a normal situation of plants that are resistant to infection, then the plant might give several crops before the need to protect them against plant fungi infection occurs. This can happen even if there are some infected plants at some distance away.

Bowers: Is it known whether or not plants which are attacked by leaf rollers contain coumarins?

Feeny: I do not know. The whole idea is quite new.

Saxena: There is also protection against predators to the level of 30–40%. So it is not only protection against insects. Sure enough these compounds do not act in isolation.

Bernays: Do you think that the field of herbaceous plants is actually more diverse than the tropical rain forest?

Feeny: I think that would be an interesting comparison. However, I believe rain forests can survive predators because some of their seeds scattered a distance away, have a chance of survival. Those near the parent tree would have no such chance. For these reasons, I think the field of herbaceous plants is quite diverse. The insects are not able to maintain their pressure on population of plants which therefore do escape some attack. In this respect, the two systems are comparable.

CHAPTER 11

The Role of Natural Products in Plant–Insects and Plant–Fungi Interaction

G. B. MARINI BETTOLO

Centro Chimica dei Recettori e delle Molecole biologicametre attive del CBR, Universita Cattolica del S. Cuore Via della Pineta Sacchetti 644, Roma 00168, Italy

CONTENTS

11.1	Introduction	188
11.2	Plant–Insect Interaction	188
	11.2.1 Toxic substances	189
	11.2.1.1 The alkaloids	189
	11.2.1.2 Pyrethrum	189
	11.2.1.3 Rotenoids	190
	11.2.1.4 Ryanodine	190
	11.2.2 Substances present in plant which affect insect behaviour	190
	11.2.2.1 Antifeedants	192
	11.2.2.2 Repellents	196
	11.2.2.3 Hormone-like substances	197
	11.2.2.4 Attractants and phagostimulants	199
	11.2.2.5 Other natural products	199
	11.2.3 Defence substances in plant resistance	200
	11.2.3.1 Quinones	201
	11.2.3.2 Phenoldienones	203
	11.2.3.3 Flavanoids and tannins	203
	11.2.3.4 Stilbenes	204
	11.2.3.5 Terpenes	204
	11.2.4 Defence substances formed in plants as a response to insect infestation	206
11.3	Plant–Fungi Interaction	207
	11.3.1 Natural resistance	209
	11.3.2 Stress substances and phytoalexins	211
11.4	Significance of the Presence in Plants and Insects of Natural Products of Similar Structure	213
11.5	Conclusion	216
11.6	Summary	217
11.7	References	218
11.8	Discussion	220

11.1   INTRODUCTION

Mankind has chosen from the thousand of plant species, a limited number, e.g. maize, wheat, barley, potato, apple tree, etc., which he grows for food instead of wandering in the forest picking wild fruits and hunting animals. This was the beginning of the history of agriculture. The next step was the improvement of the quality of these plants obtained through cross-breeding and hybridization since earlier times. The present period is characterized by the improvement of agricultural systems for higher yields necessary to support the growing population of the world.

The highly sophisticated agriculture in effect supports the increasing world population while at the same time there is increased plant attack by insects, fungi and viruses and increased weeds, resulting in crop losses. In order to overcome this, agrochemicals, generally synthetic, which act as pesticides, such as insecticides and fungicides as well as weed killers, are being applied. All this constitutes a rather complicated and unstable system, because of the progressive adaptation of the insects and fungi to the various pesticides, thus making it necessary to develop continuously new systems for crop defence.

Sometimes a new disease appears, that even the most modern techniques are inadequate for its complete control, for example as in the case of corn, coffee and tobacco diseases. In this case only new genetic varieties can resist the disease. In addition, the resultant ecological disruption makes crop protection always more difficult. Regarding future development, not enough attention has been given up to the biological systems which operate among non-cultivated species, i.e. in wild and forest areas, where natural equilibria permit the growth of plants, insects, fungi and weeds. It may thus be interesting to consider at this point the biological and chemical mechanisms which are the basis of the natural equilibria which permit the survival of many species. Possibly through a more advanced interdisciplinary study of these systems we may get some inspiration and new guidelines for action in the search for better methods of plant protection. In this review we shall examine separately the interactions between plants and arthropods and plants and fungi.

11.2   PLANT–INSECT INTERACTION

In the natural equilibria a complex of relationships exist between plants and insects. In plant life, insects are intimately associated with a number of activities beginning with pollination, where they are indispensable; some insects may on the other hand be harmful to the plant and even cause its death. In biological warfare, and mainly in the wild, harmful insects are controlled by other insects; and some plants are better protected by the presence of particular substances in their tissues. These factors prevent the excessive growth of harmful insect species. In modern agriculture and especially in monocultures, infestations become normal, plants after a long genetic selection lose many of their natural defences in order to develop other qualities. Simultaneously a disruption of the equilibria among the insect species has shifted the balance towards the injurious species.

Plant resistance to insects, one of the main factors of the survival of wild plants, can be attributed to several chemical mechanisms due to particular products formed generally in the secondary metabolism of the plant. In order to discuss the mechanisms of the plant protection we shall consider separately: toxic substances; substances acting on the insect behaviour; and defence substances formed after the infestation.

11.2.1 Toxic substances

11.2.1.1 The alkaloids The alkaloids for their general toxicity, exerted both on vertebrates and invertebrates, are present in a certain number of plants and prevent insect infestation of the plants containing alkaloids. A recent review reports much information on this subject (Levinson, 1976). One of the first pesticides used by man is nicotine, contained in *Nicotiana tabacum* L., which is strongly insecticidal although it does not protect tobacco from a large range of insects. Also other plants containing alkaloids in the parts mainly exposed to attack by insects, i.e. leaves, trunk bark or root, do not suffer from infestation. Naturally, alkaloids are not present in the plants which man has selected for ages as food for humans or animals. Their use as pesticides for plants is not advisable, because of their high toxicity, except for a few limited applications to gardening.

11.2.1.2 Pyrethrum Among the first generation of pesticides used in the past century we may recall pyrethrins. These are substances of very peculiar structure derived from cycloalkanes of three- and five-membered ring isolated from a single species *Chrysanthemum cineariaefolium*, a plant originally from Dalmatia (Casida, 1973). For their high knock-down properties and low toxicity to man, pyrethrins have found a great number of applications in the control of insects. Their sensitivity to atmospheric agents and light make them unfit for field use but extremely valuable substances for domestic use. Their cultivation has been developed in various countries especially in Kenya and Ecuador. Study of the active groups in the molecule had led to discovery of very effective synthetic products which represent the simplified molecule of pyrethrin and permethrin (Fig. 11.1). These are very active against insects and also resistant to atmospheric factors and light (Elliot, 1977). This work

Pyrethrin

Bioresmethrin

Fig. 11.1. Pyrethrin Bioresmethrin.

should be considered as one of the best examples of the role of research in this field to obtain active substances for protecting the useful plants.

11.2.1.3. Rotenoids A group of plants belonging to the sub-family of Tephrosiae and to the genera Lonchocarpus, *Derris* and *Tephrosia* contain very powerful insecticides: the rotenoids (Fig. 11.2). Rotenone and deguelin are the main components of the active species. Their toxicity for insects and also for fish and even for man is very high. This indicates that plants can have very powerful defence systems. Also in this case rotenoids are not distributed in all the species of the above genus but only in a few of them. Recently their distribution has been investigated in many species. It was found that of sixty-one species studied only twenty-seven contain rotenoids (Marini Bettolo, unpublished results). The other species contain prenylated flavo-noids, which may be considered biogenetic precursors of the rotenoids, but which are devoid of toxic activity on insects (Marini Bettolo & Delle Monache, 1975). Later, we may come back to the significance and the function of these substances in the biological equilibria. So far we must bear in mind that these plants grow in tropical areas and may be subject to the attack of insects and fungi.

11.2.1.4 Ryanodine The genus *Ryania*, plants of the Flacourtiaceae family, shrubs of South America, contain insecticidal principles in their wood which were used for controlling insects in the field. The main active constituent is ryanodine (Fig. 11.3), a complex substance whose structure has been recently elucidated (Wiesner, Valenta & Findlay, 1967). Also quassin and quassinoids from Simaroubaceae should be considered in this group. These are examples of plants with insecticide principles which have played an important role in biological equilibria. It should be underlined that the most active substances are found in tropical plants. Although of great interest, the above substances (alkaloids, pyrethrins, rotenoids, quassin and ryanodine) constitute only limited examples of plant defence mechanisms which exert only limited role in the plant–insect equilibria.

11.2.2   Substances present in plant which affect insect behaviour

To this class belong:

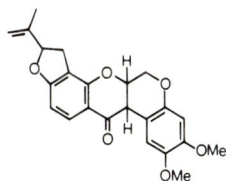

Rotenone

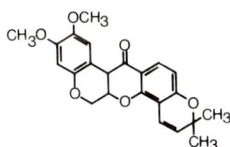

Degnelin

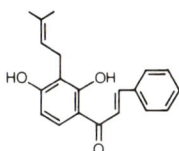

Isocordoin

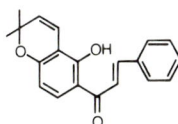

Lonchocarpin

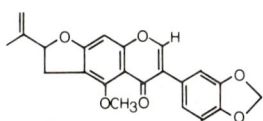

Glabrescione A

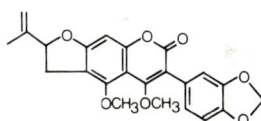

Glabrescin

FIG. 11.2. Rotenone Degnelin. Isocordoin Lonchocarpin. Glabrescione A Glabrescia.

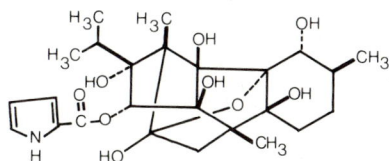

FIG. 11.3. Ryanodine.

(a) the antifeedants, which prevent the leaves or other parts of the plant from being eaten by the insect;

(b) the hormone-like substances such as phytoecdysones; juvenile hor-

$$\begin{array}{cc} NH & NH_2 \\ \parallel & \mid \\ H_2N-C-NH-O-CH_2-CH_2-CHCO_2H \end{array}$$

L - Canavanine

$$\begin{array}{c} NH_2 \\ \mid \\ H_2N-O-CH_2-CH_2-CHCO_2H \end{array}$$

L - Canaline

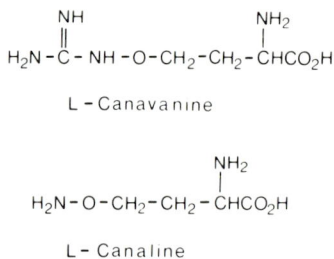

FIG. 11.4. L-Canavanine. L-Canaline.

mone (JH) analogues and precocenes which if ingested by the insect can modify their development;

(c) attractants, which through particular odours or colours attract insects and phagostimulants contained in leaves or other parts of the plants which after the first bite stimulate the insect to attack further;

(d) repellents produced by the plant with biological effects on the nervous system of the insect and characterized by their high vapour tension.

11.2.2.1 Antifeedants   The substances which prevent insects from destroying the plant, the antifeedants, belong to several chemical groups and exert their action through various mechanisms. Recent research in this field reveals the great variety of compounds having this activity and their specificity is even limited sometimes to only some insect and plant species. There are also substances that may be considered general antifeedants although this definition can never be accepted as completely valid. To this first class we can include uncommon amino acids, non-hydrolysable tannins (flavanotannins) and cyanogenetic glycosides.

Recently, the uncommon amino acids have been reviewed; they are generally found in seeds and may represent a strong defence against beetles or other insects and even to seed-eating mammals (Bell, 1977, 1980). The presence in the seeds of *Griffonia* of 5-hydroxytryptophan and in *Mucuna* of L-dioxy-phenylalanine is an interesting example of defence mechanism in these seeds against insect attack. It is well known that uncommon amino acids are rather abundant not only in seeds but also in rhizomes and in the leaves.

The presence of antifeedants in general may prevent the attack on crops and woods by dangerous pests like *Locusta* and *Prodenia*, the well-known army-worm. It has been found that the seeds of several species like *Gymonocladus dioicus* and *Dioclea megacarpa*, not attacked by bruchids, contain a high percentage of the amino acid canavanine. Canavanine (2-amino-4-guanidinoxybutyric acid) is competitive with arginine and thus exerts its toxic action against insects and mammals mainly because it is hydrolysed to L-canaline (Fig. 11.4), which is a potent neurotoxin. Recently it was found that a bruchid beetle *Caryedes brasiliensis*, the only insect which is capable of attacking

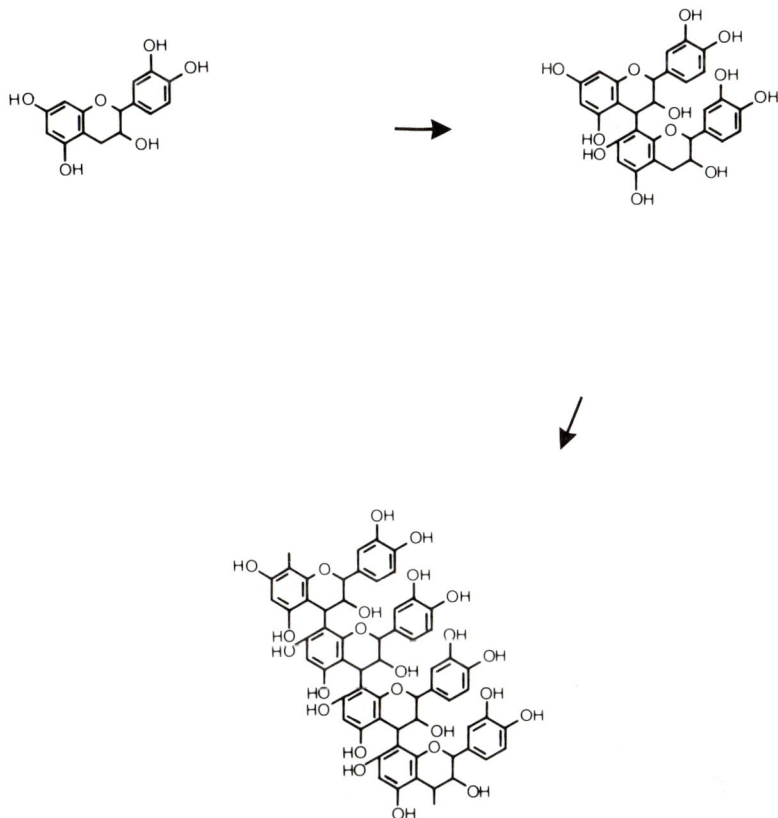

Fɪɢ. 11.5. Non-hydrolysable tannins formed by condensation of flavanic units.

Canavalia seeds, possesses an enzymatic system which allows canavaline to be deaminated to ʟ-homoserine and ammonia (Rosenthal, Dahlman & Jansen, 1978). The results of experiments on several insect types, *Locusta* (graminivorous), *Schistocerca gregaria* (polyphagous) and *Chortoicetes terminifera* (limited to grasses and legume leaves), indicate that the presence of the uncommon amino acids may reduce considerably the attack of insects against the plant parts where they are contained, acting as a general antifeedant and sometimes as a repellent.

Among the general antifeedants we can also include the tannins, that is the non-hydrolysable tannins, formed by the condensation of flavanic units (Fig. 11.5). These substances are extremely widespread in plants and their acidic character and astringent properties prevent attack by many insect species on those parts of plants like roots which contain high percentage of flavano-tannins. Another group of natural products found in many plants, the iridoids, may be also considered as general antifeedants. Bernays & De Luca

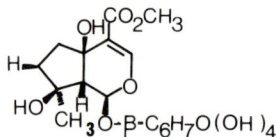

Ipolamiide

FIG. 11.6. Ipolamiide.

(1980, unpublished results) showed that ipolamiide (Fig. 11.6) is a good antifeedant for *Locusta migratoria* although less for *Schistocerca gregaria*. The bitter taste of this large group of natural products may account in part for these properties.

Some interesting antifeedants were found in limited amounts in some plant species. These products regulate the specific plant–insect interaction and although they can be considered only a very limited factor of ecological equilibria, they are most interesting as a natural model of an antifeedant that may account for structure-activity relationships. Recently, a great number of these antifeedants have been isolated and their structure determined. Work done at the ICIPE led to important results.

Azadirachtin (Fig. 11.7a) obtained from the leaves of *Azadirachta indica* (Meliacea) — the Indian neem tree, a rather common tree in the tropics ¯has proved to be one of the most powerful antifeedants known. It also acts as a systemic growth inhibitor. In Labiatae a rather strong antifeedant, ajugarin a clerodane derivative (Fig. 11.7b), was found in *Ajuga remota* (Trivedi, Komura, Kubo, Nakanishi & Goshi, 1979). Harrissonin of *Harissonia abyssinica*, a plant of East Africa, is also a very strong antifeedant (Kubo, Tanis, Lee, Miura, Nakanishi & Chapya, 1976a). Its structure is rather unusual (Fig. 11.7c) and may be related to a diterpene system, whereas *Xylocarpus molluscensis*, another Meliacea, contains an antifeedant, xylomollin an hemiacetal lactone (Fig. 11.7d) (Kubo, Miura & Nakanishi, 1976b).

From a known fly repellent, a solanacea *Nicandra physaloides*, used in Peru to this purpose, an active principle was isolated by Fraenkel in 1964 called nicandrenone (Fig. 11.8), which was recognized to be a 24-methyl steroid derivative (Begley, 1976). This substance is active as an antifeedant on insect larvae, as are also the minor substances found in the same plant, structurally related to nicandrenone. We might also mention that a Magnoliacea, *Liriodendron tulipifera*, is antifeedant for the gipsy moth larvae (*Porthetria dispar* L.) and contains two sesquiterpenes lactones, lipoferolide (Fig. 11.9) and peroxyferolide characterized by epoxy-groups (Doskotek, Feraly, Fairchild & Huang, 1976). Matsui & Muwakata (1976) have also described two lignans, epieudesmin and eudesmin, in *Parabenzoin praecox*, with antifeeding activity

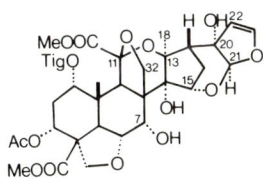

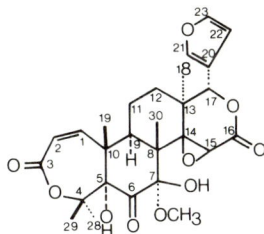

Azadirachtin

Harrisonin

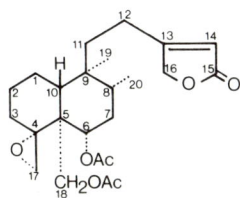

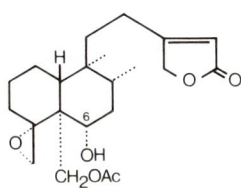

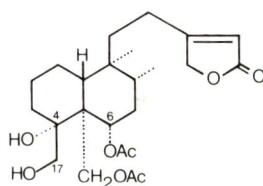

Ajuagarins

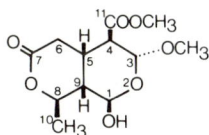

Xylomollin

FIG. 11.7. Azadirachtin. Harrisonin. Ajugarins. Xylomollin.

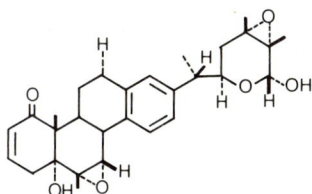

Nicandrenone

FIG. 11.8. Nicandrenone.

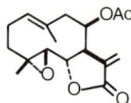

Lipiferolide

Fig. 11.9. Lipiferolide.

against the larvae of the phytophagous insect, *Spodoptera litura*, and pipere-none, another phenylpropanoid lignan in *Piper futokadzura* (Fig. 11.10).

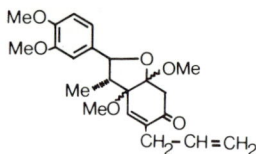

Piperenone

Fig. 11.10. Piperenone.

11.2.2.2 Repellents   It is difficult to make a sharp distinction between insect antifeedants and repellents in plants. We must bear in mind that some plants contain active principles which prevent attack and infestation by insects. The presence of particular products in the plant provides a defence system for the plant itself, and sometimes the genetic resistance of a certain species or even strain of plant is due to the existence in this variety of substances that may induce this resistance. The example of *Lupinus albus* variety, practically without alkaloids and thus well indicated as food for humans, which did not survive the attack of insects, is a clear indication of the importance of these substances in determining the resistance of the plant. This may also explain the weakness of cultivated species of plants derived through ages of selection, which may result in a species which is excellent for our use but which lacks natural defence factors.

Several terpenes exert a repellent action on insect: monoterpenes (e.g. citrals, geraniol), diterpenoids (e.g. kaurenic acid derivatives). We might mention that in some countries some plant extracts are used for their insecticidal activity as, for example, the powder of heartwood of *Juniperus recrua* of Nepal. It has been demonstrated that the active substances are thujopsene and 8-cedrene-13-ol (Fig 11.11) (Oda, Ando, Nakajma & Inoume, 1977). The presence in the leaves of many plant species of flavonotannins constitute a defence and repellent system as discussed previously. There are also other phenolic products present in plants which act as repellents to insects.

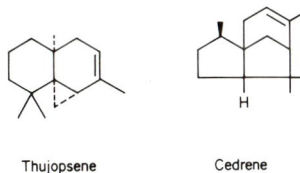

Thujopsene   Cedrene

Fig. 11.11. Thujopsene. Cedrene.

11.2.2.3 Hormone-like substances   The hormone activity considered here is that hormone system responsible for development of insects. The brain neurosecretory cells secrete an adenotropic hormone which stimulates prothoracic glands to produce ecdysone (moulting hormones: MH) and corpora allata, stimulated by the brain via the corpus cardiacum, produce juvenile hormone (JH). Ecdysone is responsible for the induction of moulting prior to the successive transformations into larval, pupal, and adult instars, while JH prevents premature metamorphosis into the adult. It has been demonstrated that excess JH may prevent development, inducing an abnormal moult which can be lethal to the insect. Similarly, ecdysones are important substances controlling the development of insect tissues.

Ecdysones are found in many plants; a recent review reports a large number of plants containing various ecdysterols (Fig. 11.12a) (Hikino & Hikino, 1975). It was shown that ecdysones are present in Pteridophyta (6 families, 44 species) Gymnosperma (6 families and 26 species) Angiosperma (22 families and 46 species) (Jacobson & Crosby, 1971). Since the presence of ecdysones in plants is so widespread they could constitute an important factor in the ecological balance between insect and plants. Moreover, ecdysones can be easily formed in plants by metabolism from other steroids (Levinson, 1972).

Juvenile hormones (Fig. 11.12b) and analogues are also present in some plants: one of the first observations is due to Carrol Williams, who noted that the waters of the Rio Negro, rich in suspended organic material (from wood decay) and thus terpenoids, do not have mosquito larvae as do the other rivers of the Amazon basin (Williams, 1969). Juvenile hormone analogues have been found in various plants and it may be inferred that they may also interfere to some extent with the control of several species of insects.

A number of chromene-derivatives, the precocenes (Fig. 11.13), isolated by Bowers and coworkers in the plant *Ageratum houstonianum*, are anti-allatotropic, that is, they prevent JH synthesis. They accelerate the development of the insect, producing dwarf animals unable to survive. This gives rise to other mechanisms of insect control. *Ageratum* and similar plants are not attacked by insects. We may wonder now if other plants contain precocenes in sufficient amounts to disrupt the insect development cycles (Bowers, Ohta, Cleere and

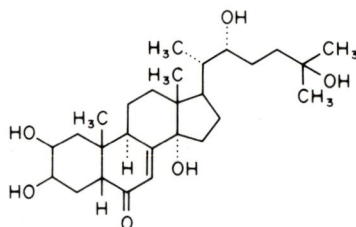

(a)   Ecdysone

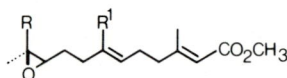

R = R¹ = Et,   JH I
R = Et, R¹ = Me, JH II
R = R¹ = Me,   JH III

(b) Juvenile hormone

Fig. 11.12. (a) Ecdysone. (b) JHI, JHII, JHIII.

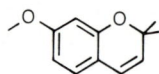   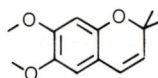

Precocene 1   Precocene 2

Fig. 11.13. Precocene 1, Precocene 2.

Marsella, 1976). Recently the isolation of precocenes was reported in some Brazilian plants (Furtado *et al.*, 1979).

We have so far little information about the presence of precocenes in other plants. We may, on the other hand, believe that some similar substances found in plants may present anti-allatotropic activity, e.g. 6-methoxy-7-geranyl-coumarin (Fig. 11.14). This substance was tested for its insecticidal activity and growth regulator effect and showed weak insecticidal activity. It is present in small quantities in a few Umbelliferae and in some Rutaceae (Torres, Delle Monache, Marini Bettolo & Cassels, 1979). It is hoped that this type of product, as well as coumarins, because of their structure, may interfere in some stage of development or affect insect behaviour. These examples show that many plants through their secondary metabolism yield products which may prevent insect infestation and attack.

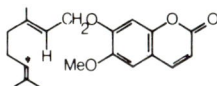

6- methoxy- 7 - geranyl-oxy- coumarin

FIG. 11.14. 6-Methoxy-7-geranyl-oxy-coumarin.

Other natural products, which are very active, have been found in a single genus and even in a single species, e.g. hypericin in St. John's Wort. Their use in the control of insects is for the moment not possible because of their toxicity or rather complex structure which prevents an easy and also economic synthesis. But the mechanisms and the properties of these substances should be considered in the new strategy of preparation of new insecticides.

11.2.2.4 Attractants and phagostimulants The relations between plant and insect are not only based on their incompatibility. This is limited to a certain number of insect species, but generally insects are necessary to the plants for their pollination and thus for their reproduction. There are some plants which are carnivorous, attract insects in order to capture them for food. There are also plants which produce fats and oils which may attract some insects from great distances. Examples are olive tree and the *Daucus oleae*.

11.2.2.5 Other natural products There are also substances which may attract insects by their smell and their taste such as the sugars, aromatic aldehydes and some terpenes. As an example of attractants for the black swallowtail (*Popilia aga*) we may quote anisic aldehyde, anethol, methyl-cavicol, present in *Citrus* sp.; attractants for *Bombyx mori* present in the leaves of Mulberry (*Morus alba* L.) are terpenes such as citral, linalol and terpenyl acetate.

Attractants of Cruciferae towards larvae of numerous insects are sulphocy-anate-glucosides (sinigrin, glucoerucin, gluconapin, etc.). Among higher fungi, the attraction exerted by *Ammanita muscaria* on flies is known. It is beyond the scope of this presentation to discuss all these substances. We can only point out that colours of the flowers attract insects, and not only those visible by humans but also those perceivable in UV, as it was demonstrated for flavones which guide the pollinating insects to the flower. The great number of aromatic aldehydes, monoterpenes and other substances, which constitute the perfume of the flowers, grasses and of other parts of plants, act as a call system for particular species of insects sensitive to the colours and the odours. A dynamic equilibrium is thus produced which regulates the plant–insect interaction. The greatest number of these substances are known; aromatic aldehydes as veratric acid, vanillic aldehyde, monoterpenes as geraniol, citral for the odours (Fig. 11.15) and flavonoids, anthocyans, polyenes and quinones for their colours. Coumarin, present in the leaves of many plants, acts at the same time as a phagostimulant and a deterrent (Matsumoto, 1962). These are normal and well-known secondary metabolites of plants. In only few cases we

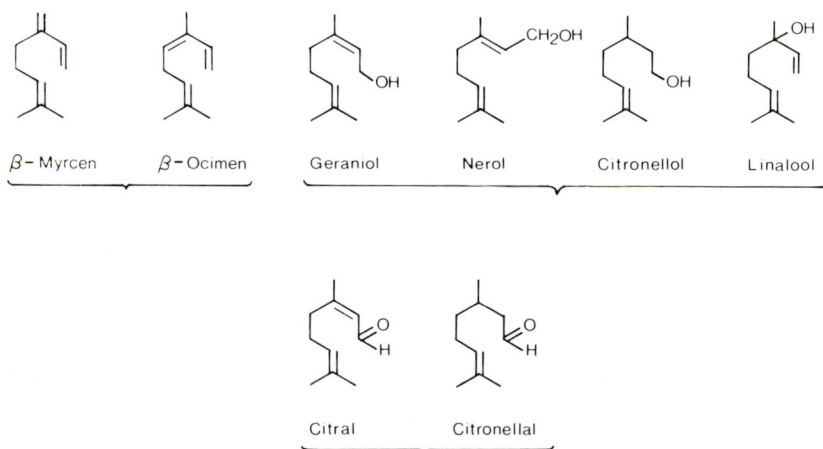

β-Myrcen   β-Ocimen   Geraniol   Nerol   Citronellol   Linalool

Citral   Citronellal

Fig. 11.15. B-Myrcen. B-Ocimen. Geraniol. Nerol. Citronellol. Linalool. Citral Citronellal.

may find some unexpected completely new product, like muscarin in *Ammanita muscaria*.

For further data one should consult the interesting and abundant literature on this subject (Torsteinson, 1960; Pavan & Quilico, 1969). On the other hand, it was necessary to point out this positive aspect of the plant–insect relationship because it is the basis of the ecology and survival of both plants and insect species. In the past it was a factor too often neglected when eradication measures for pests were taken; measures which in many areas disrupted the natural equilibria.

11.2.3   Defence substances in plant resistance

We can now consider certain constituents of plants, the chemical agents of plant resistance to insects and, less known, the plant resistance to diseases caused by fungi. A number of resistant strains contain biologically active substances, whereas non-resistant strains of the same plant do not contain them. For example, some particular strains of *Zea mays* resistant to the lepidoptera *Pyrausta bubilalis* (Hb) produce 6-methoxy-benzoxazolinone and its precursor 2,4-dihydroxy-7-methoxy, 1,4-benzoxanin-3-one. The same products are found also in resistant varieties of secale (Virtanen & Hietala, 1960).

This review must take into account not only the protection of cultivated plants or that of the forests, but also all the biological and biochemical factors which make possible the survival of the plants in the wild state in a perfect

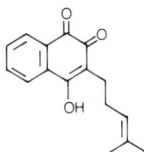

Lapachol

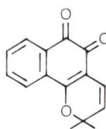

Lapachone

FIG. 11.16. Lapachol Lapachone.

ecological equilibrium. We must remember that not all insects are harmful to plants and that the injurious species are not generally harmful to all plants but only to some specific species. This means that there is a close specificity between the host plant and its pest. Gonçalves de Lima (1963), describing the antibiotic activity of some tropical woods, stresses that the resistance to decay in these woods should be attributed to a number of chemical substances mainly quinones. It is well known that many woods are rapidly attacked and destroyed by a number of insects whereas other woods are quite resistant both because of their structure and of the presence of particular substances which prevent both insect attack and fungal disease (Fortin & Poliquin, 1976).

Another example of natural defence systems in wild species (in comparison to that of the cultivated ones) is the occurrence of 2-tridecanone in a wild species of *Lycopersicon*, L. *hirsutum* f. *glabratum*. This ketone, which is active as an insecticide against lepidopterous larvae (*Manduca sexta* and *Heliothis zea*) and aphids (*Aphis gossypii*), is present in insecticidal concentration (10 µg mg^{-1} fresh foliage weight) in L. *hirsutum* the wild species and only in trace amounts (0.13 µg mg^{-1}) in L. *esculentum* the cultivated one (Williams, Kennedy, Yamamoto, Thacker & Bordner, 1980).

11.2.3.1 Quinones Naphthoquinone derivatives were found in many plant species; from the walnut, which grows in temperate countries, to the many tropical woods of the Bignonaceae family, e.g. *Tabebuia, Lapacho, Jacaranda, Tecoma*, etc. The first of these substances, lapachol, was isolated by Paterno in Rome at the end of the last century from large logs of *Lapacho* wood sent to the Chemical Institute of the University of Rome from Argentina. The structure of lapachol (Fig. 11.16) was established by Hooker. Later, a systematic study was carried on by Gonçalves de Lima in Recife on the quinones of Bignoniaceae and later by us in collaboration with the group of Recife. A particular derivative of naphthoquinone is the methyl 1-isopropyl-naphthoquinone, mansonone A, which may originate from the sesquiterpene cadalene, isolated from the heartwood of the African tree *Mansonia altissima* (Marini Bettolo, Casinovi & Galeffi, 1965). This product may give rise to other derivatives with different hydrogenation levels, mansonones D or F which contain a new oxygenated ring and may be considered derivative of

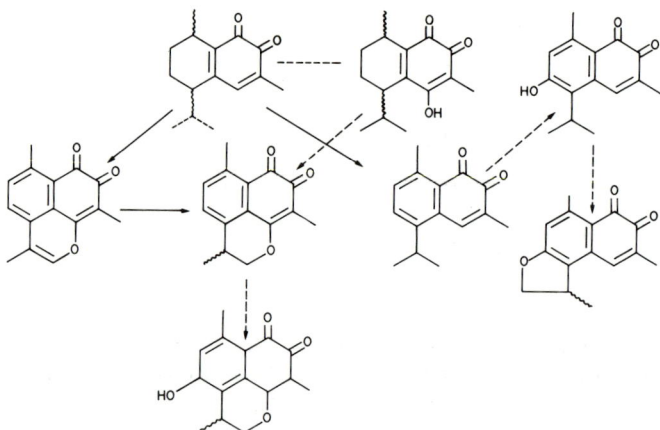

FIG. 11.17. Mansonones.

oxaphenalene (Fig. 11.17). Oxaphenalene derivatives, as biflorin isolated from the wood of *Capraria biflora* (Comin, Gonçalves de Lima, Grant, Jackmann, Keller-Schierlein & Prelog, 1963), may also represent defence substances for the plant like the other mansonones in *Mansonia* wood (Fig. 11.18). In the isolation of mansonones, it was observed that in the wood not only the quinones but also the respective hydroquinones are present. Probably in the living plant the presence of the quinone-hydroquinone system represents an effective defence against attack by insects and also fungi, as shown by the inhibitor effect exerted by the extracts of this wood on the growth of several microorganisms.

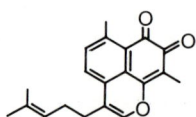

Biflorin

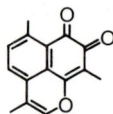

Mansonone F

FIG. 11.18. Biflorin. Mansonone F.

We must postulate that in living plants there is always an equilibrium between hydroquinone and quinone, which is shifted towards the oxidated form during extraction. Other quinones of this type, present in tropical woods, are anthranoids and principally anthraquinones, for example tectoquinone in teak wood which is one of the more resistant to insects including termites and to fungal attack. A rather complete and up-to-date account of the distribution and occurrence of quinones in plants has been developed by Thomson (1971). I wish here only to point out the existence in plants of other types of quinones which most probably account for the resistance of the plants.

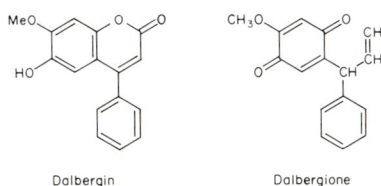

Dalbergin   Dalbergione

Fɪɢ. 11.19. Dalbergin. Dalbergione.

Dalbergiones, phenylpropanoid quinones, have been isolated and deter-mined for the first time in 1964 independently by Ollis & Gottlieb and by Gonçalves de Lima, F. Delle Monache & Marini Bettolo in plants of the Leguminosae, especially *Dalbergia* (Marini Bettolo, 1975). For their structure and structural relationships with phenyl-coumarins and isoflavones they have been named neo-flavones (Fig. 11.19). The resistance of the *Dalbergia* woods, plants distributed all over the tropical area of the world, may be attributed to these quinones and to their precursor phenyl-coumarins dalbergin and other similar compounds. A review on this subject has been made by Marini Bettolo & Delle Monache (1975b). Less common flavano-quinones are also present in several resistant woods. I may recall as an example the recent work on the root bark of *Abrus precatorius*: three new isoflavanquinones abruquinones were isolated and their structure determined (Fig. 11.20) (Lupi, Delle Monache, Marini Bettolo, Barros Costa & D'Albu-quer Que, 1979).

11.2.3.2 Phenoldienones. Other quinonoid substances found in plants consti-tute a defence for woods. We might mention the presence, in practically all species of the Celastraceae and Hippocrateaceae family, of a group of quinonoid substances which may be chemically considered triterpene phenol-dienones, such as pristimerin, celastrol and tingenone (Fig. 11.21). These products show high biological activity against bacteria and fungi and undoubtedly are an important factor in the resistance of the plants since they are distributed over the outer part of the plants. A full account of the present knowledge on this group of substances has recently been published (Marini Bettolo, 1979).

11.2.3.3 Flavanoids and tannins. Condensed flavanoids, both flavano-tannins and polycyclic flavanoids like haematoxylin, brasilin and peltogynols, constitute another system for the prevention of the attack of plants by insects and fungi. Peltogynol-mopanols distribution in various plants has been studied by Vaccaro Torracca, Galeffi, D'Albuquerque, Casinovi & Marini Bettolo (1976) and by Drewers & Roux (1966). The resistance to insect and fungal attack on these woods may be attributed to the existence of these substances which upon oxidation give rise to coloured surfaces (Fig. 11.22).

CH3O, R^1, OCH3, R^2O, OCH3, OCH3

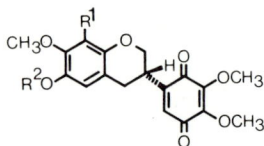

ABRUQUINONE   A   3 $R^1 = H$, $R^2 = CH_3$

ABRUQUINONE   B   4 $R^1 = OCH_3$, $R^2 = CH_3$

ABRUQUINONE   C   5 $R^1 = OCH_3$, $R^2 = H$

FIG. 11.20. Abruquinone A $R^1 = H$ $R^2 = CH_3$. Abruquinone B $R^1 = OCH_3$ $R^2 = CH_3$. Abruquinone C $R^1 = OCH_3$ $R^2 = H$.

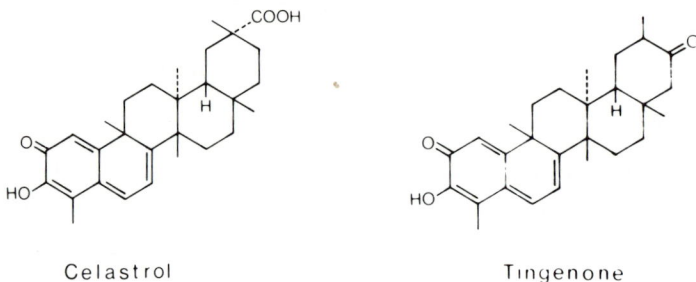

COOH   H   HO   Celastrol

H   HO   Tingenone

FIG. 11.21. Celastrol Tingenone.

11.2.3.4 *Stilbenes*. Other components of woods with strong biological activity are prenylated hydroxy stilbenes. The Iroko *Chlorophoraexcelsa* wood contains chlorophorin, and, substituted stilbenes. Other prenylated stilbenes (Fig. 11.23) were found in the roots and barks in *Lonchocarpus longistylus*, which prevent their decay from fungi and insect attack (Della Monache, Marletti Marini Bettolo, De Mello, Gonçalves & De Lima, 1977).

11.2.3.5 *Terpenes*. Many terpenes act as insect repellents and deterrents like α-pinene, present in many Coniferae. Citronella oil from the plant *Andropogon nardus*, containing mainly citron e llal, borneol and geraniol, is well known as an insect repellent in popular use. Plants of the *Geranium* sp. owe their insect-repelling activity to a mixture of terpenes.

Resistance factors for plants may be due to the presence of kaurenic acid and resin acid, according to a number of experiments on larval growth by Elliger, Zinkel, Chan & Waiss (1976). The finding was due to the observation

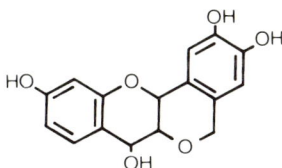

Fig. 11.22. Peltogynol (mopanol).

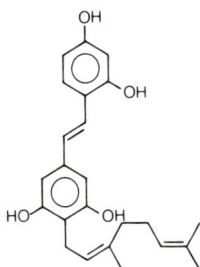

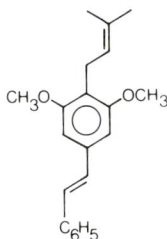

Chlorophorin   4-prenyl-3-5 dimethoxy stilbene

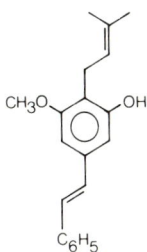

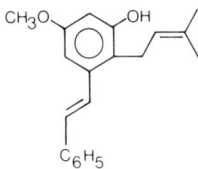

Longistylines

Fig. 11.23. Chlorophorin. 4-Prenyl-3,5-dimethoxy stilbene. Longistylines.

that the resistance of various sunflower, *Helianthus annuus*, to insect attack was proportional to their content of diterpenoid acids in the extracts. The activity could be attributed mainly to the trachylobanoic acids and kaurenoic acids present in the flowers (Fig. 22.24). The rather diffuse presence of kaurenoic acids in many plants may account for the natural resistance. Recently, kaurenoic acids were found in many plants as well as in the genus *Eupatrorium*,

Trachyloban-19-oic acid   (–)-16-kaurene-19-oic acid

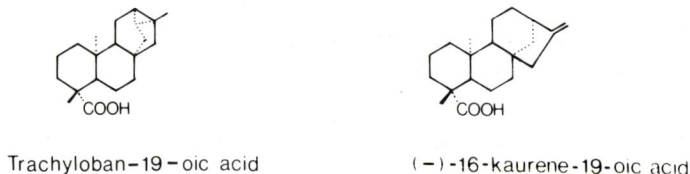

FIG. 11.24. Trachyloban-19-oic acid. (−)-16-kaurene-19-oic acid.

which is known to be free from insect infestation (Delle Monache, Mareno, Delle Monache & Marini Bettolo, 1980). Like pimaric, isopimaric, abietic and palustric acids, diterpene resins have also been considered as factors of resistance in woods mainly of the conifers (Van Buijtenen, 1972).

The occurrence of other diterpenes and oleoresins (Fig. 11.25) in tropical woods, e.g. *Copaifere multijuga*, active against various microorganisms, indicate that these products may also constitute factors of resistance for these plants (Delle Monache, D'Albuquerque, Delle Monache & Nano, 1971). Moreover, the biological activity of this type of resin acids is known through traditional and indigenous medicine for centuries.

(+) Hardwick acid   Copaiferic Acid
(methyl ester)

FIG. 11.25. (+) Hardwick acid (methyl ester). Copaiferic acid.

11.2.4   Defence substances formed in plants as a response to insect infestation

Natural resistance of plants to insects is generally due to the presence in the plant of certain secondary metabolites, as already reported. Another mechanism of resistance in the plant occurs as a result of the insect infestation. Some non-resistant plants suffer pathological transformation of their organs and tissues and die; others can produce a defence system when attacked. Although these mechanisms have not been widely studied it may be that a more detailed study of the infestation processes in plants and of resistant strains could reveal at a molecular level defence mechanisms elicited by insect attack. A closer collaboration in this field between the phytopathologist and biochemist could lead to important results. The more classical example of self-defence systems of plants towards insects is the production of nut-gall. It is known that many plants of the genus *Quercus* (Oak) are attacked by an insect of

TABLE 11.1. GALLOTANNINS (AND OTHER HYDROLYSABLE TANNINS)
PRODUCED BY DIFFERENT INSECTS

Insect species	Plant	Location
Cynips gallae tinctoriae	*Quercus infectoria*	Near and Middle East
Cynips insana West.	*Robur* genus	Near East
Cynips calicis Burgsdorff	*Quercus pedunculata* Ehr.	Hungary, Yugoslavia
	Quercus sessiflora	
Pemphigus cornicularius	*Pistacia terebinthus* L	North Africa
Pemphigus utricularius	*Pistacia terebinthus* L	Libya
Pemphigus sp.	*Pistacia vera*	Libya
Aphis genus	*Rhus semialata* Mur.	China, Japan
Eriophyles flava Trab.	*Tamarix articulata*	Magreb

the *Cynips* sp. In reacting to the insect bite, the plant forms a gall mainly constituted by an unusual product, a gallo-tannin, which for its properties prevents a further penetration by the insect (Table 11.1). The gallo-tannins structure has been revised and now we can assume the structure of differently substituted glucose esters of gallic acid (Fig. 11.26). Tannins (flavano-tannins) (Fig. 11.5) are normally present in practically all plants and they are formed especially in conditions of particular stress.

11.3   PLANT–FUNGI INTERACTION

The interaction between plant and fungi is even more complex in an ecological balance, due to the complicated chemical mechanism to which the development of plants and the growth of fungi is bound and to the environmental conditions which favour or prevent the plant–fungi interaction.

Plants may survive the diseases caused by fungi or bacteria and other microorganisms because of their natural resistance. Natural resistance to fungi is due to a number of chemical and biological factors, the latter being the presence in the plant of some substances with particular properties that inhibit fungal growth and development. Fungi, on the other hand, can develop on a plant, using the plant material as a substrate for their growth, thus causing infection. The fungi also produce, by their secondary metabolism, particular chemicals with marked and specific biological activity on the plant cells, the so-called phytotoxins. This phenomenon, which is really a chemical-mediated disease, was only studied in a limited number of cases, although rather common in nature. The complete understanding of this interaction may be of great help in the betterment of strategies for the protection of crops and forests. This warrants a thorough examination of its own and cannot be summarized here in a few words. Excellent reviews have been published in the last years (Graniti, 1972; Ballio, 1977).

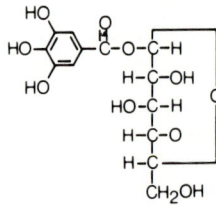

Glucogallin

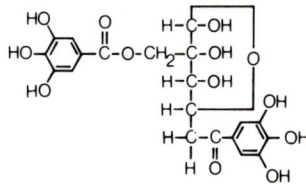

Hamamelitannin

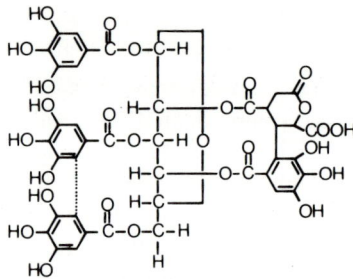

Chebulin acid

FIG. 11.26 Glucogallin Hamamelitannin. Chebulin acid.

The only point that should be mentioned is the influence of phytotoxins in some particular case of plant–fungi interaction. The other points that we wish to mention are: (1) natural resistance of plants to fungi, and (2) induced resistance of plants after fungal attack. In the first case the resistance of the plant is generally attributed to genetic factors, resistance being due to the presence in the plant of substances which prevent the growth of fungi. Non-resistant strains are generally obtained through a long natural or man-induced genetic selection, in order to obtain better fruit, seed, or in the case of forest trees, wood. These strains have lost partially or totally their defence factors. A rapid examination of the natural defence factors constitute the first part of this survey.

The second aspect has been developed recently through very interesting research, and is related to the capacity of the plant, elicited by external factors to produce through various mechanisms, new substances which are toxic to the invading agent or which can in some way control their development. These substances have been named "stress substances" or in a more limited meaning, "phytoalexins". Their presence in plants attacked by fungi is extremely interesting both from a theoretical and a practical point of view. I suggest the possibility of a special kind of defence or immunity process exerted by plants and of the possible use as fungistatics of natural products possessing new structures.

11.3.1 Natural resistance

Antibiosis is a general concept in nature far broader than the definition given at present, and thus antibiotic processes may occur between higher plants and fungi and higher plants and microorganisms. From this point of view we can examine a great number of natural products isolated from plants and consider their role in the protection of plants. The work is not always easy because when a chemist finds an interesting substance in a plant he is more interested in determining its structure than in finding a way to determine its biological activity. Moreover, the number of biological assays to which a new natural substance should be submitted varies with the feasibility of the assay and the interest of the research, and thus constitutes a limiting factor to a better understanding of natural processes. One substance can prove to be inactive as a CNS drug but active on tumour cells; another is devoid of all activity on animals but may promote growth in plants; some may show antibacterial and fungistatic properties, but have not been tested. A systematic study of new natural products thus far constitutes a great problem involving organization and great expenses. The indiscriminate biological screening of all new products (and even of extracts) in order to detect new substances in plants can be limited, according to general criteria involving chemotaxonomical considerations and traditional experiences, especially from local habits and folk medicine of indigenous culture (Marini Bettolo, 1978).

As a general policy, however, research in the natural products should at present be oriented toward ascertaining through adequate tests the role of these substances in the plant and in its environments. Gonçalves de Lima, who has developed this line for many years, in very difficult conditions in north-eastern Brazil, has investigated practically all the substances isolated

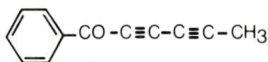

Capillin

FIG. 11.27a. Capillin.

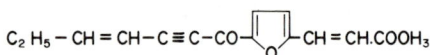

Wyerone

FIG. 11.27b. Wyerone.

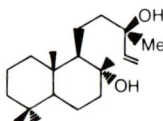

Sclareol

FIG. 11.27c. Sclareol.

from plants by biological tests and has given an account of resistance of Latino-American plants (Gonçalves de Lima, 1963). Literature indicates a number of substances present in plants which may account for their resistance to the attack of fungi and even to bacteria. Tropical woods, as earlier reported, do contain substances of quinonic or phenolic nature which prevent not only the attack of insects but also that of fungi. To this purpose we shall mention the active principle of *Capraria biflora*, biflorin which possess antibacterial and anti-fungal properties. The presence of isothiocyanate derivatives in many plants (*Tropaeolum* spp., Capparidaceae, etc.) may act also in the protection of plants against fungi.

If quinones protect wood and timber of many trees, also essential oils, rich in terpenes and phenols, prevent the attack of fungi in many species, e.g. *Eupatorium ayapana* (Compositae) and many Cyperaceae. Particular products like capillin from *Artemisia capillaris* an aromatic dialkyn-derivative shows, according to traditional medicine, strong fungicidal properties, which were confirmed in the laboratory (Fig. 11.27a). Other products with fungistatic and fungicidal activity have been reported in plants. A review was written by Wain on this subject (Wain, 1977). According to Wain, the mechanism of defence and production of phytoalexins is only a quantitative modification of a normal process of the plant where these substances are not detectable in normal conditions because of their low concentration. On the other hand in *Vicia faba* is present a furano derivative, wyerone (Fig. 11.27b), a preformed natural fungicide. Sclareol, obtained from the exudates of the healthy leaves of *Nicotiana glutinosa*, shows high protective activity against rust (Fig. 11.27c). Also fungi in the forests with their metabolites (scytalidin, hyalodendrin, etc.) may act as antagonist of other wood-rotting fungi (Strunz, Kakushima, Stillwell & Heissner, 1973). According to Wain, many of these substances are water soluble and may be found in the exudates of roots and seedlings, which explains the high resistance of these tissues to fungal and microbial attack (Wain, 1977).

Tannins have a protective action and their concentration in plant increases under stress conditions, e.g. in mangrove, when flooded and in parts exposed to microbiological attack like roots. Other substances are present in the cuticles of the leaves. Like other fungistatic substances or fungicides, natural products may also induce resistance to pathogenic fungi through the formation of enzymes which can cleave or destroy the natural substances of the fungi, as shown by De Witt-Elshove: the presence of penicillinase in antibiotic-resistant strains is an example.

11.3.2   Stress substances and phytoalexins

Phytoalexins and in general stress substances have been the object of intensive studies and research by scientists in Great Britain, Japan, Canada and Australia and a complete and up-to-date review has been published in recent years (Cruickshank, 1976).

Stress substances belong to different chemical groups. Although the main work has been done with plants of the families of Leguminosae and Solanaceae, the phenomenon is quite general although the products formed for defence may be quite different from those obtained in the cases reported above. In effect every plant genus or even family reacts to the fungal infection, mobilizing products of its metabolism and transforming them into new molecules with fungistatic properties. There is generally a connection between the structure of the stress-substances produced and the host plant depending on the particular secondary metabolism of the latter. For example, Leguminosae stress-substances, i.e. phytoalexins, are flavonoids, Solanaceae give rise to steroidal alkaloids and sesquiterpenes, Convolvulaceae to furano-derivatives like ipomaeranone. Other plants may produce stress-substances with quite different structures. The most common or the most studied, are the flavonoid phytoalexins. They may be considered derived from a C_6–C_3–C_6 system like in flavones-isoflavones, with further intramolecular condensation to give rise to pterocarpans (Dewick & Martin, 1976).

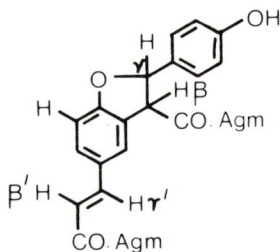

Hordatine A

FIG. 11.28

The second best studied products are sesquiterpenes mainly from Solanaceae, whereas others found in different plant families possess a terpenoid structure, mainly sesquiterpenoids. Particular substances have been found in some plant families, like hordatin in barley infected by fungi. It is a phenylpropanoid derivative (Fig. 11.28) (Stoessl, Stothers & Ward, 1976). I wish also to recall the formation of Mansonones in Ulmaceae elicited by the infection of *Ceratostomella ulmi* (Overcem & Elgermsa, 1970; Chen, Lin & Chen, 1972). In this case one can imagine that the plant synthethizes this rather active biological product from cadinene, a sesquiterpene present in the metabolism of many plants (Fig. 11.29).

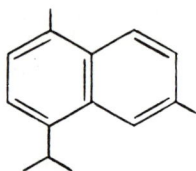

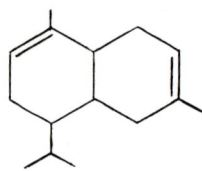

Cadalene   Cadinene

FIG. 11.29.

The study of stress substances, although very intensely pursued and developed in the last 10 years, is so far limited to a very small number of cultivated plants. Further research is now needed on the phytopathology and chemistry of the new natural fungistatic products. Even the biochemical aspect of the biosynthesis of these compounds requires further study. The mechanism of accumulation of the substance must also be examined because in order to be active a phytoalexin should be rapidly mobilized and concentrated in the points of attack. The important fact is that the substances which have proved to have fungistatic activity may control, by a self-defence mechanism, the disease caused by pathogenic fungi. This also suggests new lines of research for the control of diseases. The possible future implications of this mechanism requires a development of this study directed mainly towards wild species, where defence mechanisms are more active, because as already indicated, the studies have considered mainly plants of economic interest (potato, tobacco, tomato, legumes, etc.) and their pests.

There are great problems with industrial cultivations of coffee subject to the attack of rust *Hemileia vastatrix*. Resistant varieties have been studied in order to find differences in chemical composition and thus the protecting chemical factor but so far results are far from definitive, although higher concentrations of flavones in resistant species were observed. It would be extremely important to find an artificial mechanism through which resistant strains can be obtained. This is possibly inducing resistance by genetic procedures or by eliciting the response with the use of non-pathogenic fungi or extracts as in the case of yeasts.

11.4   SIGNIFICANCE OF THE PRESENCE IN PLANTS AND INSECTS OF NATURAL PRODUCTS OF SIMILAR STRUCTURE

The rapid review of natural products of plants, which most probably are involved in their protection from insect infestation and from fungal diseases, indicates that the structure of a number of these products is very close to that of some typical products of insects or constitute the metabolites of pathogenic fungi: an example is the occurrence in plants of insect hormones, the ecdysones already discussed. Phyto-ecdysones are identical or quite similar to the ecdysones which constitute the moulting hormone of insects (Ferrari, 1973). Several plants have been shown to contain high quantities of ecdysones like muristerone in *Ipomea* (Canonica, Danieli, Weiss-Vincze & Ferrari, 1972). The same happens in the case of juvenile hormone-like substances found in plants which exert the same activity as insect hormone.

The significance of these two groups of products both in plant and in insects is not clear whether the insect assumes these substances from the plants, or if their occurrence is due only to the fact that they both belong to the biogenetic pathway of terpenoids and steroids, common to animal and plant metabolism. Certainly there are differences in the biogenesis of ecdysones and JH in anthropods and plants, anyhow the presence of these substances in plants certainly could have an influence on insect behaviour and thus on the defence of the plants.

The presence of monoterpenes in the body of some tropical insects like in the giant Brazilian ants of *Atta* genus (Sauvas) which act as repellents towards other species, is due to the fact that the ant is in condition to obtain them from the surrounding plants (Gilbert, personal communication). This indicates that plants and insects can cooperate in their defence systems. Simple quinones like benzoquinone derivative: embelin, primin (Fig. 11.30) are

Primin   Embelin

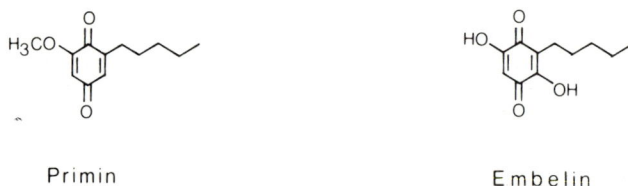

Fig. 11.30. Primin. Embelin(primula and miconia).

present in plants of different families and may exert insecticidal activity in the protection of plant. Some beetles, as shown by Schildknecht (1977), do synthesize substituted benzoquinone as a chemical weapon against other insects.

Other typical steroid derivatives, cardenolides, known for their action on the heart, have been found not only in many plants, mainly *Digitalis* sp. (Scrophulariaceae) and *Strophantus* spp. (Apocyanaceae and Asclepiadaceae), but also in a number of insects as their defensive system against mammals. Some of these insects may transfer the toxic glycoside into their own defensive system, as shown for many acridids (Reichstein, 1967). In other cases, it was found that *Chrysomelid* beetles elaborate cardenolides from plant steroid percursors. In effect these beetles secrete cardenolides which are not present in their host plants (Pasteels & Dalose, 1976). The occurrence of these substances both in plants and insects and their known defensive properties against insects indicate the extreme flexibility of natural equilibria to utilize highly potent substances for defence and survival both of plants and insects. Other examples of similar products in plants and insects can be quoted. Recently at the ICIPE in Nairobi, a new diterpene was isolated from a termite species and its structure determined, that is the Biflora-4,10 (19), 15-triene (Wiemer, Meinwald, Prestwich, Solheim & Clardy, 1979). This compound is related to biflorin, present in *Capraria biflora*, which contains a sesquiterpene system plus an isoprene unit. In the insect sesquiterpene the cyclization of the methyl group with an OH group on 5C is not present but structurally the products are indeed very close. The similarity is even more evident with the structure of the mansonones, isolated by us in *Mansonia altissima* but rather later found in other plant families. The Mansonones have an isoprene group less than biflorin, but the structure of the fundamental system is quite similar to that of the biflora-triene of termites. It can thus be inferred that all these products may derive from the sesquiterpene cadalene by further prenylation followed by cyclization (Fig. 11.31). The occurrence of active substances of closely related structure both, in plants and in insects, indicates that nature, though with slight differences, follows the same biogenetic pathway to elaborate substances which regulate the equilibria among plant and insects and among insects. These similarities apply not only to defence substances but also to repellents and attractants which are both in plants and in insects.

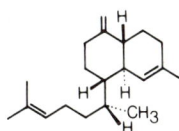

Biflora - 4 , 10(19) , 15 - triene

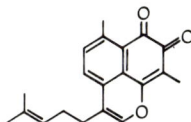

Biflorin

Fig. 11.31. Biflora-4,10(19),15-triene Mansonone A, Biflorin Mansonone E.

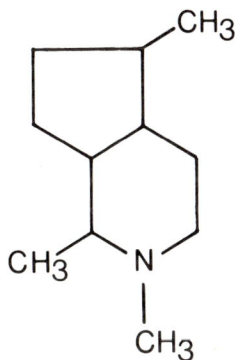

Skythantine

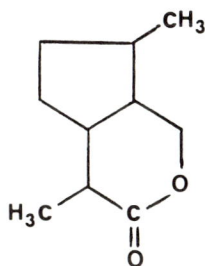

Iridomyrmecin

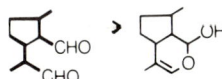

Iridodial

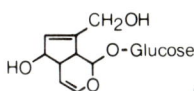

Aucubin

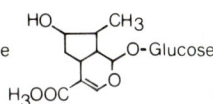

Loganin

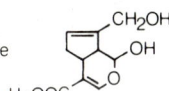

Genipin

Fig. 11.32. Iridomyrmecin Skythantine Iridodial Aucubin Loganin Genipin.

Iridoids are present in the leaves of many plants such as glycosides (aucubin, loganin) and in the form of cyclopentano-derivatives (nepetalactones) or nitrogenated iridoids (actinidine, skythantines). An iridoid (Fig. 11.32) can be found as a typical defence substance of insects; the iridomyrmecyn of the Argentine ant *Iridomyrmex humilis* (Pavan & Ronchetti, 1955; Fusco, Trave & Vercellone, 1955). A very simple product, 2-methyl-buten-2-ol, was found to form the sexual pheromones attractant of a wood coleoptera, *Ips typographus* responsible for great damage to spruce trees in Scandinavia. The most interesting feature is that the hormone, which is responsible for the attraction of the male insects and the stimulation of the females, together with two other terpene substances, is elaborated by the

2 – methyl – 3 – buten – 2 – ol

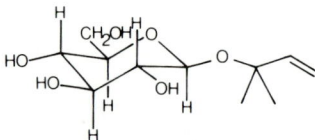

2 – methyl – but – 3 – en – 2 – yl – 1 – O – β – D – glucopyranoside

FIG. 11.33. 2-Methyl-3-buten-2-ol. 2-Methyl-but-3-en-2-yl-1-0 β-D-glucopyranoside.

insect from the terpenes myrcene and α-pinene present in the spruce tree. The use of this sexual attractant in traps in the woods has made possible the reduction of the infestation of these coleoptera in Sweden and Norway. The sexual hormone in constituted by the 2-methyl-3-buten-2-ol, the principal biological active component, and ipsdienol and *cis*-verbenol. Recently the glycoside of 2-methyl-buten-2-ol was found in an *Umbellifera* (Pinar & Martin-Lomas, 1978) and in two species of *Bergenia* (Saxifragaceae) (Esposito & Nicoletti, 1980) (Fig. 11.33).

It is not possible for the moment to advance any hypothesis on this and other facts that correlate insect chemistry to plant secondary metabolism. These few examples indicate that along the main lines of the common biogenetic pathway of higher plants, fungi and insects, secondary metabolites have a common origin, and are often structurally closely related and exert an important role in the natural equilibria.

11.5 CONCLUSION

The impressive analogies between biologically active principles in plants, insects and fungi may suggest new correlations between the host plants and insects and fungi. These researches have led to results which have already been utilized in the control of infestations and of plant diseases. It is now important to draw from these observations some guidelines for the future strategies for the protection of plants and their crops. Although very interesting and stimulating, the present results are generally preliminary and thus require further research and confirmation. One of the most important aspects of the research is the study of the self-defence mechanism of plants against arthropods and fungi. In this field it is now necessary to develop and extend further studies and to establish new correlations especially on the specificity of the interaction between host plant and insect or fungi.

Natural immunity of plants should be studied on a collaborative basis by geneticists and chemists with the aim of obtaining resistant strains without losing the quality standard of the products. A knowledge of the structures and the mechanisms of stress substances should be developed in order to obtain insecticides and fungicides, eventually systemic. In this case, the limits of their use should be governed by the specificity of the host plant and insect or fungi inter-relationship. Information about the plant immunity processes should be obtained through biochemical studies which may clarify the mechanisms of this action with the aim of studying the possibility of inducing artificial resistance in plants.

Processes, so far not completely understood, depending partly from biological, partly from chemical factors, which govern the natural equilibria in the forests, should be investigated in order to give more information about their chemical mechanisms. These will be useful in developing new systems for insect and fungi control. The importance of cultivated plants as food should not draw our attention away from the great problems of the protection of the forests, mainly of northern hemisphere (indispensable in the world ecological balance) which are often subject to severe attacks from insects and fungi because of local disruption by man of natural equilibria.

The results so far obtained along these new lines probably may be of great importance in the future for the control of pest and plant diseases, but to this end we must multiply our efforts in research, in the laboratory and in the field.

11.6 SUMMARY

Natural equilibria among plants and insects and plants and fungi indicate the existence of a particular resistance of the wild plants in respects to that of cultivated ones. The resistance of plants to arthropods can be attributed to the presence in the plant of particular products which act through various mechanisms. These products can be grouped according to the principal activity exerted: toxic substances (e.g. alkaloids, pyrethrins, rotenoids); substances acting on insect behaviour: such as antifeedants (e.g. less common amino acids, etc.); hormone-like substances (e.g. phytoecdysones, JH analogues, precocenes, etc.); attractants and phagostimulants (e.g. isoprenoids, flavones) and repellents (e.g. terpenes, quinones, phenolic derivatives); and defence substances formed after insect infestation (e.g. gallotannins).

The plant–fungi interaction is more complicated because, beyond the products present in plants which may prevent the attack of fungi (their presence constitutes the genetic resistance), some plants also develop a chemical defence mechanism as a consequence of the infectious agent producing particular products, indicated as stress substance or phytoalexins, which possess antifungal activity.

Therefore we may group these substances as: Natural antifungal products present in plants; stress substances or phytoalexins. A particular aspect of these interactions is represented by the presence both in plant and insects of chemically closely related products.

A number of examples of natural products exerting the above stated mechanisms are reported and their significance discussed both towards a better intelligence of the plant–insect and plant–fungi interaction and in regard to their possible utilization in the protection of plants.

11.7 REFERENCES

Ballio A. (1977) Phytotoxins: Chemical structure and biological activities. In *Natural Products and the Protection of Plants* (ed. Marini Bettolo), pp. 645–662. Elsevier, Amsterdam.

Begeley M. J., Crombie L., Ham P. J. & Whitney D. A. (1976) Structures of three oxygenated methyl steroids from insect-repellant plant *Nicandra physaloides*. *JCS Perkin* I, 296–304.

Bell E. A. (1977) The possible significance of uncommon amino acids in plant–vertebrate, plant–insect and plant–plant relationship. In *Natural Products and the Protection of Plants* (ed. Marini Bettolo), pp. 571–595. Elsevier, Amsterdam.

Bell E. S. (1980) *Encyclopedia of Plant Physiology*, B. 11. Sjunger Verlag-Berlin, Heidelberg.

Bowers W. S., Ohta T., Cleere J. S. & Marsella P. A. (1976) Discovery of antijuvenile hormones in plants. *Science* **193**, 542.

Bowers W. S. (1977) Antijuvenile hormones from plants; chemistry and biological activity. In *Natural Products and the Protection of Plants* (ed. Marini Bettolo G. B.), pp. 129–156. Elsevier, Amsterdam.

Canonica L., Danieli B., Weiss-Vincze & Ferrari G. (1972) Structure of muristerone and a new phytoecdysone. *JCS Chem. Comm.* 1060–1061.

Casida J. E. (1973) *Pyrethrum, the Natural Insecticide*. Academic Press, New York.

Chen F. C., Lin Y. M. & Chen A. H. (1972) Sesqiterpenes from the heartwood of Chinese elm. *Phytochem.* **11**, 1190–1191.

Comin J., Gonçalves De Lima O., Grant H. N., Jackmann L. M., Keller-Schierlein W. & Prelog V. (1963) Uber die Konstitution des Biflorin, eines o-Chinons der Diterpen-Reihe. *Helv. Chem. Acta* **46**, 409–415.

Cruickshank I. A. M. (1976) A review of the role of phytoalexins in disease resistance mechanisms. In *Natural Products and the Protection of Plants* (ed. Marini Bettolo G. B.), pp. 509–561. Elsevier, Amsterdam.

Delle Monache F., Corio E., D'Albuquerque I. L. & Marini Bettolo G. B. (1969) Diterpenes from *Copaifera multijuga*. *Ann. Chim. (Rome)* **59**, 539–551.

Delle Monache F., Cuca Suarez L. E. & Marini Bettolo G. B. (1978) Flavonoids from the seeds of six *Lonchocarpus* species. *Phytochem.* **17**, 1812–1813.

Delle Monache F., Marletti F., Marini Bettolo G. B., De Mello J. F., & Gonçalves De Lima O. (1977) Isolation and structures of longistylines from *Lonchocarpus violaceus*. *Lloydia* **40**, 201–208.

Delle Monache G., D'Albuquerque L. T., Delle Monache F. & Nano G. M. (1971)

X-Multijugenol a new sesquiterpenic alcohol with a caryophyllane carbon skeleton. *Tetrahedron Letters* 659–660.

Dello Monache G., Moreno B., Delle Monache F. & Marini Bettolo G. B. (1980) Flavones and kauronoid diterpenes from *Eupatorium tinifolium* HBK. *Farmaco* **34**.

Dewick P. M. & Martin M. (1976) Pterocarpan biosynthesis: 2 hydroxy-isoflavone and isoflavones precursors of demethylhomopterocarpan in red clover. *JCS Chem. Comm.* 637–638.

Doskotek R. W., El Feraly F. S., Fairchild E. H. & Huang C. T. (1976) Peroxyferolide: a cytotoxic Germacrynolide hydroperoxide from *Liriodendron tulipifera*. *JCS Chem. Comm.* 402–403.

Drewes S. E. & Roux D. G. (1966) Stereochemistry and biogenesis of mopanols and peltogynols from *Colophospermum mopane*. *J. Chem. Soc.* (c) 1644–1653.

Elliger C. A., Zinkel D., Chan B. G. & Weiss A. C. (1976) Diterpene acids as larval growth inhibitors. *Experientia* **32**, 1365–1366.

Elliot M. (1977) Synthetic insecticides designed from the natural pyrethrins. In *Natural Products and the Protection of Plants* (ed. Marini Bettolo G. B.), pp. 157–176. Elsevier, Amsterdam.

Esposito P. & Nicoletti M. (1980) 2-Methyl-but-3-en-2 yl-O-β-D glucopyranoside from *Bergenia crassifolia* Atti. *Ac. Naz. Lincei*, Vol. LXVII, fasc. 3, 205–206.

Ferrari G. (1973) I fitoecdisoni: scoperta, sorgenti, nuove acquisizioni. *Inf. fitopatologico* **23**, 6–24.

Fortin Y. & Poliquin J. (1976) *Natural Durability and Preservation of One Hundred Tropical African Woods*. Int. Development Research Centre, Ottawa.

Furtado H. (1979) *Proc. Int. Congress on Chagas Disease*, Rio de Janeiro.

Fusco R., Trave R. & Vercellone A. (1965) La structural dell iridominmecina. *Chim. e Ind. (Milano)* **7**, 958–959. R.K.S.

Graniti A. 1972. *Phytotoxins in Plant Diseases* (ed. Woods R. K. S., Ballio A. & Graniti A.). Academic Press, New York, London.

Gonçalves de Lima O. (1963) Antibiotic richness of Latin American plants particularly from Brazil and Mexico. *Qualitas plantarum* **9**, 257.

Gonçalves De Lima O., D'Albuquerque I. L. & Loureiro P. (1953) *An. Soc., Biol. Pernambuco* **11** (1), 3–9.

Hikino H. & Hikino Y. (1975) Phytoedysones. *Fortsch in Naturstoffe* **28**, 294–302.

Jacobson M. & Crosby D. G. (1971) *Naturally occurring Insecticides*. M. Dekker, New York.

Kubo I., Miura I. & Nakanishi K. (1976a) The structure of Xylomollin, a secoiridoid hemiacetal acetal. *J. Am. Chem. Soc.* **98**, 6704–6705.

Kubo I., Tanis S. P., Lee Y. W., Miura I., Nakanishi K. & Chapya A. (1976b) Harrissonin. *Heterocycles* **5**, 485.

Levinson H. Z. (1972) Zur evolution und Biosynthese der terpenoiden Pheromone un Hormone. *Naturwiss.* **59**, 477–484.

Levinson H. Z. (1976) The defensive role of alkaloids in insects and plants. *Experentia* **32**, 408.

Lupi A., Delle Monache F., Marini Bettolo G. B., Barros Costa D. L. & D'Albuquerque I. L. (1979) Abruquinones: new natural isoflavanquinones. *Gazz. Chim. Ital.* **109**, 9–12.

Marini Bettolo G. B. (1965) Skytanthines a new group of natural alkaloids. *Ann. Ist. Sup. Sanita* **4**, 489–500.

Marini Bettolo G. B. (1975) *Flavanic Polyphenols: International Meeting on Polyphenols*. Groupe del Polypenols, pp. 2–25.

Marini Bettolo G. B. & Delle Monache F. (1975) Flavonoids of uncommon structures. *Proc. 5th Hungarian Bioflavonoids. Symposium Akademiai Kiado Budapest*, pp. 1–13.

Marini Bettolo G. B. (1978) *Medicina Popular e Quimico-axonomia*. Univ. Fed. de Alagoas, Maceio.

Marini Bettolo G. B. (1979) Un grupo particular de substancias naturales: las phenoldienonas triterpenicas. *Rev. Latino Am. Quim.* **10**, 97–104.

Marini Bettolo G. B., Casinovi C. G. & Galeffi C. (1965) The quinones of Mansonia wood. *Lloydia* **28**, 258–265.

Marini Bettolo G. B., Casinovi C. G. & Galeffi C. (1965) A new class of quinones sesquiterpeneoid of *Mansonia altissima*. *Tetrahedron Letters* **52**, 4857–4858.

Matsui K. & Muwakata K. (1976) The structure of piperinene. *Agric. Biol. Chem.* **40** (6) 1113.

Moreno B., Delle Monache G., Delle Monache F. & Marini Bettolo G. B. (1980) Flavones and kauranoid diterpenes from *Empatorium tinifolium*. *HBK le Fanmaco*. 458–464.

Nakanishi K. (1977) Insect growth regulators from plants. *Natural Products and the Protection of Plants*. (ed. Marini Bettolo G. B.), pp. 185–195. Elsevier, Amsterdam.

Oda J., Ando N., Nakajma Y. & Inouke Y. (1977) Terpenes from *Juniperus recrua*. *Agric. Biol. Chem.* **41**, 201–.

Overcem J. C. & Elgermsa D. M. (1970) Accumulation of Mansonones E and F in *Ulmus hollandica*. *Phytochem.* **9**, 1949–.

Pavan M. & Quilico A. (1979) Prospettive di controllo degli insetti nocivi con sostanze naturali agenti sul loro comportamento. *Atti Acc. Naz. Lincei* **128**, 37–77.

Pavan M. & Ronchetti G. (1955) Ricerche chimiche biologiche sulle Iridomirmicine. *Atti. Soc. Ital. Sci. Nat.* **94**, 379.

Pasteels & Daloze D. (1976) Cardiac glycosides in the defensive secretion of Chrysomelid beetles; evidence of their production by insects. *Science* **197**, 70–72.

Reichstein T., Euw J. V., Fishelson L., Parsons J. A. & Rothschild M. (1967) Cardenolids (heart poison) in a grasshopper feeding on milkweeds. *Nature* **214**, 35–39.

Rosenthal G. A., Dahlman D. L. & Jansen D. H. (1978) L-canaline detoxification: a seed precursor's biochemical mechanisms. *Science* **202**, 528–529.

Schildknecht H. (1977) Protective substances of arthropods and plants. *Ibid.*, pp. 85–97.

Stoessl A. (1976) The antifungal factors in barley. *Can. J. Chem.* **45**.

Stoessl A., Stothers J. B. & Ward E. W. B (1976) Sesquiterpenoids stress compounds of the Solanaceae. *Phytochem.* **15**, 855–872.

Strunz G., Kakushima M., Stillwell M. & Heissner C. Y. (1973) Hyalodendrin: a new fungitoxic Epidithiodioxopiperazine produced by a *Hyalodendron* sp. *JCS Perkin* I, 2600–2602.

Thomson R. A. (1971) *Naturally Occurring quinones*, 2nd ed. Academic Press, London.

Thorsteinson A. J. (1960) Host selection in phytophagous insects. *Ann. Rev. Ent.* **5**, 193–218.

Torres R., Delle Monache F., Marini Bettolo G. B. & Cassels B. (1979) Coumarins and cinnamic acid from *Gymnophyton isatidicarpum*. *J. Nat. Products* **42**, 532–533.

Trivedi O. Komura H., Kubo I., Nakanishi K. & Goshi (1979) The ent-neo Clerodane absolute configuration of Ajugarins. *JCS Chem.* 885–886.

Vaccaro Torraca A. M., Galeffi C. L., D'Albuquerque I., Casinovi C. G. & Marini Bettolo G. B. (1976) Sulla constituzione di una nuova leucocianidina, lo pseudopeltoginolo. *Ann. Ist. Sup. San.* **2**, 328–336.

Virtanen A. I. & Hietala P. K. (1959) The structure of the precursors of benzorazolinone in rye seedlings. *Suonen Kiminstilehti* **32**B, 138.

Wain R. L. (1977) The chemical basis of plant resistance. In *Natural Products and the Protection of Plants* (ed. Marini Bettolo G. B.), pp. 483–499. Elsevier, Amsterdam.

Wiemer D. F., Meinwald J., Prestwich G. E., Solheim B. A. & Clardy J. (1979) Bifloratriene, a new diterpene from a termite soldier (Isopteratanitidae: Termitinae). *J. Org. Chem.* **45**, 191–192.

Wiesner K., Valenta Z. & Findlay J. (1967) The structure of Ryanodine. *Tetrahedron Letters* 221–225.

Williams C. M. (1969) Juvenile hormone insecticides. Nuove prospettive nella lotta contro gli insetti nocivi. *Atti Acc. Naz. Lincei* **128**, 79.

William W. G., Kennedy G. C., Yamamoto R. T., Thacker J. D. & Bordner J. (1980) 2-tridecanone a naturally occurring insecticide from wild tomato *Lycopersicon hirsutum glabratum*. *Science* 207–888.

11.8   DISCUSSION

Jacobson: I would like to offer a couple of points concerning some of the compounds that you referred to in your paper. We tested 2-tridecanol as well as 3,5,6-tridecanol on a number of lepidoptera species. We found that these compounds show antifeedant activity against first instar larvae, but after this instar, we observed no effect whatsoever. Regarding some of the quinones to which you referred, we found that 2-methylnaphthoquinones and also dihydroxynaphthoquinones were exceptional repellents for certain pest coleoptera. But 2-methyl, 5-hydroxy naphthoquinones were completely ineffective.

Marini Bettolo: I think that many things must be considered about quinones. First of all, their solubility both in the lipotropic and lyophilic solvents. Secondly, their redox potential data and thirdly, the presence of hydroquinone–quinone equilibrium system. These considerations are very important when one intends to undertake the isolation of quinones from their natural sources. The same factors are equally important in formulation and application of these compounds to insects.

Bowers: I would like to add that quinones have a wide variety of biological activity. Some display fungicidal activity. I also know of some cases where the quinones interfere with insect cuticles in a manner which leads to strange moulting behaviour.

Pryce: I also would like to comment on quinones especially on their fungicidal activity. The universal view is that quinones are more effective against fungi and that they inhibit the important enzymes that fungi require to perform their destructive roles. This view is supported by many experiments.

Bowers: I agree.

Marini Bettolo: I too think that they are very important. It is clear that more work should be done to uncover these compounds in the diverse plant species of the African continent.

Djerassi: When you say they must be considered, are you saying that they must be considered in their natural function or in the context of developing synthetic material?

Marini Bettolo: I think further basic knowledge about these compounds is desirable. With this knowledge, the quinones may possibly then act as models for developing synthetic material with superior biological activity. I have summarized our present knowledge about quinones so that those with indepth knowledge in insect physiology and biological activity may see the lines on which the quinones may be developed and structurally modified to enhance their commercial significance.

Saxena: Commenting on what Professor Djerassi said regarding choice of compounds for practical use, I would say that although some compounds may be highly effective against one insect species or another the chemical companies may not be interested in considering such products for development because of the special activity of the compound and because the companies think that such activity is not likely to improve their economies.

Marini Bettolo: Certainly, I think that before a compound is claimed to have potential for commercial exploitation, it should be subjected to the widest variety of practical tests so that full assessment of its economic potential can be done.

Bowers: I think that to look for naturally occurring biologically active compounds from plants, so that they may only act as a guide in developing models with greater value than the natural compound, is an incomplete approach as far as integrated pest management is concerned. I think that knowledge gained from extracting plants for active ingredients can be usefully employed in developing plants with much more resistance to insects than is hitherto the case. We should think on these lines as well. Surely, a two-way approach may lead to the desired solution.

Djerassi: I would do the basic toxicology first before I consider doing what you are saying. However, I agree that what you are saying is correct in theory but in practice it would take a long time (10–20 years).

Saxena: I agree with what Professor Marini Bettolo said that extensive bioassays for new products must be undertaken. Our own studies have shown that studies of compounds with established activity may uncover "new" activities with which the compounds were not associated before.

Dabrowski: The role of alkaloids, glycosides in the crop resistance to insects is already known and I would not emphasize it further. For example, some sorghum cultivars resistant to infestation contain an amount of glycosides which decrease the digestability of protein by animals. This means that 40% of the proteins of the plant are not used. This basic information should also be taken into account when thinking of the two-way approach advocated by Professor Bowers.

CHAPTER 12

Multiplicity of Insect Antifeedants in Plants

TIBOR JERMY

Research Institute for Plant Protection, H-1525 Budapest, Pf. 102, Hungary

CONTENTS

12.1	Introduction	223
12.2	Chemistry of Deterrence in Non-host Plants	225
12.3	The Search for Antifeedants of Practical Use	226
	12.3.1 The botanical occurrence of potent antifeedants	227
	12.3.2 The choice of insects for testing antifeedants	228
	12.3.3 Pure compounds or complexes of plant constituents?	229
12.4	Notes on the Non-preference Type of Resistance in Plants	230
12.5	Conclusion	230
12.6	Summary	231
12.7	Acknowledgement	231
12.8	References	232
12.9	Discussion	234

12.1 INTRODUCTION

Verschaffelt (1910) published the results of experiments with several oligophagous insect species showing that some of the non-host plants were eaten by the insects when the leaves were smeared with the saps of the host plant or with specific substances of the latter. Since then it has generally been considered that specific secondary plant substances as "token stimuli" would play a decisive role in host recognition, especially in oligophagous insects (Dethier, 1953; Fraenkel, 1953; Thorsteinson, 1953, 1958, 1960). Most authors dealing with the co-evolution of plant and phytophagous insects also claimed that secondary plant substances, which evolved as factors of resistance against insects, became "token stimuli" for some insects which succeeded in adapting to the resistant plants (Dethier, 1954; Fraenkel, 1959; Thorsteinson, 1960; Ehrlich & Raven, 1964; Whittaker & Feeny, 1971). In some oligophagous insects, special chemoreceptors were shown to respond to secondary substances of the host plant (Rees, 1969; Schoonhoven, 1967;

Wieczorek, 1976) which also seemed to support the theory of specific "token stimuli" in host selection.

However, in several oligophagous insects no specific phagostimulants could be detected. For example, Ritter (1967) and Hsiao (1976) found no secondary substances in potato leaves eliciting feeding responses in *Leptinotarsa decemlineata* while a mixture of primary plant substances strongly stimulated feeding. Similarly, Bernays & Chapman (1977) could not find specific substances in graminaceous plants as phagostimulants for the graminivorous acridid, *Locusta migratoria*. These and several other authors (see Beck, 1965; Schoonhoven, 1968; Städler, 1976) have shown that primary plant substances and their combinations may elicit normal feeding responses in both oligo- and polyphagous insects.

The mere fact that many oligophagous insect species can now be reared on diets not containing secondary substances of the relevant host plants (Singh, 1977) also indicates that feeding does not necessarily depend on the presence of "token stimuli" which can be replaced by primary substances.

Since the end of the 1950s more and more data have accumulated indicating that even in oligophagous insects the botanical distribution of substances inhibiting feeding is the main factor determining host-plant range at least in the case of herbivorous ontogenetic stages. Based on data from experiments carried out with different oligophagous species (Jermy, 1958, 1961; Gupta & Thorsteinson, 1960) the author proposed that the more an insect is oligophagous the more it is sensitive to plant metabolites occurring in the non-host plants which inhibit feeding. Thus, non-host plants are not eaten because they contain antifeedant compounds. A two-way chemoreceptor was proposed as the sensory background of this specialization emphasizing that the two ways are not equivalent since the most optimal phagostimuli, supposedly occurring in the host plants, can easily be masked by inhibitory substances while a strong enough antifeeding effect cannot be counterbalanced even with optimal phagostimuli. This is the sensory background of the use of antifeedants for controlling insect pests (Jermy, 1971). Since the great majority of non-host plants of a given insect species shows a deterrent effect, the sensory mechanism governing feeding must be sensitive to very dissimilar chemical structures (Jermy, 1966). Chapman (1974) and Vigneron (1978) published excellent reviews on the chemical nature of feeding inhibitors. However, it became more and more evident that it is inappropriate to think that phytophagous insects perceive only positive and negative chemical characteristics of the plant, i.e. phagostimulants and deterrents.

The complexity of information provided by the chemosensory mechanism can be judged by the phenomenon of induced host preference (Stride & Straatman, 1962; Jermy, Hanson & Detheir, 1968; Ma, 1972; Wiklund, 1973; Yamamoto, 1974; Phillips, 1977; Cassidy, 1978). It clearly proves that insects can distinguish not only host plants from non-host plants but also one host plant from the other. This is only possible by perceiving subtle details of the plant's chemical properties and by storing this information. Based on electrophysiological and behavioural investigations it has been assumed that the insects perceive a complex sensory picture of the plant (Dethier, 1973, 1976; Schoonhoven 1976). Recently, Dethier (1980) postulated that phytophagous insects "evolved a proliferation of receptor types, including the so-called deterrent receptors, which together are sensitive to a whole variety of secondary plant compounds, and that a central command system also evolved by which appropriate behaviour became associated with a particular sensory input".

Both the search for naturally occurring antifeedants and the selection of plant varieties for insect resistance of non-preference type should take into consideration the complexity of the "biochemical profile" of plants perceived by the insects.

12.2 CHEMISTRY OF DETERRENCE IN NON-HOST PLANTS

Many attempts have been made during the last two decades to isolate antifeedants from plants not preferred by certain or by several insect species. In most cases one or a few related secondary compounds were isolated and regarded as the antifeeding factor of the given plant species (Chuo, Hostettmann, Kubo & Nakanishi, 1977; Ikeda, Matsumura, & Benjamin 1977; Blau, Feeny, Contardo & Robson, 1978; Lee, 1978; Pettei, Miura, Kubo & Nakanishi, 1978; Russel, Sutherland, Hutchins & Christmas 1978; Warthen, Redfern, Vebel & Mills 1978; Evans & Bell, 1979; Trial & Dimond, 1979; for further references see: Chapman, 1974; Meinwald, Prestwich, Nakanishi & Kubo, 1978; Vigneron, 1978). In some cases, also, primary plant substances were found to be deterrent, e.g. nitrates in alfalfa to *Sitona cylindricollis* (Akeson, Haskins & Gorz, 1969).

However, phytochemists (Swain, 1977) estimated that the number of secondary plant chemicals equalled the number of plant species known (approximately 400,000), and they often found dozens of such compounds in each plant species (Hegnauer, 1962–1966). Therefore, it is appropriate to suppose that the "inhibitory biochemical profile" of the non-host plant, which is perceived by the insect, is in most cases constituted of several secondary plant substances presumably interacting with each other and with primary substances being additive, antagonistic or synergistic in their effect on the sensory system of the insect.

Very little has been done so far to reveal the whole spectrum or even a considerable part of compounds constituting the "inhibitory biochemical profile" of a plant species as related to a certain insect species. Adams & Bernays (1978) found that combinations of feeding deterrents were additive in their effects on the feeding behaviour of *Locusta migratoria*. A mixture of fourteen different phenolic compounds from *Sorghum bicolor* tested in naturally occurring concentrations was deterrent while individual levels of deterrence were not measurable. In seedlings of *S. bicolor* deterrence to the same insect is caused by cyanogenesis from the glycoside dhurrin and by phenolics (Woodhead & Bernays, 1978). Jacobson, Reed, Crystal, Moreno & Soderstrom (1978), by testing extracts of neem seeds and tung nuts on a series of insect species, have shown that several polar and non-polar compounds were responsible for the antifeedant properties of these plants. Jermy, Butt, Dreyer, McDonough & Rose (1980) bioassayed extracts of three plant species from the sage brush community of the North-western United States for antifeeding effect on *Leptinotarsa decemlineata* and also found that several chemical classes of compounds were active in each plant. These data clearly indicate that the "inhibitory biochemical profile" of a given plant species, as related to a certain insect species, is probably always composed of several plant constituents. Naturally, it cannot be excluded that in some instances one compound is predominant in causing deterrence.

Much more research is justified to reveal the interactions of different compounds in feeding inhibition. It can be supposed that combinations of plant substances would not only have increased antifeedant efficiency but would also be effective against a broader range of insect species.

When we tested the intensity of deterrence of a certain plant species with different insect species, all avoiding the given plant, considerable differences were found (Jermy, 1966). On the other hand, secondary plant substances isolated from non-host plants of different insect species were not necessarily deterrent for each of the latter (Schoonhoven & Jermy, 1977). Therefore, it can be concluded that the "inhibitory biochemical profile" of a given plant species is not only complex but also may vary considerably from one insect species to another.

Both the complexity of the "inhibitory biochemical profile" and its different constitution as related to different insect species determine several aspects in the strategy of developing methods for practical applications of the antifeedant principle in plant protection and in plant breeding.

12.3   THE SEARCH FOR ANTIFEEDANTS OF PRACTICAL USE

The feasibility of the use of antifeedants in pest control has been shown in several instances; nevertheless, extensive research carried out in this field during the last decade did not result in large-scale use. This is partly due to the fact that for several reasons, which are discussed by Bernays in another chapter of this book, antifeedants will rarely be effective enough as a sole control measure but could be promising when included into integrated control schemes. Since such approaches to pest control are becoming more and more widespread, the demand for suitable antifeedants is increasing. The slow development in research on antifeedants is partly due also to problems in the methodology which will be discussed later.

As natural substances are generally regarded as more acceptable for pest-control purposes than synthetic compounds, it is important to increase and coordinate efforts in order to find potent antifeedants in plants. In this connection three methodological questions arise: where do we look for most active plant substance? Is the activity range of an antifeedant predictable, i.e. what insects should be used for the bioassays? Are isolated compounds or a complex of plant constituents more promising?

12.3.1   The botanical occurrence of potent antifeedants

Very divergent opinions have been expressed concerning the question: in what types of vegetation are potent antifeedants most likely to occur? Some authors, e.g. Kubo & Nakanishi (1979), suppose that species of the tropical flora which are constantly exposed to attack by various parasites are confronted with much harsher conditions for survival than their temperate counterparts. This necessarily leads to efficient built-in defence mechanisms including substances with strong deterring effects on herbivores. In fact, tropical plants have provided substances with considerable antifeeding effect (Meinwald *et al.*, 1978).

But similar assumptions can be made also concerning plant communities at the ecological limits of primary production, e.g. under extreme arid conditions. Also here the plants must have evolved potent protection mechanisms against competitors. And indeed, plants of the sage brush (*Artemisietum*) community of the North-western United States were found to contain potent antifeedants against *Leptinotarsa decemlineata* (Jermy *et al.*, 1980).

Moreover, there are many examples of antifeedants occurring in plants of other types of vegetations. Since many different plant substances can inhibit feeding of a given insect species (Jermy, 1961, 1966; Bernays & Chapman, 1977; Schoonhoven & Jermy, 1977; Woodhead & Bernays, 1978), and since the plant species in every community are avoided by most insect species of the same community, it is evident that antifeedants must be present in the plants of every vegetation, e.g. 143 Central European plant species belonging to 43 families, when tested against 8 different insect species in the sandwich-test, were found to be deterrent in 75 to 97% depending on the insect species (Jermy, 1966). In these tests, species of some plant families were generally more deterrent than those of others. Nevertheless, any assumption on the botanical distribution of useful potent antifeedants, in relation to a given insect species, would be premature.

The intensity of the antifeedant effect may vary not only from plant species to plant species but also concerning infraspecific taxa. Phytochemists dealing with medicinal plants have shown that polychemism, i.e. infraspecific chemical differences, are a general feature in all phyla of the Plant Kingdom. This is so, for both genetic and ecological reasons (Tetenyi, 1970).

Thus, the probability of finding potent antifeedants against a certain insect species is the same for any plant community. Therefore, as sources of suitable antifeedants, local plant species available in great quantities should first of all be taken into consideration.

Because of the complexity of "inhibitory biochemical profiles" it is advisable to begin the testing of plants in bioassays using either intact plant tissues, e.g. leaves in the sandwich-test in the case of chewing insects or crude plant materials (expressed saps, dehydrated saps, powderized whole plants or organs, crude extracts, etc.) so that the whole complex of the plant constituents could be tested. Naturally, any kind of extraction, etc., changes some of the substances occurring in the intact plant; however, very unstable compounds are *a priori* inappropriate for practical use. On the other hand, chemical changes can result in antifeedants which do not occur as such in the intact plant but are produced when the tissues are damaged (Woodhead & Bernays, 1978).

As a possible source of antifeedants also waste materials of local chemical industries which process plant materials should be tested. They may contain considerable parts of the original "inhibitory biochemical profile" or new antifeedants resulting from the transformation of plant substances during processing, e.g. while the essential oils of the peppermint plant have no noticeable antifeeding effect on *Leptinotarsa decemlineata* , the water extract of the plant material remaining after steam distillation of the oils is highly active (Jermy *et al.*, 1980).

12.3.2   The choice of insects for testing antifeedants

The fact that there is no way to predict the antifeeding effect of a given plant species or plant compound relative to a certain insect species has often been neglected in antifeedant research. In most cases investigations were carried out by testing plant materials only on insect species easily reared in the laboratory without considering the insect pests against which the antifeedant should or could be used. Naturally, there are also examples where candidate target insects were used for the bioassays, e.g. at the USDA (Jacobson *et al.*, 1978; Jermy *et al.*, 1980), at the ICIPE (Meinwald *et al.*, 1978), etc.

For an earliest possible practical application it is imperative that pest species are used as test insects which are most likely to be controlled by antifeedants. Species against which an integrated control system is already available or is being developed should be preferred.

It also has to be kept in mind that geographic populations of the same insect species may vary considerably in host specificity as a result of differences in their reactions to deterrent plant chemicals, as it has been found recently with *Leptinotarsa decemlineata* (Hsiao, 1978).

A further important aspect is the possibility of changes in the insects' feeding behaviour as a consequence of mass rearing in the laboratory, e.g. Rathore & Guthrie (1973) found that larvae of *Ostrinia nubilalis*, when reared for eighty-seven generations on a meridic diet, showed reduced ability to feed on corn. Ma (1976) proved that rearing of *Spodoptera exempta* on an artificial diet resulted in basic changes of host specificity in the caterpillars. The induction of preference for the host plant, on which mass rearing is carried out in the laboratory, may also alter the insect population regarding its reactions to different antifeedants.

Such changes in the feeding behaviour of mass reared insect populations may jeopardize the reliability of laboratory screenings. Therefore, it is imperative to carry out at least preliminary field tests on the proposed target insect as soon as possible using the crude plant materials which were found to be most active in the laboratory. In most cases this has not been done, although the economy of applied research in this field demands that further investigations into isolation and identification of active ingredients, into their interactions in evoking feeding inhibition, etc., should be carried out only on materials providing good results in field tests.

12.3.3 Pure compounds or complexes of plant constituents?

From the complexity of the "inhibitory biochemical profiles" it logically follows also that more or less crude plant materials may be more effective than single purified compounds. The use of the former would be presumably also more economic than that of the latter, not to mention the synthesis of complex molecules.

No attempts have been made to find out the possibilities of building up potent antifeedants for practical use by combining several compounds or extracts of different plants. This is also worth investigating. Naturally, the use of complex plant extracts and their combination may cause even more serious problems to meet the safety requirements of human toxicology as compared to pure compounds but it is more a technical than a principal problem.

Until now the whole problem of formulation of natural antifeedants for plant protection purposes has been totally neglected although it may be crucial for successful application. The importance of the systemic action of active ingredients, the protection of non-systemic water-soluble substances against removal by rain, etc., are questions which inevitably arise and have to be solved.

12.4   NOTES ON THE NON-PREFERENCE TYPE OF RESISTANCE IN PLANTS

In a few instances the chemical background of insect resistance (non-preference) in cultivated plants has been revealed. Mostly one single compound or a group of related compounds were isolated as the main factor(s) of resistance (Buhr, Tobau & Schreiber, 1958; Schreiber, 1958; Brusse, 1962; Klun, Guthrie, Hallauer & Russell, 1970; Gorz, Haskins & Manglitz, 1972; Bordasch & Berryman, 1977; Tingey, Mackenzie & Gregory, 1978). In such cases chemical analysis of the plants could replace biological tests.

However, in this connection also, the possibly complex nature of the "inhibitory biochemical profile" should be considered. In fact there are already examples in which it turned out that the concentration of the compound(s) supposedly causing resistance did not always correlate with the resistance found in the biotest (Scriber, Tingey, Gracen & Sullivan, 1975; Tingey *et al.*, 1978). Chang & Philogene (1978) found that a large number of phenolics and other compounds cause resistance in pears to *Psylla pyricola*. Adams & Bernays (1978) revealing the additive effect of a series of compounds causing resistance of *Sorghum bicolor* to *Locusta migratoria,* rightly emphasize that the breeders should consider the whole complex of chemicals involved in non-preference.

Thus, for creating plant varieties with strong "inhibitory biochemical profiles" to a given insect pest or a group of pests, biotests under the most natural conditions are indispensable. Only such tests are able to reveal all genetic possibilities in breeding resistant varieties. A detailed chemical analysis of all the constituents of a plant which confer resistance, would provide the information for an "inhibitory biochemical profile" and could lead to a better understanding of the sensory capabilities of insects in plant recognition.

12.5   CONCLUSION

Research on the practical use of antifeedants of plant origin should consider (a) the wide capability of the chemosensory mechanism governing feeding behaviour in phytophagous insects, (b) the complexity of the "inhibitory biochemical profile" of plants deterring a given insect species, (c) the differences of these profiles in relation to different insect species, (d) the variations in feeding behaviour of different populations of the same insect species, (e) the changes of plant constituents in time and space (polychemism), and (f) the problems of formulation of antifeedants for plant protection purposes.

Research in this relatively new and promising field could be speeded up by a considerable increase of national and international co-operation. Regular exchange of crude plant materials, extracts, isolated compounds, etc., as well as of information on results of bioassays and formulation experiments, is needed.

Several of the above points apply also to the breeding of plant varieties for non-preference type of resistance against insect pests.

12.6 SUMMARY

Host plant specificity in phytophagous insects is determined mainly by the botanical distribution of plant substances inhibiting feeding since the chemosensory mechanism governing feeding behaviour is very sensitive to a wide variety of such plant constituents. Thus, plants are excellent sources of natural antifeedants for potential use in pest control. Research on antifeedants should consider: the complexity of the "inhibitory biochemical profiles" of non-host plants as perceived by the insects, the basic differences of these profiles in relation to different insect species, the variations in feeding behaviour of populations of the same insect species as well as the polychemism of plants. Candidate antifeedants should be tested on the insect species to be controlled by them. Potent antifeedants may be found in any plant community. Also waste products of chemical industries processing plant materials may contain antifeedants. Much more research is needed concerning composition and formulation of antifeedants for practical use.

Breeding of plants for non-preference type of resistance must also take into consideration the complexity of the plant's biochemical profile perceived by the insects.

12.7 ACKNOWLEDGEMENT

The author wishes to thank Dr. E. A. Bernays for critically reading the manuscript.

12.8 REFERENCES

Adams C. M. & Bernays E. A (1978) The effect of combinations of deterrents on the feeding behaviour of *Locusta migratoria. Ent. exp. appl.* **23**, 101–109.

Akeson W. R., Haskins F. A. & Gorz H. J. (1969) Sweetclover weevil feeding deterrent B: isolation and identification. *Science* **163**, 293–294.

Beck S. D. (1965) Resistance of plants to insects. *Ann. Rev. Entomol.* **10**, 207–232.

Bernays E. A. & Chapman R. F. (1977) Deterrent chemicals as a basis of oligophagy in *Locusta migratoria* (L.). *Ecol. Ent.* **2**, 1–18.

Blau P. A., Feeny P., Contardo L. & Robson D. S. (1978) Allylglucosinolate and herbivorous caterpillars: a contrast in toxicity and tolerance. *Science* **200**, 1296–1298.

Bordasch R. P. & Berryman A. A. (1977) Host resistance to the fir engraver beetle, *Scolytus ventralis* Coleoptera: Scolytidae 2. Repellency of *Abies grandis* resins and some monoterpenes. *Can. Ent.* **109**, 95–100.

Brusse M. J. (1962) Alkaloid content and aphid infestation in *Lupinus angustifolius* L. *N.Z. J. agric. Res.* **5**, 188–189.

Buhr H., Toball R. & Schreiber K. (1958) Die Wirkung von einigen pflanzlichen Sonderstoffen, insbesondere von Alkaloiden, auf die Entwicklung der Larven des Kartoffelkäfers *Leptinotarsa decemlineata* Say. *Ent. exp. appl.* **1**, 209–224.

Cassidy M. D. (1978) Development of an induced food plant preference in the Indian stick insect, *Carausius morosus. Ent. exp. appl.* **24**, 87–93.

Chang J. F. & Philogene B. J. R. (1978) Response of *Psylla pyricola*, Homoptera: Psyllidae to, and characterization of polar and lipid fractions of *Pyrus* sp. leaves. *Phytoprotection* **59**, 28–39.

Chapman R. F. (1974) The chemical inhibition of feeding by phytophagous insects: a review. *Bull. ent. Res.* **64**, 339–363.

Chuo F. Y., Hostettmann K., Kubo I. & Nakanishi K. (1977) Isolation of an insect antifeedant N-methylflindersine and several benz (c) phenanthridine alkaloids from East African plants; a comment on chelerythine. *Heterocyles* **7**, 969–977.

Dethier V. G. (1953) Host plant perception in phytophagous insects. *Trans. 9th Int. Congr. Ent.* **2**, 81–88.

Dethier V. G. (1954) Evolution of feeding preferences in phytophagous insects. *Evolution* **8**, 33–54.

Dethier V. G. (1973) Electrophysiological studies of gustation in lepidopterous larvae II. Taste spectra in relation to food-plant discrimination. *J. comp. Physiol.* **82**, 103–134.

Dethier V. G. (1976) The importance of stimulus patterns for host-plant recognition and acceptance. *Symp. Biol. Hung.* **16**, 67–70.

Dethier V. G. (1980) Evolution of receptor sensitivity to secondary plant substances with special reference to deterrents. *Am. Nat.* **115**, 45–66.

Ehrlich P. R. & Raven P. H. (1964) Butterflies and plants: a study in coevolution. *Evolution* **18**, 586–608.

Evans C. S. & Bell E. A. (1979) Non-protein amino acids of *Acacia* species and their effect on the feeding of the acridids *Anacridium melanorhodon* and *Locusta migratoria. Phytochemistry* **18**, 1807–1810.

Fraenkel G. S. (1953) The nutritional value of green plants for insects. *Trans. 9th Int. Congr. Ent.* **2**, 90–100.

Fraenkel G. S. (1959) The *raison d'être* of secondary plant substances. *Science* **129**, 1466–1470.

Gorz H. J., Haskins F. A. & Manglitz G. R. (1972) Effect of coumarin and related compounds on blister beetle feeding in sweetclover. *J. econ. Ent.* **65**, 1632–1635.

Gupta P. D. & Thorsteinson A. J. (1960) Food plant relationships of the diamond-back moth, *Plutella maculipennis* Curt. I. Gustation and olfaction in relation to botanical specificity of the larva. *Ent. exp. appl.* **3**, 241–250.

Hegnauer R. (1962–1966) *Chemotaxonomie der Pflanzen*, Vols. 1–4. Birkhäuser Berlag, Basel.

Hsiao T. H. (1976) Chemical and behavioural factors influencing food selection of *Leptinotarsa* beetle. *Symp. Biol. Hung.* **16**, 95–99.

Hsiao T. H. (1978) Host plant adaptations among geographic populations of the Colorado potato beetle. *Ent. exp. appl.* **24**, 437–447.

Ikeda T., Matsumura F. and Benjamin D. M. (1977) Chemical basis for feeding adaptation of pine sawflies *Neodiprion rugifrons* and *Neodiprion swalnei*. *Science* **197**, 497–499.

Jacobson M., Reed D. K., Crystal M. M., Moreno D. S. & Soderstrom E. L. (1978) Chemistry and biological activity of insect feeding deterrents from certain weed and crop plants. *Ent. exp. appl.* **24**, 448–457.

Jermy T. (1958) Untersuchungen über Auffinden und Wahl der Nahrung beim Kartoffelkäfer (*Leptinotarsa decemlineata* Say). *Ent. exp. appl.* **1**, 197–208.

Jermy T. (1961) On the nature of the oligophagy in *Leptinotarsa decemlineata* Say (Coleoptera: Chrysomelidae). *Acta Zool. Acad. Sci. Hung.* **7**, 119–132.

Jermy T. (1966) Feeding inhibitors and food preference in chewing phytophagous insects. *Ent. exp. appl.* **9**, 1–12.

Jermy T. (1971) Biological background and outlook of the antifeedant approach to insect control. *Acta Phytopath. Acad. Sci. Hung.* **6**, 253–260.

Jermy T., Butt B. A., Dreyer D. L., McDonough L. M. & Rose A.F. (1980) Antifeedants for the Colorado potato beetle. I. Antifeeding constituents of some plants from the sagebrush community. *Insect Sci. Appl.* **1**, 237–242.

Jermy T., Hanson F. E. & Dethier V. G. (1968) Induction of specific food preference in lepidopterous larvae. *Ent. exp. appl.* **11**, 211–230.

Klun J. A., Guthrie W. D., Hallauer A. R. & Russel W. A. (1970) Genetic nature of the concentration of 2,4-dihydroxy-7-methoxy-2H-1,4-benzoxazin-3 (4H)-one and resistance to the European corn borer in a diallele set of eleven maize inbreds. *Crop Sci.* **10**, 87–90.

Kubo I. & Nakanishi K. (1979) Some terpenoid Insect Antifeedants from Tropical Plants. In *Advances Pestic. Sci.*, Part 2 (ed. Geissbuhler, H.) pp. 284–294. Pergamon Press, Oxford.

Lee Y. W. (1978) Potent insect antifeedants from East African plants, pp. 108–136. Doct.thesis, Columbia University.

Ma W. C. (1972) Dynamics of feeding responses in *Pieris brassicae* Linn. as a function of chemosensory input: a behavioural, ultrastructural and electrophysiological study. *Meded. Landbouwhogeschool Wageningen* **72-11**, 1–162.

Ma W. C. (1976) Experimental observations on food-aversive responses in larvae of *Spodoptera exempta* Wlk. (Lepidoptera: Noctuidae). *Bull. ent. Res.* **66**, 87–96.

Meinwald J., Prestwich G. D., Nakanishi K. & Kubo I. (1978) Chemical ecology: Studies from East Africa. *Science* **199**, 1167–1173.

Pettei M. J., Miura I., Kubo I. & Nakanishi K. (1978) Insect antifeedant sesquiterpene lactones from *Schkuhria pinnata*: the direct obtention of pure compounds using reverse phase preparative liquid chromatography. *Heterocycles* **11**, 471–479.

Phillips W. M. (1977) Modification of feeding "preference" in the flea-beetle, *Haltica lythri* (Coleoptera: Chrysomelidae). *Ent. exp. appl.* **21**, 71–80.

Rathore Y. S. & Guthrie W. D. (1973) European corn borer: leaf-feeding damage on field corn by larvae reared for 87 generations in a meridic diet. *J. econ. Ent.* **66**, 1195–1196.

Rees C. J. C. (1969) Chemoreceptor specificity associated with choice of feeding site by the beetle, *Chrysolina brunsvicensis* on its foodplant, *Hypericum hirsutum*. *Ent. exp. appl.* **12**, 565–583.

Ritter F. J. (1967) Feeding stimulants for the Colorado beetle. *Meded. Rijksfac. Landbouwwet.* **32**, 291–305.

Russel G. B., Sutherland O. R. W., Hutchins R. F. N. & Christmas P. E. (1978) Vestitol: a phytoalexin with insect feeding-deterrent activity. *J. Chem. Ecol.* **4**, 571–579.

Schoonhoven L. M. (1967) Chemoreception of mustard oil glucosides in larvae of *Pieris brassicae*. *Proc. Kon. Ned. Akad. Wtsch.* C, **70**, 556–563.

Schoonhoven L. M. (1968) Chemosensory bases of host plant selection. *Ann. Rev. Entomol.* **13**, 115–136.

Schoonhoven L. M. (1976) Feeding behaviour in phytophagous insects: on the complexity of the stimulus situation. *Colloques Internationaux du C.N.R.S.* No. 265, 391–398.

Schoonhoven L. M. & Jermy T. (1977) A behavioural and electrophysiological analysis of insect feeding deterrents. In *Crop Protection Agents — Their Biological Evaluation* (ed. McFarlane N. R.), pp. xvii + 638. Academic Press, New York.

Schreiber K. (1958) Über einige Inhaltsstoffe der Solanaceae und ihrer Bedeutung für die Kartoffelkäferresistenz. *Ent. exp. appl.* **1**, 28–37.

Scriber J. M., Tingey W. M., Gracen V. E. & Sullivan S. L. (1975) Leaf-feeding resistance to the European corn borer in genotypes of tropical (low-DIMBOA) and U.S. inbred (high-DIMBOA) maize. *J. econ. Ent.* **68**, 823–826.
Singh P. (1977) *Artificial Diets for Insects, Mites, and Spiders*, pp. xi + 594. IFI/Plenum, New York.
Städler E. (1976) Sensory aspects of insect plant interactions. *Proc. 15th Int. Congr. Ent.* 228–248.
Stride G. O. & Straatman R. (1962) The host plant relationship of an Australian swallowtail *Papilio aegeus*, and its significance in the evolution of host plant selection. *Proc. Linn. Soc. N.S.W.* **87**, 69–78.
Swain T. (1977) Secondary compounds as protective agents. *Ann. Rev. Plant Physiol.* **28**, 479–501.
Tétényi P. (1970) *Infraspecific Chemical Taxa of Medicinal Plants*, p. 225. Akadémiai Kiadó, Budapest.
Thorsteinson A. J. (1953) The chemotactic responses that determine host specificity in an oligophagous insect, *Plutella maculipennis* Curt. *Can. J. Zool.* **31**, 52–72.
Thorsteinson A. J. (1958) Acceptability of plants for phytophagous insects. *Proc. 10th Int. Congr. Ent.* **2**, 599–602.
Thorsteinson A. J. (1960) Host selection in phytophagous insects. *Ann. Rev. Ent.* **5**, 193–218.
Tingey W. M., Mackenzie J. D. & Gregory P. (1978) Total foliar glycoalkaloids and resistance of wild potato species to *Empoasca favae* Harris. *Am. Potato J.* **55**, 577–585.
Trial H. & Dimond J. B. (1979) Emodin in buckthorn-feeding deterrent to phytophagous insects. *Can. Ent.* **111**, 207–213.
Verschaffelt E. (1910) The cause determining the selection of food in some herbivorous insects. *Proc. Kon. Ned. Acad. Wtsch.* **13**, 536–542.
Vigneron J. P. (1978) Substances antiappétantes d'origin naturelle. *Ann. Zool. Ecol. Anim.* **10**, 663–694.
Warthen J. D., Redfern R. E., Uebel E. C. & Mills G. D. (1978) An antifeedant for fall armyworm larvae from neem seeds. USDA, ARR-NE-1, pp. iv + 9.
Whittaker R. H. & Feeny P. P. (1971) Allelochemics: chemical interactions between species. *Science* **171**, 757–770.
Wieczorek H. (1976) The glycoside receptor of the larvae of *Mamestra brassicae* L. (Lepidoptera: Noctuidae). *J. comp. Physiol.* **106**, 153–176.
Wiklund C. (1973) Host plant suitability and the mechanism of host selection in larvae of *Papilio machaon*. *Ent. exp. appl.* **16**, 232–242.
Woodhead S. and Bernays E. A. (1978) The chemical basis of resistance of *Sorghum bicolor* to attack by *Locusta migratoria*. *Ent. exp. appl.* **24**, 123–144.
Yamamoto R. T. (1974) Induction of hostplant specificity in the tobacco hornworm, *Manduca sexta*. *J. Insect Physiol.* **20**, 641–650.

12.9 DISCUSSION

Bernays: What is the importance of the complex profiles of unpalatable plants? You mentioned some work on cypress. I think that it is very important if synergism is found. If you can show the synergism which I have unsuccessfully tried to do, then the potential of this work is very great. Did you find any evidence for synergism?

Jermy: We have not yet done anything to establish whether or not synergism is involved in this exercise. However, I agree with you that the presence of synergism would increase the potential of this work. The complex profiles show that if we had a set of plants to protect, one profile would protect against one insect species, while the other would be useful in protecting the other species. Even if the compounds are not synergists they can, however, enlarge the profiles to confer deterrency against other insect species.

Bernays: I brought out this question because I think synergism confers about 10 times activity in some of these plants and so it is very important.

Marini Bettolo: In addition to that, I think a very important consequence comes out from the presentation of Dr. Jermy if water-soluble fraction is also considered in looking for antifeedants from the plants. This approach may lead to the discovery of more antifeedant compounds. In addition to this, I should like to ask you which is the antifeedant for the Colorado beetle which you mentioned several times? Has it been isolated and identified?

Jermy: I did not refer to a specific feeding deterrent for the Colorado beetle. A wide range of compounds has been tested and I referred to the range.

Otieno: Do you think that these antifeedants have any prospects for commercial exploitation.

Jermy: Yes, I think they do. I would also stress that prospects for exploitation by local industries are good. Because as Dr. Saxena said, a stage has now been reached when active crude plant extracts should be considered for commercial developments, using very simple industrial processes. In this regard, it is also important to emphasize Professor Djerassi's feelings about the safety factor involved in using crude products. I admit that the active components of these extracts could be harmful. However, I do not think that it is wise to condemn the extract on the strength of this suspicion. Adequate tests evaluation should be done before condemning such active plant extracts.

Djerassi: The question of environmental safety is as important as the need to increase food production by minimizing food losses due to pests. I think that new active compounds must undergo a full assessment to determine their impact on the environment. Only when they are found to respect the laws governing environmental safety, should they be considered for commercial development.

Whitehead: I do not see anything wrong with spraying crude extracts on a crop to deter insects from feeding on it long before the crop is ready for harvest. Indeed, with corn, you protect the corn against the armyworm long before the ears ever ripen. I think this practice reduces the possibility of concentrating the extract on the corn and thus minimizes risks to the consumer.

Djerassi: I am not against the use of natural products. In fact I have spent my entire life working on natural products. What I am saying is that an extract or compound is not necessarily safe to use simply because it is natural or comes from a natural origin, despite any strong economic arguments which might prevail. I argue that each active compound or extract natural or not, must be thoroughly evaluated before it is recommended for use in pest management.

Jermy: Arguing on the issue of safety, I would say that some species of plants could develop compounds which are strongly active against some insects over a period of time. Because of this, there is some chance that these compounds could be environmentally safe when used commercially. But this is only speculative.

Djerassi: I should like to make a suggestion now regarding this discussion. I believe that the questions we are discussing are relevant to several other papers, like that of Dr. Bernays which deals with issues similar to those we are discussing now. I suggest, therefore that we discuss them together after the presentation of these other papers.

CHAPTER 13

Antifeedants from Clausena anisata (Willd.) Hook F. Ex Benth (Rutaceae)

T. GEBREYESUS and A. CHAPYA

*The International Centre of Insect Physiology and Ecology (ICIPE)
P.O. Box 30772, Nairobi, Kenya*

CONTENTS

13.1	Introduction	237
13.2	Materials and Methods	238
13.3	Results and Discussion	238
	13.3.1 Isolation	238
	13.3.2 Characterization	239
13.4	Summary	240
13.5	Acknowledgement	240
13.6	References	240
13.7	Discussion	241

13.1 INTRODUCTION

The need to look for alternative methods of insect control in the face of the serious problems created by the use of insecticides and the realization that plants which have to survive in the presence of pests must have a built-in mechanism to protect themselves against insect attack, have prompted us to initiate a programme of screening African plant extracts for antifeeding activity. This programme has already led to the isolation of quite a number of antifeedants described elsewhere (Kubo & Nakanishi, 1979).

This paper describes the isolation and characterization of two antifeedants from *Clausena anisata* (Rutaceae), a tree which grows 6–10 ft high and is common in dry areas, savanna bush and forest edges (Dale &. Greenway, 1962). The plant is used by various tribes throughout Africa for different medical purposes. It is reportedly used as a heart tonic, anthelminthic, parasiticide, purgative and for the treatment of rheumatism, malaria,

237

influenza and other ailments (Watt & Breyer-Brandwijk, 1962; Kokwaro, 1976).

13.2   MATERIALS AND METHODS

The plant was collected in Chiromo, Nairobi, and identified by the Botany Department of the University of Nairobi and by the East African Herbarium. The bark was cut into small pieces and extracted at room temperature with petroleum ether 60–80°C. The solvent was removed under vacuum and the residue tested for antifeeding activity using African armyworm, *Spodoptera exempta*, larvae in a leaf disc bioassay (Kubo & Nakanishi, 1977). The extract was dissolved in acetone and a 2-cm diameter leaf disc of corn, *Zea mays*, dipped in it for about 2 sec, dried and offered to larvae which had previously been starved for a minimum of 2 h. Leaf discs treated similarly with only the solvent served as control. The amount of leaf material eaten was estimated every hour for 2 hr, the time it usually took the larvae to eat all the control discs.

Column chromatography was carried out using silica gel 60, 70–270 mesh ASTM (Macherey Nagel & Co., 516 Duren, Germany). 20 × 20–cm plates coated with silica gel containing fluorescent indicator UV_{254} were used for preparative thin-layer chromatography and were developed in acetone-hexane mixtures. Gas chromatography was performed using a Hewlett Packard 402 instrument. The ultraviolet spectra were taken in ethanol on a Perkin-Elmer 402 spectrophotometer. The infrared spectra were measured either in carbon tetrachloride solution or as a mull in nujol on a Perkin-Elmer 720 instrument. Proton magnetic resonance (PMR) spectra were obtained in deuterochloroform on a Hitachi Perkin-Elmer R-24A 60-MHz instrument. The mass spectra were determined using a Finnigan 1015D mass spectrometer.

13.3   RESULTS AND DISCUSSION

13.3.1   Isolation

The petroleum ether extract was chromatographed and eluted with increasing percentages of diethyl ether in hexane. Two fractions were found to be active as antifeedants. These fractions were purified by preparative thin-layer chromatography and recrystallization from acetone–hexane mixtures. The two pure compounds so obtained showed feeding activity at 100 and 500 ppm.

13.3.2 Characterization

The more active of the two compounds had a melting point of 98–99.5°C. Its ultraviolet (UV) spectrum showed absorptions at 218, 249 and 299 nm (log 4.42, 4.20, 3.91). The infrared (IR) spectrum showed absorptions at 1740, 1625, 1585 and 1250 cm^{-1} which suggests an α-β unsaturated lactone. The PMR spectrum showed a singlet of six protons at 1.73δ, a doublet of two protons at 4.98δ, another doublet of one proton at 5.56δ and a complex of five olefinic and aromatic protons in the 6–8δ region.

The mass spectrum showed a very small molecular ion at m/e 270 with prominent peaks at m/e 201, 173, 145, 117, 89 and 69. This represents loss of 69 (C_5H_9) from the molecular ion followed by consecutive losses of 28 mass units. These physical and spectral data agree with that of the known furanocoumarin, imperatorin (Fig. 13.1). The identity was further confirmed by gas and thin-layer chromatographic comparison with an authentic sample.

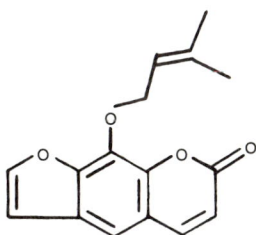

Fig. 13.1. Imperatorin.

The less active antifeedant had a melting point of 132.5–134°C. The UV spectrum showed absorptions at 227, 268 and 346 nm (log 4.26, 4.30, 4.00). Its IR spectrum showed absorptions at 1740 cm^{-1} corresponding to an α-β unsaturated lactone and at 1618 cm^{-1} corresponding to conjugated carbon–carbon double bond. The PMR spectrum showed a singlet of six protons at 1.46δ, another singlet of three protons at 3.85δ and a complex of five olefinic and aromatic protons in the 5.5–7.8δ region. The mass spectrum showed a molecular ion at m/e 258 with prominent peaks at m/e 243, 228, 200, 172, 144, 129 and 115 corresponding to consecutive losses of two methyl groups followed by loss of 28 mass units. These physical and spectral data correspond with that of the known coumarin, xanthoxyletin (Fig. 13.2). Further confirmation of the structure was obtained by gas and thin-layer chromatographic comparison with an authentic sample.

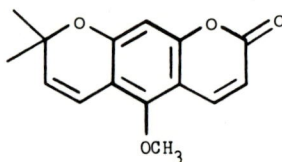

OCH₃

FIG. 13.2. Xanthoxyletin.

A number of coumarins, imperatorin and xanthoxyletin among them, have been isolated from the roots of *C. anisata* (Mester, Szendrei & Reisch, 1977; Okorie, 1975) but no biological activity has been reported for any of them.

13.4 SUMMARY

Two antifeedants against the African armyworm, *Spodoptera exempta*, larvae have been isolated from the petroleum ether extract of *Clausena anisata* (Rutaceae) using a leaf disc bioassay. The compounds have been identified as the coumarins imperatorin and xanthoxyletin from spectral data as well as by comparison with authentic samples.

13.5 ACKNOWLEDGEMENT

We are grateful to Drs. D. Dreyer and Von I. Mester for samples of imperatorin and xanthoxyletin.

13.6 REFERENCES

Dale I. R. & Greenway P. J. (1962) *Kenya Trees and Shrubs*, pp. 481–482. Buchanan's Kenya Estates Ltd., Nairobi.
Kokwaro J. O. (1976) *Medicinal Plants of East Africa*, pp. 195–196. East African Literature Bureau, Nairobi.
Kubo I. & Nakanishi K. (1977) Insect antifeedants and repellents from African plants. In *Host Plant Resistance to Pests* (ed. Paul, A.), pp. 165–178. ACS Symposium Ser. No. 62, American Chemical Society.
Kubo I. & Nakanishi K. (1979) Some terpenoid insect antifeedants from tropical plants. In *Advances in Pesticide Science*, Part 2 (ed. Geissbuhler H.), pp. 284–294. Pergamon Press, Oxford and New York.
Mester I., Szendrei K. & Reisch J. (1977) Inhaltsstoffe aus *Clausena anisata* (Willd.) Oliv. (Rutaceae). I. Cumarine aus der Wurzelrinde. *Planta Medica* **32**, 81–85.
Okorie D. A. (1975) A new carbazole alkaloid and coumarins from roots of *Clausena anisata*. *Phytochemistry* **14**, 2720–2721.

Watt J. M. & Breyer-Brandwijk, M. G. (1962) *The Medicinal and Poisonous Plants of Southern and Eastern Africa*, 2nd ed., pp. 917–918. E. & S. Livingstone Ltd., Edinburgh and London.

13.7   DISCUSSION

Jacobson: If you're not already aware of it you might be interested to know the work of Novak in Czechoslovakia who found that a crude extract of *Clausena anisata* leaves was very repellent to ticks.

Gebreyesus: We are actually testing some of these compounds for tick repellency as well. What we have not been able to establish is the tick bioassay. We are using the tendency of ticks to climb up glass plates. We are trying this in the laboratory and are having some difficulties. For one thing, the ticks appear to behave one day and not the next. We are not very sure what induces them to climb one day and not the next. There are, of course, factors such as temperature and humidity affecting them. However, we are trying and have found repellency in certain plants. But we could not repeat this on a regular basis.

Marini Bettolo: I think it's very interesting that it has been shown that these coumarins have this activity. I should like to ask if you have noticed an influence of light during your experiments, because coumarins can be extremely sensitive to light and even when in contact with DNA and similar compounds they give photoaddition products. I think it is worth while to consider this aspect.

Gebreyesus: If you mean in the bioassay; no! But as I said these are crystalline solids and one of the coumarin compounds appears to have no activity whatsoever. I have not extracted back to see if it is the same compound or whether it has been changed by light. We have not had any experience of not being able to repeat the experiment with the same compound.

Chapya: I want to add something to what Professor Marini Bettolo said. This non-active furanocoumarin if kept for several weeks, shows two spots by TLC?

Marini Bettolo: Yes, these compounds can be very reactive.

Gebreyesus: As I've said I have not back-extracted the leaves to see if the compound, in this case xanthoxyletin, is in the same form or not. If there are breakdown products, they might be responsible for the antifeeding activity. I have not extracted the treated leaves to see if there are breakdown products.

Marini Bettolo: Anyhow, the spirit of my question was just to underline that the mechanism of activity may involve some change due to the reactivity of the coumarins on the substrate.

Gebreyesus: It certainly is possible. In the case of blister beetles it has been shown that the *trans*-o-hydroxycinnamic acid present in the plant has no activity as a feeding deterrent whereas the *cis*-isomer does, and this is postulated to be so because the *cis*-isomer forms a coumarin and this is responsible for the activity.

Marini Bettolo: True, but these are not simple coumarins, rather chromano-coumarins.

CHAPTER 14

Attractants, Arrestants, Feeding and Oviposition Stimulants in Insect–Plant Relationships: Application for Pest Control

ERICH STÄDLER

Eidg. Forschungsanstalt, CH-8820 Wädenswil, Switzerland

CONTENTS

14.1	Introduction	243
14.2	Methods and Difficulties	244
	14.2.1 Attractants and arrestants	244
	14.2.2 Feeding stimulants	245
	14.2.3 Oviposition stimulants	245
	14.2.4 Electrophysical recordings	246
	14.2.5 General consideration of isolation and identification techniques	246
14.3	Conclusions on the Role of Attractants, Arrestants and Stimulants in Insect–Plant Relationships	247
14.4	Application for Pest Control	248
	14.4.1 Attractants	248
	14.4.2 Feeding stimulants	250
	14.4.3 Mass trapping	250
	14.4.4 Disguising non-host plants as hosts	250
	14.4.5 Breeding resistant plants	250
14.5	Conclusions	251
14.6	Summary	251
14.7	Acknowledgements	252
14.8	References	252
14.9	Discussion	255

14.1 INTRODUCTION

The aim of this contribution is to review briefly our present methodology and knowledge about plant chemicals increasing or stimulating plant attack by insects and to evaluate their possible applications. It is important to realize that chemicals acting as repellents and deterrents (in the sense of Dethier, Barton Browne & Smith, 1960) are at least as important for host selection (for ref. see Bernays, Jermy in this volume). In addition it has to be stressed that plant chemicals seldom determine insect behaviour alone but always act in concert with non-chemical plant characteristics such as colour, shape and texture. Thus if we limit the scope of this contribution to host-plant compounds which increase the attack of specific plants by insects, we have to be aware of the fact that this is artificial and can only partially explain the insect–plant relationship occurring in the real environment.

Specific difficulties can arise from the "labelling" of chemicals in terms of end results of a chain of behaviour sequences. The use of the terminology introduced by Dethier *et al.* (1960) requires a careful analysis of the behaviour sequences involved in host selection. Kennedy (1977a,b, 1978) suggested the use of an even more detailed description of the behaviour reactions influenced by chemicals. However, no terminology solves all the problems of classification of active compounds, such as the fact that a specific chemical can have different effects on different insects. An example is the toxic activity of furanocoumarins occurring in Umbelliferae on armyworms (Berenbaum, 1978) and the stimulation of oviposition by the same compounds in the carrot fly (Städler, unpubl.)

The same compound may also influence different behaviour sequences in the same animal such as attraction and oviposition stimulation in the cabbage fly (Finch, 1978; Nair & McEwen, 1976). Therefore the use of any terminology does imply both a detailed analysis of the behavioural sequence and the specification of the circumstances as already pointed out by Dethier *et al.* (1960).

14.2   METHODS AND DIFFICULTIES

Any investigation of plant chemicals which potentially may influence insect behaviour has to depend on a reliable bioassay method. Obviously, it is important to use an assay which is simulating as close as possible the natural environment and providing an accurate measure of the specific behaviour sequence. Furthermore, the ideal assay should involve as little cost and labour as possible. Since the two aims are usually contradictory a compromise must be found which should ideally be based on the complete analysis of the relevant behaviour. In the following text, methods and recent examples of investigation of behaviourally active compounds are listed.

14.2.1   Attractants and arrestants

An excellent review of the methodology and appreciation of the difficulties involved has been presented by Kennedy (1977b). The author stresses the fact that olfactometers do not usually distinguish as well between behavioural as between chemical components of insect behaviour. In his own words he suggests the following solution, "all of the known types of response, except chemotaxis, could be assayed unambiguously by presenting the insect with temporal instead of spatial changes of odour stimulation in a uniform arena". A useful tool for this approach is the wind tunnel as has recently been proven in the studies of the olfactory orientation (anemotaxis) of the cabbage fly by Hawkes & Coaker (1979) and the Colorado beetle by Visser, Van Straten & Maarse (1979). However, this does not mean that every insect will behave as nicely as certain moths (Kennedy & Marsh, 1974) in response to pheromones. The ideal bioassay obviously is different for each species of insect.

14.2.2   Feeding stimulants

A very useful review and investigation of the different methods applied and the interpretation of the results obtained has been presented by Cook (1976). In addition to these techniques it seems worth mentioning that glass-fibre filter papers have recently been found acceptable as an inert substrate for testing of feeding stimulants by Städler & Hanson (1976) and Woodhead & Bernays (1978). Like other activities, feeding can be automatically monitored. This has been demonstrated by Jones (1979) for *Pieris brassicae* larvae using an electronic linear transducer.

For the study of the choice behaviour of aphids, assays have been developed which are based either on the number of settled insects (Greenway, Griffiths & Lloyd, 1978) or on the consumption of diet (Arn & Cleere, 1971). The sucking activity itself can be monitored and correlated to patterns of electric signals (Tjallingii, 1978).

14.2.3   Oviposition stimulants

Oviposition has been studied with a variety of assays reflecting the different behaviours and substrates observed in different insects. Recent examples of assays for insects ovipositing into the soil have been presented by Nair & McEwen (1976), Städler (1971/1972), Swailes (1967) and Zohren (1968). Females using leaf surfaces as oviposition site have been tested by Heinicke (1978), Matsumoto & Thorsteinson (1968), Röttger (1978, 1979), Städler (1974), Vernon, Borden, Pierce & Oehlschlager (1977). Haisch & Levinson (1980) bio-assayed fruit volatiles influencing oviposition in the cherry fruit fly using artificial fruits developed by Prokopy & Boller (1970). In some investigations including our own (Städler, 1971/1972) no clear discrimination between oviposition stimulation or attractant and arrestant activity was made. Thus it may be that so-called oviposition stimulants are attractants and arrestants or a combination of all the three. This calls for more detailed investigations of the behaviour sequences leading to oviposition, as has been attempted by Städler (1977), involving the development of new or additional bioassay techniques.

14.2.4   Electrophysiological recordings

Provided the chemoreceptive sensilla involved in the perception of host-plant chemicals have been identified, electrophysiological recordings (review about techniques: Hansen, 1978; Städler 1976) can be used to explain the behaviour of pest insects in terms of sensory perception. These techniques are ideal to test many compounds in minimal quantities and have proven to be powerful techniques in the analysis of pheromones (e.g. Arn, Städler & Rauscher, 1975). Since stimulus quantities are also usually a limiting factor in the analysis of plant extracts, it seems logical to apply electrophysical techniques. No systematic study has been performed yet. The obvious reason for this fact is the present view that chemoreception of plant chemicals is far more complex than the perception of pheromones (Städler, 1980). It has indeed proven to be difficult to relate sensory input to insect behaviour because the sensory information going to the brain is coded in, across fibre patterns which are, as the perceived natural stimuli, rather complex (Dethier, 1973, 1974a). This conclusion is based on well-studied contact chemoreceptors of Lepidoptera larvae which perceive the composition of plant saps during feeding. It seems that the chemistry of the undamaged plant, the plant surface which plays a crucial role in insect-plant relations, is less complex (Chapman, 1977; Klingauf, Nocker-Wenzel & Rottger, 1978). Consequently, it can be argued that the codes of chemoreceptors involved in the perception of undamaged plants, such as olfactory receptors, are less complex too (Städler, 1980). At present the limited available evidence seems to support this hypothesis: so Schoonhoven (1974), Ma & Visser (1978) and Dethier (1980) were able to record clearly different sensory responses to host and non-host plant odours. These studies hold promise that electrophysiological techniques may be powerful tools in the analysis of insect attractants of plant origin. In order to test non-volatile, apolar compounds stimulating contact chemoreceptors the present techniques are less suited (Blaney, 1974; Städler, 1976). This could be an incentive to develop new recording techniques adapted to the analysis of plant compounds and extracts of unknown composition.

14.2.5   General consideration of isolation and identification techniques

All mentioned assays are affected and limited by the variability of two biological "partners": host plant and insect. This fact has to be considered in detail in order to obtain reproducible results. The sources of variability have been discussed recently by Städler (1976) and detailed investigations have been presented by Dethier (1974b), Chapman & Bernays (1977) and Schoonhoven (1976).

A difficulty often encountered in the isolation and identification of active plant compounds has been the dissipation of activity during purification procedures. Different factors can be mentioned which may be involved in this problem such as synergistic, masking (inhibitory) effects between individual fractions or compounds and the change of the extracts due to destruction during fractionation procedures (Hedin, Maxwell & Jenkins, 1974; Städler & Woggon, unpubl.). Further problems such as small amounts of contaminations have been discussed by Greenway *et al.* (1978).

Difficulties during isolation of plant compounds can lead to another approach of investigation which is based on our growing knowledge of the chemotaxonomy of plants (Hegnauer, 1973). Providing the relevant compounds are available the technique of screening of all plant compounds with a high chemotaxonomic value (that means occurring only in the group of host plants of the insect studied) can be very useful (e.g. Schoonhoven, Tramper & Van Drongelen, 1977; Van Drongelen, 1979). Such results may help to design improved isolation techniques which cannot be omitted and which are indispensable for the understanding of the chemical ecology of plant–insect relations.

14.3 CONCLUSIONS ON THE ROLE OF ATTRACTANTS, ARRESTANTS AND STIMULANTS IN INSECT–PLANT RELATIONSHIPS

Recently, several reviews on our present knowledge have been presented: Schoonhoven (1972), Hedin *et al.* (1974), Kogan (1976) and Städler (1976, 1980). All authors came basically to the same conclusions:

1. All known plant chemicals have the potential of influencing insect behaviour by increasing (acting as attractant, arrestant, stimulant) plant attack. (This is true even for alkaloids — Bernays, personal communication.) Allelochemics do play a primary role in most insect–plant relations.

2. Ovipositing females are recognized as the crucial "agents" in most insect–plant relations. Remarkably, we know least about this important first step in plant attack. Reasons for this are our general lack of knowledge and the difficulty of designing assays and of observing oviposition behaviour. Further it is obvious that we know relatively little about the important plant surface chemistry (Chapman, 1977, Klingauf *et al.*, 1978).

3. In no case could conclusive evidence be presented that a single plant compound can explain host selection in an insect. Even the often cited insects associated with crucifers do respond to other chemicals than glucosinolates and isothiocyanates such as flavonolglucosides (Nielsen, Larsen & Sorensen, 1979).
4. No example of a plant–insect relation is known in which all plant compounds involved in host selection could be identified. Obviously for a few insects such as the boll weevil (Hedin, Thompson & Guildner, 1973; Hedin, 1976), the silkworm (Ishkawa, Hirao & Arai, 1969; Kato, 1978) and the locust (Chapman & Bernays, 1977; Cook, 1977) an impressive number of compounds have been identified responsible for different behavioural reactions. But the search for additional important compounds is by no means finished. The aim of all studies is to explain host selection completely in terms of plant chemicals influencing insect behaviour. This appears difficult even in the best studied examples.

Taking into account these conclusions and the possibilities and limitations of present methods, an attempt is made to point out present and future possible applications of our knowledge of the chemical ecology of insect plant interactions.

14.4 APPLICATION FOR PEST CONTROL

14.4.1 Attractants

Examples of host-plant-derived attractants which are used to increase the efficiency and specificity of visual traps in order to monitor the occurrence and density of pest insect populations are listed in Table 14.1. Surprisingly this list is rather small considering the number of pest insects which should be monitored in order to apply the principles of integrated pest management. The reason for this lack of application of plant odours for trapping could be manifold and may lead to suggestions for future research:

1. Apparently only in a few economic pest insects has host selection been studied in detail. For some insects the identified attractants have not yet been applied to the field such as in the case of the codling moth (Sutherland, Wearing & Hutchins, 1977). Obviously more attention should be given to the host selection of pest insects.
2. Plant volatiles may, despite our current beliefs, play only a minor role in the location of the host plant. Instead, visual cues may be much more important even though the literature provides fewer examples for this as has been stressed by Prokopy & Owens (1978). This would mean that

TABLE 14.1 EXAMPLES OF PRACTICAL FIELD APPLICATION OF
HOST-PLANT DERIVED COMPOUNDS FOR TRAPPING OF INSECT
PESTS

Insect species	Plant compound and source	Literature
Pest of cruciferae	Isothiocyanates from seeds of cruciferae	Görnitz (1953, 1957) Feeny *et al.* (1970) Finch *et al.* (1980)
Onion fly Delia (*Hylemya antiqua*)	Propylmercaptan from onions	Matsumoto & Thorsteinson (1968) Städler (unpubl.)
Carrot fly (*Psila rosae*)	Phenylpropanoids from Umbelliferae	Guerin & Städler (1980)
Boll weevil (*Anthonomus grandis*)	Cotton seed oil bait: (+) α-pinene (+) limonene (−)-β-caryophyllene (−)-β-caryophyllene oxide (+)-β-bisabolol	Hedin *et al.* (1973) Hedin (1976)
Lepidoptera	Phenylacetaldehyde from flower of *Araujia sericofera*	Cantelo & Jacobson (1979)
Bark beetles	Terpenes from barks in combination with pheromones	Wood (1977) Payne (1979)

the examples in Table 14.1 are exceptions which seem premature but points to the fact that visual and olfactory stimuli together should be studied in future investigations.

3. The identified chemicals may not be available in sufficient amounts and purity, may not be stable enough or may be too expensive for use in traps designed for practical application.

4. As the examples in Table 14.1 show, some insects are attracted by single pure compounds. However, this may not be the case in other species like the Colorado beetle which is only attracted by a complex mixture of volatiles which seems to be typical of its host plant (Visser *et al.*, 1979). Since pure compounds do seldom occur naturally it can be assumed that complex host-plant odours may be more attractive in general. Such natural mixtures do not have to be a limit for practical application. It can be envisaged that crude plant or seed extracts may be both more attractive and economically more feasible.

At present sex pheromones are much more popular and their application does far outnumber host-plant chemicals as attractants. But since most pheromones available are sex-specific, attracting only males, it is not surprising that it has proved to be difficult to correlate the number of trapped insects with pest population density and crop damage. In this respect traps using attractants originating from host plants hold much more promise as we have been able to show in the case of the carrot fly (Guerin & Städler, 1980, Städler, Freuler & Gfeller, unpubl. OILB report 1978).

14.4.2   Feeding stimulants

The most universal feeding stimulant of plant origin, sucrose, has been widely used in baits and bait sprays. One of the successful applications is the use of bait sprays in combination with the parapheromone methyleugenol against different fruit flies (Chambers, 1977). Several types of commercial baits incorporate sucrose and wheat germ as a feeding stimulant for cut worms, tipulid larvae and male crickets. Protein baits, which do not have to be of plant origin, can act both as attractants and feeding stimulants, the active principle probably being not protein but compounds associated with proteins or products of their decay. Sucrose baits have also been used successfully for the estimation of population densities (Mulkern, Records & Carlson, 1978). Baits may be also ideal agents for pest control based on rural resources and low technological input. Only a few of the potential applications have been exploited yet and it can be assumed that a better knowledge of general feeding stimulants may lead to more effective bait formulations providing useful inexpensive tools for integrated pest control.

14.4.3   Mass trapping

Provided a highly efficient attractant and trap can be found, attempts can be made to reduce pest populations by mass trapping. This approach seems to be viable for some bark beetles using traps and baited trees combining aggregation pheromones with host-plant terpenes (Wood, 1977). In the case of the cabbage fly, Finch (1977) estimated that the efficiency of the present traps would have to be increased by a factor of 5 to 10, in order to achieve control of this insect.

14.4.4   Disguising non-host plants as hosts

Host-plant specific compounds could theoretically be used to attract and release feeding and oviposition on non-host plants naturally protected by deterrents or toxins. Since it seems very unlikely that the host crop will lose its attractivity care would have to be taken to reduce its apparency by using a delayed sowing time or by the application of repellents or deterrents. Future investigations will show if this possibility has any chance of application.

14.4.5   Breeding resistant plants

Breeding of less preferred plants may be accomplished by a reduction of

attractants, arrestants and stimulants. Past and present breeding programmes did and do usually not involve the investigation of basic mechanisms of
plant resistance. Maxwell (1977) mentioned several reasons for our lack of
knowledge about the chemistry of plant resistance: the most important being
that resistance is chemically and by consequence also genetically very
complex. This did favour the "concept" of intentionally ignoring the basic
aspects of resistance and pushing ahead with "success-assured" applied
research. At present no practical chemical assay methods have been developed based on identified attractants or stimulants which could be used as
a primary screening device for detecting less susceptible plants. Despite the
above-mentioned difficulties there exist, however, good reasons to investigate
the chemical basis of resistance as pointed out by Maxwell (1980) and others.
A promising first step to use the content of attractant chemicals in a breeding
programme has been presented by Ellis, Hardman, Crisp, Johnson & Cole
(1976) who explain the preference of the cabbage fly for certain varieties and
developmental stages in terms of their content of mustard oils. In summing up,
it seems that we have to know much more about the basis of plant resistance
and the chemistry of insect–plant relationships in order to apply this to
breeding programmes.

14.5 CONCLUSION

At present the investigation of plant volatiles acting as attractants,
arrestants or stimulants seem to hold most promise for future applied research
because this knowledge may be used to design or improve traps and baits
which are essential for the estimation of populations and the control of some
insects in integrated pest management. However, volatiles are only a part of
the stimuli mediating plant–insect relations. Therefore it is essential that we
try to understand all factors influencing host selection and aggregation on food
sources in order to be able to apply our knowledge to "biorational" ways of
pest control.

14.6 SUMMARY

The influence of attractants, arrestants, feeding and oviposition stimulants
in insect–plant relation is briefly reviewed and two examples (*Manduca sexta*
larvae and *Psila rosae* females) are presented in more detail including some
sensory aspects of these relations. The study of these attractants, arrestants
and stimulants is difficult as plant stimuli are extremely complex. There is

good evidence that insects reacting to plants perceive and are guided by a multitude of compounds (including repellents and deterrents). Plant components in isolation on the other hand may influence very different behavioural responses. This complexity in plant–insect relations is also reflected in the responses of sensory cells to plant stimuli.

Our present limited knowledge about plant chemicals which influence insect behaviour can already be applied for pest control. Several examples are presented for the successful application of plant odour components in traps (insects attacking crucifers, carrot fly, bark beetles). The traps are used in the integrated pest control by setting population thresholds (as estimated by the number of pest insects/trap) for insecticide applications.

In contrast to the application of repellent or deterrent chemicals for masking host plants, no attempts have been made yet to disguise non-hosts as host plants with attractants, arrestants or stimulants. This may be due to the fact that our knowledge on non-volatile stimulants is rather limited even though these compounds may be crucial for host selection in general. As more information is gathered on host selection and the natural chemicals involved, we may be able to apply it not only for the development of species-specific traps and the breeding of resistant host-plant varieties, but also for novel applications.

14.7 ACKNOWLEDGEMENTS

Thanks are due to the convenors for the organization of the symposium, Drs. E. A. Bernays, P. Feeny and T. Jermy for discussions, Dr. P. Guerin for improving and Michelle Flury for typing this manuscript.

14.8 REFERENCES

Arn H. & Cleer J. S. (1971) A double-label choice-test for the simultaneous determination of diet preference and ingestion by the aphid *Amphorophora agathonica*. *Ent. exp. appl.* **14**, 377–387.
Arn H., Städler E. & Rauscher S. (1975) The electroantennographic detector — a selective and sensitive tool in the gas chromatographic analysis of insect pheromones. *Z. Naturforsch.* **30c**, 722–725.
Berenbaum M. (1978) Toxicity of a furanocoumarin to armyworms: a case of biosynthetic escape from insect herbivores. *Science* **201**, 532–534.
Blaney W. M. (1974) Electrophysiological responses of the terminal sensilla on the maxillary palps of *Locusta migratoria* (L.) to some electrolytes and non-electrolytes. *J. exp. biol.* **60**, 275–293.
Cantelo W. W. & Jacobson M. (1979) Phenylacetaldehyde attracts moths to bladder flower and to blacklight traps. *Environ. Ent.* **8**, 444–447.
Chambers D. L. (1977) Attractants for fruit fly survey and control. In *Chemical Control of Insect Behaviour* (eds. Shorey H. H. & McKelvey J. Jr.), pp. 327–344. John Wiley & Sons, N.Y.

Chapman R. F. (1977) The role of the leaf surface in food selection by Acridids and other insects. *Coll. Int. CNRS Comportement des insectes et milieu trophique* **265**, 133–149.

Chapman R. F. & Bernays E. A. (1977) The chemical resistance of plants to insect attack. In *Produits naturels et la protection des plantes* (ed. Marini-Bettolo G. B.), pp. 603–633. Pont. Acad. Sci. Scripta varia 41.

Cook A. G. (1976) A critical review of the methodology and interpretation of experiments designed to assay the phagostimulatory activity of chemicals to phytophagous insects. *Symp. Biol. Hung.* **16**, 47–54.

Cook A. G. (1977) Nutrient chemicals as phagostimulants for *Locusta migratoria* L. *Ecol. Ent.* **2**, 113–121.

Dethier V. G. (1973) Electrophysiological studies of gustation in Lepidopterous larvae. II. Taste spectra in relation to food plant discrimination. *J. comp. Physiol.* **82**, 103–134.

Dethier V. G. (1974a) The specificity of the labellar chemoreceptors of the blowfly and the response to natural foods. *J. Insect Physiol.* **20**, 1859–1869.

Dethier V. G. (1974b) Sensory input and the inconstant fly. *Experimental Analysis of Insect Behaviour* (ed. Barton Browne L.), pp. 21–31. Springer, Berlin.

Dethier V. G. (1980) Responses of some olfactory receptors of the eastern tent caterpillar (*Malacosoma americanum*) to leaves. *J. chem. Ecol.* **6**, 213–220.

Dethier V. G., Barton Browne L. & Smith C. N. (1960) The designation of chemicals in terms of the responses they elicit from insects. *J. econ. Ent.* **53**, 134–136.

Drongelen Van W. (1979) Contact chemoreception of host plant specific chemicals in larvae of various *Yponomeuta* species (Lepidoptera). *J. comp. Physiol.* **134**, 265–279.

Ellis P. R., Hardman J. A., Crisp P., Johnson A. G. & Cole R. A. (1976) Resistance of brassicas and radish to cabbage root fly. *Ann. Rep. Natl. Vegetable Res. Sta. for 1975* **26**, 88–90.

Feeny P., Paauwe K. L. & Demong N. J. (1970) Flea beetles and mustard oils: host plant specificity of *Phyllotreta cruciferae* and *P. striolata* adults (Coleoptera: Chrysomelidae). *Ann. Ent. Soc. Am.* **63**, 832–841.

Finch S. (1977) Effect of secondary plant substances on host-plant selection by the cabbage root fly. *Coll. Int. CNRS. Comportement des insectes et milieu trophique* **265**, 251–267.

Finch S. (1978) Volatile plant chemicals and their effect on host plant finding by the cabbage root fly (*Delia brassicae*). *Ent. exp. appl.* **24**, 350–359.

Finch S., Freuler J. & Städler E. (1980) Trapping *Hylemya brassicae* adults. *IOBC/WPRS Bull.* III/1, 11–17.

Görnitz K. (1953) Untersuchungen uber in Kruziferen enthaltene Insekten-Attraktivstoffe. *Nachrichtenbl. Dtsch. Pflanzenschutzd.* N.F. **7**, 81–95.

Görnitz K. (1957) Ueber die Reaktion einiger an Cruciferen lebenden Insektenarten auf attraktive Duft-und Farbreize. *Ber. 100 jahrfeier Dtsch. Ent. Ges. Berlin* 188–198.

Greenway A. R., Griffiths D. C. & Lloyd S. L. (1978) Response of *Myzus persicae* to components of aphid extracts and to carboxylic acids. *Ent. exp. appl.* **24**, 169–174.

Guerin P. & Städler E. (1980) Carrot fly olfaction: behavioural and electrophysiological investigations. In *Olfaction and Taste* VII (ed. van der Starre, 4), p. 95. Information Retrieval, London.

Haisch A. & Levinson H. Z. (1980) Influences of fruit volatiles and coloration on oviposition of the cherry fruit fly. *Naturwiss.* **67**, 44.

Hansen K. (1978) Insect chemoreception. In *Taxis and Behaviour* (*Receptors and Recognition*, Ser. B, Vol. 5) (ed. Hazelbauer G. L.), pp. 232–292. Chapman & Hall, London.

Hawkes C. & Coaker T. H. (1979) Factors affecting the behavioural responses of the adult cabbage root fly, *Delia brassicae*, to host plant odour. *Ent. exp. appl.* **25**, 45–58.

Hedin P. A. (1976) Seasonal variations in the emission of volatiles by cotton plants growing in the field. *Environ. Ent.* **5**, 1234–1238.

Hedin P. A., Maxwell F. G. & Jenkins J. N. (1974) Insect plant attractants, feeding stimulants, repellents, deterrents and other related factors affecting insect behaviour. In *Proc. Summer Institute on Biological Control of Plants, Insects and Diseases* (eds. Maxwell F. G. & Harris F. A.), pp. 494–527. Univ. Mississippi, Jackson.

Hedin P. A., Thompson A. C. & Gueldner R. C. (1973) The boll weevil-cotton plant complex. *Toxicol. Environ. Chem. Rev.* **1**, 291–351.

Heinicke D. H. K. (1978) Untersuchungen uber den Einfluss von Pflanzeninhalts-stoffen auf das

Verhalten der Fritfliege (*Oscinella frit* L.) und ihrer Maden. Dissertation Hohenheim BRD, 219 pp.

Hegnauer R. (1973) *Chemotaxonomie der Pflanzen.* Birkhäuser, Basel.

Ishikawa S., Hirao T. & Arai N. (1969) Chemosensory basis of host plant selection in the silkworm. *Ent. exp. appl.* **12**, 544–554.

Jones C. G. (1979) An automatic feeding detector (AFD) for use in insect behaviour studies. *Ent. exp. appl.* **25**, 112–115.

Kato M. (1978) Phenols as indispensible components of the synthetic diet of the silkworm, *Bombyx mori. Ent. exp. appl.* **24**, 485–490.

Kennedy J. S. (1977a) Olfactory responses to distant plants and other odour sources. In *Chemical control of Insect Behaviour* (eds. Shorey H. H. & McKelvey J. Jr.), pp. 67–91. John Wiley & Sons, N.Y.

Kennedy J. S. (1977) Behaviourally discriminating assays of attractants and repellents. In *Chemical Control of Insect Behaviour* (eds. Shorey H. H. & McKelvey J. Jr.), pp. 215–229. John Wiley & Sons, N.Y.

Kennedy J. S. (1978) The concepts of olfactory "arrestment" and "attraction". *Physiol. Ent.* **3**, 91–98.

Kennedy J. S. & Marsh D. (1974) Pheromone-regulated anemotaxis in flying moths. *Science* **184**, 999–1001.

Klingauf F., Nöcker-Wenzel K. & Röttger U. (1978) Die Rolle peripherer Pflanzenwachse für den Befall durch phytophage Insekten. *Z. Pflkrankh. Pflschutz* **85**, 228–237.

Kogan M. (1976) The role of chemical factors in insect/plant relationships. *Procl XVth Int. Congr. Entomology*, 211–227.

Ma W.-C. & Visser J. H. (1978) Single unit analysis of odour quality coding by the olfactory antennal receptor system of the Colorado beetle. *Ent. ecp. appl.* **24**, 520–533.

Matsumoto Y. & Thorsteinson A. J. (1968) Effect of organic sulfur compounds on oviposition in onion maggot *Hylemya antiqua* Meigen (Diptera: Anthomyiidae). *Appl. Ent. Zool.* **3**, 5–12.

Maxwell F. G. (1977) Host-plant resistance to insects–chemical relationships. In *Chemical Control of Insect Behaviour* (eds. Shorey H. H. & McKelvey J. Jr.), pp. 299–304. John Wiley & Sons, N.Y.

Maxwell F. G. (1980) Future opportunities and directions. In *Breeding Plants Resistant to Insects* (eds. Maxwell F. G. & Jennings P. R.), pp. 535–538. John Wiley & Sons, N.Y.

Mulkern G. B., Records J. C. Jr. & Carlson R. B. (1978) Attractants and phagostimulants used for control and estimation of grasshopper populations. *Ent. exp. appl.* **24**, 550–561.

Nair K. S. S. & McEwen F. L. (1976) Host selection by the adult cabbage maggot, *Hylemya brassicae* (Diptera: Anthomyiidae): Effect of glucosinolates and common nutrients on oviposition. *Can. Ent.* **108**, 1021–1030.

Nielsen J. K., Larsen L. M. & Sorensen H. (1979) Host plant selection of the horseradish flea beetle *Phyllotreata armoraciae* (Coleoptera: Chrysomelidae): Identification of two flavonol glycosides stimulating feeding in combination with glycosinolates. *Ent. exp. appl.* **26**, 40–48.

Payne T. L. (1979) Pheromone and host odour perception in bark beetles. In *Neurotoxicology of Insecticides and Pheromones* (ed. Narahashi T.), pp. 27–57. Plenum N.Y.

Prokopy R. J. & Boller E. F. (1970) Artificial egging system for the European cherry fruit fly. *J. econ. Ent.* **63**, 1413–1417.

Prokopy R. J. & Owens E. D. (1978) Visual generalist — visual specialist phytophagous insects: Host selection behaviour and application to management. *Ent. exp. appl.* **24**, 609–620.

Röttger U. (1978) Blattwachskomponenten als Schlüsselreize der Eiablagen von *Pegomya betae* Curt. *Mitt. Dtsch. Ges. allg. angew. Ent.* **1**, 22–28.

Röttger U. (1979) Untersuchungen zur Wirtswahl der Rübenfliege *Pegomya betae* Curt. (Diptera, Anthomyiidae). I. Olfaktorische Orientierung zur Wirts-pflanze. *Z. ang. Ent.* **87**, 337–348.

Schoonhoven L. M. (1972) Secondary plant substances and insects. In *Structural and Functional Aspects of Phytochemistry* (eds. Runeckles V. C. & Tso T. C.). *Rec. Adv. Phytochem.* **5**, 197–224.

Schoonhoven L. M. (1974) Studies on the shootborer *Hypsipyla grandella* (Zeller) (Lep: Pyralidae). XXIII. Electroantennogram (EAG) as a tool in the analysis of insect attractants. *Turrialba* **24**, 24–28.

Schoonhoven L. M., (1976) On the variability of chemosensory information. *Symp. Biol. Hung.* **16**, 261–266.
Schoonhoven L. M., Tramper N. M. & Van Drongelen W. (1977) Functional diversity in gustatory receptors in some closely related *Yponomeuta* species (Lep.). *Neth. J. Zool.* **27**, 287–291.
Städler E. (1971/1972) Ueber die Orientierung und das Wirtswahlverhalten der Möhrenfliege, *Psila rosae* F. (Diptera: Psilidae). *Z. angew. Ent.* **69**, 425–438; **70**, 29–61.
Städler E. (1974) Host plant stimuli affecting oviposition behaviour of the eastern spruce budworm. *Ent. exp. appl.* **17**, 176–188.
Stadler E. (1976) Sensory aspects of insect plant interactions. *Proc. XVth Int. Congr. Entomology* 228–248.
Städler E. (1977) Host selection and chemoreception in the carrot rust fly (*Psila rosae* F., Diptera Psilidae): Extraction and isolation of oviposition stimulants and their perception by the female. *Coll. Int. CNRS Comportement des insectes et milieu trophique* **265**, 357–372.
Städler E. (1980) Chemoreception in arthropods: sensory physiology and ecological chemistry. In *Animal and Environmental Fitness* (ed. Gilles R.), pp. 223–241. Pergamon, London.
Städler E. & Hanson F. E. (1976) Influence of induction of host preference on chemoreception of *Manduca sexta*: Behavioural and electrophysiological studies. *Symp. Biol. Hung.* **16**, 267–273.
Sutherland O. R. W., Wearing C. H. & Hutchins R. F. N. (1977) Production of α-farnesene, an attractant and oviposition stimulant for codling moth, by developing fruit of ten varieties of apple. *J. chem. Ecol.* **3**, 625–631.
Swailes G. E. (1967) A substrate for oviposition by the cabbage maggot in laboratory cultures. *J. Econ. Ent.* **60**, 619–620.
Tjallingii W. F. (1978) Electronic recording of penetration behaviour by aphids. *Ent. exp. Appl* **24**, 721–730.
Vernon R. S., Bordon J. H., Pierce Jr. H. D. & Oehlschlager A. C. (1977) Host selection by *Hylemya antiqua* laboratory bioassay and method of obtaining host volatiles. *J. chem. Ecol.* **3**, 359–368.
Visser J. H., Van Straten S. & Maarse H. (1979) Isolation and identification of volatiles in the foliage of potato, *Solanum tuberosum*, a host plant of the colorado beetle, *Leptinotarsa decemlineata*. *J. chem. Ecol.* **5**, 11–23.
Wood D. L. (1977) Manipulation of forest insect pests. In *Chemical Control of Insect Behaviour* (eds. Shorey H. H. & McKelvey J. Jr.), pp. 369–384. John Wiley & Sons, N.Y.
Woodhead S. & Bernays E. A. (1978) The chemical basis of resistance of *Sorghum bicolor* to attack by *Locusta migratoria*. *Ent. exp. appl.* **24**, 123–144.
Zohren E. (1968) Laboruntersuchungen zu Massenzucht, Lebensweise, Eiablage und Eiablage-verhalten der Kohlfliege *Chortophila brassicae* Bouché (Diptera, Anthomyiidae). *Z. angew. Ent.* **62**, 139–188.

14.9 DISCUSSION

Bowers: Is the *trans*-asarone a component of the carrot?
Städler: Yes.
Bowers: Does it have any toxicity to the fly? Have you tried that directly?
Städler: We tried it for olfactory response but we didn't apply it directly to the fly. Not all of these phenyl propanoids occur in the carrot, but the carrot fly larvae seem able to cope with them.
Bowers: Is the *cis*-isomer present?
Städler: I do not know, but I am sure it is in the literature. In EAG studies the *trans*-isomer is much more active than the *cis*.
Bowers: Some work has been done in India with the *cis*-isomer as an insect growth regulator.

Feeny: Is it your impression that the tarsal chemoreceptors on the tarsal hairs do not actually penetrate through the leaf surface? Are they responding to chemicals on the surface of the leaf only? I have never looked at this aspect.

Städler: I think it would be the case because you can bend the tarsal hairs very easily. I don't see how they could go through the waxy layer. The leaf surface has been looked at by scanning electron microscopy to see if marks due to tapping or scraping were visible. One was barely able to see the little marks of the hooks on the tarsae, which are much stronger than the tarsal hairs.

Feeny: Logically, then, the glucosinolate must be on the leaf surface?

Städler: Yes, we were able to make extracts, but even if you do it very carefully you are never sure that you did not damage the leaf. So I was never completely convinced of my experiments. In oviposition bioassay, both water and methanol extracts were active.

Feeny: I am prepared to believe it's on the leaf surface, but how does a polar compound like that adhere to the waxy surface of leaves particularly in Europe where it rains a lot?

Städler: Well, one could consider that the compounds are washed off the upper leaf surface but are retained on the lower surface. It would be interesting to find out the concentration of these compounds on the leaf surface in relation to the sensitivitity of the chemoreceptors which are very sensitive. But this would, in fact, be very difficult to do.

Bernays: We have found that cinergrin is present on the surface of cabbage leaves in up to 5% of the total concentration. So there is plenty there.

Feeny: Was that in a long dry spell?

Graham-Bryce: How was that measured?

Bernays: We had two techniques but the most important one was by surface extraction with a 3-second dip in chloroform.

Graham-Bryce: Chloroform is a very severe extractant.

Bernays: There was no chlorophyll extracted.

Städler: It is a standard technique.

Graham-Bryce: Yes. I am aware that it's a standard technique but of the solvents used, chloroform is a very severe one and even with a very short dip you can't be sure that you're not washing out as well as stripping off.

Bernays: Because we are so interested in this problem we now have a student doing EM work and she has been looking at the surface structure before and after washing and certainly the cell membranes on the epidermis are not disrupted by 10-sec dips.

Feeny: May I ask either Dr. Städler or Dr. Bernays or both of them, what seems to me to be a very important question. To what extent do you think secondary compounds are actually physically or chemically bound to the outside of the leaf and to what extent are they a sort of sap that gets washed off and replaced? Is it possible that there are "islands" on the surface that can bind polar compounds for short periods of time?

Städler: It could be because the flies are moving over the leaf surface a good deal and they cover a lot of surface area and during that time would come into close contact with the stomata or areas around them where you could imagine there are more compounds present than the other areas of the leaf surface.

Bernays: We have shown in wheat leaves that there is a fair amount of fluctuation and the water-soluble compounds are in much higher concentration in the distal half of the leaves on the surface than in the proximal half. Certainly it varies.

Whitehead: For armyworm there are preferred areas of attack on the maize leaf; they tend to eat certain areas in preference to others. Perhaps there are "islands" that can be found.

Graham-Bryce: I have a question which I think relates to the prospects for practical application. Perhaps I misunderstood you, but in talking about the bioassay, particularly of the chemoreceptors, I think you stressed the importance of using an individual insect. Saying that the actual coding and not just the response itself could vary between individuals.

Städler: It's just an hypothesis to explain some of our failures.

Graham-Bryce: If that's correct, I presume it has quite interesting implications because it seems to me not like the normal quantitative difference in response you would get, say, to a pesticide, but a qualitative difference which would mean you need different mixtures to get a given response.

Städler: It's my experience that you do come across enormous differences in response, which is why I devised this hypothesis. But it still remains to be tested.

Bernays: We had some interesting results in locusts, because in bioassays for antifeedants we use very large numbers and, for example, one locust in a 100 would not be deterred by a quite strong deterrent. However, the next day the same locust may react normally. It's not like a fixed response.

Graham-Bryce: It seems to depend on what one might call the "Mood" or something like that. But could this also relate to the fractionation question in the bioassay?

Städler: So far our bioassay has always worked with a population of 100 females. So I have an average, but I sometimes wonder if I should not switch my technique.

Feeny: If you had a field of crucifers, say cabbages, that was subject to infestation by cabbage fly, what would happen if you distributed across the whole field in some form or other, allyl isothiocyanate?

Städler: Probably you would attract many more flies than normal. Stan Finch tried something like that with cabbage fly. He tried putting out as many traps as had been used in the lab, to see how effective it would be for mass trapping. He used 100 traps per hectare^{-1} to show significant trapping. This number is not very good for practical application but he is optimistic that he will find an attractant more potent than allylisothiocyanate. I am less optimistic I must say. Now in your experiment there is evidence these flies are attracted over very large distances, and this would mean you would except higher infestation.

Feeny: In fact we found this sometime ago, but the results were never published because they were so negative. One of my students did this — she put out various colour treatments, conventional pesticides, blanks, mustard oils, etc., in various combinations. The heaviest damage was to plots which had traps containing mustard oils. In other words, the traps were doing a good job trapping the insects but they were attracting so many additional insects that the plots were devastated. At that point I switched projects!

CHAPTER 15

Antifeedants in Crop Pest Management

E. A. BERNAYS

Centre for Overseas Pest Research, College House, Wrights Lane,
Kensington, London W8 5SJ, England

CONTENTS

15.1 Introduction   259
15.2 The Natural Occurrence of Antifeedants   259
15.3 The Search for Antifeedants   260
    15.3.1 The initial bioassay   262
    15.3.2 Plant factors   264
15.4 Realistic Field Trials   265
    15.4.1 Field persistence of sprayed antifeedants   266
    15.4.2 Logistics   267
15.5 Summary   268
15.6 References   268
15.7 Discussion   269

15.1 INTRODUCTION

The use of antifeedants in pest-management programmes has enormous intuitive appeal. It satisfies the need to protect specific crops while avoiding damage to non-target organisms, so that the potential value is very great, but so too are the pitfalls. These problems must be thoroughly examined so that the practical side may be discussed in a climate of reality.

15.2 THE NATURAL OCCURRENCE OF ANTIFEEDANTS

It is now widely recognized that host-plant selection by phytophagous insects is based to a large extent on the distribution of repellent and deterrent chemicals in the plant kingdom. Thus, for example, Jermy (1966) showed that for a number of different kinds of insects, with over 100 plant species, those not

259

TABLE 15.1 PALATABILITY OF 100 PLANT SPECIES TO EIGHT
PHYTOPHAGOUS INSECTS AND THE PERCENTAGE OF THOSE WHICH
CONTAIN FEEDING INHIBITORS (JERMY, 1966)

Insects	A % plants unacceptable	B % of A containing feeding inhibitors
Coleoptera		
Tanymecus dilaticollis	57	75
Phyllobius oblongus	80	91
Leptinotarsa decemlineata	97	94
Cassida nebulosa	96	97
Phytodecta fornicata	98	97
Lepidoptera		
Hyphantria cunea	72	78
Plutella maculipennis	98	95
Hymenoptera		
Athalia rosae	80	97

eaten, or eaten only in small amounts usually contained deterrents (Table 15.1). Similar results have been obtained at the Centre for Overseas Pest Research in another very extensive survey (Bernays, Chapman, McDonald & Salter, 1976). Furthermore, approximately 200 so-called secondary plant chemicals have now been tested for antifeedant effects on grasshoppers and locusts. The majority of these reduce feeding in the graminivorous species *Locusta migratoria* while many fewer affect the polyphagous *Schistocerca gregaria* (Table 15.2). As a general rule polyphagous species are less sensitive to feeding deterrents.

There are now many examples of particular insect species being prevented from feeding by a chemical or variety of chemicals derived from plants. This is no longer surprising and it illustrates that the natural array of feeding deterrents is large and likely to be enormous. More and more people are becoming involved, chemists identifying more and more chemicals and entomologists describing more and more antifeedant effects, but where is this leading in practical terms? Can these results really be translated into methods of crop protection? For the most part we do not know. The extraordinary fact is that, although antifeedant spraying of natural products were tried in the United States in the 1930s, we are scarcely any further advanced now in practical terms, though so many chemicals have been identified. Why have we not advanced? We should examine the essential steps for progress in this field, in an attempt to find the faults and problems, and thereby to seek a remedy.

15.3   THE SEARCH FOR ANTIFEEDANTS

Some of the first discoveries of antifeedant effects were simply made by

TABLE 15.2 THE EFFECT OF EIGHT DIFFERENT PLANT COMPOUNDS
ON THE FEEDING BEHAVIOUR OF FIVE DIFFERENT INSECT SPECIES
WHEN PRESENTED IN APPROXIMATELY NATURAL
CONCENTRATIONS

Compound	Insect species				
	L.m.	L.d.	D.k.	P.b.	S.g.
Tannic acid	—	—	—	0	0
Sinigrin	—	—	—	+	0
Salicin	—	—	0	0	+
Caffeine	—	—	—	—	0
Conessine	—	0	—	—	0
Atropine	—	0	0	—	0
Amygdalin	0	0	—	0	0
Azadirachtin	—	—	0	—	—

L.m. *Locusta migratoria* (Bernays & Chapman, 1977) (Bernays, unpub.).
L.d. *Leptinotarsa decemlineata* (Schoonhoven & Jermy, 1977).
D.k. *Dysdercus koenigii* (Schoonhoven & Derksen-Koppers, 1973).
P.b. *Pieris brassicae* (Ma, 1969) (Schoonhoven & Jermy, 1977).
S.g. *Schistocerca gregaria* (Bernays, unpub.).
— = deterrent.
0 = no effect.
+ = phagostimulatory.

chance: thus organo-metal fungicides were found by a number of people to reduce insect damage (Ascher & Rones, 1964) and a few insecticides have demonstrable antifeedant effects (Jermy & Matolcsy, 1967). Such chemicals have obvious disadvantages, and the approach generally is to search for very resistant plants, test extracts of these, and then examine the extracts for active chemicals. In 1932 Metzger & Grant tested 500 such chemicals against *Popillia japonica* but their results were not very encouraging. On the other hand, Pradhan *et al.* (1962) drew attention to the value of extracts of neem tree, *Azadirachta indica*, in preventing feeding by the desert locust. It has been the best find in the field of naturally occurring antifeedants so far. It is effective against a large number of insects, but particularly the desert locust. Using the same approach, bigger efforts are now being put into finding active secondary plant chemicals.

Even with this somewhat empirical behavioural approach we should be clear exactly what the search is for. Chapman (1974) and Schoonhoven & Jermy (1977) consider that searching for single antifeedant compounds against a large number of pests would be fruitless because of the interspecific variability in responses to secondary plant compounds. Such variability is illustrated in Table 15.2 with a few selected compounds and insects. There are very different spectra of antifeedant compounds for the four different oligophagous species, while the polyphagous *S. gregaria* is insensitive except to the highly deterrent azadirachtin.

TABLE 15.3 THE EFFECT OF A VARIETY OF SECONDARY PLANT
COMPOUNDS ON FEEDING BEHAVIOUR OF *SCHISTOCERCA GREGARIA*
AND *LOCUSTA MIGRATORIA*.[a] EFFECTS ARE RECORDED AT
APPROXIMATELY NATURAL CONCENTRATIONS[b]

Chemical group	*Schistocerca gregaria*			*Locusta migratoria*		
	+	0	—	+	0	—
N-containing compounds	26	64	10	2	48	50
Mono- and sesquiterpenes	0	90	10	0	15	85
Triterpenoids	22	61	17	0	18	82
Organic acids	55	45	0	40	40	20
Phenolics	27	63	10	0	48	52

[a]Bernays & Chapman (1977), Bernays (unpub.).
[b]Based on over 250 results.
The table shows the percentage of compounds in each chemical group which are phagostimulatory (+), without effect (0), or deterrent (—).

Such interspecific differences have been shown for many more species now
and do point to the need to search selectively for specific antifeedants against a
particular pest. These results, and those of others such as All & Benjamin
(1976), indicate that specific feeders, that is to say pests with a limited host
range, give more profitable results. With the two contrasted locust species in
Table 15.3, the number of effective antifeedants for the restricted feeder
L. migratoria is more than 4 times greater than for the polyphagous feeder
S. gregaria. Moreover, the chemicals are generally effective at much lower
concentration. Another interesting feature is that some compounds deterrent
to *L. migratoria* are actually phagostimulants for *S. gregaria*. However, the
anomaly of azadirachtin — a uniquely potent antifeedant against *S. gregaria* —
means that such polyphagous species should not be neglected.

The search for effective compounds might be accelerated if attention is
concentrated on particular chemical groups. The results from Table 15.2
suggest that the terpenoid groups may be a profitable area in which to
concentrate. Azadirachtin is triterpenoid and it has been shown recently that
the neem tree also has more than twelve other related compounds, at least two
of which have antifeedant activity against some insects (Warthen, 1979). It
may be noted that many of the antifeedants identified by Munakata (1977)
against *Spodoptera litura* are diterpenoid. Again, the array of antifeedant
sesquiterpene lactones in the Compositae (Mabry & Gill, 1979) suggests that
this may be a fruitful chemical group to concentrate on.

So far, our knowledge is rather too fragmentary to pick out narrow chemical
groups for special attention, but it could in the future be a useful approach, as
it has been with the pyrethroid insecticides.

15.3.1   The initial bioassay

However well controlled the laboratory bioassay may be, it is best

TABLE 15.4. THE DETERRENT EFFECT (AS % INHIBITION OF FEEDING)
ON *LOCUSTA MIGRATORIA* OF THREE DIFFERENT PLANT CHEMICALS,
TESTED ON EITHER WHEAT FLOUR WAFERS (BERNAYS & CHAPMAN,
1977) OR ON WHEAT LEAVES

Chemical and concentration (% dry weight)	Per cent inhibition on feeding compared with untreated	
	Wafers	Leaves
Halostachine 0.025	98	48
Perloline 0.25	76	25
Tannic acid 5	64	4

performed with the test chemical applied to a host plant. This is so because in most cases we do not know the full importance of the host-plant phagostimulants and other factors which interact, and these may override inhibitory factors rather more strongly than do even the best artificial diets. For example, three chemicals tested against *L. migratoria* were much less effective on the host plant than on high nutrient wheat flour wafers (Table 15.4).

Choice tests are not relevant except with a mobile insect, although this is the commonest bioassay method. When an insect can choose either control or treated leaves, it will never become starved, with the result that chemicals under test will appear more effective than they would if there were no choice (Bernays & Chapman, 1975). A similar problem arises with very small field experiments, where even a slight non-preference may cause the insect to move, if it can, onto an adjacent control plant, grossly enhancing the effect of an antifeedant. Consideration should also be given to carrying out bioassays on antifeedants applied to growing plants, since cutting leaves can greatly alter palatability (Bernays, Chapman, Caffery, Modder & Leather, 1977), and for realistic tests, some food deprivation should be imposed. A shortage of food and particularly of water will lead to the breakdown of antifeedant effects in some cases, and it is chemicals which remain effective under such conditions, which could then qualify for field testing.

There is the problem of habituation of the insect to the deterrent chemical as discussed by Schoonhoven & Jermy (1977). After careful experiments with inorganic inhibitors, these authors considered that this phenomenon was unlikely to be important. On the other hand, Gill (1972) has shown that *S. gregaria* exposed to azadirachtin for 9 days on an artificial medium were less deterred from feeding at the end even though given uncontaminated green food for 4 h each day, so that they were not starved (Table 15.5). The possibility of adaptation or habituation is clearly very important.

TABLE 15.5 HABITUATION OF ADULT *SCHISTOCERCA GREGARIA* TO
THE DETERRENT ACTION OF AZADIRACHTIN (GILL, 1972)

	Per cent inhibition of feeding on artificial medium	
	8×10^{-2} mg cm^{-2}	1.6×10^{-1} mg cm^{-2}
Day 1	94	99
2	77	99
3	83	98
4	10	89
5	33	75
6	39	75
7	8	33
8	0.5	34
9	29	43

Electrophysiological studies indicate that some chemicals such as heavy metals and the recently isolated natural product warburganol, seriously affect the taste sensilla (Chapman, 1974; Ma, 1977). Habituation may occur less readily on these compounds and a neurophysiological approach to this problem might be most productive.

15.3.2 Plant factors

Phytotoxicity is a potential problem. Early testing could save a lot of wasted effort, if the antifeedant should be phytotoxic to the important crop. Some plants are particularly sensitive to organo-metals but little is known about the effects of natural products in this sphere. We are unlikely to avoid this problem since allelopathic effects have been found in a number of ecological situations, where a secondary plant chemical leached from one species adversely affects growth of another (Muller & Chou, 1972). Phenolics and monoterpenoids have so far been implicated, but it is likely that other chemicals will be deleterious, since the plant producers themselves usually sequester such materials in vacuoles, latex, hairs or other special glands, or on the surface wax (Muller, 1976), away from the processes of primary metabolism. If the materials are not sequestered, special mechanisms of tolerance are required (Hegnauer, 1976).

Related to this is the need for sprayed antifeedants to be taken up by the plants, as pointed out by others in this field. If they are not systemic, then new growth becomes extra-vulnerable, since it provides the insects with a palatable food close at hand; thus the use of the antifeedant may be limited to periods of very slow plant growth. On the other hand, if a foreign, biologically active natural product is systemic, it has a greater chance of being phytotoxic or being exposed to biochemical alteration by the plant. Two reports of systemic uptake without phytotoxicity relate to neem (Gill & Lewis, 1971;

Radwanski, 1977) although their results have proved difficult to repeat. Perhaps the plant biochemists could give pointers to kinds of chemicals most likely to be taken up by the plant but not altered.

15.4 REALISTIC FIELD TRIALS

While chemists in numerous laboratories put effort into the task of separating, identifying and studying compounds involved, entomologists test more insect species but tend to lack the vision or the resources to run meaningful field trials. Without such field trials, the bioassays have little value. The problem, of course, is that field trials are difficult and expensive and need extremely careful planning. Extracts of the neem tree *Azadirachta indica* have been shown in the laboratory to be deterrent to over fifty species of phytophagous insects (Warthen, 1979), but field trials with convincing results are hard to find. In seven cases cited by Ketkar (1976), there was no effect on the pests concerned, while in a number of other cases the positive effect was either marginal, undocumented, anecdotal or based on an unreal situation such as single potted plants within field cages.

Ladd, Jacobson & Buvill (1978), carried out tests of neem seed extracts against Japanese beetle (*Popillia japonica*) using plots consisting of four to five plants. The results were good with little damage to the treated plants, and totally destroyed control plants. The beetles were very mobile and of course quickly moved off the treated plants onto the control ones, but what would have been the result if a large area of the vegetation had been treated? Some feeding occurred on the treated plants and if no alternative food had been readily available, would the crop have been protected? Certainly deterrency is to some extent reduced by starvation, and this must be considered in the experimental design. The trial reported is thus a field bioassay, but not a convincing field trial. Unless we eventually test areas as large as those we may wish to protect, we may get results which are more encouraging than the real situation would give, and although large-scale trials are so much more costly, it could be cost effective in the long run to carry them out at an early stage. The more mobile the insect, the larger the test plots should be.

A further requirement for realistic trials is an adequate pest population at the time of trials. This may seem obvious, but a fall in pest incidence at the time the trial was mounted is one of the reasons for some ambiguous results to date. The only critical, larger scale and well-documented field trials are those done with the organo-tins, for example against the Colorado potato beetle in Europe (Jermy, 1961; Murbach, 1967), against *Neodiprion* species by All & Benjamin (1976) and against *Spodoptera litura* by Kamel, Mitri, Abo & Zake (1970). Unfortunately organo-metals have many disadvantages but the work was extremely important: sprayed antifeedants could protect a crop.

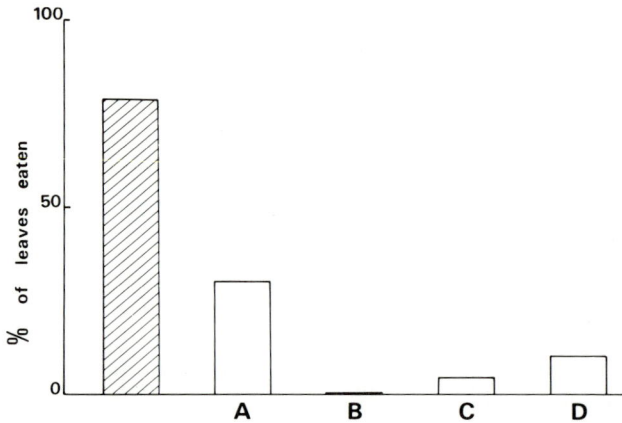

FIG. 15.1. A comparison of feeding on leaves of cassava by *Zonocerus variegatus* in the laboratory, after treatment with different amounts of neem extract. Hatched = control untreated leaves, A = 0.2% dry plant material in water, laboratory dipped and tested with starved insects, B = as A but with 2% dry material, C = 2% field sprayed 6 days before testing, D = 2% field sprayed 12 days before testing.

15.4.1   Field persistence of sprayed antifeedants

Pradhan *et al.* (1962) claim that crops would be protected from attack by desert locust using neem extract sprays, which last 2 weeks if there is no rain. This problem of washoff is not the only one, as there have been many failed field trials in suitable weather conditions — most of them never fully reported. In the experiments by Ladd *et al.* with Japanese beetle (1978), laboratory feeding tests showed much greater success 1 week after spraying than occurred in field trials. For the latter, spraying every 3 days was necessary. Similarly, in COPR trials with *Zonocerus variegatus* in Nigeria, field-sprayed leaves were fairly effective after 12 days when brought in for laboratory testing (Fig. 15.1) but field trials gave much less protection. A difference was found between sprayed and unsprayed treatments at the beginning of invasion, but the sprayed treatment suffered almost as much damage 2 weeks later in spite of three more spray applications (Fig. 15.2).

In these examples, the problem is not just a matter of chemical instability, since leaves sprayed and exposed in the field are effectively protected when brought into the laboratory; rather there must be some feature of the insect in the field environment which makes it more ready to tolerate the unpalatable. Perhaps there is an increased water requirement for the insect under field conditions, or perhaps there is an important behavioural component such as social facilitation of feeding, or perhaps there might be habituation to the antifeedant chemicals. We do not know what the cause of the discrepancy is

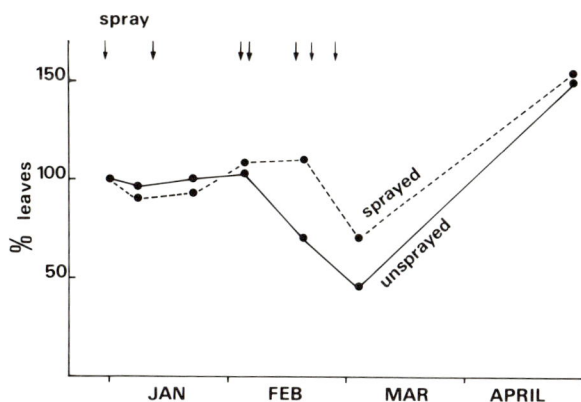

Fɪɢ. 15.2. Defoliation (as per cent of leaves present at the beginning of spraying) of cassava by *Zonocerus variegatus* on plots which were untreated or sprayed with 2% neem extract, in Nigeria. Arrows represent time of spraying, invasion of the insects occurred at the end of January, 1976.

but it must be acknowledged and it must be investigated, if we are to overcome it.

15.4.2 Logistics

If a natural product is shown to be effective in large-scale field trials, is it then practical to use it? Does one need forests of producer plants? Can the material concerned be reasonably easily synthesized? Much of this problem relates to cost, which can hardly be judged at the present time, for it will in the future be relative cost, together with safety, and not absolute cost which is to be balanced against alternative methods of crop protection. It is only when convincing field trials can be demonstrated that costing will be taken seriously. If we believe that the enormous potential of natural deterrents can ever be realized in applied antifeedants, then all must be centred on the demonstration of effectiveness. For this we need a most thoughtful approach to the strategies employed and a full appreciation of the problems. As each year passes we should either be obtaining the field results or else have the knowledge and courage to say "this approach has no future". We need more time, but I suspect we are near to this latter point and that effort may be better spent in helping crop plants to help themselves, namely by breeding for pest resistance and the increase in antifeedants already *in situ* in the plant and, in the ideal situation, of their being systemic and restricted to the crop itself.

15.5 SUMMARY

The appeal of using antifeedants in pest management systems is set against the rather limited progress in this field to date. The search for antifeedants present in plants resistant to predation by insects has led to the description of various deterrents but the problems of developing a commercially effective, cheap product have so far beset pest managers. However, suggestions which might minimize these problems are made, but the basic question of the practicality of using antifeedants in the field are reiterated.

15.6 REFERENCES

All J. N. & Benjamin D. M. (1976) Potential of antifeedants to control larval feeding of selected *Neodiprion* sawflies (Hymenoptera: Diprionidae). *Can. Ent.* **108**, 1137–1144.

Ascher K. R. S. & Gerta Rones (1964) Fungicide has residual effect on larval feeding. *Int. Pest. Control* **6**, 6-9.

Bernays E. A. & Chapman R. F. (1975) The importance of chemical inhibition of feeding in host-plant selection by *Chorthippus parallelus* Zetterstedt. *Acrida* **4**, 83–93.

Bernays E. A. & Chapman R. F. (1977) Deterrent chemicals as a basis of Oligophagy in *Locusta migratoria*. *Ecol. Ent.* **2**, 1–18.

Bernays E. A., Chapman R. F., McCaffery A. M., Modder W. W. D. & Leather E. M. (1977) The relationship of *Zonocerus variegatus* and cassava (*Manihot esculenta*). *Bull. Ent. Res.* **67**, 391–404.

Bernays E. A., Chapman R. F. McDonald J. & Salter J. E. R. (1976) The degree of oligaphagy in *Locusta migratoria*. *Ecol. Ent.* **1**, 223–230.

Chapman R. F. (1974) The chemical inhibition of feeding by phytophagous insects. *Bull. Ent. Res.* **64**, 339–363.

Gill J. S. (1972) Studies on insect feeding deterrents with special reference to the fruit extracts of the neem tree, *Azadirachta indica* A. Juss. Ph.D. thesis, University of London.

Gill J. S. & Lewis C. T. (1971) Systemic action of an insect feeding deterrent. *Nature (Lond.)* **232**, 402–403.

Hegnauer R. (1976) Accumulation of secondary products and its significance for biological systematics. In *Secondary Metabolism and Coevolution* (eds. Luckner M., Mothes M. & Nover L.), pp. 45–76. Deutsche Akademie der Naturforscher Lepoldina Halle, Leipzig.

Jermy T. (1961) The rejective effect of some inorganic salts on the Colorado beetle (*Leptinotarsa decemlineata* Say) adults and larvae. *Növényyéd. Kut. Intér. Évk.* **8**, 121–130 (in Hungarian).

Jermy T. (1966) Feeding inhibitors and food preference in chewing phytophagous insects. *Ent. exp. appl.* **9**, 1–12.

Jermy T. & Matolcsy G. (1967) Antifeeding effect of some systemic compounds on chewing phytophagous insects. *Acta phytopath. Acad. Sci. Hung.* **2**, 219–224.

Kamel A. A. M., Mitri S. H., Abo El Ghar M. & Zake M. M. (1970) Field trials for the use of two promising antifeedants against the cotton leaf worm. *Bull. ent. Soc. Egypt Econ. Ser.* **4**, 35–41.

Ketkar C. M. (1976) Utilization of neem (*Azadirachta indica* Juss.) and its by-products. Report of Khadi & Village Industries Commission, Bombay, India.

Ladd T. L., Jacobson M. & Buriff C. R. (1978) Japanese beetles: extracts from neem tree seeds as feeding deterrents. *J. Econ. Ent.* **71**, 810–813.

Ma W. C. (1969) Some properties of gustation in the larva of *Pieris brassicae*. *Ent. exp. appl.* **12**, 584–590.

Ma W. C. (1977) Alterations of chemoreceptor function in armyworm larvae (*Spodoptera exempta*) by a plant-derived sesquiterpenoid and by sulphydryl reagents. *Physiol. Ent.* **2**, 199–207.

Mabry T. J. & Gill J. E. (1979) Sesquiterpene lactones and other terpenoids. In *Herbivores: Their Interaction with Secondary Plant Metabolites* (eds. Rosenthal G. A. & Janzen D. H.), pp. 502–538. Academic Press, N.Y.

Metzger F. W. & Grant D. H. (1932). Repellancy to the Japanese beetle of extracts made from plants immune to attack *Tech. Bull US Dept. Agric.* no. 299–21 pp.

Muller E. (1976) Principles in transport and accumulation of secondary products. In *Secondary Metabolism and Coevolution* (eds. Luckner M., Mothes K. & Nover L.). Deutsche Akademie der Naturforscher Leopoldina Halle, Leipzig.

Muller C. H. & Chou, Chang-Hung (1972) Phytotoxins: an ecological phase of phytochemistry. In *Phytochemical Ecology* (ed. Harborne J. B.), pp. 201–216.

Munakata K. (1977) Insect antifeedants of *Spodoptera litura* in plants. In *Host Plant Resistance to Pests* (ed. Hedin, P. A.), pp. 185–196. American Chemical Society, Washington.

Murbach R. (1967) Effet en plein champ de fongicides a base de fentin-acetate, de manebe et d'oxychlorure de cuivre sur la densite de population du doryphore de la pomme de terre (*Leptinotarsa decemlineata* Say). *Schweiz. Landwirt. Forsch.* 6, 345–357.

Pradhan S., Jotwani M. G. & Rai B. K. (1962) The neem seed deterrent to locusts. *Indian Fmg* 12 (8), 7–11.

Radwanski S. (1977) Neem tree 3: Further uses and potential uses. *Wld Crops & Livestock* 29, 167–168.

Schoonhoven L. M. & Derksen-Koppers I. (1973) Effects of secondary plant substances on drinking behaviour in some Heteroptera. *Ent. exp. appl.* 16, 141–145.

Schoonhoven L. M. & Jermy T. (1977) A behavioural and electrophysiological analysis of insect feeding deterrents. In *Crop Protection Agents — Their Biological Evaluation* (ed. M. R. McFarlane). pp. 133–145. Academic Press, N.Y.

Warthen J. D. (1979) *Azadirachta indica*: a source of insect feeding inhibitors and growth regulators. *USDA Agricultural Rev. & Man.* ARM-NE-4, 21 pp.

15.7 DISCUSSION

Djerassi: I refer to the dashes in some of your slides. Do they mean that nothing has been done?

Bernays: It really means that, the advantage is not there.

Djerassi: Why is a chemical sometimes more active when it is in the plant than when you spray it on the plant?

Bernays: Activity in the plant may be aided by other simple compounds which may be present in the plant at various concentrations and none of which by themselves are at all toxic to insects. For the antifeedant alone to confer the same degree of protection to the plant, more of it must be sprayed on to the plant. The case of resistant sorghum is different. We know that there are twenty-five different chemicals which are important; only one of them (i.e. thiocyanate) is by itself active against insects *per se*. We also know that it is a combination of these substances which has made resistance so important.

Djerassi: Suppose a person offered to support your research both financially and in terms of manpower, which crop would you pick as a target for spraying and which particular antifeedant would you try? Just approximately how long do you think it would take to get results?

Bernays: I would choose sorghum. I think it would take 10 years to get results.

Djerassi: Why would you choose sorghum?

Bernays: Because it is an important crop grown in several semi-arid tropics. Furthermore, a lot of development work has been done on sorghum.

Graham-Bryce: As far as I know, there have been some resistant varieties of plants which have been proclaimed. It has not been deemed necessary to submit these to regulatory authorities because the authorities might, it has been argued, call for a systematic and scientific study of the plants, which would, in the opinion of many people, create new problems.

Djerassi: The moment you submit some natural chemical or biochemical constituent to field study, you invite several rational questions about risk and safety of your material. For this reason, it is even more risky in terms of invoking these sentiments to start proclaiming the advantages of crude plant extracts or plant material before exhaustive field tests have been done.

Graham-Bryce: It seems to me that the only way to explain the basis of resistance, especially in the more resistant varieties, is to know the identities of all the constituents responsible for the resistance phenomenon.

Djerassi: Yes I agree. At Zöecon we have started to examine the chemical and biochemical characters which plants use to evolve resistance. Our approach involves genetic engineering techniques and is therefore very difficult. However, I think 10 years seems to be a reasonable period to expect fruitful results.

Bowers: I also think that the issue of resistance in plants should be studied systematically. In this way, we shall gain insight into the phenomenon. The chemist should elucidate the constituents of the complex mixtures of compounds found in resistant plants, and the pure plant constituents so obtained should be incorporated into young plants by the plant breeder. I think we need to blend chemical and plant breeding techniques to uncover the mystery surrounding the phenomenon of plant resistance.

Bernays: I fully agree with that. Knowing the basis of resistance seems to me to offer the best chance for success in trying to protect food and fibre crops and their products from attack by pests.

Feeny: I want to emphasize that this approach is superior to trying to induce resistance in plants by spraying the plants with antifeedants or some other insect-active compounds which, in the first place, were obtained from the plants themselves through long arduous processes. In addition, spraying on to the plant is wasteful, and as has been said, likely to raise questions of safety and risks to the environment.

Jotwani: I disagree with Professor Bowers' view that a blend of chemical and plant breeding techniques is the answer to understanding the phenomenon of resistance in plants. As shown in my paper to this workshop, there are a number of other factors which can confer resistance and which need to be understood too in their own right. It is now known that the level of resistance which we bring about by breeding strategy is inferior to that evolved by the plant itself.

Tahori: Dr. Pryce posed a question this morning about whether there have been any antifeedant compound marketed. There has been one antifeedant compound developed by an American scientist, Donald White. In field trials using tobacco plants, this product showed excellent antifeedant properties against the Southern US armyworm *Spodoptera frugiperda*. Further development of this compound failed because the product was later shown to be a carcinogen.

Bowers: Development of biologically active compounds to the marketing stage has been a problem. The US Department of Agriculture has tried several times doing tests on every conceivable activity. But even then, success has not been impressive.

Feeny: I would like to comment on Tahori's remark about the antifeedant compound which was never developed. It seems to me that the transnational companies producing secondary compounds are not coordinated. I also think that they do not collaborate. Otherwise, they would have rescued the antifeedant product from oblivion. They would have pooled their effort to solve the resistance problem as well.

Djerassi: I think such collaboration or coordination might complicate the lives of these organizations.

Saxena: I wish to add that very major advances have been made in our effort to solve the resistance problem. For example, we have succeeded in incorporating in a number of rice plants multiple resistance to a number of insect pests.

Feeny: I still think that spraying antifeedants onto plants to induce resistance in them might jeopardize the environment.

Bernays: Most antifeedants are not toxic. People might not be affected by spraying them in the field.

Feeny: I think it is worth while to concentrate on developing repellents for specialized pests.

Bernays: Yes. I would tend to agree with that.

CHAPTER 16

Biochemical Composition of Bean Varieties Resistant and Susceptible to Mexican Bean Beetle (Coleoptera: Coccinellidae)

A. K. RAINA

International Centre of Insect Physiology and Ecology (ICIPE)
P.O. Box 30772, Nairobi, Kenya

and

S. S. LEE

Virginia State University, Petersburg, Virginia, USA

CONTENTS

16.1	Introduction	273
16.2	Materials and Methods	274
16.3	Results	274
16.4	Discussion of results	276
16.5	Summary	276
16.6	Acknowledgement	277
16.7	References	277
16.8	Discussion	277

16.1 INTRODUCTION

Mexican bean beetle, *Epilachna varivestis* Mulsant, is a serious pest of vegetable beans, *Phaseolus vulgaris* and *P. lunatus*, and more recently achieved great importance as a pest of soybeans, *Glycine max* (Turnipseed & Kogan, 1976). Both adults and larvae feed on bean foliage; completely defoliating the crop under heavy infestations. Host plant resistance has been proposed and investigated as a tool for the management of the Mexican bean beetle. Raina, Benepal & Sheikh (1978) screened 141 snapbean, *P. vulgaris*, and 20 limabean varieties as moderately resistant to adult feeding.

Mexican bean beetle is an oligophagous species that feeds almost exclusively on plants of the genus *Phaseolus*. Lippold (1957) suggested that

273

olfaction was of minor importance in host selection by Mexican bean beetle but gustatory stimulants, possibly glycosides, played an important role. Consequently the cyanogenic glucoside, phaseolunatin was reported as a phagostimulant for Mexican bean beetle (Klingenburg & Bucher, 1960; Nayar & Fraenkel, 1963).

In order to determine factors responsible for reduced adult feeding on resistant bean varieties, biochemical analysis of a resistant and a susceptible variety from each of the two groups of beans was performed.

16.2   MATERIALS AND METHODS

Plants of two resistant varieties "Regal" (snapbean) and "Baby Fordhook" (Limabean) and two susceptible varieties "Morses Pole 191" (snapbean) and "Fordhook 242" (Limabean) were grown in a greenhouse in plastic pots filled with commercial potting soil. Plants of two trifoliate leaf stage were used for chemical analysis.

The first trifoliate leaves from ten plants of each variety were excised and each leaflet weighed. One leaflet from each leaf was extracted with 80% acetone for chlorophyll determination. The second leaflets were dried in an oven at 45°C for two days and macerated. This was then used to determine the protein content by micro-Kjeldahl method. The third leaflets were individually immersed in 80% boiling alcohol and the leaf extract used for determination of total carbohydrate, reducing sugars, total phenolic compounds and free amino acids. Beckman 121 M amino acid analyser with system AA integrator was used for amino acid analysis. The data was statistically analysed.

16.3   RESULTS

The results of the analysis are presented in Tables 16.1 and 16.2. The most striking difference between the resistant and the susceptible varieties was reflected in the chlorophyll content of the leaves. Almost twice as much chlorophyll was present in the susceptible varieties as in the resistant. Similar difference was evident in the carbohydrate and reducing sugar content of susceptible and resistant varieties of limabeans. Although their amount was higher in case of the susceptible snapbean than the resistant variety the difference was significant only in the case of reducing sugars. The total phenolic compounds and % protein was significantly higher only in the case of susceptible snapbean, the differences in the case of limabeans being insignificant.

TABLE 16.1 BIOCHEMICAL COMPOSITION[a] OF LEAVES OF SNAP AND
LIMABEANS, RESISTANT AND SUSCEPTIBLE TO MEXICAN BEAN
BEETLE (MBB)

Variety name	Type	MBB rating	Chlorophyll $mg\ g^{-1}$ of leaf tissue	Carbohydrate $mg\ g^{-1}$ of leaf tissue	Reducing sugars	Phenolic compound	% Protein
Regal	Snap	Resistant	0.98	2.77	1.15	2.14	30.3
Morses pole 191	Snap	Susceptible	2.62	3.03	1.74	2.91	37.8
Baby ford-hook	Lima	Resistant	1.85	3.38	1.45	1.45	32.8
Fordhook 242	Lima	Susceptible	2.34	6.23	3.18	1.45	35.0
LSD at 5% level			0.478	0.852	0.517	0.329	3.678

[a]Based on average of ten samples, from ten individual plants.

A comparison of the free amino acids showed that threonine, serine,
glutamic acid and cysteine were significantly higher in both resistant varieties
whereas arginine and isoleucine were lower compared to the susceptible
varieties.

TABLE 16.2 QUANTITATIVE ESTIMATION OF FREE AMINO ACIDS IN
LEAVES OF BEAN VARIETIES, RESISTANT AND SUSCEPTIBLE TO THE
MEXICAN BEETLE

Amino acids	Moles of amino acid ml^{-1} leaf extract			
	Regal	Morses pole 191	Baby fordhook	Fordhook 242
Lysine	13	15	9	18
Histidine	6	7	35	17
Arginine	0	Trace	0	8
Aspartic acid	274	275	286	232
Threonine	132	67	108	77
Serine	246	155	398	193
Glutamic acid	571	503	617	523
Proline	55	46	86	136
Glycine	56	Trace	Trace	13
Alanine	19	14	17	13
Half cysteine	197	139	143	90
Valine	23	30	63	71
Methionine	Trace	Trace	Trace	Trace
Isoleucine	9	15	9	45
Leucine	9	25	9	Trace
Tyrosine	Trace	Trace	Trace	Trace
Phenylalanine	Trace	Trace	Trace	Trace

16.4   DISCUSSION OF RESULTS

After an insect arrives on a plant host and takes its first bite, whether or not it will continue feeding depends on whether proper phagostimulation occurs. Some feeding stimulants have been shown to be host-specific substances, but others like glucose, sucrose and amino acids are present in varying amounts in almost all plants (Thorsteinson, 1960).

Sugars have been reported to be of considerable significance in the regulation of feeding in phytophagous insects. Augustine, Fisk, Davidson, La Pidus & Cleary (1964) bioassayed thirteen sugars and fifteen amino acids and reported that only sucrose and to a lesser extent its hexose components induced feeding by the Mexican bean beetle. Van Duyan (1971) found that resistant soybean varieties contained less reducing sugar than susceptible varieties. Thorsteinson (1960) reported negligible response by *Pleutella xylostella* to sucrose alone at any concentration and a weak response to 0.1% sinigrin but both combined evoked appreciable feeding response. Nayar and Fraenkel (1963) reported that the cyanogenic glycoside, phaseolunatin acts as a phagostimulant for the Mexican bean beetle larvae at lower concentration but at higher concentrations it acts as a deterrent or inhibitor. More extensive feeding in case of the susceptible varieties in the present study can be attributed to the presence of higher amounts of both carbohydrate and reducing sugars. This is perhaps further complimented by a higher chlorophyll content of the susceptible varieties which were darker green in colour than the resistant ones. Tood, Parker & Gaines (1972) reported a greater number of Mexican bean beetle on soybeans with increased leaf protein. However, they observed that plants with higher protein also had darker-green leaves. Tester (1977) reported that susceptible cultivars of soybean contained about 25% more soluble proteins and amino acids than the resistant cultivars.

Beck and Hanec (1958) and Thorsteinson (1960) reported that the amino acids, alanine, serine and α-amino-n-butyric acid evoked feeding response in certain insects. In the present study only isoleucine and to some extent arginine were present in higher levels in susceptible varieties. The relatively higher amounts of threonine, serine, glutamic acid and cysteine in both resistant varieties can not be linked with reduced feeding.

16.5   SUMMARY

Based on field and greenhouse screening of 131 snapbean, *Phaseolus vulgaris* L., and 20 limabean, *P. lunatus* L., varieties for resistance to Mexican bean beetle, *Epilachna varivestis* Mulsant; one snap and two limabean varieties were reported as moderately resistant to adult feeding.

Biochemical analysis of a resistant and a susceptible variety from each of the two groups of beans was performed. Susceptible varieties of both snap and limabeans had significantly higher amount of chlorophyll. The susceptible variety of limabeans had almost twice as much carbohydrate and reducing sugars as the resistant variety, but the differences were not significant for snap beans. Total phenolic compounds and protein was higher in the susceptible snap beans. Both resistant varieties were higher in the free amino acids; threonine, serine, glutamic acid and cysteine but lower in isoleucine and arginine. Possible role of these constituents in reduced feeding by Mexican bean beetle of the resistant varieties is discussed.

16.6 ACKNOWLEDGEMENT

Part of the research was supported by Co-operative State Research Service Grant No. 216-15-10.

16.7 REFERENCES

Augustine M. G., Fisk F. W. Davidson R. H., La Pidus J. B. & Cleary R. W. (1964) Host-plant selection by the Mexican bean beetle, *Epilachna varivestis*. *Ann. Ent. Soc. Am.* **57**, 127–134.

Beck S. D. & Hanec. W. (1958) Effect of amino acids on feeding behaviour of the European corn borer, *Pyrausta nubilalis* Hübn. *J. Insect Physiol* **2**, 85–96.

Klingenburg M. & Bucher J. (1960) Biological oxidations. *Ann. Rev. Biochem.* **29**, 669–708.

Lippold P. C. (1957) The history and physiological basis of host specificity of the Mexican bean beetle, *Epilachna varistestis* Muls. (Coleoptera: Coccinellidae). Ph.D. thesis, University of Illinois, Urbana. 146 pp.

Nayar J. K. & Fraenkel G. (1963) The chemical basis of the host selection in the Mexican bean beetle, *Epilachna varivestis* (Coleoptera: Coccinellidae). *Ann. Ent. Soc. Am.* **56**, 174–178.

Raina A. K., Benepal P. S. & Sheikh A. Q. (1978) Evaluation of bean varieties for resistance to Mexican bean beetle. *J. Econ. Ent.* **71**, 313–314.

Tester C. R. (1977) Constituents of soybean cultivars differing in insect resistance. *Phytochemistry* **16**, 1899–1901.

Thorsteinson A. J. (1960) Host selection in phytophagous insects. *Ann. Rev. Ent.* **5**, 193–218.

Todd J. W., Parker M. B. & Gaines T. P. (1972) Population of Mexican bean beetles in relation to leaf protein of nodulating and non-nodulating soybeans. *J. Econ. Ent.* **65**, 729–731.

Turnipseed S. G. & Kogan M. (1976) Soybean entomology. *Ann. Rev. Ent.* **21**, 247–283.

Van Duyan J. W. (1971) Investigations concerning host plant resistance to the Mexican bean beetle, *Epilachna varivestis* Mulsant, in soybeans, *Glycine max* L. Merril. Ph.D. dissertation. Clemson Univ., Clemson S.C. 210 pp.

16.8 DISCUSSION

Marini Bettolo: I should like to ask you about the amino acid content. Has it been done after hydrolysis or are there free amino acids also?
Raina: No, these are all free amino acids.

Saxena: Many of the factors you have elucidated are considered as varietal resistance against sorghum shootfly. My point is that they used to be considered to confer resistance in rice varieties, especially the presence of trichomes, silica and lignification. I have indicated that these factors have no role or, I should say, they have very little role to play in rice, very simple experiments like rubbing off the trichomes could very easily settle the question whether they are involved in resistance or not. For example, in pigeon pea, when the trichome hairs are rubbed off the oviposition response of the moth remain the same.

    Plants do not produce silica. Silica comes from the soil with nutrients and is then deposited. If we take a susceptible plant and plant it in a silica-rich field there probably will be deposition of silica in the leaves. I do not know how silica got involved in explaining varietal resistance problem.

Raina: I think I agree with you on both aspects, particularly on trichomes. You might probably hear something more from Dr. Reed. I think the problem is very difficult in sorghum. The trichomes are extremely small and shaving them off might also alter the leaf surface in other ways. Trichomes by themselves, however, could be associated with some other characters. For silica, I accept the very early report by Blum and others. I don't think there is any concrete evidence for its involvement.

CHAPTER 17

Antifungal Compounds and Disease Resistance in Plants

J. A. CALLOW

Department of Plant Sciences, University of Leeds, Leeds LS2 9JT, UK

CONTENTS

17.1	Introduction	279
17.2	Host–Parasite Recognition	280
	17.2.1 Elicitors, suppressors and specificity factors	283
	17.2.1.1 Non-specific glucan elicitors	283
	17.2.1.2 Non-specific glycoprotein elicitors	283
	17.2.1.3 Specificity factors	283
	17.2.1.4 Host receptors	284
17.3	Triggering of Resistance	284
17.4	Resistance Mechanisms, the Control of Fungal Growth	285
	17.4.1 Pre-formed inhibitors	285
	17.4.2 Phytoalexins	288
17.5	Potential Use of Naturally Occurring Antifungal Agents in Disease Control	291
17.6	Summary	292
17.7	Acknowledgements	292
17.8	References	293
17.9	Discussion	294

17.1 INTRODUCTION

Plants, in common with all other living organisms, have evolved immune or resistance mechanisms of various types by which they are able to counteract the advance of foreign organisms, the result being that a given parasite can usually only infect a distinct range of host plants. Disease, therefore, is generally the exception rather than the rule despite the fact that plants are

279

surrounded by many potential pathogens (Callow, 1977). The mechanisms by which plants are able to resist infection are diverse and complex, but in the context of the aims of this meeting it is pertinent to examine those aspects of resistance which are thought to depend on the presence or *de novo* synthesis of antibiotic agents. Conventionally two forms of resistance are distinguished: "non-host resistance", where plants are clearly outside the range of hosts a parasite can infect, and "host resistance", resulting from genetic modification to the host, rendering it resistant to pathogens that would otherwise infect it (Day, 1974). This distinction may be exemplified in a simple way by considering the rice blast pathogen, *Pyricularia oryzae*. This fungus will infect rice cultivars but not most other plants, tomato, beans, corn, etc. The latter plants therefore constitute "non-hosts". On the other hand, there are cultivars of rice into which have been bred genes for resistance to rice blast, such varieties constituting "resistant hosts".

17.2 HOST–PARASITE RECOGNITION

Resistance may be divided mechanistically into two phase as shown on Fig. 17.1. The initial, determinative phase of resistance is seen as one in which the host plant has the ability to detect and "recognize" the potential invader as "non-self", the result being a triggering or expression of the various components constituting the resistance mechanism. Failure of the host to recognize the pathogen as "non-self", or recognition as "self" results in susceptibility and disease.

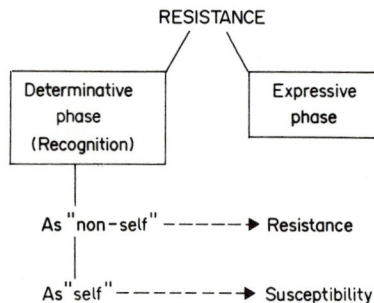

Fig. 17.1. The two phases of resistance to microorganisms.

The ability to recognize and discriminate between "self" and "non-self" is a fundamental property of all organisms and, in general, proceeds via complementary molecular mechanisms located at the cell surface (Callow, 1977; Sequiera, 1979). In the case of host plant/microorganism interactions there is now evidence which supports the view that plants have the ability to

Plant Pathogen

Non-specific recognition by all non-host species	Non-recognition or suppression of non-specific recognition

NON-HOST RESISTANCE   BASIC COMPATIBILITY

Introduction of R gene to recognise specific features of the pathogen

CULTIVAR RESISTANCE

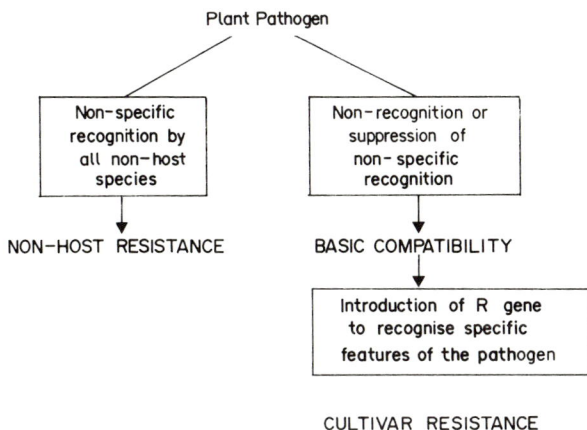

Fig. 17.2. Possible origins of resistance and compatibility (after Heath, 1981).

recognize components on the surface of the invading pathogen, the act of recognition triggering metabolic changes in the host cells, including mRNA and enzyme synthesis, the sum of which results in death of the potential pathogen, or in the case of mutualistic symbionts such as *Rhizobium*, acceptance of the microorganism (Sequiera, 1979; Callow, 1981). Non-specific recognition of such surface components by the majority of plants might thus be viewed as the basis of at least some aspects of non-host resistance. If this is the case, then host susceptibility, i.e. susceptibility in those plants constituting host species, might therefore be explained by the ability of successful, or compatible pathogens to suppress or divert the general, non-specific recognition, so that the fungus is recognized as "self", resulting in disease. Evidence supporting this hypothesis has recently been presented by Doke, Garas & Kuc (1979).

Race- and varietal-specific forms of resistance have probably evolved from systems of more general resistance and induced susceptibility through plant breeding. Resistance (R) genes are incorprated into "host species" from wild relatives permitting the new cultivar to recognize the pathogen as "non-self" once more, resulting in resistance (Fig. 17.2). Callow (1977) has discussed how such a pattern of man-guided evolution has resulted in gene-for-gene systems with multiple gene loci and multiple alleles. Close examination of the genetics of resistance and susceptibility in systems of varietal specificity demonstrates the presence of the so-called "gene-for-gene" relationship, illustrated in Fig. 17.3. Ellingboe (1976) has presented cogent arguments to suggest that the complementary genes in host and parasite conferring specificity have their specific interaction for the incompatible relationship, i.e. recognition is for resistance. Person & Mayo (1974) termed the combination R:A a "stop-signal" since it triggers incompatibility. A model incorporating this concept is shown in Fig. 17.4. Thus, on this basis, specific, induced host resistance has been

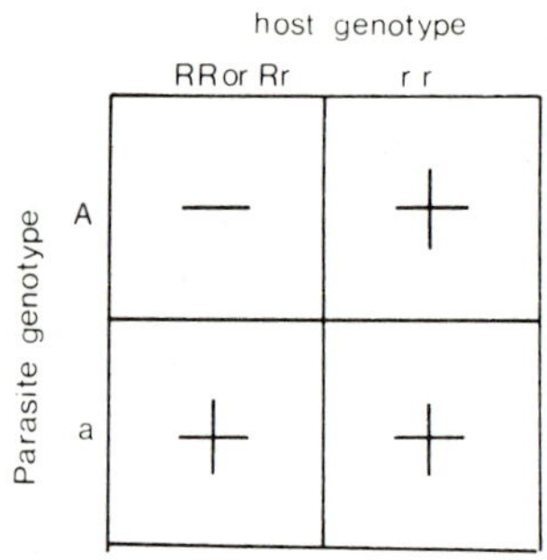

J. A. Callow

host genotype

Fig. 17.3. The quadratic check in gene-for-gene interaction. R and r represent host genes for resistance (dominant) and susceptibility (recessive). A and a represent parasite genes for avirulence (dominant usually) and virulence (recessive). Compatible and incompatible responses are denoted by + and − respectively.

developed on a system of non-specific or general resistance and induced susceptibility. A number of questions may now be posed, for example:

1. What types of fungal surface molecule may be recognized by the host?
2. What is the nature of the host receptors responsible for recognizing these parasite "signals"?
3. How are the processes of resistance triggered following the act of recognition?
4. What is the actual mechanism by which fungal growth is arrested in the incompatible response?

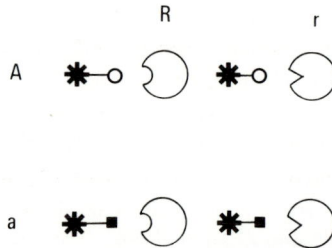

Fig. 17.4. Simple lock-key interpretation of the molecular implications of the gene-for-gene concept. Products of host resistance/susceptibility alleles are shown as surface-localized receptors with defined binding-site configurations. Products of parasite avirulence/virulence alleles are shown as molecules with a substantially common structure but specific recognized termini. Only in the case R/A can correct binding and recognition take place, triggering the resistance response.

17.2.1 Elicitors, suppressors and specificity factors

17.2.1.1 Non-specific glucan elicitors. Extensive studies (reviewed by Albersheim & Valent, 1978) have shown that plants can recognize polydisperse, high molecular weight, branched β1-3 glucans isolated from culture media or cell walls of a number of parasitic fungi and saprophytes. The glucans appear to be structural polymers of the fungal cell walls. Their activity is manifest by the rapid accumulation of phytoalexins in the treated plant. The glucans are highly active, as little as 0.2 μg eliciting phytoalexin accumulation. The smallest effective glucan molecule is an oligomer with nine glucose residues, produced by partial hydrolysis of the isolated polysaccharide. The activity of these elicitors is neither cultivar, nor race or fungal species-specific.

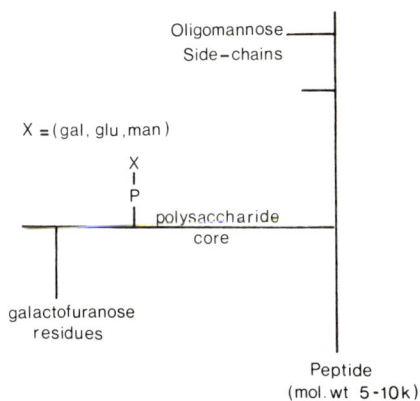

```
                          Oligomannose
                          Side-chains


          X = (gal, glu, man)
                          X
                          |
                          P
                          |  polysaccharide
                             core


     galactofuranose
      residues

                              Peptide
                          (mol. wt 5-10k)
```

Fig. 17.5. Schematic representation of the major structural features of the glycopeptide secreted by *Fulvia fulva* (after Dow & Callow, 1979a).

17.2.1.2 Non-specific glycoprotein elicitors. Dow & Callow (1979a) partially characterized a peptidophosphoglucomannan from culture filtrates and cell walls of the tomato leaf mould pathogen, *Fulvia fulva* (syn. *Cladosporium fulvum*) (Fig. 17.5). This glycoprotein has a number of effects when applied to tomato tissue or cells (Dow & Callow, 1979b), but one of its activities is a rapid elicitation of phytoalexins, not just in tomato, but in plants which are not hosts for this pathogen. The activity of the elicitor is sensitive to periodate, indicating a primary role for carbohydrate residues, and the oligomannose side chains appear to be particularly important for full activity.

17.2.1.3 Specificity factors. Three research groups have reported the isolation of fungi cell wall fractions which may be effective in determining varietal specificity. Keen (1978) reported the isolation of glycoproteins from cell walls of *Phytophthora megasperm sojae*, which when applied to soyabean elicited the accumulation of the phytoalexin, glyceollin, in a variety- and species-specific manner. In contrast, Wade & Albersheim (1979) also isolated cell wall glycoproteins from the same fungus but these did not possess phytoalexin elicitor activity. However, when applied to soyabean plants, these glycoproteins were able to protect treated plants against subsequent infection by the pathogen, *P. megasperma sojae*, and protection was race- and variety-specific. The degree of correspondence between these two observations remains to be determined.

Specific resistance in the interaction between potato and *P. infestans* is determined by the elicitation in the host of a hypersensitive response involving host cell death and terpenoid phytoalexin accumulation. These responses can also be elicited in a non-specific way by β1-3 glucans from the fungal cell wall, and the key to specificity here appears to lie in the ability of the compatible races of the fungus to suppress the hypersensitive reaction. Doke *et al.* (1979) have described additional glucan components of the cell wall which have suppressor activity.

17.2.1.4 Host receptors. A common feature of the various molecules already described is that they contain carbohydrate residues and, at least in some cases, the saccharide moieties have been shown to be important in the activity of the molecules. By implication, therefore, the corresponding host receptors responsible for recognizing the fungal molecules may be surface-localized carbohydrate-binding proteins known to exist in plants in various forms as "lectins". At the present time, however, such receptors have not been characterized. The evidence that reception of parasite signals is a surface-localized event is also largely circumstantial, resting on the rapidity of response and the fact that all the native parasite molecules putatively involved are macromolecules, and hence are probably unable to enter host cells (Callow, 1977). Dow & Callow (1979b) were able to demonstrate that the glycoprotein elicitor of *Fulvia fulva* (Fig. 17.5) does bind reversibly to isolated tomato cells, and the elicitor from *Phytophthora infestans* will agglutinate and kill potato leaf protoplasts (Peters, Cribbs & Stelzig, 1978).

17.3   TRIGGERING OF RESISTANCE

If it is confirmed that macromolecular host–parasite recognition phenomena in host–parasite interactions are located at the surface of host cells, then there are no fundamental difficulties in postulating mechanisms by which such a surface event may then be transduced to trigger resistance. There is ample evidence to show that surface recognition in other systems results in the release of "secondary messengers" on the cytoplasmic face of the membrane. Interaction of such molecules, such as cAMP, with the cell genome may then result in genome activation. In the case of varietal-specific resistance responses accompanied by phytoalexin synthesis, there is evidence for *de novo* mRNA synthesis (Yoshikawa, Masago & Keen, 1977). Also, phytoalexin induction through elicitor treatment has been shown to be accompanied by *de novo* synthesis of enzymes on the biosynthetic routes leading to the phytoalexin (Dixon & Lamb, 1979). In the case of *Helminthosporium sacchari*-sugarcane interactions, binding of the carbohydrate-based toxin to host surface receptors does induce conformational changes in other membrane proteins (Strobel 1976). However, as yet, the triggering effect of elicitor or specificity factor activity has yet to be examined.

17.4 RESISTANCE MECHANISMS, THE CONTROL OF FUNGAL GROWTH

The actual means by which fungal growth is controlled in a resistance response are diverse and complex (Callow, 1977, 1981). However, those mechanisms which appear to depend on the presence in the plant of molecules which are inhibitory to fungal growth and development will be discussed. Such molecules may be either present in the plant before infection (pre-formed or constitutive) or synthesized *de novo* on infection (phytoalexins).

17.4.1 Pre-formed inhibitors

Whilst there is still much debate on the metabolic significance of secondary compounds in plants, there is now substantial evidence to support a primary ecological role for these compounds as protective agents against various parasites and predators. However, the evidence is more convincing when considered for the interactions between plants and insects than plants and parasitic microorganisms. There are those that take the extreme view that in few cases, if any, that have convincingly been demonstrated, resistance of a living plant tissue is due to the presence in the plant of a particular inhibitory compound. Now, it is relatively easy to demonstrate that plants do contain compounds which are fungi- or bacterio-static such as cyanogenic glycosides, steroid glycoalkaloids, phenol glucosides and tannins, isothiocyanates, cyclic hydroxamates, etc. (for reviews see Overeem, 1976; Schonbeck & Schlosser, 1976). It is interesting that many of these compounds are also effective against animal predators (Rosenthal & Janzen, 1979). To attempt to review the role of such compounds in resistance would be impossible in the present article. Rather, the criteria by which a convincing role for secondary metabolites may be demonstrated will be discussed, as illustrated by the steroid glycoalkaloid tomatine, a saponin with antifungal activity. Such criteria have also been discussed and emphasized by Wood (1967) and Schonbeck & Schlosser

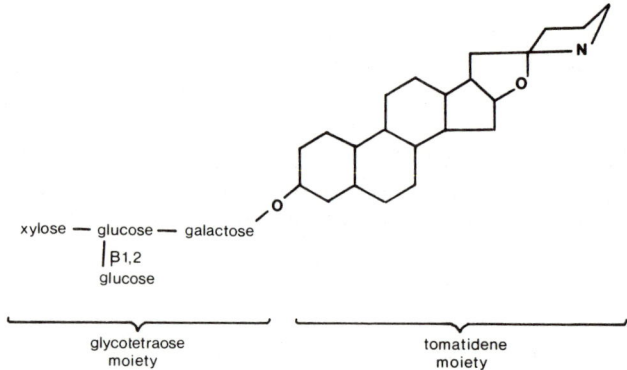

xylose — glucose — galactose
        | β1,2
        glucose

glycotetraose   tomatidene
  moiety   moiety

FIG. 17.6. The steroidal glycoalkaloid saponin, tomatine.

(1976). Saponins are glycosides containing one of four aglycone moieties: alkaloid, triterpene, furostanol or spirostanol. The distinct hydrophobic–hydrophilic asymmetry possessed by the saponins permits reaction with sterols in cellular membranes leading to cell lysis and this is generally taken as the basis for their biological activity. The bitter taste of many saponins may also act as a feeding deterrent to insects (Applebaum & Birk, 1979). Breeding programmes have been devised for the production of legume pasture plants that accumulate high levels of saponin in roots to deter root-damaging insects, and low levels in the shoots to improve nutritional value to herbivores (Applebaum & Birk, 1979). The effect of this on a possible weakening of the chemical barriers to fungal attack is not recorded.

The primary criterion necessary to establish at least a circumstantial role of any secondary compound in disease resistance is that the compound should exhibit strong antibiotic activity *in vitro* at concentrations relevant to the effective concentration of compound at the actual site of contact between host and parasite. Saponins have pronounced antifungal properties with LD_{100} values in the range 15–200 μg cm^{-3} for many different fungi (Schonbeck & Schlosser, 1976). Since saponin concentrations in certain plant tissues may be as several thousand μg cm^{-3} of plant extract, or gram^{-1} fresh weight, the saponins are clearly prime candidates as agents of chemical defence.

In the specific example of tomatine (Fig. 17.6) the antifungal activity demonstrated depends on the assay technique employed. Dow & Callow (1978) showed that tomatine inhibited hyphal extension of the leaf mould pathogen, *Fulvia fulva* (ED_{50} 15 μg cm^{-3}), but that toxicity was dependent on both pH and nutrient status of the medium. In shake cultures, with high concentration of carbon source, van Dijkman (1972) failed to detect any effect of tomatine on the same fungus in terms of biomass yield. Within the tomato plant, tomatine is present at highest concentration (up to 3 mg cm^{-3}) in young leaves and green fruits (Schonbeck & Schlosser, 1976). In young leaf tissue, tomatine may occupy 0.5% by dry weight, equivalent to a concentration of approximately 1 mM assuming even distribution of the compound in vacuole and cytoplasm. It is apparent, therefore, that gross tissue damage, such as that caused by a biting insect or some necrotrophic fungi, would release high concentrations of the alkaloid. Whether the compound is then effective depends on a number of factors, but particularly pH and the activity of the invading pathogen to degrade the tomatine to a non-toxic form. Thus, resistance of tomato fruit to non-pathogenic forms of *Alternaria* spp. has been attributed to an increase in the pH of the infection site to 6.5–8.0, at which tomatine is maximally effective, and to the absence of an enzyme system capable of deglycosylating tomatine to inactive forms. Pathogenic forms of *Alternaria* were able to carry out such degradation (Schonbeck & Schlosser, 1976). A similar correlation between pathogenicity and ability to degrade tomatine has been noted for *Septoria* spp. (Arneson & Durbin, 1968). Susceptibility of tomato fruit to fruit-rot pathogens may be due to the fact that rotting lowers the pH of pulp to pH 3–4, under which conditions tomatine is protonated and largely inactive. In contrast, tomatine is present in low concentrations (50 µg cm^{-3}) in root tissue and Kern (1952) concluded that tomatine played no important role in the resistance of roots to *Fusarium oxysporum*. However, not all fungal parasites of tomato exhibit widespread tissue degradation. Some, for example *Fulvia fulva*, exert minimal cellular damage, in this case growing between host mesophyll cells. Here, a postulated role for tomatine in resistance would require leakage of the saponin from the cells at effective concentrations of the order 15 µg cm^{-3}. It may be relevant, therefore, that the glycopeptide phytoalexin elicitor of *F. fulva* described earlier will also induce changes in host cell permeability, causing ions and other low molecular weight compounds, including tomatine, to leak out of the cell (Dow & Callow, 1979b). Whether the concentration of tomatine around the hyphae would be sufficiently high to cause growth inhibition is not known; the appropriate methods for such an estimation have yet to be devised. It is also likely that the effective concentration of tomatine in the middle lamellae and intercellular spaces where the fungus grows is less than the actual chemical concentration. The relatively low pH at such sites caused by the presence of acidic cell wall polymers would reduce the effective concentration of the unprotonated form of tomatine.

The evidence implicating tomatine and many other metabolites in resistance is therefore largely circumstantial. More direct approaches through the use of mutants is urgently required. Defago & Luescher (1978) induced mutants of a number of fungi with lower sterol content in their membranes. Such mutants were less sensitive to tomatine. An unrelated example of the type of genetic approach required is that of Couture, Routley & Dunn (1971), and cited in Day (1974). Leaves of *Zea mays* contain a glucoside of the cyclic hydroxamate DIMBOA (2,4,-dihydroxy-7-methoxy-1,4-benzoxazin-3-one). In this form the compound is not antifungal. On wounding the aglucone, DIMBOA is released which is toxic to many fungi and insects. Recessive corn mutants with very low levels of DIMBOA showed reduced, but not totally abolished specific resistance to *Helminthosporium turcicum* suggesting that DIMBOA is only one line of defence against this fungus.

17.4.2 Phytoalexins

The defence system of plants, like other organisms contains facultative, i.e. induced, components in addition to the standing components, and it is noteworthy that those concerned with plant–microorganism interactions have developed a more thorough understanding of facultative defence compared with those concerned with animal pests.

If the evolution of defence mechanisms is considered from the point of view of a cost–benefit analysis or optimal defence strategy (Rhoades, 1979; McKey, 1979), then since defences are costly in terms of energy required to synthesize the defensive agents, it would seem to be advantageous to the plant to possess mechanisms by which commitment to defence is increased when organisms are subject to attack and decreased when the pathogen is absent (estimates of the "costs" of defence in plants are difficult to obtain but in recent experiments Smedegaard-Petersen [personal communication] reported a 7% reduction of yield in barley plants repeatedly inoculated with an incompatible race of powdery mildew). In fact there is now substantial literature to show that under conditions of stress there is a tendency on the part of plants to expand or activate many secondary metabolite pathways. One such stress is the attempted invasion by microorganisms through which phytoalexins are induced. The triggering nature of recognition mechanisms has already been discussed.

The subject of phytoalexins has been extensively reviewed elsewhere (e.g. Deverall, 1976; Kuc, 1976; Van Etten & Pueppke, 1976). Phytoalexins may be defined as low molecular weight antimicrobial compounds that are both synthesized by, and accumulate in plants after exposure to microorganisms. The 100 or so compounds characterized fall into a limited number of chemical categories (Fig. 17.7). The antifungal activity exhibited varies within the range 1–50 μg cm^{-3} for ED$_{50}$ values *in vitro*, depending on conditions, and their effect on the pathogen is generally one of fungistasis. The ability to produce phytoalexins is widespread in the plant kingdom and is not specific to pathogen type, i.e. the same phytoalexin may be produced by inoculation with a range of different pathogens. Phytoalexins are also induced by a variety of biotic and abiotic compounds, wounding, UV light, water stress, etc. Their

phenolic (phenanthrene)
orchinol , orchid

phenolic (isocoumarin)
carrot

phenolic (pterocarpan)
glyceonin , soyabean

phenolic (phenylpropanoic)
coniferyl alcohol , flax

phenolic (stilbene)
resveratrol , grapevine

polyacetylenid
wyerone acid (broad bean)

sesquiterpenoid
rishitin , potato

terpene
momilactone

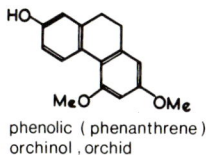

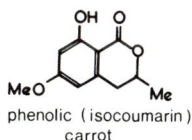

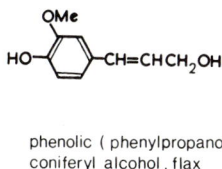

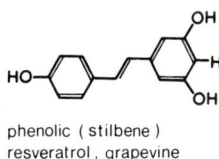

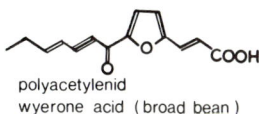

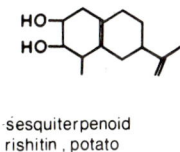

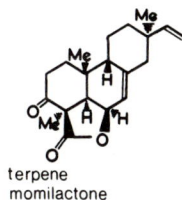

Fig. 17.7. Some phytoalexin structures.

production may take place in living cells as a result of active metabolism, *de novo* RNA and protein synthesis (Yoshikawa, Masago & Keen, 1977; Dixon & Lamb, 1979), or they may constitute the products of cells undergoing necrosis.

Phytoalexins have been implicated both in varietal-specific forms of resistance and general resistance or immunity. In the case of the former, the ability to synthesize phytoalexins is not restricted to genetically resistant cultivars, i.e. genes for resistance do not themselves code for enzymes directly involved in the biosynthetic routes leading to phytoalexins. As shown in Fig. 17.8 the rate, and time at which phytoalexins accumulate differs in the compatible and incompatible associations and in some cases the correlation between phytoalexin accumulation and inhibition of fungal growth has

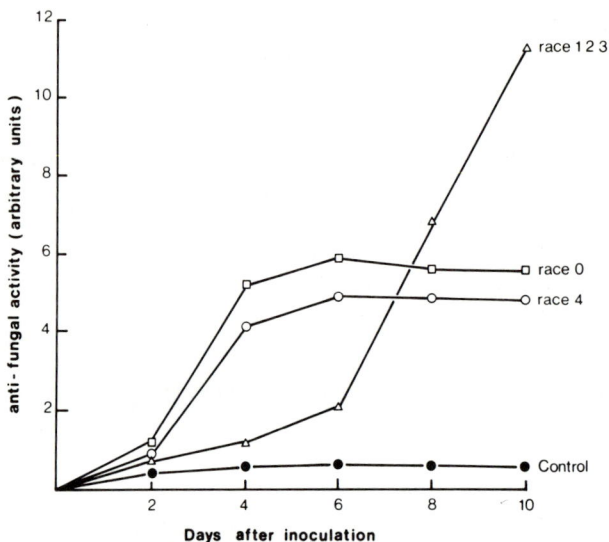

Fig. 17.8. Time course of antifungal activity in tomato plants, cv. Vetomold, inoculated with avirulent (0 and 4) and virulent (1,2,3) races of *Fulvia fulva*, and uninoculated controls. Antifungal activity was measured by examining the effect of methanolic extracts of tissue on fungal germ tube extension and is expressed in arbitrary units (Dow & Callow, unpublished data).

suggested a causal relationship (Yoshikawa, Yamauchi & Masago, 1979; Keen & Littlefield, 1979). A simple interpretation of the role of genes for resistance is that they code for the ability of a plant to quickly recognize parasite signals, triggering phytoalexin synthesis either directly or indirectly. Reference has already been made to the role of elicitors and other macromolecules in the evolution of host–parasite recognition systems.

In the case of general immunity, a further consideration, over and above the ability of the parasite to elicit phytoalexins in the host, is sensitivity of the microorganism to the phytoalexin. Generally, fungi pathogenic to a given species are less sensitive to the phytoalexin than non-pathogens through their ability to degrade the phytoalexin to a non-toxic form. However, as in the case of pre-formed metabolites, the evidence implicating phytoalexins in disease resistance is largely circumstantial. Extraction techniques estimate only the gross tissue level without taking into account microlocalization in the vicinity of the fungal hyphae and the correlation between cessation of fungal growth and accumulation of the phytoalexin is imprecise in all but a few examples. Unfortunately, the use of mutants of fungi resistant to phytoalexins has yet to be successful in producing unequivocal answers.

17.5 POTENTIAL USE OF NATURALLY OCCURRING ANTIFUNGAL
AGENTS IN DISEASE CONTROL

What are the practical consequences, if any, of the studies on natural antifungal compounds discussed in this article? Can compounds such as phytoalexins be used either directly, or indirectly as models for synthetic agents in biorational disease control? At the present time this author is not aware of a single parallel between the use of natural antifungal agents and the potent natural insect control agents such as pyrethrum. The vast range of phytoalexins and pre-formed antifungal agents suffer from the following drawbacks when it comes to consider them as control agents.

1. Specific antibiotic activity tends to be low compared with the most potent synthetic fungicides. Thus, ED_{50} values for inhibition of spore germination *in vitro* are generally in the range 5–25 µg cm^{-3} for most natural inhibitors and activity tends to fungistatic rather than fungicidal, whereas fungicidal activity (LD_{99} values) for modern synthetic agents can be as low as 1–5 µg cm^{-3} (Marsh, 1977; Fawcett & Spencer, 1978).
2. Most phytoalexins and pre-formed agents are not systemic in the plant and fail to show chemotherapeutic activity. Many agents with high or moderate antifungal activity *in vitro* possess negligible fungicidal activity *in vivo* (Fawcett & Spencer, 1978).
3. Many phytoalexins are phytotoxic.
4. The chemical synthesis of most phytoalexin structures appears to be difficult.

An alternative strategy would be to induce the formation of antifungal compounds *in vivo* rather than exogenous application. An obvious candidate as an inducing agent would be the highly active β-glucan elicitors described earlier. However, problems associated with their use would be: (1) not systemic; (2) easily broken down in the field; (3) difficult to synthesize, although fermentation technology could be utilized to produce large amounts; (4) the induction of the phytoalexins inside the plant might lead to phytotoxicity problems. In addition, a natural system of energy-saving, facultative defence is, in effect, converted to one of high-cost constitutive defence.

A more suitable model compound is one which is able to potentiate or amplify natural porcesses of resistance when required, i.e. when the plant is challenged by the pathogen. Recent work by Cartwright, Langcake, Pryce, Leworthy & Ride (1977) has identified such an agent. The dichlorocyclopropane compound WL 28325 induced systemic protection in rice against the rice blast pathogen, *Pyricularia oryzae*, when applied as a root drench. This protection took the form of rapid host cell death, melanin pigment accumulation and momilactone phytoalexin accumulation, much as in the natural resistant response to a virulent race. WL 28325 did not, however, exhibit any direct fungicidal activity *in vitro* but rather appears to act by sensitizing the host so that on subsequent infection, the natural processes of resistance arc accelerated. Thus, the plant only invokes its resistance mechanisms when required. It might therefore be suggested that the development of compounds like WL 28325, active perhaps at the recognitional phase of resistance, forms the most realistic opportunity for future attempts to control disease by manipulating natural defences. The design and study of such compounds will be facilitated when our understanding of the molecular biology of this aspect of host–parasite relationships is improved.

17.6 SUMMARY

Plant defence against insect predation is often considered separately from the defence against parasitic microorganisms despite the fact that it is now clear that the mechanisms of defence have elements in common. A comprehensive view of plant defence therefore requires that the various hypotheses of optimal defence theory, developed primarily for plant–herbivore interactions, be evaluated in a more general way.

Resistance in plants to microorganisms, like that in man or other animals, is based on two groups of mechanisms. Firstly, there are physical and chemical barriers to keep infectious agents out of the plant. Secondly, and superimposed on these constitutive mechanisms, there may be active or induced responses leading to the *de novo* synthesis of compounds which inhibit the development of the potential parasite.

Whilst it is relatively easy to demonstrate that many plants contain compounds inhibitory to the growth of pathogens, the evidence implicating these compounds in resistance in a causal role is slight. The criteria to satisfy such a role are discussed by reference to secondary metabolites, in particular the steroid glycoalkaloid, tomatine. Facultative, or induced defence involves synthesis of phytoalexins, chemically diverse secondary compounds which exhibit fungi- or bacteriostatic activity. The triggering of the *de novo* synthesis of such compounds following host–parasite recognition will be discussed. Finally, possibilities for the use of natural resistance mechanisms in chemical control of plant disease will be examined.

17.7 ACKNOWLEDGEMENTS

The author gratefully acknowledges grants from The Agricultural Research

Council and Ministry of Overseas Development. The author also wishes to thank Dr. M. C. Heath for permission to examine an unpublished paper.

17.8 REFERENCES

Albersheim P. & Valent B. S. (1978) Host–pathogen interactions in plants. *J. Cell Biol.* **78**, 627–643.

Applebaum S. W. & Birk Y. (1979) Saponins. In *Herbivores: their interaction with secondary plant metabolites* (eds. Rosenthal G. A. & Janzen D. H.), pp. 539–566. Academic Press, New York.

Arneson P. A. & Durbin R. D. (1968) Studies on the mode of action of tomatine as a fungitoxic agent. *Plant Physiol.* **43**, 683–686.

Callow J. A. (1977) Recognition, resistance and the role of plant lectins in host–parasite interactions. *Adv. Bot. Res.* **4**, 1–49.

Callow J. A. (1982) Fungal infection. In *The Molecular Biology of Plant Development* (eds. Smith H. & Grierson D.) Botanical Monog. **18**, Blackwells Sci., Oxford.

Cartwright D., Langcake P., Pryce R. J., Leworthy D. P. & Ride J. P. (1977) Chemical activation of host defence mechanisms as a basis for crop protection. *Nature* **267**, 511–512.

Couture R. M., Routley D. G. & Dunn G. M. (1971) Role of cyclic hydroxamic acids in monogenic resistance of maize to *Helminthosporium turcicum*. *Physiol. Plant Pathol.* **1**, 515–521.

Day P. (1974) *Genetics of Host–Parasite Interaction*. W. H. Freeman and Co., San Francisco.

Defago G. & Luescher D.(1978) Induced mutants for studying the role of saponins as resistance factors. *Proceedings 3rd International Congress of Plant Pathology, Munich, 1978*, p. 218.

Deverall B. J. (1976) Current perspectives on research in phytoalexins. In *Biochemical Aspects of Plant–Parasite Relationships* (eds. Friend J. & Threlfall D. R.), pp. 207–223. Academic Press, London.

Dixon R. A & Lamb C. J. (1979) Stimulation of *de novo* synthesis of L-phenylalanine ammonia-lyase in relation to phytoalexin accumulation in *Colletotrichum lindemuthianum* elicitor-treated cell suspension cultures of French bean (*Phaseolus vulgaris*). *Biochim. Biophys. Acta* **586**, 453-463.

Doke N., Garas N. & Kuc J. (1979) Partial characterisation and aspects of the mode of action of a hypersensitivity-inhibiting factor (HIF) isolated from *Phytophthora infestans*. *Physiol. Plant Pathol.* **15**, 127–140.

Dow J. M. & Callow J. A. (1978) A possible role for α-tomatine in the varietal-specific resistance of tomato to *Cladosporium fulvum*. *Phytopath. Z.* **92**, 211–216.

Dow J. M. & Callow J. A. (1979a) Partial characterisation of glycopeptides from culture filtrates of *Fulvia fulva* (Cooke) Ciferri (syn. *Cladosporium fulvum*), the tomato leaf mould pathogen. *J. Gen. Microbiol.* **113**, 57–66.

Dow J. M. & Callow J. A. (1979b) Leakage of electrolytes from isolated leaf mesophyll cells of tomato induced by glycopeptides from culture filtrates of *Fulvia fulva* (Cooke) Ciferri (syn. *Cladosponium fulvum*). *Physiol Plant Pathol.* **15**, 27–34.

Ellingboe, A. H. (1976) Genetics of host-parasite interactions. In *Physiological Plant Pathology* (eds. Heitefuss R. & Williams P. H.), pp. 761–778. Springer-Verlag, Berlin, Heidelberg, New York.

Fawcett C. H. & Spencer D. M. (1978) Plant chemotherapy with natural products. *Ann. Rev. Phytopathol.* **16**, 403–418.

Keen N. T. (1978) Elicitation and sites of formation of phytoalexins and induced resistance. *Proceedings 3rd International Congress of Plant Pathology, Munich, 1978*, p. 213.

Keen N. T. & Littlefield L. J. (1979) The possible association of phytoalexins with resistance gene expression in flax to *Melampsora lini*. *Physiol Plant Pathol.* **14**, 265-280.

Kern H. (1952) Uber die Beziehungen zwischen dem alkaloidgehalt versheidener Tomatensorten und ihrer Resistenz gegen *Fusarium oxysporum*. *Phytopath. Z.* **19**, 351–382.

Kuc J. (1976) Phytoalexins. In *Physiological Plant Pathology* (eds. Heitefuss R. & Williams P. H.), pp. 632–652. Springer-Verlag, Berlin, Heidelberg, New York.

McKey D. (1979) The distribution of secondary compounds within plants. In *Herbivores; their interaction with secondary plant metabolites* (eds. Rosenthal G. A. & Janzen D. H.), pp. 55–133. Academic Press, New York.

Overeem J. C. (1976) Pre-existing antimicrobial substances in plants and their role in disease resistance. In *Biochemical Aspects of Plant–Parasite Relationships* (eds. Friend J. & Threlfall D. R.), pp. 195-206. Academic Press, London.

Person C. & Mayo G. M. E. (1974) Genetic limitations on models of specific interactions between a host and its parasite. *Can. J. Bot.* **52**, 1339–1347.

Peters B. M., Cribbs D. H. & Stelzig D. A. (1978) Agglutination of plant protoplasts by fungal cell wall glucans. *Science* **201**, 364–365.

Rhoades D. F. (1979) Evolution of plant chemical defence against herbivores. In *Herbivores; their interaction with secondary plant metabolites* (eds. Rosenthal G. A. & Janzen D. H.), pp. 3-54. Academic Press, New York.

Rosenthal G. A. & Janzen D. D. (1979) *Herbivores; their interaction with secondary plant metabolites.* Academic Press, New York.

Schönbeck F. & Schlösser E. (1976) Preformed substances as potential protectants. In *Physiological Plant Pathology* (eds. Heitefuss R. & Williams P. H.), pp. 653–678. Springer-Verlag, Berlin, Heidelberg, New York.

Sequeira L. (1979) Recognition between plant hosts and parasites. In *Host–Parasite Interfaces* (ed. Nickol B. B.), pp. 71–84. Academic Press, New York.

Strobel G. A. (1976) Toxins of plant pathogenic bacteria and fungi. In *Biochemical Aspects of Plant–Parasite Relationships* (eds. Friend J. & Threlfall D. R.), pp. 135–159. Academic Press, London.

Van Dijkman A. (1972) Natural resistance of tomato to *Cladosporium fulvum*. A biochemical study. Ph.D thesis, State University Utrecht, The Netherlands.

Van Etten H. D. & Pueppke S. G. (1976) Isoflavonoid phytoalexins. In *Biochemical Aspects of Plant–Parasite Relationships* (eds. Friend J. & Threlfall D. R.), pp. 239–289. Academic Press, London.

Wade M. & Albersheim P. (1979) Race-specific molecules that protect soybeans from *Phytophthora megasperma* var. *sojae. Proc. Nat. Acad. Sci. USA* **76**, 4433–4437.

Wood R. K. S. (1976) *Physiological Plant Pathology.* Blackwell, Oxford, Edinburgh.

Yoshikawa M., Masago H. & Keen N. T. (1977) Activated synthesis of poly (A)-containing messenger RNA in soyabean hypocotyls infected with *Phytophthora megasperma* var. *sojae. Physiol. Plant Pathol.* **10**, 125–138.

Yoshikawa M., Yamauchi K. & Masago H. (1979) Biosynthesis and biodegradation of glyceollin by soyabean hypocotyls infected with *Phytophthora megasperma* var. *sojae. Physiol Plant Pathol.* **14**, 157–169.

17.9 DISCUSSION

Bernays: I am interested to know exactly where these receptors are. Are they on the surface or underneath the cuticle?

Callow: The limited amount of evidence that we have is that they are surface localized. They have to be essentially because the molecules that are effective are very large and could not penetrate the surface.

Bernays: So they are sitting around under the wax?

Callow: Oh no. When I talk about the leaf surface, I am talking about inside the leaf surface tissue.

Pryce: I really can't agree with you about phytoalexins not serving as possible models for chemical synthesis. I think they can be models. The other thing is that you said they were not very useful because they were too water insoluble, whereas I think some of them are, in fact, quite soluble. For example, polyphenolics from grape vines.

Callow: My comments about their limited appeal is based on an extensive series of studies which really achieved very little in designing and synthesizing modified phytoalexins as effective control agents.

Djerassi: Could one ask of you why this cyclopropylcarboxylic acid is not a feasible practical agent? As a very simple organic molecule I would think that it is very easy to synthesize.

Pryce: There were various commercial reasons why we didn't develop this compound. It was under development for a long time before it was dropped. I would say the same commercial restraints may not apply to another company.

Raina: In your expression of the gene-for-gene hypothesis you have indicated that the resistant gene and the avirulent gene of the pathogen have to be compatible to produce resistance.

Callow: In most cases, yes.

Raina: The disease is normally caused by a virulent gene and not the avirulent gene. What happens in these cases?

Callow: The host plant is unable to recognize the parasite and therefore the mechanisms for resistance are not induced or brought into play.

Raina: So resistance in these cases will only be applicable to avirulent strains?

Callow: That's what the geneticists tell us.

Graham-Bryce: In relation to the glucans, is there any information on the minimum requirements in terms of recognition?

Callow: The slide I showed is only a tentative model for the minimum component of these glucans which can be recognized and still be effective.

Graham-Bryce: That does pose considerable problems, then, in going for a smaller molecule. Turning to the lectins, you said there is very little information about this interaction, the nature of the receptors, etc. Is it considered that the glutinating properties of the lectins are unrelated to their action?

Callow: In terms of what I've been discussing, yes. But in terms of a different aspect, namely binding of bacteria to the leaf cell surface preventing them from multiplying, there may be quite a different function of the different lectins.

Graham-Bryce: I would also like to comment that in this type of interaction, this type of investigation, one should not just look for its usefulness in terms of suggesting new fungicides, because particularly in this lectin type of work, there can be great value in strain identification of pathogens which in certain cases can be very important to devising strategies, which may or may not include fungicides. One can differentiate in terms of lectin response between different strains of pathogen.

Callow: Yes, you can use the lectins which are particularly common in seeds. The sort of lectin component we are looking for is a much more subtle thing which is part of the membrane structure perhaps, or which may not have more than one binding site.

Graham-Bryce: I wasn't just thinking of seed but of actual tissue which you can use for strain differentiation.

Feeny: Where did you get your figure of 7% for the energy saved by a plant having phytoalexins rather than static defences?

Callow: This comes from very recent work not yet published. I only became aware of it myself about 2 weeks ago. It's from some work carried out in Denmark on infections in barley, where the variation in yield was measured on infection of both resistant and susceptible cultivars of barley. The difference in yield was approximately 7% compared to the controls.

Feeny: Did they at the same time measure approximately how much biomass of secondary compounds were produced in the stimulated plants?

Callow: No. I think the reason is that one has no knowledge of the secondary compounds that might be affected.

Feeny: That's going to be very interesting information and many people are trying to establish this.

Callow: It's assuming of course that there are secondary compounds involved, which may not necessarily be the case.

Otieno: One of your reasons for dismissing these phytoalexins is that they are too toxic to work with.

Callow: I'm not actually dismissing them, but they don't seem to be attractive.

Otieno: It was one of the reasons you gave, anyway. I find it amazing that a lot of effort should be spent on developing compounds like these cyclopropane carboxylic acids which you say act by sensitizing the plant to produce phytoalexins. What are your comments on that?

Callow: Well, the phytoalexin is only being formed when the plant is challenged by a parasite, that is, it is only being produced when needed. It is not being formed on immediate application of the sensitizing chemical. In other words, these chemicals are assisting, potentiating, if you like, the natural disease-resistance process.

Djerassi: And what is the actual mechanism? Is it increasing the production of phytoalexins?

Callow: Nobody knows that yet!

Saxena: What would be your approach in breeding resistance to pathogens?

Callow: I have no approach since I'm not a plant breeder. However, it seems to me that what the plant breeder needs to know is the actual expression of the gene for resistance and we do not know this in any single case. If we did have this knowledge it would seem to me that the art of the plant breeder would be very much improved.

CHAPTER 18

Nematocidal Natural Products

KATSURA MUNAKATA

Department of Agricultural Chemistry, Nagoya University, Nagoya, 464, Japan

CONTENTS

18.1	Introduction	299
18.2	Screening for Nematocidal Activity	301
18.3	Nematocidal Polyacetylenes from *Carthamus tinctorius* L.	301
18.4	Odoracin and Odratrin, Nematocidal Constituents from *Daphne odora*	303
18.5	Nematocidal Fatty Acids from *Japonica* Thumb.	305
18.6	Nematocidal Substance from *Angelica pubescens*	305
18.7	Conclusion	308
18.8	Summary	308
18.9	References	308
18.10	Discussion	309

18.1 INTRODUCTION

The control of plant–parasitic nematodes by chemical means is very limited mainly to the use of halogenated hydrocarbons, but the continuous use of the nematocides has caused several environmental problems, and research needs are aimed at establishing means of alternative nematode control. Some plant species possess characteristics which may be detrimental to populations of specific nematodes. When the active principles of the nematocidal plants are elucidated chemically, the chemical structure teaches us the mother structure of new nematocidal substances. Marigolds (*Tagetes electa* L.) have been found to suppress populations of *Meloidogyne*, *Pratylenchus* and other nematode genera.

1

$\equiv C-CH=CH_2$

2

$CH_3-(C\equiv C)_5-CH=CH_2$

3

OH

4

COOH

5

OH
OH

6

R_1
R_2 OH $N-CH_3$

7 $R_1=R_2=CH_2\!\!<^{O-}_{O-}$

8 $R_1=R_2=OCH_3$

CH_3O OH $N-CH_3$

9

10

11

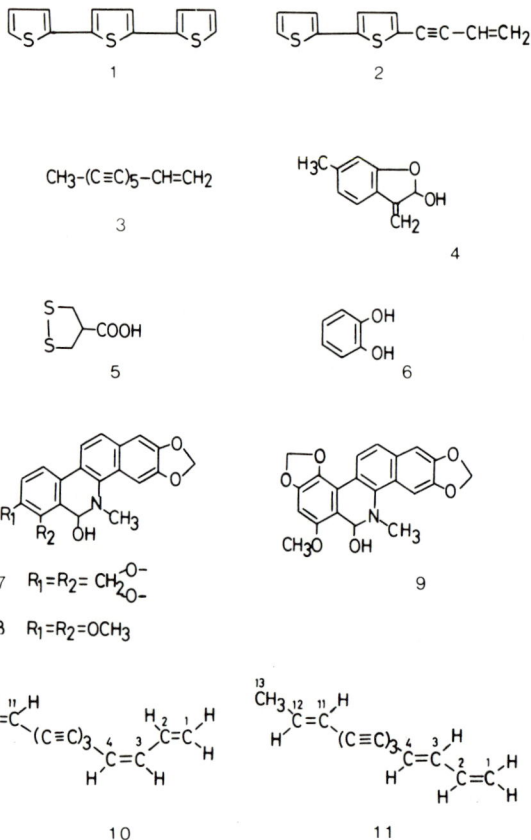

Fig. 18.1, nos. 1–11

After the investigation of the nematocidal principles of marigold (*Tagetes electa* L.), Uhlenbrook & Bijloo (1956, 1958) reported that the nematocidal substances in the roots of marigolds are α-terthienyl (Fig. 18.1, 1) and 5-(3-buten-1-ynyl)-2,2′-bithienyl (Fig. 18.1, 2). These compounds have nematocidal activities towards *Ditylenchus dispaci, Anguina tritici, Heterodera rostochiensis* and *Pratylenchus penetrans*. They also synthesized the related compounds and tested their nematocidal activities and found that the active structure needed has the bithienyl group.

Gommers (1973) discovered two nematocidal principles from the roots of *Helenium* hybrid "Moreheim Beauty" and identified their chemical structures to be 1-tridecanene-3,5,7,9,11-pentayne (Fig. 18.1, 3) and 2,3-dihydro-3-methylene-6-methyl benzofuran (Fig. 18.1, 4) after the bioassay of *Pratylenchus penetrans*. Asparagus (*Asparagus officinalis*) is resistant to *Trichodorus christiei*, and contains a highly toxic glycoside in the roots, stems and leaves, but the chemical structure of the toxic principle is not reported (Rohde & Jenkins, 1958). By a bioassay using *P. curvitus* it was revealed that one of the nematocidal substances in the root of *Asparagus* is asparagusic acid (Fig. 18.1, 5) (Takasugi, Yachida, Anetai, Masamune & Kigasawa, 1975), which inhibited *Heterodera rostochiensis*, *Meloidogyne hapla*, *P. curvitus* at 55 ppm in water. The nematocidal principle in the root exudates of *Eragrostis curvula* against *Meloidogyne* species elucidated to be a high concentration of pyrocatechol (Fig. 18.1, 6) (Scheffer, Kickuth & Visser, 1962).

After the observation of nematode proof of *Bocconia cordata*, Onda, Takiguchi & Hirata (1965) have investigated nematocidal substances in the roots of *B. cordata* against *Rhabditis* sp. and *Panaglolaims* sp. and extracted and identified three alkaloids, two of them were sanguinarine (Fig. 18.1, 7) and chelerythrin (Fig. 18.1, 8) and the third was an unknown substance named bocconine. The chemical structure of bocconine was later elucidated to be as shown in Fig. 18.1, 9 (Onda *et al.*, 1965).

As pointed out by Rohde (1972) and Avivardi (1971), many antihelmintic chemicals from plants for nematode parasites of man and domestic animals such as santonin, ascaridole, finicinic acid and embelin are toxic to animal-parasitic nematodes. An extensive survey for nematocidal substances from plants against rice white-tip nematode, *Aphelenchoides besseyi*, has been conducted at the laboratory of Pesticide Chemistry Nagoya University, Nagoya, Japan. Here, nematocidal substances from four plants (*Carthamus tinctorius* L., *Daphne odoro* Thumb., *Angelica pubescens*, *Iris japonica* Thumb.) are reported and discussed.

18.2 SCREENING FOR NEMATOCIDAL ACTIVITY

Immersion test was used for bioassay (Kogiso, Wada & Munakata, 1976a): to an ethanol solution (0.04 ml) of a test compound was added 1 ml of 0.3% solution of the detergent (tween 20) and emulsified. Rice-tip nematode was reared for 1 month on the slant of a fungus, *Alternaria citri*. Two hundred nematodes in 2 ml of distilled water and emulsified sample solution (1 ml) were transferred into a Syracuse watch glass and incubated at 25°C for 24 h or for 5 days. Then the nematodes, dead (straight) or alive (winding), were counted under a dissecting microscope.

18.3 NEMATOCIDAL POLYACETYLENES FROM
CARTHAMUS TINCTORIUS L.

In preliminary experiments, benzene and methanol extracts of the flowers, stems and roots of the safflower plant, *Carthamus tinctorius* L., were examined by a bioassay for nematocidal activity to *Aphelenchoides besseyi*, and only benzene

```
                              Flowers  (IO kg)

                              Benzene  20 (twice)

                              evaporation

                    Benzene | extracts  (63 g)

                              Silicic  acid  column  chromatogram


    n-hexane              Benzene                        EtOAc
    ─────────
          Silicic  acid  column  chromatogram

          n-hexane

          fractionation


    I            II            III          IV              V

   3.52          1.85g        1.353 g      50 mg          85 mg

                              Silicic  acid  column  chromatogram (230 g)
                              n-pentane

   Fraction                   fractionation
   number
          1-10         11-14        15-19      20-23          24-35

                7.9 mg        14.7 mg

                              Preparative  high-speed  liquid  chromatogram
                              (C-18  CORASIL)

                              30 %  water  in  methanol


          Component - I            II
          20 mg                   I mg
```

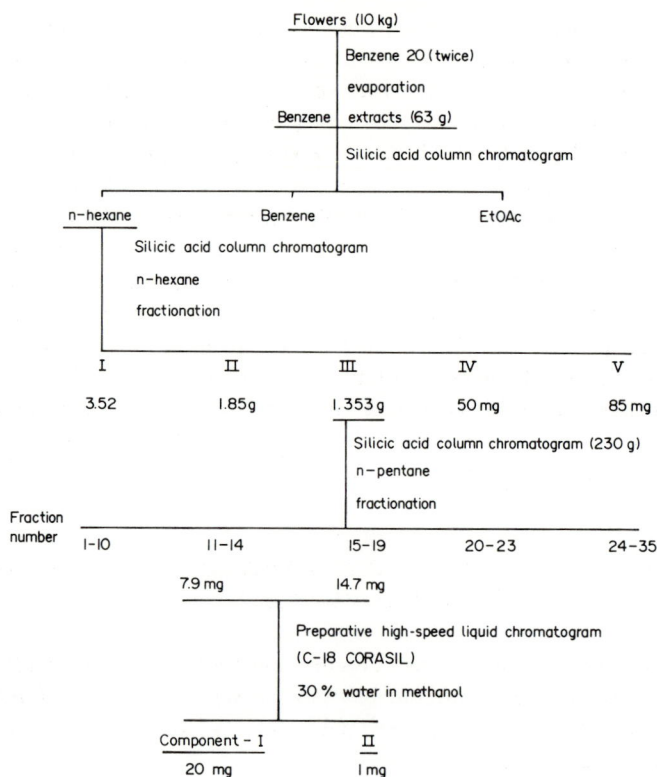

Scheme 18.1
Protocol for extraction of Safflower

extracts of the flowers and roots showed nematocidal activity. The benzene extracts of the flowers were then subjected to a silicic acid column chromatography and eluted with n-hexane, benzene and ethyl acetate. Only the n-hexane eluate showed nematocidal activity and was rechromatographed on silicic acid with n-hexane (Scheme 18.1). The fraction III showed nematocidal activity at 20 ppm. During the isolation of the active fraction, we noticed that active components were sensitive to light, oxygen and temperature. Thus, silicic acid column chromatography of the fraction III was carried out in a dark room and the eluate were carefully evaporated below 0°C under reduced pressure. The active fraction, Nos. 11–14 and 15–19, showed only one spot on thin-layer chromatography, but two peaks on a gas-liquid chromatography. The combined fraction was not separated into components by PTLC or GLC. This difficulty of separation was overcome by preparative high-speed liquid chromatography on reverse-phase column, BONDAPAK C_{18}/COR-ASIL under dark condition and the components I and II were obtained pure. Chemical structures of I and II were elucidated as 3-*cis*,11-*trans* trideca-1,3,11-triene-5,7,9-triyne (Fig. 18.1, 10) and 3-*trans*, 11-*trans*-trideca-1,3,11-triene-5,7,9-triyne (Fig. 18.1, 11), respectively, by the Mass, NMR, IR and UV spectra (Kogiso, Wada & Munakata, 1976b,c; Sörensen & Sörensen, 1958; Bohlman, 1958). Nematocidal activities of chemical structures (Fig. 18,1, 10 and 11) to rice white-tip nematode were observed by immersion test under dark condition. The nematocidal activity (of Fig. 18.1, 11) was 95% nematode mortality at 1 ppm concentration after 46 h. However, the main components (Fig. 18.1, 10) showed 80% nematode mortality at 10 ppm after 48 h. It is very interesting that these geometrical isomers showed significantly different activity. Many polyacetylenes in Compositae plants have been reported to have their growth inhibitory properties on other plants as allelopathic substances (Kobayashi, Marutomo, Shibata & Yamashita, 1974). Similar polyacetylenes have been obtained from fungi belonging to Basidiomycetes (Turner, 1971). These fungal polyacetylenes might have some effects on the population of nematodes.

18.4 ODORACIN AND ODRATRIN, NEMATOCIDAL CONSTITUENTS FROM *DAPHNE* ODORA

In further studies on the nematocidal substances in higher plants, we discovered that the benzene extract of the roots of a sweet daphne, *Daphne odora* Thumb. (Thymelaeaceae), showed nematocidal activity to rice white-tip nematode in immersion test.

Nematocidal crude extract (70 g) was obtained from the dry root (9.4 kg) of *D. odora* by benzene extraction. The extract was subjected to silicic acid (Mallinkrodt) column chromatography using isopropanol-benzene. The nematocidal eluate with benzene containing 3% isopropanol was rechromatographed using acetone-benzene to afford fractions containing odoracin (Fig. 18.2, 12) and odratrin (Fig. 18.2, 13). Successive preparative thin-layer chromatography with n-hexane-ethyl acetate gave odoracin (215 mg) and odoratrin (36.1 mg).

The molecular formula of odoracin (Fig. 18.2, 12), $C_{37}H_{44}O_{10}$, was determined by high-resolution mass spectrometry. Acetylation of odoracin with acetic anhydride and pyridine afforded diacetate, and methanolysis of odoracin with methanolic potassium hydroxide gave desbenzoyl derivative. This desbenzoyl derivative gave ethyl 2,4-decadienoate by acid catalysed ethanolysis in sealed tube. This fact showed that desbenzoyl derivative contained an orthoester group of 2,4-decadienoic acid in its structure. The plane structure of odoracin was elucidated by carefully comparing the proton NMR data of this compound with those of the structurally related diterpenes, huratoxin (Fig. 18.2, 14) (Sakata, Kawazu & Mitsui, 1971), daphnetoxine (Fig. 18.2, 15) (Stout, Balkenhol, Poling & Hickernell, 1970), and Mezerein

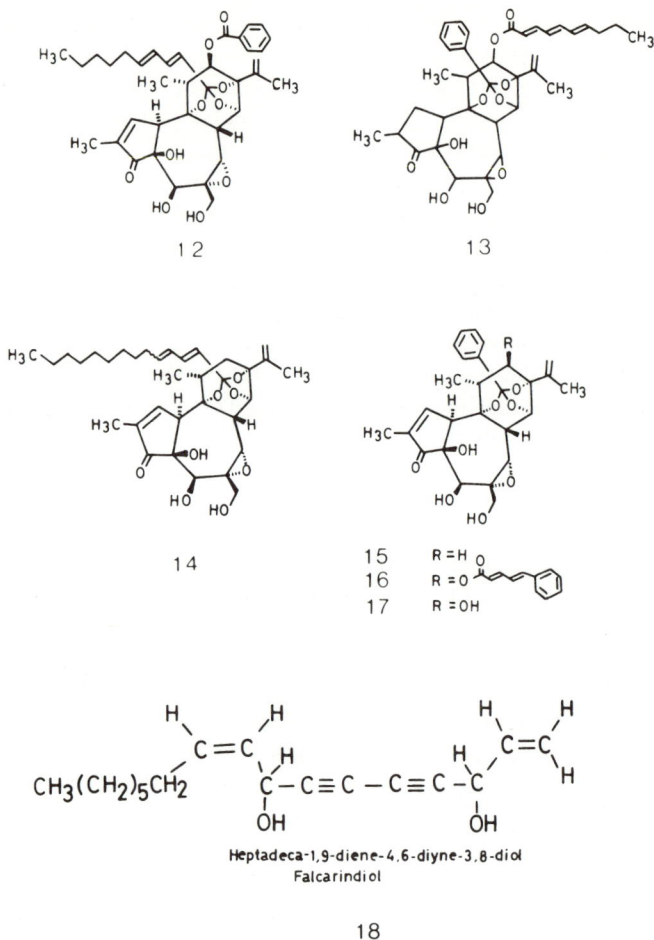

H_3C ... CH_3 ... OH ... HO HO

1 2

H_3C ... CH_3 ... OH ... HO HO

1 3

H_3C ... CH_3 ... OH ... HO HO

1 4

H_3C ... R ... CH_3 ... OH ... HO HO

15   R = H
16   R = O
17   R = OH

$CH_3(CH_2)_5CH_2$ $\overset{H}{\underset{}{}}C=C\overset{H}{\underset{}{}}\overset{H}{\underset{OH}{C}}-C\equiv C-C\equiv C-\overset{H}{\underset{OH}{C}}C=C\overset{H}{\underset{H}{}}$

Heptadeca-1,9-diene-4,6-diyne-3,8-diol
Falcarindiol

18

FIG. 18.2, nos. 12–18

(Fig. 18.2, 16) (Ronlan & Wickberg, 1971), previously isolated from *Hura crepitans* L. (Euphorbiaceae), and *D. mezereum* L., respectively. The absolute configuration of odoracin (Fig. 18.2, 12) was determined by comparing the circular dichroism datum of desbenzoyl derivative with those of daphnetoxin (Fig. 18.2, 17) (Ronlan & Wickberg, 1971). Surprisingly odoracin was identical with gnidilatidin isolated from *Gnidia latifolia* Gilg. by Kupchan, Shizuri, Summer, Haynes, Leighton & Shickles (1974). Another nematocidal substance, odoratrin (Fig. 18.2, 13), had the same molecular formula as odoracin. Odoratrin gave diacetate on acetylation, but by solvolysis with methanolic potassium hydroxide afforded a parent alcohol and methyl 2,4,6-decatrienoate. The configuration of methyl 2,4,6-decatrienoate was determined to be all *"trans"* by careful double resonance experiments in proton NMR. The plain structure of odoratrin (Fig. 18.2, 13) was elucidated by comparing proton NMR and carbon 13 NMR spectra of odoratrin with those of odoracin, but the absolute configuration has not been established yet.

Nematocidal activities of odoracin, odoratrin and their derivatives were estimated. Odoracin and odoratrin showed 70% and 96% nematode mortality at 1 ppm after 5 days, respectively. But diacetate and isopropylidene derivatives of odoracin and odoratrin were inactive. It was remarkable for the desbenzoyl derivative of odoracin to retain the same level of activity with odoracin and odoratrin.

Diterpenes having orthoester groups are also known to possess piscicidal (Sakata *et al.*, 1971; Ohigashi, Katsumata, Kawazu, Koshimizu & Mitsui, 1974), antileukemic (Kupchan *et al.*, 1974) and irritant activity (Zayed, Adolf, Hafez & Hecker, 1977).

18.5   NEMATOCIDAL FATTY ACIDS FROM *IRIS JAPONICA* THUMB.

In preliminary experiments, the benzene extracts of the roots showed strong nematocidal activity. The benzene extracts (14 g) were then subjected to silicic acid column chromatography and eluted with n-hexane, mixtures of hexane-EtOAc and methanol. The 5% EtOAc-hexane and 10% EtOAc-hexane eluate showed strong activity. The 10% EtOAc-hexane eluate (III) was rechromatographed on PTLC with 10% EtOAc-benzene and separated into four fraction (A,B,C,D). The fraction D showed strong nematocidal activity.

Fraction D dissolved in ether was extracted with 2% NaOH aqueous solution and separated into acidic fraction and neutral fraction. The acidic fraction showed strong nematocidal activity. The acidic fraction was separated by HPLC (Lichrosorb RP-2,dioxane-MeOH-water) and gave three acidic substances (Acid I, Acid II, Acid III). The spectral data of Acid I, Acid II and Acid III were identical with those of myristic acid, palmitic acid and linoleic acid, respectively. Because simple fatty acids (myristic acid and palmitic acid) showed strong nematocidal activity, nematocidal activities of various fatty acids were tested. Undecanoic acid (C-11) showed the strongest nematocidal activity (80–100% nematode mortality at 20 ppm after 24 h). Moreover, 2-undecylenic acid showed stronger activity (70–80% mortality at 10 ppm) (Table 18.1 and Scheme 18.2).

18.6   NEMATOCIDAL SUBSTANCE FROM *ANGELICA PUBESCENS*

The fresh roots (1.8 kg) of *Angelica pubescens* were extracted with methanol.

TABLE 18.1 NEMATOCIDAL ACTIVITIES OF FATTY ACIDS[a]

		200 ppm	50 ppm	20 ppm	10 ppm
Valeric acid	C-5	—	—	—	—
Caproic acid	C-6	—	—	—	—
Peragonic acid	C-9	+++	++	++	—
Capric acid	C-10	+++	++	++	—
Undecanoic acid	C-11	+++	+++	+++	+
11-Undecylenic acid	C-11	+++	+++	+++	+
2-Undecylenic acid	C-11	+++	+++	+++	++
Lauric acid	C-12	+++	+++	++	—
Tridecanoic acid	C-13	+++	+++	+	—
Myristic acid	C-14	+++	++	—	—
Palmitic acid	C-16	++	—	—	—
Heptadecanoic acid	C-17	—	—	—	—
Stearic acid	C-18	—	—	—	—
Linoleic acid	C-18	—	—	—	—
Elaidic acid	C-18	—	—	—	—
Arachidic acid	C-20	—	—	—	—

[a]80–100% nematode mortality +++, 60–80% nematode mortality ++.
30–60% nematode mortality +, 0–30% nematode mortality — for 48 h.

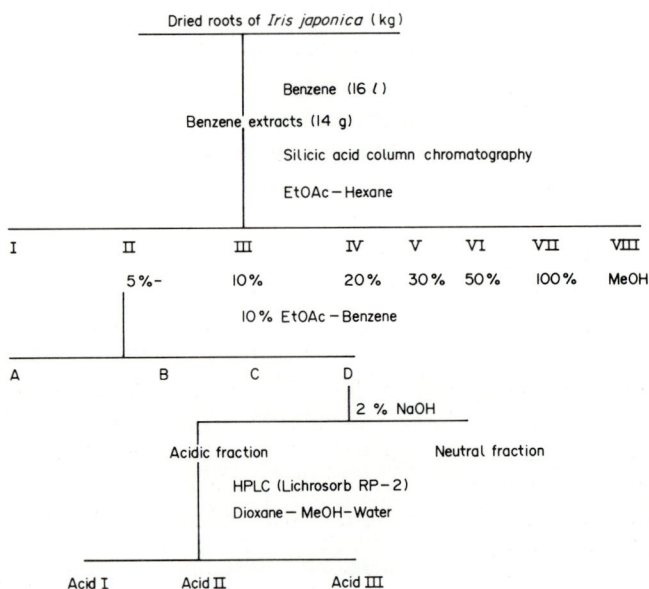

Scheme 18.2

The methanol solution was evaporated under reduced pressure. Obtained aqueous solution was adjusted to pH 2 with HCl solution and extracted with EtOAc. The EtOAc solution was evaporated to dryness. The extract was

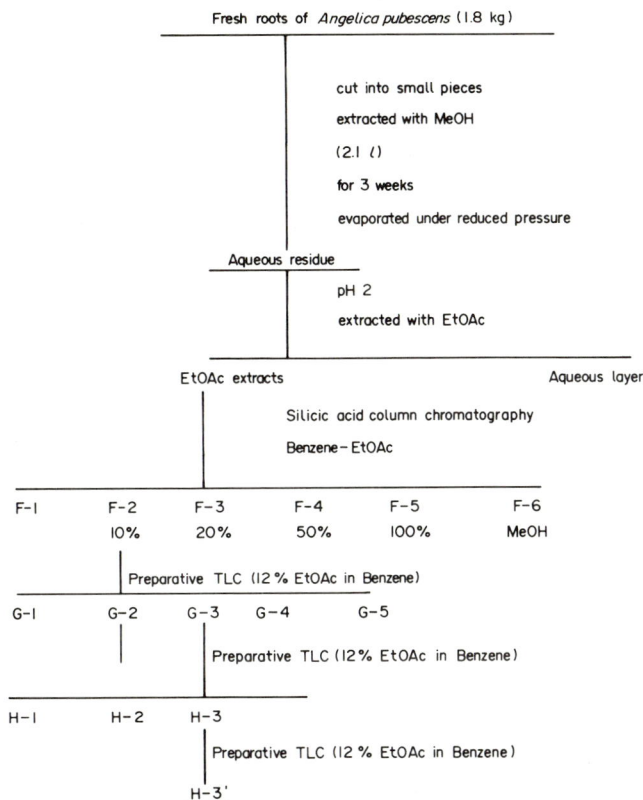

Fresh roots of *Angelica pubescens* (1.8 kg)

cut into small pieces
extracted with MeOH
(2.1 *l*)
for 3 weeks
evaporated under reduced pressure

Aqueous residue

pH 2
extracted with EtOAc

EtOAc extracts   Aqueous layer

Silicic acid column chromatography

Benzene-EtOAc

F-1	F-2	F-3	F-4	F-5	F-6
	10%	20%	50%	100%	MeOH

Preparative TLC (12 % EtOAc in Benzene)

G-1   G-2   G-3   G-4   G-5

Preparative TLC (12% EtOAc in Benzene)

H-1   H-2   H-3

Preparative TLC (12 % EtOAc in Benzene)

H-3'

Scheme 18.3

subjected to silicic acid column chromatography using EtOAc-benzene. The nematocidal eluate with benzene containing 10% EtOAc was rechromatographed by PTLC using 10% EtOAc-benzene to afford fractions containing nematocidally active substance (H–3') (Scheme 18.3). The chemical structure of H–3' was determined from spectral data. Molecular formula of H–3' was determined by high resolution mass spectra as $C_{17}H_{24}O_2$. H–3' showed IR absorption at 3610, 3380 cm^{-1} (hydroxyl group) and 2250 cm^{-1} (carbon–carbon triple bond) and 1645, 990 cm^{-1} (double bond), UV absorption at 258, 244, 242 nm (conjugated triple bonds). Carbon 13 NMR of H–3' showed the existence of disubstituted double bond and terminal vinyl group [δ ppm, 135.8 (d), 133.1 (d), 128.1 (d), 116.4 (t)] and terminal methyl group, [14.0 (q)], two secondary alcohol [62.7(d), 57.9(d)], four acetylenic carbon [79.5(s), 78.0(s), 70.2(s), 68.7(s)] and methylene group [31.6, 29.0, 27.4, 22.3(t)]. From the NMR, C-13 NMR, IR and UV spectral data, the chemical structure of H–3' was determined as heptadeca-1,9-diene-4,6-diyne-3,8-diol. The spectral data of H–3' was identified with those of Falcarindiol (Bohlman, Rohde & Zdero, 1966; Malcom 1978).

18.7 CONCLUSION

Several natural products from plants possess nematocidal activities. These compounds probably play a role as resistance factors, protecting the plants against nematode attack, and teach us the mother structure of nematocidal substances. Based on our experience with isolation and identification of nematocidal substances from plants, the following may be pointed out:

1. The chemical structures of nematocidal substances from plants extend very widely from phenolics, polyacetylenes, alkaloides, sesquiterpenes, diterpenes, to sulphur-ring compounds. It must be very difficult to generalize "structure-activity relationships" among natural nematocidal substances, because a big decrease in nematocidal activities (as in the case of polyacetylenes from *Carthamus tinctorius* and *Angelica pubescens*, and odoracin and its derivatives from *Daphne odora*, and fatty acids from *Iris japonica*) is apparent.
2. Naturally occurring nematocidal substances may be systemic in plants and decomposable in the environment.
3. Nematocidal natural products can be used as genetic index substances in selective breeding programmes for nematode-resistant plant varieties.
4. There are few investigators on the nematocidal substances from plants, and those researches offer a good opportunity to meet new compounds in the field of natural product chemistry.

18.8 SUMMARY

During the research for nematocidal substances in higher plants, four species of plants, *Carthamus tinctorius* L. (Compositae), *Daphne odora* Tumb. (Thymelaeceae), *Iris japonica* Thumb. (Iridaceae), and *Angelica pubescens* Max. (Apiaceae) were found to show nematocidal activity.

The isolation of the nematocidal components was guided by the toxicities against rice white-tip nematode, *Aphelencoides besseyi*, in immersion test. Two nematocidal polyacetylenes, 3-*cis*, 11-*trans*-trideca-1,3,11-triene-5,7,9-triyne and 3-*trans*, 11-*trans*, 11-*trans*-trideca-1,3,11-triene-5,7,9-triyne were isolated from flowers of *C. tinctorius*. Two nematocidal diterpenes having an ortho-ester group, named Odoracin and Odratrin, were isolated from roots of *D. odora*. One nematocidal polyacetylene (Falcarindiol) was isolated from roots of *A. pubescens*.

The research results of nematocidal substances from plants are reviewed and discussed.

18.9 REFERENCES

Avivardi C. (1971) Studies on the effects of nine Iranian anthelmintic plant extracts on the root-knot nematode *Meloidogyne incognita*. *Phytopath. Z.* **71**, 300–308.

Bohlman F., Rohde K. M., Zdero C. (1966) Polyacetylenverbindungen, CXVIII, Über neue polyine mit C_{17}-Kette. *Chem. Ber.* **99**, 3552–3558.

Gommers F. J. (1973) A nematocidal principle in Compositae. *Meded. Landbouwhoogesch., Wageningen* **73**-17, 1–71.

Kobayashi A., Marutomo S., Shibata Y. & Yamashita K. (1974) Abstracts of papers. *Annual meeting of Agric. Chem. Soc. of Japan, Tokyo, April 1974*, p. 161.

Kogiso S., Wada K. & Munakata K. (1976a) Odoracin, as nematocidal constituent from *Daphne odora. Agric. Biol. Chem.* **40**, 2119–2120.

Kogiso S., Wada K. & Munakata K. (1976b) Isolation of nematocidal polyacetylenes from *Carthamus tinctorius* L. *Agric. Biol. Chem.* **40**, 2085–2089.

Kogiso S., Wada K. & Munakata K. (1976c) Nematocidal polyacetylenes, 3Z,11E-and 3E,11E-trideca-1,3,11-triene-5,7,9-triyne from *Carthamus tinctorius* L. *Tetrahedron Lett.* 109–110.

Kupchan S. M., Shizuri Y., Summer W. C. Jr., Haynes H. R., Leighton A. P. & Shickles B. R. (1974) Isolation and structural elucidation of new potent antileukemic diterpenoid esters in *Gnidia* species. *J. Org. Chem.* **41**, 3850–3852.

Malcolm S. M. (1978) Falcarindiol; An antifungal polyacetylene from *Aegopodium podagraria. Phytochemistry* **17**, 1002.

Ohigashi H., Katsumata H., Kawazu K., Koshimizu K. & Mitsui T. (1974) Piscicidal constituent of *Excoecaria agallocha. Agricl Biol. Chem.* **38**, 1093–1095.

Onda M., Takiguchi T. & Hirata M. (1965) Studies on the constituents of *Bocconia cordata. Nippon Nogeikagaku Kaishi* **39**, 168–170.

Rohde R. A. (1972) Expression of resistance in plants to nematodes. *Ann. Rev. Phytopath.* **10**, 233–252.

Rohde R. A. & Jenkins W. R. (1958) Basis for resistance of *Asparagus officinalis* var *altilis* L. to the stubby-root nematode *Trichodorus christiei* Allen. *Bull. Md. Agric. Exp. Sta.* **A-97**, 1–19.

Ronlan A. & Wickberg B. (19780) The structure of Mezerein, a major toxic principle of *Daphne mezereum* L. *Tetrahedron Lett.* 1261 1261.

Sakata K., Kawazu K. & Mitsui T. (1971) Studies on a piscicidal constituent of *Huracraptans. Agric. Biol. Chem.* **35**, 2113–2126.

Scheffer F., Kickuth R. & Visser H. (1962) Root excretions from *Trifolium pratense, Medicago sativa,* and *Eragrostis curvula. Z. Pflanzenernaehr Dueng Bodenk* **97**, 26–40.

Seinhorst J. W. (1961) Plant–nematode interrelationships. *Ann. Rev. Microbiol.* **15**, 177–196.

Stout G. H., Balkenhol W. G., Poling M. & Hickernell G. L. (1970). The isolation and structure of Daphnetoxin, the poisonous principle of *Daphne* species. *J. Am. Chem. Soc.* **92**, 1070–1071.

Takasugi M., Yachida Y., Anetai M., Masamune T. & Kegasawa K. (1975) Identification of asparagusic acid as a nematicide occurring naturally in the roots of *Asparagus. Chem. Lett.* 43–44.

Turner W. B. (1971) *Fungal Metabolites,* pp. 66–70. Academic Press, London and New York.

Uhlenbrook J. H. & Bijloo J. D. (1958) Isolation and structure of a nematocidal principle occurring in *Tagetes* roots. *Int. Plant Prot. Congress* (4th), *Proc. L.* 579–580 (Hamburg).

Zayed S., Adolf W., Hafez A. & Hecker E. (1977) New highly irritant 1-alkyldaphnane derivatives from several species of Thymelaeaceae. *Tetrahedron Lett.* 3481–3482.

18.10 DISCUSSION

Marini Bettolo: You have shown that the fractionation of *Carthamus tinctorius* had led to the isolation of polyacetylenes and fatty acids which are active against nematodes. Have you any evidence of the activity and the role of Carthamin (a glycoside of carthamidine) isolated in the early 1940s from *C. tinctorius*? This substance is a flavoquinone and may be of a certain interest, because of the known biological activity of the quinones.

Munakata: No! this has not been tried.

Callow: I note with interest your demonstration of falcarindiol as a nematocide. The major phytoalexin accumulating in our tomato extracts from the inoculation of pathogen is also falcarindiol.

Bowers: There are very closely related compounds to the falcarindiols found in carrots which have a fungicidal activity.

Munakata: We have tried only for nematocidal activity.

Djerassi: Let me ask a general question. Is it really feasible to use any of these polyacetylenes as practical agents; whether it is nematocidal, fungicidal or anything else? Aren't they really too unstable for practical use?

Munakata: As I said, the fatty acids usually show very strong activity, so genetically it might be possible to increase, for example, the production of fatty acids which confer nematocidal protection.

Casida: You mentioned odoracin and odoratrin. What type of mammalian toxicity do your compounds have? Did you check that?

Djerassi: I think I can answer that one for him. He had only 1 mg of one and 20 mg of the other and he must have used some of that for structure determination. I doubt if there was much left for mammalian toxicity studies.

Casida: True, but sometimes such toxicity levels go down to only 10 μg.

Munakata: No full-scale studies have been carried out!

Casida: Do any synthetic analogues of asparaguisic acid show any activity?

Munakata: The natural compound showed the strongest activity in fact.

Djerassi: How stable is this compound?

Munakata: Very stable.

Djerassi: Can this cyclic disulphide be used as a practical nematocide?

Munakata: It might be possible to use it.

Djerassi: Have many analogues of this disulphide been made?

Munatake: Yes, a number have been made.

Pryce: There are also some disulphide analogues of nereistoxin, one of which is a commercial insecticide.

Djerassi: That is from a marine source.

Pryce: Yes, but what I am saying is that the disulphide moiety is not a problem.

Section D

Mode of Action of Phytochemicals Affecting Insects

CHAPTER 19

Phytochemical Action on Insect Morphogenesis, Reproduction and Behaviour

W. S. BOWERS

*New York State Agricultural Experiment Station and Department of Entomology
Entomology-Plant Pathology Laboratory, Cornell University, Geneva, N.Y. 14456, USA*

CONTENTS

19.1	Introduction	313
19.2	Natural Insecticides	314
	19.2.1 Insect growth regulators	316
	19.2.2 Antijuvenile hormones (Precocenes)	318
19.3	Pheromones	318
19.4	Conclusion	321
19.5	Summary	321
19.6	References	321

19.1 INTRODUCTION

Botanists and entomologists have long entertained a common belief that insects and plants, in their evolutionary interactions, have developed strategy and counter-strategy in their mutual struggle for survival. For their part, plants have developed both passive and active means of protection against insect predation. Thick cuticle waxes, sticky hairs, thorns and numerous other morphological features are easily identified as antiherbivore mechanisms (Fig. 19.1). Less obvious, but certainly as important, plants have developed a variety of chemical repellents, antifeedants and toxicants designed to protect their roots, stems, foliage and flowers from the voracious appetites of plant-feeding insects. At the same time we recognize that many plants have developed attractive chemical secretions designed to draw beneficial insects to their flowers for the purpose of pollination. Indeed plants and insects can be

Fɪɢ. 19.1. *Solanum mammosum*, an insect-resistant plant.

seen to have entered into co-operative agreements of benefit to each of the participants. In certain *Acacia* communities ants are allowed to build their nests formed from the very tissues of the *Acacia* bush, and the ants in turn keep it free of other herbivorous parasites. While these commensal relationships provide exciting glimpses into the variety and beauty of nature, the agricultural scientist is of necessity preoccupied by the chemical nature of the natural plant protectants.

The use of insecticides in the protection of our public health and in support of our agricultural endeavours has provided freedom from famine and pestilence undreamed of in previous eras. Nevertheless, the hazards of certain toxicants have made us re-examine our approaches to the development of new agrochemicals. To this end agricultural scientists are beginning to examine the chemical basis of plant resistance to insects as a starting point in the development of safer and more selective insecticides.

19.2   NATURAL INSECTICIDES

The Persians are recorded to have used the powdered flowers of the pyrethrum plant to kill insects since the mid-nineteenth century (McLaughlin, 1973). Certainly the insecticidal use of this plant predates all records. Only comparatively recently have organic chemists and entomologists turned their

Pyrethrin I

Nicotine   Anabasine

Rotenone

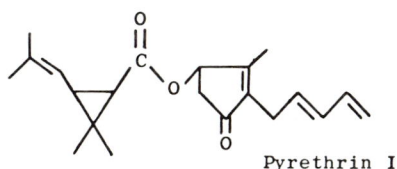

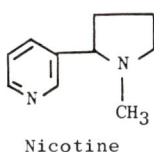

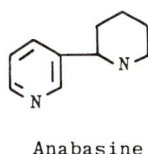

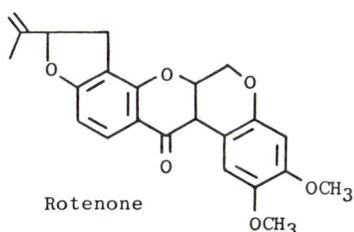

Fig. 19.2. Plant-derived insecticides.

attention to the pyrethrum plant isolating and identifying the natural ingredients (Fig. 19.2), and then synthesizing more potent analogues (for review see Elliott & Janes, 1977). Other insecticides of antiquity include the use of extracts of various solanaceous plants which are now known to contain nicotine, anabasine (Fig. 19.2) and other toxic alkaloids (Schmeltz, 1971). Rotenone (Fig. 19.2), a fish poison employed throughout the tropics, was found to be an excellent insecticide, and moreover was demonstrated to have a very low toxicity to warm-blooded animals (Fukami & Nakajima, 1971). Rotenone, and several of its natural analogues extracted from various *Derris* species, serves as a reminder today that compounds with efficient insecticidal action need not be highly toxic to man and domestic animals. The complicated structure of rotenone and the difficulty of efficient synthesis prevented its development as a modern insecticide. Neverthless, a successful cottage industry concerned with the extraction of rotenone from plants as a natural insecticide was once a flourishing business. On re-examination one finds scant attempt to identify the toxic nature of rotenone or to attempt to synthesize simpler analogues of it. The rapid development of chlorinated hydrocarbons and other more cheaply produced toxicants no doubt prevented in-depth research on the rotenoids.

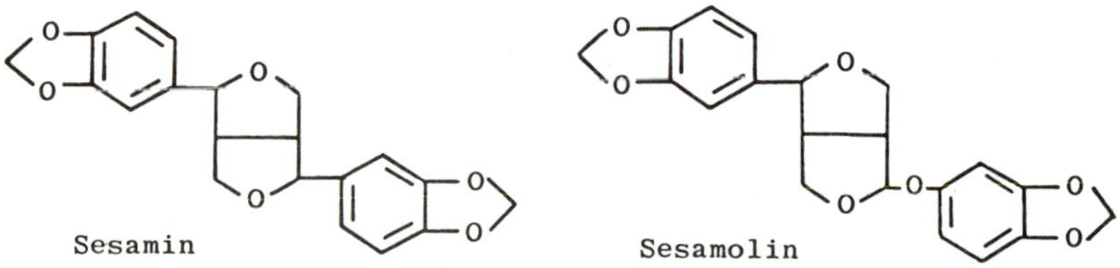

Sesamin   Sesamolin

Fig. 19.3. Plant-derived insecticide synergists.

It seems appropriate to mention that not only have plants developed a variety of toxicants, but insects in their evolutionary progress learned to overcome certain of these toxicants through metabolism. Plants respond in kind by developing inhibitors of the enzymes which metabolize the primary toxicant. It was discovered by early investigators (Eagleson, 1942) working with the natural pyrethrins that coformulation with sesame oil increased the toxicity of the pyrethrins and, on examination, two compounds (Fig. 19.3) were isolated from sesame oil which were found to account for the insecticide synergism. Metabolism studies revealed that the compounds sesamin and sesamolin inhibited microsomal enzymes in insects which otherwise were able to metabolize the plant toxicants. More recently, a thorough investigation of the extract from the pyrethrum flower (Doskotch & El-Feraly, 1969) has revealed the presence of natural synergists which heretofore had gone unnoticed. Plants have doubtless evolved dual and mutually supportive mechanisms of chemical protection against herbivorous predators. The isolation and identification of sesamin and sesamolin prompted the industrial development of the more efficient insecticide synergists piperonyl butoxide and sesamex. Thus, plants have served to provide both insecticidal as well as synergist models which were successfully optimized into more useful agricultural products.

19.2.1   Insect growth regulators

From basic studies in insect physiology we now recognize that insects regulate their growth, development, reproduction, diapause and behaviour by the use of several hormones. Of these hormones, the juvenile hormones are perhaps the best known and their physiological actions have been characterized in the greatest detail (Fig. 19.4). The development of immature insects depends upon the presence of the juvenile hormone (Wigglesworth, 1934, 1935) to allow growth and increase in size without radical change in form. Finally, the insect must change to the adult breeding stage and the remarkable changes which take place are called metamorphosis. This process must take place in the absence of juvenile hormone. Scientists found that if insects

BOWERS, et al., 1965 JH III
JUDY, et al., 1973

MEYER, et al., 1968 JH II

ROLLER, et al., 1967 JH I

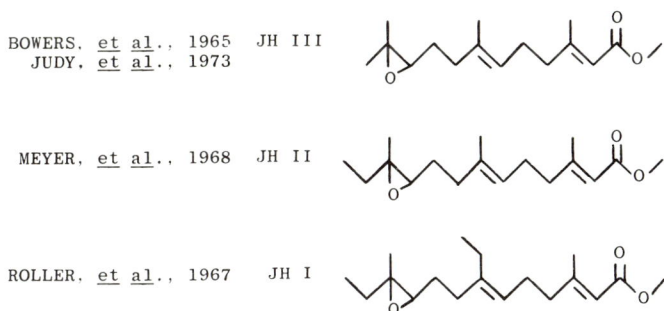

Fig. 19.4. Natural juvenile hormones.

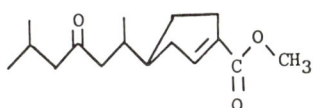

Juvabione

I   II

Juvocimenes

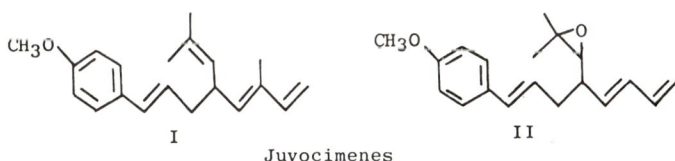

Fig. 19.5. Insect juvenile hormone mimics from plants.

undergoing metamorphosis were treated with juvenile hormone (Williams, 1956) they moulted into intermediates which were half immature and half adult, and which died without gaining reproductive competence. The identification of the juvenile hormones allowed industrial optimization of these structures into highly efficient synthetic hormone mimics, and a few of these are now used in the control of mosquitoes and manure-breeding flies. Wider application of hormonal technology to the control of agricultural pests was prevented by the fact that the juvenile hormones only control insects in the latter stages of development. Since the immature stages are those which do most of the plant destruction, the use of juvenile hormone or its analogues in their control comes too late. However narrow the use of juvenile hormone, it was of interest to find that a number of plants contain compounds with juvenile hormone activity. The balsam fir was shown to contain compounds called juvabione (Fig. 19.5), which is clearly homologous with the natural juvenile hormones (Bowers, Fales, Thompson & Uebel, 1966). More recently, two compounds (juvocimene 1 and 11, Fig. 19.5) isolated from the herb sweet basil (Bowers & Nishida, 1980) possessed juvenile hormone activity of a higher order than the natural hormones of the insects themselves. These compounds incorporate a number of structural features related to the juvenile hormones and yet retain an aromatic nucleus. The occurrence of compounds with juvenile hormone activity in plants led to the exciting speculation that plants have developed a variety of chemical defences based upon interfering with or mimicking the insect endocrine system.

19.2.2   Antijuvenile hormones (Precocenes)

Recently, compounds have been discovered in plants which have antijuvenile hormone activity (Bowers, 1976; Bowers, Ohta, Cleere & Marsella, 1976). These compounds, called precocenes, destroy the gland known as the corpus allatum (Bowers & Martinez-Parto, 1977; Unnithan, Nair & Bowers, 1977) which produces the juvenile hormone in insects. The precocenes are able to destroy the corpus allatum following oxidation within the gland to reactive intermediates which alkylate crucial cellular elements. The result is destruction of the gland and invasion of it by connective tissue (Unnithan, Nair & Syed, 1980). In the absence of the juvenile hormone, immature insects halt their voracious feeding and moult precociously into tiny adults (Fig. 19.6). Since adult insects require juvenile hormone for the development of their ovaries, the precocious adults are completely sterile. Moreover, treatment of normal adult insects with the precocenes results in their being sterilized permanently (Fig. 19.7). Certain insects which normally diapause (Bowers, 1976) by turning off the flow of their juvenile hormone are induced to go into diapause when treated with the precocenes. Such insects never awaken and eventually die. The ability to induce diapause in an insect at a time when it is ordinarily programmed to feed is of manifest utility.

Unfortunately the precocenes do not affect all insects in all ways. The entire range of antijuvenile hormone actions of the precocenes appears to be confined to hemimetabolus insects. Although certain adult holometabolus species can be sterilized, their immature stages appear to be refractory to the action of the precocenes. Continued optimization of the precocenes by synthesis (Bowers, 1977) may result in the development of analogues effective against holometabolus insects.

19.3   PHEROMONES

Insects possess a variety of chemicals useful in communication. Chemical secretions may signal an interest in mating, serve to attract siblings to a food

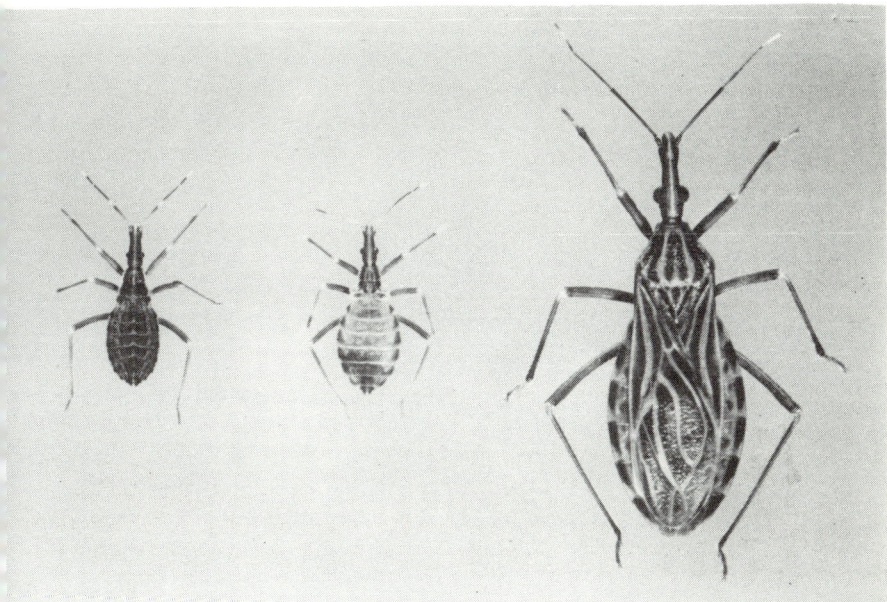

FIG. 19.6. Induction of precocious metamorphosis in *Rhodnius prolixus*. Normal third instar nymph (left). Third instar precocious adult (middle); and normal adult (right).

source, mark territorial boundaries, signal the approach of predators and parasites, and in some cases act as overt defensive weapons. In most cases it appears that insects synthesize their pheromones from simple biochemicals. However, certain insects have been found to sequester plant compounds and in turn use them in communication. In a few unique instances it would appear that plants may produce secondary chemicals which are identical with or mimic certain insect pheromones. It is uncertain at this point whether the production of insect pheromone-like compounds by plants are used for defensive purposes. Nevertheless, pheromonal activity identical with that of natural insect pheromones can be demonstrated in the laboratory. D-bornyl acetate isolated from needle and cone oil distillates of spruce and fir possess sex attractant and sex-excitant activity against the American cockroach (Bowers & Bodenstein, 1971). Similarly, germacrene D isolated from the common goldenrod (Tahara, Yoshida, Mizutani, Kitamura & Takahashi, 1975) also possesses potent sex stimulatory activity for American cockroaches. Following the identification of the natural sex attractant of this cockroach (Persoons, Verwiel, Ritter, Talman, Nooijen & Nooijen, 1976), it could be demonstrated that the nucleus of the natural attractant was based upon the germacrene structure (Fig. 19.8). At this time no explanation can be offered for the identical activity of D-bornyl acetate.

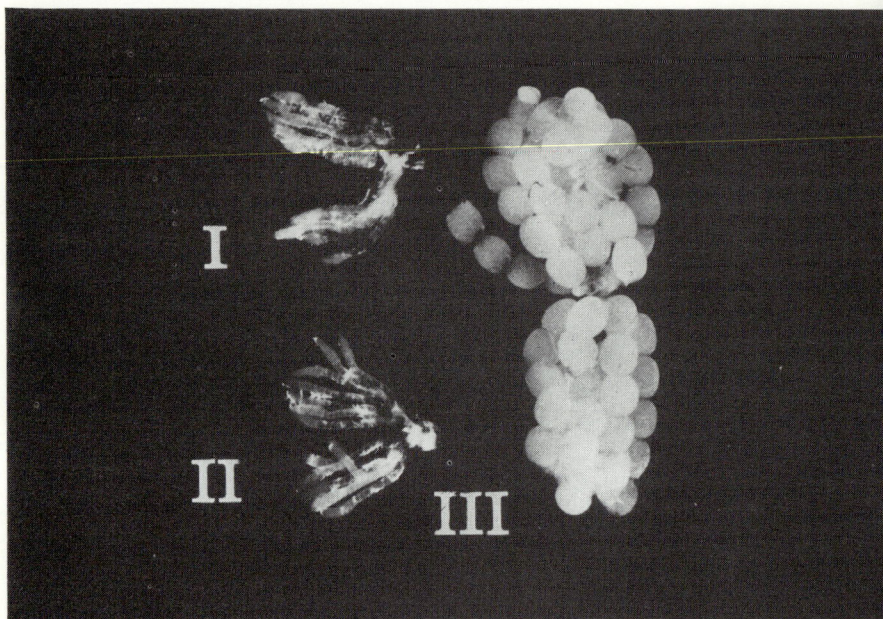

FIG. 19.7. Ovaries of precocious adult (I). Sterilized normal adult and normal developed ovaries of the cotton stainer, *Dysdercus cingulatus*.

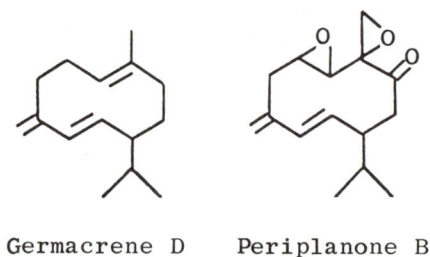

Germacrene D   Periplanone B

FIG. 19.8. Germacrene D, Periplanone B.

Aphids signal the attack of a predator by the production of an alarm pheromone from specialized structures called cornicles located on the tip of their abdomen. The structures of two of the aphid alarm pheromones have been identified as *trans*-β-farnesene and germacrene A (Fig. 19.9) (Bowers, Nault, Webb & Dutky, 1972; Bowers, Nishino, Montgomery, Nault & Nielson, 1977). Both of these compounds occur in plants, although their significance in possible protection from aphids is uncertain. Since aphids respond to their alarm pheromone by running from or jumping off of their host plant to escape the attacking predator or parasite, it would seem possible that

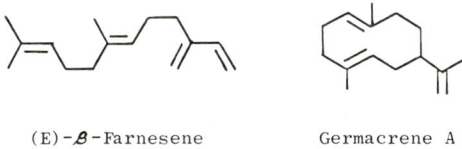

(E)-*β*-Farnesene   Germacrene A

FIG. 19.9. (E)-β-Farnesene. Germacrene A.

an alarm pheromone might be used commercially to repel foraging aphids or to cause them to leave desirable food plants.

While it is tempting to speculate that plants may have developed defensive pheromonal technologies against herbivorous insects, such protection is difficult to demonstrate and the possession of such pheromonal chemicals in plants may be altogether incidental and have no direct protective value. This is obviously a fruitful area for investigation.

19.4 CONCLUSION

The selection of chemicals for the protection of the public health and agriculture in the future clearly must be developed in such a way that the resulting compounds are highly specific to the pest species and are without toxicity to man and domestic animals. Plants are a vast natural resource of compounds developed to resist insect and plant pathogen attack. Although many of the toxicants in plants are also poisonous to man and domestic animals, plants have developed a great many unique and selective approaches to interfering with specific aspects of insect growth, development, reproduction, diapause and behaviour. We must address our best efforts to the isolation and identification of these compounds, and their optimization into eventual biorational protectants.

19.5 SUMMARY

Basic investigations into the chemistry of insect–plant interactions reveal the presence of secondary plant compounds targeted to interfere with specific aspects of insect physiology and biochemistry. Many such plant defensive chemicals are non-toxic to man and domestic animals and may serve as molecular models useful in the development of safe and selective insecticides of the future.

19.6 REFERENCES

Bowers W. S. (1976) Discovery of insect antiallatotropins. In *The Juvenile Hormones* (ed. Gilbert L. I.), pp. 394–408, Plenum Press. New York.

Bowers W. S. (1977) Anti-juvenile hormones from plants: Chemistry and biological activity. In *Natural Products and the Protection of Plants* (ed. Marini Bettolo G. B.), Proc. Pontifical Acad. Sci., Scripta Varia 41, Vatican City, Italy.

Bowers W. S. & Bodenstein W. G. (1971) Sex pheromone mimics of the American cockroach. *Nature* **232**, 259–261.

Bowers W. S., Fales H. M., Thompson M. J. & Uebel (1966) Juvenile hormone: identification of an active compound from the balsam fir. *Science* **154**, 1020–1022.

Bowers W. S. & Martinez-Pardo R., (1977) Antiallatotropins — inhibition of corpus allatum development. *Science* **197**, 1369–1371.

Bowers W. S., Nault L. R., Webb R. E. & Dutky S. R. (1972) Aphid alarm pheromone: isolation, identification and synthesis. *Science* **177**, 1121–1122.

Bowers W. S. & Nishida R. (1980) Juvocimenes: Potent juvenile hormone mimics from sweet basil. *Science* **209**, 1030–1032.

Bowers W. S., Nishino C., Montgomery M. E., Nault L. R. & Nielson M. W. (1977) Sesquiterpene progenitor Germacrene A: An alarm pheromone in aphids. *Science* **196**, 680–681.

Bowers W. S., Ohta T., Cleere J. S. & Marsella P. A. (1976) Discovery of insect anti-juvenile hormones in plants. *Science* **193**, 542–547.

Bowers W. S., Thompson M. & Uebel E. C. (1965) Juvenile and gonadotropic hormone activity of 10,11-epoxy-farnesenic acid methyl ester. *Life Sci.* **4**, 2323–2331.

Doskotch R. W. & El-Feraly F. S. (1969) Isolation and characterization of (+)-sesamin and -cyclopyrethrosin from pyrethrum flowers. *Can. J. Chem.* **47**, 1139–1142.

Eagleson C. (1942) Sesame oil as a synergist for pyrethrum insecticides. *Soap and Sanit. Chem.* **18**, 125–127.

Elliott M. & Janes N. F. (1977) Chemistry of the natural pyrethrins. In *Pyrethrins* (ed. Casida J. E.). Academic Press, San Francisco, CA.

Fukami H. & Nakajima M. (1971) Rotenone and the rotenoids. In *Naturally Occurring Insecticides* (eds. Jacobson M. and Crosby, D. G.), pp. 77–95. Marcel Dekker, Inc., New York.

Judy K. J., Schooley D. A., Dunham L. L., Hall M. S., Bergot B. J. & Siddall J. B. (1973) Isolation, structure and absolute configuration of a new natural insect juvenile hormone from *Manduca sexta*. *Proc. Natl. Acad. Sci., USA* **70**, 1509–1513.

McLaughlin G. A. (1973) History of Pyrethrum. In *Pyrethrum* (ed. Casida J. E.). Academic Press, San Francisco, CA.

Meyer A. S., Schneiderman H. A., Hanzman E. & Ko J. H. (1968) The two juvenile hormones from the cecropia moth. *Proc. Natl. Acad. Sci., USA* **60**, 853–860.

Persoons C. J., Verweil P. E. J., Ritter F. J., Talman E., Nooijen P. J. F. & Nooijen W. J. (1976) Sex pheromones of the American cockroach, *Periplaneta americana*: A tentative structure of periplanone B. *Tetrahedron Lett.* 2055–2058.

Roller H., Dahm K. H., Sweeley C. C. & Trost B. M. (1967) Die struktur des juvenilhormons. *Angew. Chem.* **79**, 190–191.

Schmeltz I. (1971) Nicotine and other tobacco alkaloids. In *Naturally Occurring Insecticides* (eds. Jacobson M. & Crosby D. G.), pp. 99–131. Marcel Dekker, Inc., New York.

Tahara S., Yoshida M., Mizutani J., Kitamura C. & Takahashi S. (1975) A sex stimulant to the male American cockroach in Compositae plants. *Agric. Biol. Chem.* **39**, 1517–1518.

Unnithan G. C., Nair K. K. & Bowers W. S. (1977) Precocene-induced degeneration of the corpus allatum of adult females of the bug *Oncopeltus fasciatus*. *J. Insect Physiol.* **23**, 1081–1094.

Unnithan G. C., Nair K. K. & Syed A. (1980) Precocene-induced metamorphosis in the desert locust, *Schistocerca gregaria*. *Experientia* **36**, 135–236.

Wigglesworth V. B. (1934) The physiology of ecdysis in *Rhodnius prolixus* (Hemiptera) 11. Factors controlling moulting and metamorphosis. *Quart. J. Micr. Sci.* **77**, 191–222.

Wigglesworth V. B. (1935) Functions of the corpus allatum of insects. *Nature* **136**, 338.

Williams C. M. (1956) The juvenile hormone of insects. *Nature, Lond.* **178**, 212–213.

CHAPTER 20

The Mode of Action of Pro-allatocidins

*Agricultural Research Council, Unit of Invertebrate Chemistry and Physiology,
University of Sussex, Brighton, Sussex, BNI 9RQ, England*

CONTENTS

20.1 Introduction   323
20.2 Lethal Bioactivation of Precocenes   325
20.3 Tissue and Species Specificity of Precocenes   330
20.4 Chemical Reactivity of Precocene-Epoxides   335
20.5 Structure/Activity Optimization   338
20.6 The Cancer Risk from Precocenes   342
20.7 Conclusion   344
20.8 Summary   348
20.9 Acknowledgements   348
20.10 References   349
20.11 Discussion   353

20.1 INTRODUCTION

The precocenes and certain of their analogues are the only known compounds, whether of natural or synthetic origin, which cause profound morphogenetic effects in larvae and adults of certain sensitive insect spp. through the induction of specific cellular necrosis in the corpora allata (CA). The CA is the physiological source of juvenile hormone whose principal, but not exclusive, roles are the maintenance of the larval condition up until metamorphosis, and then the positive stimulation of gonadal development, with associated phenomena, in the final adult stage, particularly the female. Clearly, these chromene derivatives are of exceptional interest both as experimental probes in fundamental research (offering the possibility of specific non-surgical allatectomy) and as possible model compounds from which might be developed a conceptually new type of insect growth regulator, the "fourth generation insecticide" (Bowers, Ohta, Cleere & Marsella, 1976). It has been widely recognized for many years that insect growth regulators (IGRs) of the juvenile hormone analogue type (e.g. Methoprene[R]) are of inherently little value for the control of phytophagous pests, since they are without effect on the larval feeding stages which cause crop damage, and only operate during a narrow time-window prior to metamorphosis. In contrast, any compound which permanently inactivates (or destroys) the CA can terminate somatic growth or egg production, and associated feeding, at almost any stage in the life cycle. There may be a delay of only one or two moult cycles before these effects are manifest (Bowers *et al.*, 1976; Brooks, Pratt & Jennings, 1979a) so that, at least in principle, they could be efficiently deployed against both larval and adult pests. In this respect, growth regulators of the precocene type have a better chance than those of the juvenile hormone type, in withstanding objective comparison with conventional insecticides of the neurotoxic/neurotransmitter type. The precocenes themselves are unlikely to find use as field control chemicals: they are relatively photo-labile, and they are only effective against a limited range of insect species, and have rather low specific activities. It is agriculturally noteworthy that the Lepidoptera are largely insensitive to precocenes.

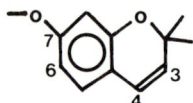   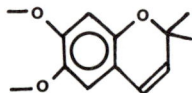   

     precocene I   precocene II   C 16 juvenile hormone

Following the dramatic discovery of precocenes (Bowers, 1976), there have been several attempts to elucidate their mode of action with a view to understanding their species-limited action, and low specific activity. Our progressively better understanding of their action is reflected in the changing fashions of their nomenclature over the last few years, which now warrants clarification. The original term "antijuvenile hormone" (Bowers *et al.*, 1976) implies that precocenes antagonize juvenile hormones. In fact, the prescriptive diagnosis of allatocidal activity of precocenes (the acute reversibility of precocene action by simultaneous administration of juvenile hormones or analogues) demands the absence of such a negative interaction. The term "antiallatotropin" (Bowers & Martinez-Pardo, 1977) was coined at a time when some authors believed that precocenes blocked a stimulatory transaction of the brain on the CA (mediated by the unidentified allatotropin), but no direct evidence has been forthcoming. It now appears correct to interpret the observed effects of precocene treatment on brain neurosecretory cells in *Oncopeltus fasciatus* (Unnithan, Nair & Kooman, 1978) as being a feed-back consequence of allatal necrosis, rather than its cause. The term "antiallatal agent" (Pratt & Bowers, 1977) seems to be correct, for it implies a direct action of precocenes on the CA. Recent studies in our laboratory (Brooks *et al.*, 1979a; Pratt, Jennings, Hamnett & Brooks, 1980) suggest that precocenes are necrogenic in the CA because of their *in situ* massive bio-oxidation to a highly reactive epoxide which is the ultimate cytotoxin. Accordingly, we have adopted the term "pro-allato-cidins" for precocenes (Hamnett, Pratt, Ottridge & Alexander, 1981a) by analogy with the use of the term "pro-carcinogen" to classify, for example, poly-nuclear aromatic hydrocarbons, which are bioactivated to ultimate carcinogens. Other well-known oxidative bioactivations, but of a non-cell-specific nature, are the insecticidal conversions of aldrin to dieldrin, and of malathion to malaoxon (Brooks, 1979).

In this paper, we attempt to summarize present knowledge on the mode of action of precocenes, and to use this to develop guidelines for further studies aimed at defining the bases for species-restricted activity and low specific activity in those species which are sensitive. We reaffirm (Pratt *et al.*, 1980) that the epoxidase in the CA must be a focal point for studies of both species specificity and structure/activity optimization. We also draw cautionary parallels between the modes of action of precocenes and of poly-aromatic hydrocarbon, (PAH) the mammalian pro-carcinogens.

20.2   LETHAL BIOACTIVATION OF PRECOCENES

At the present time, the only evidence which indicates directly the crucial metabolic activation of precocenes comes from studies of the metabolism of precocene-I by CA from adult female *Locusta migratoria* (Pratt *et al.*, 1980), although the demonstrated antagonism of precocenes, *in vivo*, by two methylenedioxyphenyl analogues (mixed function oxidase inhibitors) in larval *L. migratoria* and *Oncopeltus fasciatus* is strong supporting evidence for a wider application of the hypothesis (Brooks *et al.*, 1979a; Brooks, Hamnett, Jennings, Ottridge & Pratt, 1979b). We chose *L. migratoria* for these experiments according to a set of pragmatic dictates (Pratt & Weaver, 1978): (a) it provides a good example of chemical allatectomy (albeit at high doses, *vide infra*) with no known obfuscation by general toxic or lethal reactions at effective allatocidal doses, (b) the CA of the

adult female are large (up to 30 nl) which greatly facilitates biochemical
analysis, in comparison with the glands from larval stages, and (c) the
CA synthesize a known juvenile hormone (C_{16} JH) whose synthesis can
be readily monitored *in vitro* by short-term radiochemical methods
already developed in our laboratory (Pratt & Tobe, 1974; Pratt, Tobe,
Weaver & Finney, 1975; Hamnett & Pratt, 1978; Feyereisen R., Pratt
G. E. & Hamnett A. F., unpublished). The latter dictate marks a clear
distinction in favour of the better studied Orthoptera or Dictyoptera,
rather than the more sensitive Hemiptera *O. fasciatus* and *Rhodnius
prolixus*, whose juvenile hormone has not yet been chemically identified,
so we have no relevant knowledge of the biochemistry of their CA.

The CA of adult female *L. migratoria* rapidly metabolize radio-labelled
precocene-I *in vitro* (up to 400 pmol per gland pair h^{-1}) and
simultaneously effect extensive covalent modification of cellular macromolecules (Pratt *et al.*, 1980). Both metabolic conversions are similarly
inhibited by a mixed function oxidase inhibitor modelled on the natural
substrate of the CA epoxidase (methyl farnesoate), and to the same
extent as the conversion of methyl farnesoate to C_{16} JH. Furthermore,
the quantitative distributions of microsomal enzyme activities responsible for precocene-I metabolism, methyl farnesoate epoxidation (both
NADPH dependent) and NADPH-cyt_c reductase, on linear sucrose
density gradient analyses of CA homogenates, are virtually identical
(Feyereisen & Pratt, unpublished). Finally, the 3,4-dihydrodiols of
precocene-I, produced both by incubation of CA with precocene-I and
by chemical hydrolysis of synthetic precocene-I 3,4-epoxide, contain a
high proportion of *cis* isomer, indicative of a carbonium ion intermediate
during epoxide solvolysis (Pratt *et al.*, 1980; Jennings & Ottridge, 1979;
Hamnett, Ottridge, Pratt, Jennings & Stott, 1981b). In contrast to the
CA, wherein the 3,4-dihydrodiol is the sole extractable metabolite,
preliminary studies on isolated non-target tissues and whole insects
reveal a complex spectrum of metabolites and relatively little macromolecular labelling (Pratt, Jennings & Hamnett, 1981). This augments
published studies on the metabolism of precocene-II by several other
spp. of intact insect and isolated non-target tissues (Ohta, Kuhr &
Bowers, 1977; Soderlund, Messequer & Bowers, 1980; Bergot, Judy,
Schooley & Tsai, 1980) and suggests that the bulk of the primary
metabolites of precocenes are produced by peripheral, non-target
tissues, and consist mostly of O-demethylated and 3,4-dihydrodiol
products, which are subsequently conjugated prior to excretion. There
may be important differences between species in respect of both the
principle pathways and the intensity of peripheral detoxification. These
findings are summarized qualitatively in Fig. 20.1, which highlights the
dual opportunities for "disarming" the precocene molecule: (a) by

Flow sheet for Precocene I

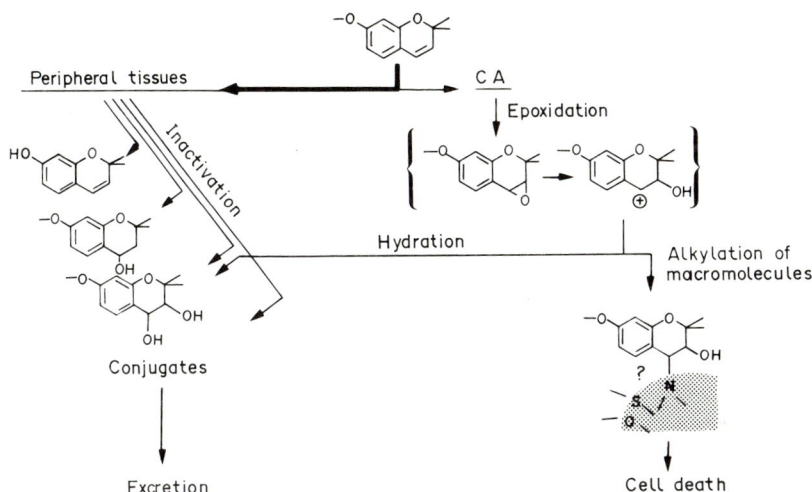

Fig. 20.1. Simplified metabolic flow-sheet for precocenes in insects, based upon the findings of Ohta *et al.* (1977); Pratt *et al.* (1980); Soderlund *et al.*, (1980); and Bergot *et al.* (1980). The chemical nature of the bond formed between precocene moieties and macromolecules is unknown, and might involve hydroxyl, thiol, amino/imino or phosphate groups. The 3,4-diols arise by epoxidation followed by non-enzymic hydration (Burt *et al.*, 1978; Soderlund *et al.*, 1980). The 6- or 7-hydroxy metabolites probably arise by oxidative O-demethylation (Bergot *et al.*, 1980; Pratt *et al.*, 1980b) of the possible chromanols, only the 3-chromanol (not shown) has been found thus far and the mechanism of its formation is presently obscure (Soderlund *et al.*, 1980). The balance of these metabolites varies appreciably between different species, as may be the proportion of precocene which interacts with the Ca.

detoxification of the precocene by oxidative metabolism in the very much larger mass of peripheral tissue (e.g. fat body, mid-gut, Malpighian tubules), and (b) by neutralization of the carbonium ion formed spontaneously from the biosynthesized epoxide, by water or some other consumable nucleophile. We have shown that ortho-phosphate anion at neutral pH is an effective scavenger for the precosyl carbonium ion (Hamnett *et al.*, 1981a), as expected from work with other carbonium ion generators in vertebrate systems (Gamper, Tung, Straub, Bartholomew & Calvin, 1977), and it is to be expected that thiols such as reduced glutathione would also be effective (Jollow, Mitchell, Zampaglione & Gillette, 1974; Sarrif, McCarthy, Nesnow & Heidelberger, 1978) although this has yet to be demonstrated in the case of precocene epoxide. The epoxide might also suffer inactivating reduction to the 3-chromanol by reaction with cellular nicotinamide adenine dinucleotide phosphate (Yang & Gelboin, 1976). Our failure to find evidence of such polar adducts in the *L. migratoria* CA suggests that the sensitive tissue is not protected by high intracellular levels of detoxifying nucleophiles, and this may contribute towards tissue sensitivity. The position of peripheral non-target tissues in this respect deserves closer investigation.

These studies provide a clear insight into the nature of the interaction between precocenes and the CA of one sensitive insect, but they must be placed on the correct quantitative basis if we are to accept them as representative of the mode of action *in vivo*. This requires knowledge of the circulating titres of precocene during an effective exposure. This in turn requires knowledge of the effective dose for a particular species and its stage of development. Previous studies showing the inhibitory effects of precocenes in CA *in vitro* are not directly interpretable, for want of this information (Pratt & Bowers, 1977; Muller, Masner, Kalin & Bowers, 1979). Therefore, we found the effective doses of precocene-I to both fourth instar and adult female *L. migratoria*, and then used radio-labelled material to determine the heamolymph titres of precocene following effective and sub-effective doses (Pratt *et al.*, 1981).

Most of the founding studies on the dose/effectiveness of precocenes and their synthetic analogues have centred on the induction of precocious metamorphosis in larval stages, which provides an indirect indication of chemical allatectomy, quite suited to routine screening (Bowers, 1977; Chenevert, Perron, Paquin, Robitaille & Wang, 1980; Brooks *et al.*, 1979b). It appears that those doses which afford the complete induction of prothetely are those which induce complete necrosis of the CA (Pener, Orshan & deWilde, 1978; Unnithan *et al.*, 1980). We find a sensitivity of fourth instar *L. migratoria* to precocenes I and II which is quite similar to that found in other laboratories (Brooks *et al.*, 1979a; Pedersen, 1978; Chenevert *et al.*, 1980). In the case of precocene-I, applied or injected to fourth instars, 70 µg affords complete prothetely and 7–10 µg gives no observable effect. The corresponding peak titres of precocene in the haemolymph, attained after ½–1 h, are $1–2 \times 10^{-4}$ M precocene-I for a few hours appears to induce irreversible cell death in the fourth instar CA (Pratt *et al.*, 1981). The situation in the adult female is slightly more complicated, from the point of view of setting up the bioassay. We have measured not only a convenient overall expression of the effects of chemical allatectomy (length of terminal oocytes), but also two direct parameters: the volume of the CA 7 days after treatment, and their enzymic competence to utilize exogenous farnesenic acid to support maximum C_{16} JH biosynthesis (i.e. O-methyl transferase and 10,11-epoxidase enzymes) (Pratt & Tobe, 1974; Tobe & Pratt, 1976). Animals were injected prior to the onset of vitellogenesis (day 5) and sacrificed on day 12, towards the end of vitellogenesis in the first wave of oocytes in control animals. We found a markedly biphasic response in terms of CA volume, and the enzyme activity which varied coordinately with it. The CA were diminished approximately 70% after doses of 100 μg, without there having been any significant effect upon the course of vitellogenesis. Complete suppression of vitellogenesis was obtained at doses of 600–800 μg, which also caused complete necrosis of the CA. Apparently, much of the synthetic competence of the CA can be lost by exposure to the lower doses, without incurring any manifestation of hormonal insufficiency in the first wave of oocytes. This could be interpreted as evidence of "excess capacity" in the CA, but we urge caution here, since these preliminary studies give no evidence of the pharmacological time-course, which may be complex. Other studies, *in vitro*, have shown that exposure to low levels of precocene-I actually stimulates spontaneous juvenile hormone synthesis in these CA, in contrast to the acute inhibition of synthesis in CA from *Periplaneta americana* at high levels of precocene-II (Pratt & Bowers, 1977); this interesting effect, which would be expected to complicate the dose/response relationship in any indirect bioassay, is currently the subject of study in our laboratory. Leaving this complexity aside, the allatocidal dose of precocene-I to adult female *L. migratoria* is 600–800 μg, by injection, which affords a peak haemolymph titre of $1–2 \times 10^{-5}$ M precocene (Pratt *et al.*, 1981). Clearly, the adult CA are approximately 10 times more sensitive to precocene than the larval CA. The fact that rather similar doses per unit fresh body weight are required to allatectomize fourth instars and adults *in vivo* suggests a much greater detoxifying capacity in the adult.

As already reported (Pratt *et al.*, 1980) exposure of adult female *L. migratoria* CA to $1–2 \times 10^{-5}$ M precocene-I *in vitro* results in extensive labelling of cellular macromolecules, which we presume to be the instrument of cell death. Although the biochemical progress of cellular pathology in the CA is currently unexplored, the evidence so far is suggestive of a non-selective shotgun attack on the cell: certainly, a large number of cellular proteins and sub-cellular organelles suffer covalent modification. It is of interest, although inexplicable at present, that the epoxidase itself appears to be relatively immune to precocene action, even during exposure for several hours at high concentration (4×10^{-4} M). Recent studies, in vertebrate systems, with several cytotoxins, including alkylating agents, have suggested that the induction of calcium permeability in the cytomembrane (perhaps a result of the alkylation of some vital membrane component(s) may provide an important amplification of the cytotoxic effect (Schanne, Kane, Young & Farber, 1979). Cytological studies on the precocene-induced demise of CA from larval *L. migratoria* have shown that a pathological permeability to the complex colloidal dye Evans Blue sets in about 4 h after exposure to precocene (Schooneveld, 1979), which further indicates an important loss of membrane integrity. In this connection it may be relevant to note that many CA, including those of *L. migratoria*, contain much endoplasmic reticulum (smooth or rough) and large surface areas resulting from cellular interdigitation (Cassier, 1979). These extensive membrane boundaries to the CA cells, being the site of methyl farnesoate and precocene epoxidation, and apparently the natural exit for juvenile hormone into the extracellular spaces (Tobe & Pratt, 1974, 1976), may afford above-average sensitivity to alkylating agents. Clearly, much more work is needed to support a detailed picture of the biochemical pathology of CA during precocene-induced necrosis.

20.3 TISSUE AND SPECIES SPECIFICITY OF PRECOCENES

We do not yet know enough about the balance of metabolism in different tissues from a wide range of insect spp. to be able to produce an exclusive categorical model of precocene susceptibility. However, the available evidence does provide clues to at least some factors of likely importance. Considering first the question of tissue specificity, it must be recognized that no detailed pharmacological study has been carried out on any of the spp. which show "classical chemical allatectomy", i.e. those in which the only superficially observable effect is the induction of juvenile hormone deficiency, which is experimentally reversible by juvenile hormone therapy. A widely embracing biochemical study might well reveal a spectrum of other, but "sub-clinical", lesions. Slama (1978) has cautioned that precocenes may well have antifeedant effects in some insects, although we know of no evidence that this is true in the two species most studied to date, *Oncopeltus fasciatus* and *Locusta migratoria*. More alarmingly, precocenes exert lethal effects on many insects, sometimes at doses less than one order of magnitude greater than those required to elicit chemical allatectomy (Chenevert *et al.*, 1980), but the mechanisms are completely unstudied. Hopefully, a detailed toxicological study will be forthcoming, which will at least indicate whether a specific bioactivation is common to both insecticidal and allatocidal actions of precocenes. Until that information is available, we may only speculate on why the CA of certain spp. happen to be the most crucially sensitive in the body. Simple logic suggest that the specific activity of an appropriate epoxidase (enzyme units per unit weight cellular protein), and absence of detoxifying systems (alternative oxidases or consumable nucleophiles), may be important factors. The extremely high chemical reactivity of precocene epoxide (Hamnett *et al.*, 1981b) makes it unlikely that epoxide hydrases could have a significant protective effect in non-target tissues (they are absent in the CA), as already deduced (Burt,

Kuhr & Bowers, 1978). The especial sensitivity of the CA may simply result from the high degree of biochemical specialization of this tissue towards the synthesis and rapid release of a harmless, and relatively unreactive epoxide (juvenile hormone). It may be helpful to summarize these thoughts in the form of a provisional checklist of features which must be satisfied if a precocene is to be an effective antiallatal agent:

1. Physical access to the CA is assured, whether from the gastro-intestinal tract or cuticle (via the haemolymph) or tracheolar supply. (There is presently no direct evidence for the importance of a tracheolar route, but the sensitivity of some Hemiptera and Orthoptera to contact/vapour exposure (Bowers, 1976; Masner, Bowers, Kalin & Muhle, 1979; Unnithan *et al.*, 1980) suggests that this possibility must not be overlooked.)

2. Under circumstances where the CA are exposed via the haemolymph circulation, the extent of peripheral metabolism must be sufficiently low as to maintain effective titres of precocene. This applies whether or not peripheral metabolism is potentially harmful (epoxidative) or directly detoxifying (O-demethylation, hydroxylation, conjugation, etc.).

3. The CA have, as part of their natural anabolic enzymic machinery, a high concentration of P_{450} epoxidase capable of accepting precocenes as lethal alternative substrates (*vide infra*).

4. The CA are biochemically ill-equipped to detoxify the carbonium ion resulting from spontaneous dative-bond catalysed opening of the oxirane ring in precocene-3,4-epoxide. (There is as yet no comparative information of the pathways of metabolism of precocenes in CA from sensitive and insensitive spp.)

A discussion of species specificity of precocenes proceeds naturally from these preliminary considerations of tissue specificity, but with similar uncertainty. Studies on insects from Hemiptera, Orthoptera, Dictyoptera, Coleoptera, Diptera, Hymenoptera, Isoptera and Lepidoptera (and a species of soft tick) have afforded results ranging from undetectable, through ill-defined to clear-cut chemical allatectomy (Tarrant & Cupp, 1978; Chenert *et al.*, 1980; Bowers *et al.*, 1976; Masner *et al.*, 1979; Pound & Oliver, 1979; Walker, 1978; Dietz, Hermann & Blum, 1979; Kelly & Fuchs, 1978; De Loof, Van Loon & Vanderrost, 1979; Cupp, Lok & Bowers, 1977; Mackauer, Nair & Unnithan, 1979; Pener *et al.*, 1978). Whilst some insects respond well to precocenes, there are some spp. in which it is difficult, perhaps impossible, to find clear evidence of an antiallatal effect. Both the fundamental and applied scientist will be curious to know whether or not this distinction has an amenably quantitative basis, which might be overcome by modified methods of application or exposure, or by structure/activity optimization through the use of synthetic analogues. Regrettably, we are unable to answer this question satisfactorily, and attempting to do so merely outlines several gaps in our fundamental knowledge of CA biochemistry, which we outline below.

We first discuss the possible importance of application methods and structure/activity relationships to species specificity. It is now quite clear that the mode of application of precocenes has a marked effect upon their efficiency; the method of contact/fumigation first routinely employed by Bowers and co-workers has proved to be the most widely efficacious, which may be encouraging from the point of view of selected pest-control applications. To take some examples from our own work: larval *Rhodnius prolixus* respond well to precocene-II by jar or filter-paper deposits (cf. Tarrant & Cupp, 1978), but if the operator "misses" the paper and instead applies the equivalent dose directly to the insect, no effect is observed. Increasing the topical dose beyond the equivalent vapour-effective dose (20 μg cm^{-2} per 10 insects) incurs insecticidal rather than antiallatal responses. Fourth instar *Locusta migratoria* 2–6 h old, are equally sensitive to precocenes by oil injection or by topical abdominal application in acetone, whereas 5-day-old adult females are completely insensitive (up to insecticidal levels) by the topical route. Unlike *L. migratoria*, larval *Schistocerca gregaria* are very insensitive to precocenes by either topical or injection routes (Chenevert *et al.*, 1978; Pratt, unpublished results), whereas Unnithan *et al.* (1980) found comparatively good sensitivity to precocene-II by fumigation. The efficiency of fumigation may also depend upon the volatility of the particular pro-allatocidin, that of precocene-I being considerably greater than that of precocene-II (unpublished results from this laboratory) and whether or not an oily reservoir is employed. These findings are sufficient to indicate that care should be taken in interpreting a negative result from a single type of laboratory bioassay (Brooks *et al.*, 1979b) and, conversely, make it possible that more spp. may be sensitive to precocenes than is presently apparent. One is reminded of analogous studies on the sensitivity of several insects to the juvenilizing effects of juvenile hormone mimics, in which it was found that too rapid an introduction of the active principle into the insect circulation sometimes resulted in greatly lost efficiency (Gilbert & Schneiderman, 1960; Wigglesworth, 1969; Cruickshank & Palmere, 1971; Patterson & Schwarz, 1977). The same may apply to precocenes and their analogues, suggesting that improved fumigation techniques could be beneficial. At the present time there are few examples from structure/activity studies of marked interspecific differences in analogue sensitivity (Brooks *et al.*, 1979b), but this may simply reflect the relatively limited number of combination of analogue and species tested.

The paucity of comparative information on the biochemistry of both CA and precocenes, relative to the entire insect Class, provides no firm basis of fundamental knowledge upon which to base predictions or explanations of the antiallatal activity of precocenes in different insects. More constructively, what we do know about the mode of action of precocenes in *L. migratoria* may be of use in helping us to decide what lines of fundamental inquiry might best yield that required basis. It may be helpful to consider the position according to a three-dimensional model, in which the overall effect of the precocene is a balance between allatocidal action, peripheral detoxifying metabolism and (unspecified) insecticidal action (Fig. 20.2). The least well-defined components of this simple model are the radiations towards insecticidal action, the

Action of Precocenes

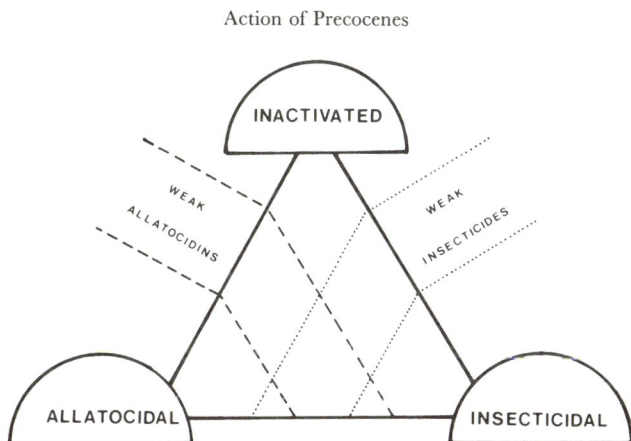

Fig. 20.2. A classical model to represent the three possible interactions of putative proximal allatocidins within the insect. Judging by the quantities of precocenes required to affect different species, they can be classified as weakly allatocidal and weakly insecticidal. Structure optimization aims to achieve more efficient and powerful interaction with the CA, whilst diminishing the rate of inactivation in peripheral tissues. The mechanisms of insecticidal action of precocene is unknown.

mechanisms of which is presently quite unknown. We need to know whether insecticidal action involves (a) epoxidation of precocenes in tissues other than the CA, and (b) all or just some other tissues (e.g. CNS). Although it is only limited to indirect evidence, our finding that a variety of compounds of this type, but with and without allatocidal activity, have rather similar insecticidal activities in *L. migratoria* (Brooks, Pratt & Ottridge, unpublished results), suggests that the insecticidal activity of precocenes may be independent of their ability to suffer oxidation to a highly reactive epoxide. However, the subjects need further study. From the point of view of the possible use of precocene analogues for pest control, insecticidal action may only be disadvantageous if it correlates with some other, undesired, property (e.g. vertebrate or pest-predator toxicity), and this we cannot yet exclude.

The model indicates that there are three important components which might vary quite independently of each other, depending upon the insect spp. (including the stage in the life cycle), the nature of the precocene analogue under study and the mode of administration. Shifting the balance towards the desired pole (allatocidal action) can be achieved either by strengthening that interaction, or by diminishing the other two interactions. Some spp. may be insensitive to precocenes because of enhanced oxidative detoxification in non-target tissues (Burt *et al.*, 1978), perhaps associated with polyphagy (Krieger, Feeny & Wilkinson, 1971). The importance of peripheral detoxification might depend upon the route of administration of the pro-allatocidin. It seems inherently unlikely that this detoxification could be blocked by co-administration of metabolic inhibitors, without simultaneously preventing the allatocidal action of precocenes (Brooks *et al.*, 1979a). Some spp. may be insensitive to precocenes because of a weak interaction with the CA. This might arise through (a) a greater substrate specificity of the epoxidase, leading to low rates of lethal bioactivation of precocenes, or (b) a markedly reduced level of epoxidase activity in the CA, affording the same consequence, or (c) the existence of biochemical or ultrastructural systems which protect against the damaging effects of the reactive epoxide. These considerations suggest a list of fundamental studies which might illuminate the questions of tissue and species specificity;

1. The gross rate of metabolism of precocenes in different spp., having regard for the quantity and route of administration.
2. The resultant titres of precocene in the haemolyph of different spp., having regard for the quantity administered, and the effective titre found in sensitive spp.
3. The epoxidase activity in the CA of different spp. (enzyme units/unit gland volume), measured in either intact CA or microsomal preparations, with a physiological substrate.
4. The gross rate of metabolism of precocenes by isolated CA from different spp., relative to (3) and with regard to the balance between macromolecular labelling, dihydrodiol formation and other possible metabolites.

There appear to be strong practical restrictions to some of these undertakings. For example, it may be impracticable to attempt (1) and (2) in the case of fumigation studies, or (3) and (4) in the case of larval CA from small insects, or for spp. in which the physiological substrate of the epoxidase is unknown (e.g. Hemiptera). Nevertheless, the residue of studies which we can presently see to be possible should indicate whether it is interspecific variation in the biochemistry of the CA or peripheral tissues, or both, which is mainly responsible for the varying tissue and species specificity of precocenes throughout the Insect Class. This knowledge might facilitate structure/activity optimization.

The very limited knowledge presently available on the comparative biochemistry of CA is at least consistent with the view that there may be important differences in the epoxidase levels of CA from different spp. It is known that the CA from adult females of several spp. (drawn from the Orthoptera, Dictyoptera and Coleoptera) readily utilize exogenous farnesenic acid, or its analogues (Pratt, Hamnett, Finney & Weaver, 1978; Hamnett & Pratt, 1978), *in vitro* to promote elevated rates of C_{16} JH biosynthesis (Pratt & Tobe, 1974; Pratt, Tobe & Weaver, 1975; Weaver, Pratt, Hamnett & Jennings, 1980; Pratt *et al.*, 1980). Although the evidence is not entirely definitive, it seems most likely that the farnesenic acid is first estertified (O-methyl transferase) and then epoxidized (methyl farnesoate-10,11-epoxidase). Different CA behave differently when "stimulated" with optimum concentrations of farnesenic acid, apparently because of differences in levels of activity of O-methyl transferase and 10,11-epoxidase enzymes. CA from *Periplaneta Americana* and *Tenebrio molitor* (both insensitive to precocenes *in vivo*) show limited epoxidase activity, leading to accumulation of methyl farnesoate within the glands, whereas CA from *Schistocerca gregaria* and *Locusta migratoria* (both sensitive to precocenes *in vivo*) show much higher epoxidase activities which are not driven to saturation by the O-methyl transferase (Pratt *et al.*, 1975; Weaver *et al.*, 1980; Pratt G. E., unpublished results). Therefore, it will be most interesting to extend this survey to other sensitive and insensitive spp.

20.4 CHEMICAL REACTIVITY OF PRECOCENE-EPOXIDES

The exact chemical reactivity of precocene epoxides was only established after their careful synthesis was achieved (Soderland, *et al.*, 1980; Jennings & Ottridge, 1979). Prior to these signal advances, only qualitative predictions were available, based upon general chemical knowledge (Brooks, 1977), specific attempts to synthesize a precocene epoxide by conventionally successful methods (Bergot *et al.*, 1980) and the deduced S_N1 character of the solvolysis transition state of putatively biosynthesized epoxide (Pratt *et al.*, 1980). An early attempt to isolate precocene-II epoxide as a reactive metabolite (Ohta *et al.*, 1977), apparently led to wrong interpretation because of the unexpectedly high reactivity of this compound (Soderland *et al.*, 1980). Natural product chemists in the juvenile hormone field have known for many years of the difficulties in isolating these hormones from natural sources, largely because of the chemical instability of the epoxide ring (Meyer, 1971), yet it is of the essence of the allatocidal action of precocenes that their epoxides are almost incomparably more reactive than the natural hormones. It is therefore timely to review briefly some of the salient features governing the reactivity of epoxides, to provide the basis for considering ways of possibly optimizing the structure of precocene analogues, and for assessing their possible harmful actions in non-target organisms.

Substitution reactions with Precocenes

Fɪɢ. 20.3. Broad classification of epoxides according to their solvolytic reactivity, after Dewar & Dougherty (1975), indicating the importance of different chemical substituent groups on the mechanism, and direction of nucleophile attack. The most reactive epoxides undergo Sɴ1 reactions, whilst the least reactive undergo Sɴ2 attack at the least substituted carbon atom (anti-M). The examples are classified according to their reaction in protic solvent under mild acid catalysis, and can exhibit different mechanisms under other conditions. a, Precocenes-I 3,4-epoxide; b, benzo (α) pyrene 7,8-dihydrodiol 9,10-epoxide; c, C_{18} juvenile hormone; d, epichlorhydrin.

Figure 20.3 shows a broad classification of epoxides, according to Dewar & Dougherty (1975) (*contra* Mumby & Hammock, 1979). The purpose is to emphasize the difference between highly reactive potentially cytotoxic epoxides (with ±E or −E substituents greatly stabilizing an Sɴ1 intermediate) and the two classes of relatively unreactive alkyl-substituted epoxides. However, it is important to note that the boundaries are not immutable in the case of any one compound, depending upon the conditions in solution. Thus precocene-I-3,4-epoxide, whose carbonium ion is well stabilized by the *p*-anisolyl moiety isoconjugant with benzyl anion, can be pushed towards significant Sɴ2 reaction at neutral pH, by quite minor quantitative replacement of water by aprotic solvents (Hamnett *et al.*, 1981b). The epoxide of the biologically inactive iso-precocene-1 is approximately 4000 times less reactive than precocene-I-epoxide (the putative carbonium ion now being attached at an inactive position on the −E substituent) and shows a mixture of SN1 and SN2 products in aqueous solution, but the balance is sensitive to pH and to the concentration of weak acid (general) catalysts such as orthophosphate (Hamnett *et al.*, 1981a). The phenomenon is well known from many studies on the reaction of epoxides having ±E (aryl) substituents, which are the ultimate carcinogenic metabolites of polynuclear aromatic hydrocarbons (PAHs) (Tsang W- S. & Griffin, 1979).

Clearly there are graded spectra of epoxide reactivity both between different epoxides, and for individual compounds in different solution conditions. It remains to be seen how this latter aspect relates to the possible partitioning of biosynthesized epoxides between aqueous (general-acid catalyst containing) and hydrophobic compartments in living cells. A recent report on the additional of benzo (α) pyrene diolepoxide to adenine residues in DNA has carefully intimated the probable importance of micro-environment in determining the reactivity of epoxides in complex biochemical milieux (Jeffrey, Gryzeskowiak, Weinstein, Nakanishi, Roller & Harvey, 1979).

It would be helpful to have biologically valid parameters for biosynthesized epoxides which could serve as useful predictors of cytotoxicity or mutagenicity. However, in the case of carcinogenic epoxides, it may well be the case that additional molecular properties, which may promote interaction with the genetic material (by hydrophobic interaction with stacked bases in nucleic acid) are as important as the reactivity of the epoxide once at its crucial target (Horning, Thenot & Helton, 1978). In comparison, it seems likely that the interaction between precocene-epoxides and the living components of the CA is less sophisticated, but one should keep in mind the possibility that more structurally elaborate analogues of precocenes (if realized) might be more selective in their alkylation of CA macromolecules, and perhaps thereby be more efficient. This is a quite unexplored area at present. For the time being we must fall back on some simple chemical yardstick, with which to assess the likely importance of the degree of epoxide reactivity towards allatocidal effectiveness. Although it must be remembered that the more reactive epoxides are subject to "general acid catalysis" at near neutral pH by a wide variety of weak acids or conjugate acids (Whalen, Ross, Montemarano, Thakker, Yagi & Jerina, 1979), which destabilize the epoxide ring by formal protonation or hydrogen-bonding, probably of considerable physiological importance, it has recently become the practice to compare reactivities at neutral pH in the absence of such catalysts (the so-called k_o rate) (Becker, Janusz & Bruice, 1979). Low concentrations of ethylenediaminetetra acetate (about 10^{-4} M) afford satisfactory pH buffering without incurring appreciable catalysis (Hamnett *et al.*, 1981a). Under these conditions, precocene-I epoxide has a half-life of 14 sec at 25°C, which makes it about 100 times more reactive than the highly carcinogenic antibenzo (α) pyrene diol-epoxide. The epoxide of the biologically inactive iso-precocene-I is very unreactive under these conditions (Table 20.1) with a

TABLE 20.1 CLASSIFICATION OF PRECOCENE-I AND RELATED
COMPOUNDS ACCORDING TO THEIR PHYSIOLOGICAL ACTIVITY,
AND THE CHEMICAL REACTIVITY OF THEIR 3,4-EPOXIDES

Compound	Antiallatal activity $ED_{50}(\mu g)^a$	Substrate for CA epoxidase	Relative reactivity of 3,4-oxide[b]
Precocene-I	40	yes	1.0
Precocene	50	presumably yes	unknown, 0.1–1. ?
Iso-precocene-I	inactive	presumably yes	0.00025
Methylenedioxy-precocene	inactive (antagonistic)	presumably not (inhibits)	3.04

[a]Determined by topical application bioassay to 2–6-h-old fourth instar *Locusta migratoria*, "inactive" compounds were tested up to their lethal dose.
[b]Based upon measurements of pseudo-first-order rate constants for hydrolysis at pH 7.0 (25°C), with/without 40% tetra-hydrofuran, as described by Hamnett *et al.* (1981 a b).

half-life of about 100 min similar to that of, for example, phenanthrene 9,10-oxide (Bruice, Bruice, Dansett, Selander, Yagi & Jerina, 1976). Two questions of obvious germanity are:

1. is the very high reactivity of precocene-epoxide essential for allatocidal action, or is there some other reason why iso-precocene-I shows no pro-allatocidal activity?, and
2. is the quantitative similarity in reactivities of precocene epoxides and a selection of ultimate cytotoxins and carcinogens of medical importance, reasonable ground for suspecting that precocene-like compounds may have unacceptable toxic actions against a wide spread of non-target organisms? We shall consider these points under separate headings.

20.5   STRUCTURE/ACTIVITY OPTIMIZATION

The discovery of precocenes (Bowers, 1976) afforded an exciting vision of possibly new types of insect-control chemicals. We wish to know whether or not we can progress from the structure of precocenes, using available knowledge of their mode of action and of the biochemistry of insect CA, to design more effective compounds which might realize this potential goal. As we have already indicated above, our fundamental knowledge is insufficiently comprehensive to fully support such a rational approach, and would seem to provide only a tentative conceptual framework within which to discuss the possibilities (see Fig. 20.2). Clearly, the balance between allatocidal action and peripheral detoxification is of paramount importance, but it seems premature to do more than merely introduce the subject at present, for want of basic comparative data. However, there may be some useful points to be made regarding the interaction within the CA.

The historical development of the related field of mammalian carcinogenic epoxides has moved between intensive study of the pathways of metabolism of the pro-carcinogens, and the application of chemical theories attempting to prescribe the carcinogenicity of the parent PAH according to the calculated reactivity of the putative epoxide formed therefrom (Lehr, Schaefer-Ridder & Jerina, 1977; Manatt, Beerman & Oesch, 1979). As with any system in which a pro-toxicant is converted to an ultimate toxicant by metabolic bio-activation, the physico-chemical properties of the pro-toxicant must simultaneously satisfy two separate criteria:

1. it is subject to the crucial biotransformation (i.e. it is an efficient substrate for the appropriate enzyme(s), and
2. the resultant metabolite has the required chemical reactivity.

It is worth recalling that these two requirements are specified by sets of molecular properties which only partially intersect. The former has an important contribution from the more physical attributes of the molecule (owing to the importance of spatial and hydrophobic qualities in determining the binding of substrates to enzyme reactive sites), whereas the latter is more concerned with varying chemical influences on the functional groups being modified or generated during the crucial bio-activation. So, in the case of pro-allatocidins, we must consider both their acceptability as substrates for the CA epoxidase, and the particular reactivity of the ensuing epoxide. Table 20.1 indicates how several compounds related to precocenes might be fitted into this dual classification, by a combination of facts and predictions. Clearly, there is a real need for more definitive observations on this limited set of examples, before the scheme can be regarded as having been properly tested.

The "Bay region theory" describing the mammalian carcinogenicity of PAHs (Jerina, Yagi, Lehr, Thakker, Schaefer-Ridder, Karle, Levin, Wood, Chang & Conney, 1978) shows how increasing the size of the poly-aromatic system increases the reactivity of the ultimate dihydrodiolepoxides, through the application of simple Perturbational Molecular Orbital theory (Dewar & Dougherty, 1975). The precocenes achieve greater reactivity in a smaller molecule because of the much greater stabilization of the S_N1 carbonium ion by the $-E$ p-anisolyl grouping (being iso-conjugant with a carbanion in PMO theory), than by a $\pm E$ aryl grouping of similar size. At the present time, all compounds with known allatocidal activity contain the $-E$ substituent moiety in the form of 7-alkoxy or 7-dialkyl amino group attached to the chromene ring (Bowers 1977; Brooks, 1979; Chenert *et al.*, 1980). We have tentatively concluded that the inactivity of iso-precocene-I *in vivo* may be taken as evidence of a requirement to generate epoxides of very high reactivity in order to realize allatocidal action (Table 20.1); what is implicit in this reasoning is that iso-precocene-I is expected to be a suitable substrate for the CA epoxidase. Clearly, this assumption demands early experimental verification. From the point of view of designing more effective precocene analogues, the need for a −E substituent relative to the epoxidized double bond (if indeed it is an absolute requirement) would not seem to incur any major problems of synthetic chemistry or rapid detoxification, perhaps quite the opposite. Nevertheless, it would be helpful to know the required instability of the ultimate epoxide, and hence the need for −E substitution. It is unfortunate that "pencil and paper" methods for calculating the NBMO coefficients in carbonium ions with −E substituents are not routinely available yet. On the other hand, a sizeable body of evidence on the reactivity of ±E (aryl) substituted epoxides, reinforced by simple PMO calculations (Dewar & Dougherty, 1975; Jerina *et al.*, 1978), indicates that a correctly positioned poly-aryl system of considerable size would be needed to afford the same degree of reactivity as that found in precocene-I epoxide. Such a large molecule might be unacceptable to the CA epoxidase.

We can say very little at present regarding the constraints upon precocene analogue structure which may be imposed by the substrate specificity of the CA epoxidase. It seems most probable that the natural substrates adopt a prescribed configuration on the active site (at least in respect of the olefinic terminus), because the epoxidations show complete stereochemical specificity (Nakanishi, Schooley, Koreeda & Dillon, 1971; Judy, Schooley, Dunham, Hall, Bergot & Siddall, 1973a; Hammock & Mumby, 1978; Judy, Schooley, Hall, Bergot & Siddall, 1973b; Judy, Schooley, Troetschler, Jennings, Bergot & Hall, 1973c; Jennings, Judy, Schooley, Hall & Siddall, 1975). The enzyme does not discriminate markedly between the sidechain homologues of methyl farnesoate (Pratt & Tobe, 1974) and so is not responsible for determining the biosynthetic balance of juvenile hormone homologues, which is determined at early steps in the biosynthetic pathway (Schooley, Judy, Bergot, Hall & Siddall, 1973; Lee, Schooley, Hall & Judy 1978). The only other evidence which can be brought to bear is indirect, coming from studies on the susceptibility of methyl farnesoate epoxidase to inhibition by typical mixed function oxidase inhibitors (Hammock & Mumby, 1978), which suggest a rather broad structural requirement for effective interaction. What is required at this stage is unequivocal evidence that methyl farnesoate (or alternative natural substrates) and precocenes are alternative substrates for the same enzyme in the CA, followed by direct tests of the substrate specificity of the epoxidase(s) using analogues modelled on both methyl farnesoate and precocenes. It is usually helpful and reassuring to have a conceptual basis for such exploratory designs, but no categorical model is yet available for the CA epoxidase. Figure 20.4 illustrates one of the several models available, being marginally favoured by the present author. It is fixed upon the premise that

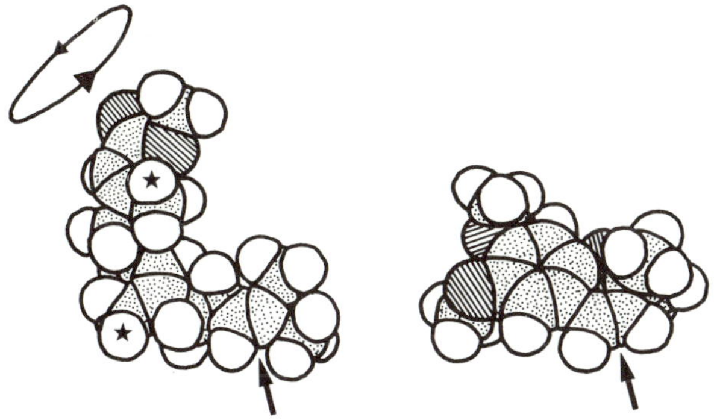

a   b

FIG. 20.4. Line drawings of photographs of space-filling molecular models (CPK) of methyl farnesoate (a) and precocene-II (b), showing the rather poor structural similarity. The methyl farnesoate molecule has been configured intuitively so that the 10, 11 double bond which is enzymitically epoxidized is isospatial with the 3, 4 double bond of precocene (arrows), and so that its 6,7 double bond is co-planar with the benzo ring of precocene. The 3- and 7-side-chain methyl groups in methyl farnesoate are identified by asterisks, that at the 7 position, being the site of attachment of an additional methyl group in C_{18} juvenile hormone, which then occupies the same space as the 6-methoxyl group in precocene-II. The model is purely speculative, and incidentally makes no prediction as to the configuration of the carbonyl terminus of methyl farnesoate *vis à vis* the reactive side of the epoxidase; Hammock & Mumby (1978) have found that the unnatural 2Z isomer of methyl farnesoate is a good substrate for the CA epoxidase in one species. Some implications of this model should be amenable to direct testing (see text).

the 3,4 and 10,11 double bonds in precocene and methyl farnesoate respectively must be iso-spatial on the reactive site of the epoxidase, and that the rest of the two molecules will show a greater, rather than random, congruence in three-dimensional space. Since the precocene molecule has only slight conformational flexibility, this effectively defines (in part) a limited number of configurations of the natural substrate, of which one is illustrated. This representation implies that considerable extension of the 7-alkoxy group in precocenes should be possible, and it is noteworthy that several chain-extended analogues have already been tested and found to have good activity (Bowers, 1977; Brooks G. T., unpublished results). Perhaps the most tangible assumption in this model is the one which implies a specific stereochemistry for biosynthesized precocene epoxide, and hence one absolute configuration at C3 in the dihydrodiols resulting from spontaneous hydrolysis. We hope that the current study of this stereochemistry in our laboratory will help to eliminate a number of presently available options. Once this is achieved, it should be possible to move forward with greater confidence towards an unambiguous dissection of the enzymic and chemical requirements for activity of a pro-allatocidin of the precocene type. In particular we would like to be able to delineate the space-filling options more precisely, and to assess the separate importance of substituent groups on the nucleophillic reactivity of the 3,4 double bond in precocenes during epoxidation, the energy level of the epoxide ring and the degree of stabilization of the resultant carbonium ion. The function of the gem 2-dimethyl groups is particularly enigmatic at present, it being known only that they are important for biological activity, particularly in larval *Oncopeltus fasciatus* (Bowers, 1977). If the recently synthesized 2-trifluoromethyl analogue (Camps, Coll, Messeguer & Pericas, 1980) should prove to have lower biological activity than the iso-spatial parent precocene, this would suggest the chemical importance of a balancing electron induction into the C3, 4π systems, rather than a spatial role in substrate binding to the epoxidase.

In summary, our present knowledge permits only the broadest of guidelines to be applied to structure/activity optimization, and these are based largely upon general knowledge of seemingly relevant chemobiodynamics and chemical enzymology, rather than definitive quantitative data on the particular systems. Further screening of precocene analogues in whole animal bioassays, whilst immediately available, cannot identify the causes of inactivity, and hence cannot provide wholly reliable clues for further improvement. It would seem that unambiguous interpretation of available structure/activity data requires a fairly large injection of fundamental information, along the following lines: (a) how the physiology/biochemistry of different insect spp., matrixed with differing biochemical susceptibilities in precocene analogues, determines the degree of exposure of the CA to the pro-toxicant, and (b) how stringent, and how separable the biological and chemical constraints are upon a successful interaction once the pro-allatocidin has reached the CA.

20.6   THE CANCER RISK FROM PRECOCENES

The present evidence suggests that precocenes are allatocidal, in sensitive spp. of insect, because they are metabolized in the CA at a very high rate per weight tissue to a potent alkylating agent. Whilst the biochemical details of the ensuing cellular pathology, dramatically signalled by observations of dying CA in the electron microscope (Schooneveld, 1979; Unnithan, Nair & Bowers, 1977; Leichty & Sedlak 1978), are as yet unreported, the large extent of macromolecular labelling observed in *Locusta migratoria* CA *in vitro* (Pratt *et al.*, 1980) suggests that cell death may result from a "blitzkrieg" upon the cell by a relatively non-selective alkylating agent. This hypothesis, which awaits experimental verification, implies that the element of tissue (or animal) specificity of precocenes resides in the intensity of their conversion to a powerful alkylating agent, and not in the target selectivity of that alkylating agent. The question naturally arises of whether a low level of formation of precocene epoxides in some "non-target" tissues might significantly increase the incidence of mutation of tumorigenesis, with obvious implication for the possible use of such chemicals as pest-control agents. These are matters in which our laboratory is not an authority, therefore we only raise them for consideration.

There are many types of carcinogen, and many types of cytotoxic/mutagenic alkylating agent, of human importance. Purely on grounds of chemical structure, the precocenes show greatest similarity to the carcinogenic PAHs, but they are of appreciably lower molecular weight than the most active PAHs such as benzo (α) pyrene. Figure 20.5 illustrates the close biochemical parallel between the activation of precocene in the CA, and the conversion of benzo (α) pyrene to one of its most mutagenic metabolites, the 7,8-dihydrodiol-9,10-epo-

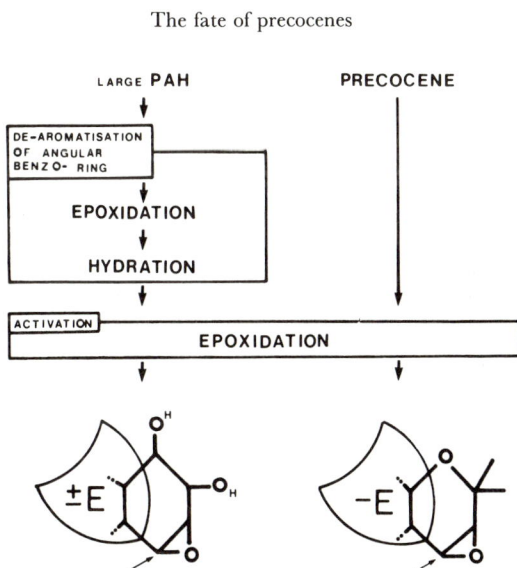

Fig. 20.5. Comparison of the metabolic steps involved in the bioactivation of carcinogenic polynuclear aromatic hydrocarbons (PAHs) and allatocidal precocenes, to highly reactive epoxides. The most reactive epoxides of PAHs are formed by a series of three enzymic conversions (Jerina *et al.*, 1978), which are in balance with a variety of other metabolic transformations leading to less reactive compounds. The first two reactions serve to break the aromaticity of a terminal, angular benzo ring. It may be important that the first-formed epoxide has sufficient chemical stability to permit reaction with epoxide hydrase enzyme (second step), otherwise an internal NIH rearrangement diverts in into a less reactive "dead-end" phenol. The second-formed epoxide may be highly activated by the ±E substituent (the remainder of the PAH), but now reacts only by nucleophillic substitution at the stabilized carbonium ion (small arrow), which makes it a powerful alkylating agent. The precocene molecule requires no such metabolic preamble: a single epoxidation suffices to generate a very powerful alkylating agent.

xide, in vertebrate tissues (Buening, Wislocki, Levin, Yagi, Thakker, Akagi, Koreeda, Jerina & Conney, 1978). One can see that the precocenes are in a sense pre-adapted for bioactivation to potent alkylating agents, in comparison with PAHs, since the pyran ring requires no preliminary metabolism to break its aromaticity prior to epoxidation. In view of the occurrence of epoxidizing mixed function oxidases in the nuclear membranes of vertebrate cells (Mukhtar, Elmamlouk & Bend, 1979; Fahl, Jefcoate & Kasper, 1978; Thomas, Korzeniowski, Bresnick, Bornstein, Kasper, Fahl, Jefcoate & Levin, 1979) we hold the carcinogenicity of precocenes to be worthy of study.

Precocene-II has been subjected to preliminary screening by Ames test, using both frame-shift and base-substitution sensitive mutants of *Salmonella typhimurium*, with and without rat liver microsomes: no clear-cut evidence for mutagenesis was obtained at doses of up to 250 $\mu g \, ml^{-1}$ (Hubbard S. & Brooks G. T., unpublished results). These negative results appear encouraging, but there may be a need for caution in applying the results of such a contrived test to the possible activity of the compound in the intact animal. It is known that the observed relative activities of different PAHs and their metabolites can vary widely in different laboratory tests (Tsang & Griffin, 1979) and there may be several reasons for this. In the case of the Ames test, the bio-activation if effected by a microsomal system outside the bacterium, the ultimate mutagen experiences a long statistical diffusion path prior to interaction with the bacterial genome. Possibly, this feature of the test may discriminate against epoxides of very high reactivity (e.g. precocene-I epoxide) in a way which is more severe than that encountered in the intact eukaryotic cell. More optimistically, precocene epoxides may be too reactive in the aqueous compartments of cells to gain access to the genetic material, but this speculation is no grounds for complacency. We therefore urge caution in the continued laboratory use of these compounds, until the human risks have been thoroughly assessed.

20.7 CONCLUSION

At the present time the only hypothesis, with experimental support, which attempts to define the mode of action of the precocenes, is that outlined above. It should be remembered that most of the direct evidence comes from observations on just one species of insect, *Locusta migratoria*. However, until shown otherwise, it seems reasonable to suggest that all precocene-sensitive insects suffer chemical allatectomy because of an important conversion of precocenes to cytotoxic epoxides in their CA. It remains to be seen whether insensitive insects are so because their peripheral metabolism does not permit a sufficiently high titre of precocenes to appear in the haemolymph, or because their CA are inherently less sensitive to precocenes, perhaps because of lower specific activities or stricter substrate specificities of their epoxidase.

The precocenes provide an example of potential new types of insect growth regulator, which were discovered by random screening of natural products (Bowers, 1976), not by human invention based upon advanced fundamental knowledge (Casida, 1976). Fundamental studies, of the kind outlined here, are helpful both in facilitating structure/activity optimization, and in predicting possible areas of toxicological concern. In addition to those compounds already synthesized and subjected to limited activity testing (Bowers, 1977; Brooks *et al.*, 1979b; Chenevert *et al.*, 1980), there may be many other analogues which have been screened by the agrochemical industry, to whose structures and activities we are not yet privy. Further studies on the substrate specificity of the CA, in a wider range of insects, might provide a basis for the rational design of more effective precocene analogues, within the context of general theories of chemobiodynamics. The fact that precocenes share a common type of enzymic bioactivation with a host of carcinogenic xenobiotics, suggests that we should bear in mind the possibility that they, or their analogues, may have undesirable properties from the ecological/medical point of view. Further studies are required.

The action of the precocenes points towards the insect CA as a potential "weak link", or possible target for the action of selective insecticides or growth regulators. The CA is an essential endocrine gland and it appears to have a highly dedicated biochemistry involving a limited range of anabolic and oxidative enzymes, with little or no enzymic provision for detoxification of harmfully reactive compounds. The prothoracic glands, source of α-ecdysone, may be similar in this respect. There are relatively few examples of chemicals (other than immunochemicals) which induce tissue-specific necrosis in animals, but it is interesting to note (Schooneveld, 1979) that precocenes in the insect CA (Pratt *et al.*, 1980), metyrapone in the testis of some vertebrates (Del Conte, 1979), and 6-hydroxydopamine in adrenergic nerve terminals (Rotman, 1977) are all examples which exert their various effects via an interaction with oxidative enzyme(s). Horning *et al.* (1978) have summarized, "The fact that numerous organic compounds that are essentially non-toxic, so long as their structure is preserved, can be converted into cytotoxic, teratogenic, mutagenic or carcinogenic compounds by normal biologic oxidative metabolism in animals or humans is now well established". Thus there is a *prima facia* general case for focusing attention upon the oxidative steps in the juvenile hormone biosynthetic pathway, particularly if these are likely to involve enzymes whose specificity or other molecular properties may be special to the CA.

The biosynthesis of all known juvenile hormones proceeds via mevalonate and iso-prenyl units to farnesyl (or homo-farnesyl, in the case of C_{17} and C_{18} hormones) moieties, with bio-oxidation steps at the conversion of farnesol (from its pyrophosphate) to farnesenic acid, and the epoxidation of methyl farnesoate to the final hormone. Although neither of these reactions are unique to the insect CA, they are much more dominant conversions there than in any other known tissue. Thus, farnesenic acid is only a minor by-product from sterol synthesis in vertebrate tissues, and the methyl farnesoate epoxidase clearly differs from most mixed-function oxidases, since it is not closely associated with epoxide hydrases (as part of a detoxifying system) and also appears to differ from the squalene epoxidase in the pathway of sterol biosynthesis in non-insectan organisms (Yamoto & Block, 1970; Polito, Popjak & Parker, 1972). The discovery of the mode of action of precocenes is stimulating further study of the methyl farnesoate epoxidase, as outlined above. However, nothing is known at present about the enzymic conversion of farnesol to farnesenic acid in the CA. Reibstein, Law, Bowlus & Katzenellen-bogen (1976) speculated that farnesol dehydrogenase might be a nicrotina-mide adenine dinucleotide-linked enzyme, and did obtain low levels of radio-labelled incorporation from farnesyl pyrophosphate into farnesenic acid with homogenates from *Manduca sexta* CA incubated at pH 5–5.2, but did not demonstrate the co-factor requirement. Apparently, neither farnesol nor its pyrophosphate are utilized by intact CA *in vivo* (Metzler, Dahm, Meyer & Roller, 1971), which may have discouraged study of this enzymic conversion. The possibility that exogenous compounds have severely restricted access to key intracellular enzymes in the biosynthetic pathway may also apply to earlier steps such as the 2-hydroxy, 2-methyl-glutaryl coenzyme A reductase and prenyl transferases (Schooley, Judy, Bergot, Hall & Jennings, 1976) which are involved in the production and utilization of the unique homo-isoprenoid moieties in those insects which synthesize C_{17} and C_{18}JH (Lee *et al.*, 1978). Attempts to block the action of these early enzymes either *in vivo* or *in vitro*, using analogues of mevalonic acid (Matolcsy, Varjas & Bordas, 1975), chlofibric acid (Pratt & Finney, 1977; Hammock, Kuwano, Ketterman, Scheffrahn, Thompson & Sallume, 1978) and methyl 2-fluorophenylacetate (Pratt G. E. & Clements A. N., unpublished), have thus far been rewarded with negative results. This may be a problem of access, rather than enzymic susceptibility. Although there is a report from one laboratory of exogenous mevalonic acid interacting efficiently with the biosynthetic enzymes in intact CA from several spp. (Dahm, Bhaskaran, Peter, Shirk, Seshan & Roller, 1976), a rather careful radioisotopic study on *Manduca sexta* CA (Schooley *et al.*, 1976) showed an interaction efficiency of only about 1%. The general importance of cytomembrane integrity during organ studies *in vitro* may be relevant here (Pratt, Weaver & Hamnett, 1976), and it may be of only academic interest to inquire the interaction of such low molecular weight water-soluble compounds with CA whose cytomembranes have been awarded some unnatural degree of permeability. In contrast, more lipophyllic compounds such as farnesenic acid analogues (Pratt & Weaver,

1978), mixed function oxidase inhibitors (Pratt & Finney, 1977; Pratt *et al.*, 1980) and precocene-I (Pratt *et al.*, 1980), appear to interact readily with the O-methyl transferase and 10,11-epoxidase, at least in some insects. Figure 20.6 summarizes the presently known strong interactions between exogenous

The fate of juvenile hormones in precocene treated insects

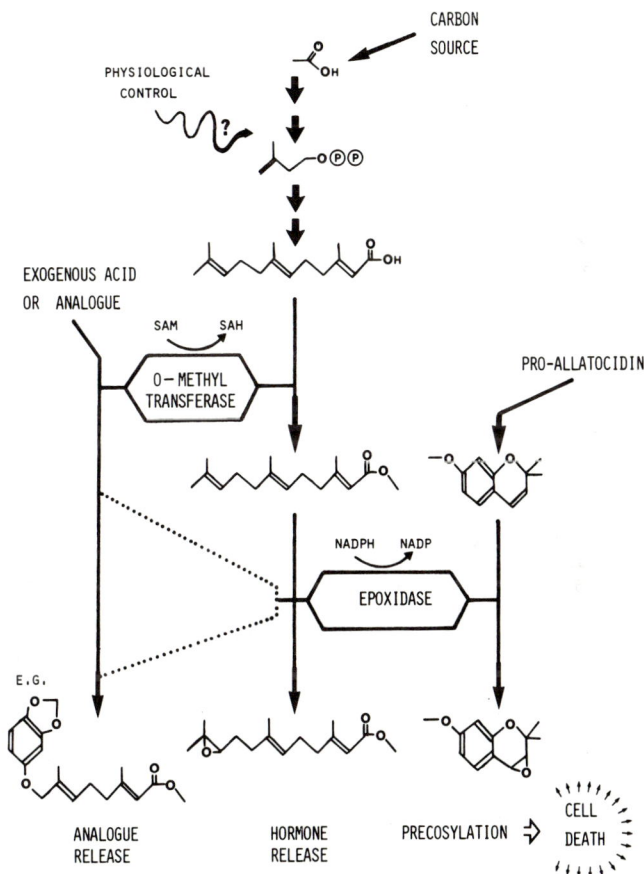

FIG. 20.6. Simplified version of the biosynthetic pathway for C_{16} juvenile hormone, highlighting the presently known sites where exogenous chemicals or influences (physiological control by the brain) are able to interact efficiently with the enzymic machinery. See text for details and references.

compounds and the juvenile hormone biosynthetic pathway. Although all aspects of this pathway are worthy of further study, including the possibility of reactive affinity-label inhibitors of the O-methyl transferase (Pratt & Weaver, 1978), the precocenes are the only known CA antagonists which cause total atrophy, and they do so by a mechanism which is still being explored.

20.8 SUMMARY

Precocenes are lethally converted to highly reactive epoxides by high levels of mixed function oxidases in the corpora allata of sensitive spp., e.g. *Locusta migratoria*. The corpora allata are biochemically specialized for the synthesis of epoxides (juvenile hormones). A current study of this pharmacology requires knowledge of the circulating titres of precocenes during effective exposure, and these show that the glands from adult females are about 10 times more sensitive than those from larvae. The putative reactive epoxide labels (precosylates) many polypeptides in the cell, and most cytoplasmic organelles: it appears to act like a scatter-gun.

Biological action of pro-allatocidins requires first that they gain admission and are transported in adequate proportion to the glands. It is then necessary that (a) they serve as substrates for the activating enzyme (epoxidase in the case of precocenes) and (b) that the metabolite is of sufficient reactivity to induce cellular lesions. Thus the epoxide of the 3,4-methylenedioxy analogue of precocene is slightly more reactive than precocene-I epoxide, yet the parent compound is inactive; rather it antagonizes precocenes. That is because it inhibits the epoxidase. The epoxide of the 7-methoxy positional isomer of precocene-I (*iso*-P-I) is 4000 × less reactive than P-I epoxide, and shows significant S$_N$2 reaction under physiological conditions; this may suffice to explain the lack of biological activity of the parent chromene. The chemical reactivity of the epoxides can be largely predicted from PMO theory, but knowledge of the substrate specificity of the epoxidases is lacking, and the subject of current research. This may facilitate structure/activity predictions.

There are close parallels between the action of precocenes and many polyaromatic hydrocarbon carcinogens and cytotoxins. The pyran ring of chromenes is pre-adapted, being already "de-aromatized". The carcinogenicity of precocene analogues should be kept under scrutiny.

20.9 ACKNOWLEDGEMENTS

I am grateful to my colleagues, particularly G. T. Brooks, R. Feyereisen, A. F. Hamnett and R. C. Jennings, for their helpful discussions and contribution toward the cited work upon which this article is based, and to Professor A. W. Johnson FRS, for his continuing encouragement. B. J.

Bergot and D. M. Soderlund kindly afforded sight of articles in press.

20.10 REFERENCES

Becker A. R., Janusz J. M. & Bruice T. C. (1979) Solution chemistry of the syn- and anti-tetrahydrodiol epoxides, the syn- and anti-tetrahydrodimethoxy epoxides, and the 1,2- and 1,4-tetrahydroepoxides of naphthalene. *J. Am. Chem. Soc.* **101**, 5679–5687.

Bergot B. J., Judy K. J., Schooley D. A. & Tsai L. W. (1980) Precocene II metabolism: comparative *in vivo* studies among several species of insect, and structure elucidation of two major metabolites. *Pest. Biochem. Physiol.* **13**, 95–104.

Bowers W. S. (1976) Discovery of insect anti-allatotropins. In *The Juvenile Hormones* (ed. Gilbert L. I.), pp. 394-408. Plenum Press, New York.

Bowers W. S. (1977) Anti-juvenile hormones from plants: chemistry and biological activity. *Pontif. Acad. Scient. Scripta Varia* **41**, 129–142.

Bowers W. S. & Martinez-pardo R. (1977) Anti-allatotropins: Inhibition of corpus allatum development. *Science* **197**, 1369–1371.

Bowers W. S., Ohta T., Cleere J. S. & Marsella P. A. (1976) Discovery of insect anti-juvenile hormones in plants. *Science* **193**, 542–547.

Brooks G. T. (1977) Action and inaction of certain non-anticholinesterase insecticides. *Proc. Brit. Crop Protection Conf.: Pests and Disease* **3**, 731–740.

Brooks G. T. (1979) The metabolism of xenobiotics in insects *Progress in Drug Metabolism* (eds. Bridges J. W. & Chasseaud L.), pp. 151–213. John Wiley & Son, Chichester, England.

Brooks G. T., Hamnett A. F., Jennings R. C., Ottridge A. P. & Pratt G. E. (1979b) Aspects of the mode of action of precocenes on milkweed bugs (*Oncopeltus fasciatus*) and locusts (*Locusta migratoria*). In *Proc. Brit. Crop Prot. Conf.: Pests & Diseases*, pp. 273–279.

Brooks G. T., Pratt G. E. & Jennings R. C. (1979a) The actions of precocenes in milkweed bugs (*Oncopeltus faciatus*) and locusts (*Locusta migratoria*). *Nature* **281**, 570–572.

Bruice P. Y., Bruice T. C., Dansette P. M., Selander H. G., Yagi H. & Jerina D. M. (1976) Comparison of the mechanisms of solvolysis and rearrangement of K-region vs. non-K-region oxides of phenanthrene. Comparative solvolytic rate constants of K-region and non-K-region arene oxides. *J. Am. Chem. Soc.* **98**, 2956–2973.

Buening M. K., Wislocki P. G., Levin W., Yagi H., Thakker D. R., Akagi H., Koreeda M., Jerina D. M. & Conney A. H. (1978) Tumorigenicity of the optical enantiomers of the diastereomeric benzo (α) pyrene 7,8-diol-9,10-epoxides in newborn mice: exceptional activity of (+)-7β, 8α-dihydroxy-9α,10α-epoxy-7,8,9,10-tetrahydrobenzo (α) pyrene. *Proc. Natl. Acad. Sci. USA* **75**, 5358–5361.

Burt M. E., Kuhr R. J. & Bowers W.S. (1978) The metabolism of precocene-II in the cabbage looper and european corn borer. *Pest. Biochem. Physiol.* **2**, 300–303.

Camps F., Coll J., Messeguer A. & Pericas M. A. (1980) Fluorinated chromenes. Synthesis of 6,7-dimethoxy-2-methyl-2-trifluoromethyl-2*H*-chromene. *J. Heterocylic Chem.* **17**, 207–208.

Casida J. E. (1976) Prospects for new types of insecticide. In *The future for Insecticides: needs and prospects* (eds. Metcalf R. L. & McKelvey J. J.), pp. 349–366. John Wiley & Sons, New York, London, Sydney & Toronto.

Cassier P. (1979) The corpora allata of insects. *Int. Rev. Cytol.* **57**, 1–73.

Chenevert R., Perron J. M., Paquin R., Robiteille M. & Wang Y. K. (1980) Activity of precocene analogues on *Locusta migratoria migratorioides* (R. & F.). *Experientia* **36**, 379–380.

Cruickshank P. A. & Palmere R. M. (1971) Terpenoid amides as insect juvenile hormones. *Nature* **233**, 488–489.

Cupp E. W., Lok J. B. & Bowers W. S. (1977) The developmental effects of 6,7-dimethoxy-2,2-dimethyl chromene on the pre-imaginal stages of *Aedes aegypti* (Diptera: Culicidae). *Ent. Exp. Appl.* **22**, 23–28.

Dahm K. H., Bhaskaran G., Peter M. G., Shirk P. D., Seshan K. R. & Roller H. (1976) On the identity of the juvenile hormone in insects. In *The Juvenile Hormones* (ed. Gilbert L. I.), pp. 19–47. Plenum Press, New York.

Del Conte E. (1979) Ultrastructural aspects of degradation and necrosis of Leydig cells in lizards by effect of metyrapone. *Gen. Comp. Endocrinol.* **37**, 101–110.

De Loof A., Van Loon J. & Vanderroost C. (1979) Influence of ecdysterone, precocene and compounds with juvenile hormone activity on induction, termination and maintenance of diapause in the parasitoid wasp, *Nasonia vitripennis*. *Physiol. Entomol.* **4**, 319–328.

Dewar M. J. S. & Dougherty R. C. (1975) *The PMO Theory of Organic Chemistry*. Plenum Press, New York, 576 pp.

Dietz A., Hermann H. R. & Blum M. S. (1979) The role of exogenous JH I, JH III and anti-JH (precocene-II) on queen induction of 4.5-day-old worker honey bee larvae. *J. Insect Physiol.* **25**, 503–512.

Fahl W. E., Jefcoate C. R. & Kasper C. B. (1978) Characteristics of benzo (α) pyrene metabolism and cytochrome P_{450} heterogeneity in rat liver nuclear envelope and comparison to microsomal membrane. *J. Biol. Chem.* **253**, 3106–3113.

Gamper H. B., Tung A. S. -C., Staub K., Bartholomew J. C. & Calvin M. (1977) DNA strand scission by benzo (α) pyrene diol epoxides. *Science* **197**, 671–674.

Gilbert L. I. & Schneiderman H. A. (1960) The development of a bioassay for the juvenile hormone of insects. *Trans. Am. Micro. Soc.* **79**, 38–67.

Hammock B. D., Kuwano E., Ketterman A., Scheffrahn E. H., Thompson S. N. & Sallume D. (1978) Acute toxicity and developmental effects of analogues of ethyl α-(4-chlorophe-noxy)-α-methylpropionate on two insects, *Oncopeltus fasciatus* and *Tenebrio molitor*. *J. Agric. Food Chem.* **26**, 166–170.

Hammock B. D. & Mumby S. M. (1978) Inhibition of epoxidation of methyl farnesoate to juvenile hormone III by cockroach corpus allatum homogenates. *Pest. Biochem. Physiol.* **9**, 39–47.

Hamnett A. F., Ottridge A. P., Pratt G. E., Jennings R. C. & Scott K. M. (1981b) The hydrolysis of precocene-I epoxide: the putative biologically active form of the insect cytotoxin: precocene-I. *Pesticide Science* **12**, 245–254.

Hamnett A. F. & Pratt G. E. (1978) Use of automated capillary column radiogas chromatography in the identification of insect juvenile hormones. *J. Chromatog.* **258**, 387–399.

Hamnett A. F., Pratt G. E., Ottridge A. P. & Alexander B. A. J. (1981a). The role of HPLC in the elucidation of the mechanisms of hydrolysis of an organ specific cytotoxin, precocene I epoxide and some inactive analogues. *Biomedical and Biomedical Applications of HPLC* (eds. Hawk G. & Dekker M.). New York.

Horning E. C., Thenot J-P. & Helton E. D. (1978) Toxic agents resulting from the oxidative metabolism of steroid hormones and drugs. *J. toxicol. env. Health* **4**, 341–361.

Jeffrey A. M., Grzeskowiak K., Weinstein I. B., Nakanishi K., Roller P. & Harvey R. G. (1979) Benzol (α) pyrene-7,8-dihydrodiol 9,10-oxide. Adenosine and deoxyadenosine adducts: structure and stereochemistry. *Science* **206**, 1309–1311.

Jennings R. C., Judy K. J., Schooley D. A., Hall M. S. & Siddall J. B. (1975). The identification and biosynthesis of two juvenile hormones from the tobacco budworm moth (*Heliothis virescens*). *Life Sciences* **16**, 1033–1040.

Jennings R. C. & Ottridge A. P. (1979) Synthesis of precocene-I epoxide (2,2-dimethyl--3,4-epoxy-7-methoxy-2*H*-1-benzopyran). *Chem. Commun.* 920–921.

Jerina D. M., Yagi H., Lehr R. E., Thakker D. R., Schaefer-Ridder M., Karle J. M., Levin W., Wood A. W., Chang R. L. & Conney A. H. (1978) The Bay-region theory of carcinogenesis by polycyclic aromatic hydrocarbons. In *Polycyclic Hydrocarbons and Cancer*, Vol. I (eds. Gelboin H. V. & Ts's P. O. P.), pp. 173–188. Academic Press Inc., New York, San Francisco, London.

Jollow D. J., Mitchell J. R., Zampaglione N. & Gillete J. R. (1974) Bromebenzene induced liver necrosis. Role of glutathione and evidence of bromobenzene-3,4-epoxide as the hepatotoxic metabolites. *Pharmacology* **11**, 151–169.

Judy K. J., Schooley D. A., Dunham L. L., Hall M. S., Bergot B. J. & Siddall J. B. (1973a) Isolation, structure and absolute configuration of a new natural insect juvenile hormone from *Manduca sexta*. *Proc. Natl. Acad. Sci. USA* **70**, 1059–1513.

Judy K. J., Schooley D. A., Hall M. S., Bergot B. J. & Siddall J. B. (1973b) Chemical structure and absolute configuration of a juvenile hormone from grasshopper corpora allata *in vitro*. *Life Sciences* **13**, 1511–1516.

Judy K. J., Schooley D. A., Troetschler R. G., Jennings R. C., Bergot B. J. & Hall M. S. (1973c)

Juvenile hormone production by corpora allata of *Tenebrio molitor in vitro*. *Life Sciences* **16**, 1059–1066.

Kelly T. J. & Fuchs M. S. (1978) Precocene is not a specific anti-gonadotropic agent in adult female *Aedes aegypti*. *Physiol. Entomol.* **3**, 297–301.

Krieger R. I., Feeny P. & Wilkinson C. F. (1971) Detoxication enzymes in the guts of caterpillars: an evolutionary answer to plant defenses? *Science* **172**, 579–581.

Lee E., Schooley D. A., Hall M. S. & Judy K. J. (1978) Juvenile hormones biosynthesis: homomevalonate and mevalonate synthesis by insect corpus allatum enzymes. *J. Chem. Soc. Chem. Commun.* 290–292.

Lehr R.E., Schaefer-Ridder M. & Jerina D. M. (1977) Synthesis and reactivity of diol epoxides derived from non-K-region *trans*dihydrodiols of benzo (α) anthracene. *Tetrahedron Letts.* **6**, 539–542.

Leichty L. & Sedlak B. J. (1978) The ultrastructure of precocene-induced effects on the corpora allata of the adult milkweed bug, *Oncopeltus fasciatus*. *Gen. Comp. Endocrinol.* **36**, 433–436.

Mackauer M., Nair K. K. & Unnithan G. C. (1979) Effect of precocene-II on alate production in the pea aphid, *Acyrthosiphon pisum*. *Can. J. Zool.* **57**, 856–859.

Manatt S. L., Beermann D. & Oesch F. (1979) On the potential carcinogenic and mutagenic character of benzobiphenylenes. *Tetrahedron Letts.* **39**, 3691–3694.

Masner P., Bowers W. S., Kalin M. & Muhle T. (1979) Effect of precocene-II on the endocrine regulation of development and reproduction in the bug, *Oncopeltus fasciatus*. *Gen. Comp. Endocrinol.* **37**, 156–166.

Matolcys G., Varjas L. & Bordas B. (1975) Inhibitors of steroid biosynthesis as potential insect anti-hormones. *Acta Phyto. Acad. Scient. Hung.* **10**, 455–463.

Metzler M., Dahm K. H., Meyer D. & Roller H. (1979a) On the biosynthesis of juvenile hormone in the adult cecropia moth. *Z. Naturforsch.* **26b**, 1270–1276.

Meyer A. S. (1976) The juvenile hormones of the cecropia silkmoth. A *Chronicle. Mitteil. Schweiz. Ent. Gesell.* **44**, 37–63.

Mukhtar H., Elmamlouk T. H. & Bend J. R. (1979) Epoxide hydrase and mixed-function oxidase activities of rat liver nuclear membranes. *Arch. Biochem. Biophys.* **192**, 10–21.

Muller P. J., Masner P., Kalin M. & Bowers W. S. (1979) *In vitro* inactivation of corpora allata of the bug *Oncopeltus fasciatus* by precocene-II. *Experientia* **35**, 704–705.

Mumby S. M. & Hammock B. D. (1979) Stability of epoxide-containing juvenoids to dilute aqueous acid. *J. Agric. Food Chem.* **27**, 1223–1228.

Nakanishi K., Schooley D. A., Koreeda M. & Dillon J. (1971) Absolute configuration of the C_{18} juvenile hormone: application of a new circular dichroism method using tris-(dipivaloyl-methanato)-praseodymium. *Chem. Commun.* 1235–1236.

Ohta T., Kuhr R. J. & Bowers W. S. (1977) Radiosynthesis and metabolism of the insect anti-juvenile hormone, Precocene-II. *J. Agric. Food Chem.* **25**, 478–481.

Patterson J. W. & Schwarz M. (1977a) Chemical structure, juvenile hormone activity and persistence within the insect of juvenile hormone mimics for *Rhodnius prolixus*. *J. Insect. Physiol.* **23**, 121–129.

Pedersen L-E. K. (1978) Effects of anti-juvenile hormone (precocene-I) on the development of *Locusta migratoria* L. *Gen. Comp. Endocrinol.* **36**, 502–509.

Pener M. P., Orshan L. & DeWilde J. (1978) Precocene-II causes atrophy of corpora allata in *Locusta migratoria*. *Nature* **272**, 350–353.

Polito A., Popjak G. & Parker T. (1972) Artificial substrates in squalene and sterol biosynthesis. *J. Biol. Chem.* **247**, 3463–3470.

Pound J. M. & Oliver J. H. (1979) Juvenile hormone: Evidence of its role in the reproduction of ticks. *Science* **206**, 355–357.

Pratt G. E. & Bowers W. S. (1977) Precocene-II inhibits juvenile hormone biosynthesis by cockroach corpora allata, *in vitro*. *Nature* **246**, 548–550.

Pratt G. E. & Finney J. R. (1977) Chemical inhibitors of juvenile hormone biosynthesis *in vitro*. In *Crop Protection Agents* (ed. McFarlane N. R.), pp. 113–132. Academic Press Inc., London.

Pratt G. E., Hamnet A. F., Finney J. R. & Weaver R. J. (1978) The biosynthesis of cecropia (C_{18}) juvenile hormone and several analogues by isolated corpora allata of the American cockroach (*Periplaneta americana*) during incubation with labelled methionine and analogues of farnesenic acid. *Gen. Comp. Endocrinol.* **34**, 113.

Pratt G. E., Jennings R. C. & Hamnett A. F. (1981) Metabolism of precocene-I in *Locusta migratoria*. In *Proc. Int. Conf. Reg. Insect Dev. and Behav., Wroclaw, Poland* **2**, 347–356.
Pratt G. E., Jennings R. C., Hamnett A. F. & Brooks G. T. (1980) Lethal metabolism of precocene-I to a reactive epoxide by locust corpora allata. *Nature* **284**, 320–323.
Pratt G. E. & Tobe S. S. (1974) Juvenile hormones radiobiosynthesised by corpora allata of adult female locusts, *in vitro*. *Life Science* **14**, 575–586.
Pratt G. E., Weaver R. J. & Hamnett A. F. (1976) Monitoring of juvenile hormone released by superfused corpora allata of *Periplaneta americana*. In *The Juvenile Hormones* (ed. Gilbert L. I.), pp. 164–178. Plenum Press, New York.
Pratt G. E., Tobe S. S. & Weaver R. J. (1975b) Relative oxygenase activities in juvenile hormone biosynthesis of corpora allata of an African locust (*Schistocerca gregaria*) and American cockroach (*Periplaneta americana*). *Experientia* **31**, 120–122.
Pratt G. E., Tobe S. S., Weaver R. J. & Finney J. R. (1975a) Spontaneous synthesis and release of juvenile hormone by isolated corpora allata of the locust *Schistocerca gregaria* and cockroach *Periplaneta americana*. *Gen. Comp. Endocrinol.* **26**, 478–484.
Pratt G. E. & Weaver R. J. (1978) The rate of juvenile hormone biosynthesis by insect corpora allata, *in vitro*. In *Comparative Endocrinology* (ed. Gaillard P. J. & Boer H. H.), pp. 503–506. Elsevier/North Holland Press, Amsterdam.
Reibstein D., Law J. H., Bowlus S. B. & Katzenellenbogen J. A. (1976) Enzymatic synthesis of juvenile hormone in *Manduca sexta*. In *The Juvenile Hormones* (ed. Gilbert L. I.), pp. 131–146. Plenum Press, New York.
Rotman A. (1977) The mechanism of action of neurocytotoxic compounds. *Life Sciences* **21**, 891–900.
Sarrif A. M., McCarthy K. L., Nesnow S. & Heidelberger C. (1978). Separation of glutathione S-transferase activities with epoxides from the mouse liver h-protein, a major polycyclic aromatic hydrocarbon-binding protein. *Cancer Res.* **38**, 1438–1443.
Schanne F. A. X., Kane A. B., Young E. E. & Farber J. L. (1979) Calcium dependence of toxic cell death: a final common pathway. *Science* **206**, 700–702.
Schooley D. A., Judy K. J., Bergot H. J., Hall M. S. & Jennings R. C. (1976) Determination of the physiological levels of juvenile hormones in several insects and biosynthesis of the carbon skeleton of the juvenile hormones. In *The Juvenile Hormones* (ed. Gilbert L. I.), pp. 101–117. Plenum Press, New York.
Schooley D. A., Judy K. J., Bergot J. B., Hall M. S. & Siddall J. B. (1973) Biosynthesis of the juvenile hormones of *Manduca sexta*: labelling pattern from mevalonate, propionate and acetate. *Proc. Natl. Acad. Sci. USA* **70**, 2921–2925.
Schooneveld H. (1979) Precocene-induced necrosis and haemocyte-mediated breakdown of corpora allata in nymphs of the locust *Locusta migratoria*. *Cell Tissue Res.* **203**, 25–33.
Slama K. (1978) The principles of antihormone action in insects. *Acta entomol. bohemoslav.* **75**, 65–82.
Soderlund D. M., Messeguer A. & Bowers W. S. (1980) Precocene-II metabolism in insects: synthesis of potential metabolites and identification of initial *in vitro* biotransformation products. *J. Agric. Food Chem.* **28**, 724–731.
Tarrant C. A. & Cupp E. W. (1978) Morphogenetic effects of precocene-II on the immature stages of *Rhodnius prolixus*. *Trans. Roy. Soc. Trop. Med. Hygiene* **72**, 666–668.
Thomas P. E., Korzeniowski D., Bresnick, Bornstein W. A. Kasper C. B., Fahl W. E., Jefcoate C. R. & Levin W. (1979) Hepatic cytochrome P-448 and epoxide hydrase: enzymes of nuclear origin are immunochemically identical with those of microsomal origin. *Arch. Biochem. Biophys.* **192**, 22–26.
Tobe S. S. & Pratt G. E. (1974) Dependence of juvenile hormone release from corpus allatum on intra-glandular content. *Nature* **252**, 474–476.
Tobe S. S. & Pratt G. E. (1976) Farnesenic acid stimulation of juvenile hormone biosynthesis as an experimental probe in corpus allatum physiology. In *The Juvenile Hormones* (ed. Gilbert L. I.), pp. 147–163. Plenum Press, New York.
Tsang W-S. & Griffin G. W. (1979) *Metabolic Activation of Polynuclear Aromatic Hydrocarbons*. Pergamon Press, Oxford, New York, Toronto, Sydney, Paris & Frankfurt, 125 pp.
Unnithan G. C., Nair K. K. & Bowers W. S. (1977) Precocene-induced degeneration of the corpus allatum of adult females of the bug *Oncopeltus fasciatus*. *J. Insect Physiol.* **23**, 1081–1094.

Unnithan G. C., Nair K. K. & Kooman C. J. (1978) Effects of precocene II and juvenile hormone III on the activity of neurosecretory A-cells in *Oncopeltus fasciatus*. *Experientia* **34**, 411.
Unnithan G. C., Nair K. K. & Syed A. (1980) Precocene-induced metamorphosis in the desert locust *Schistocerca gregaria*. *Experientia* **36**, 135–136.
Walker W. F. (1978) Mating behaviour in *Oncopeltus fasciatus* (Dallas): effects of diet, photoperiod, juvenoids and precocene-II. *Physiol. Ent.* **3**, 147–155.
Weaver R. J., Pratt G. E., Hamnett A. F. & Jennings R. C. (1980) The influence of incubation conditions on the rates of juvenile hormone biosynthesis by corpora allata isolated from adult females of the beetle *Tenebrio molitor*. *Insect Biochem.* **10**, 245–254.
Whalen D. L., Ross A. M., Montemarano J. A., Thakker D. R., Yagi H. & Jerina D. M. (1979) General acid catalysis in the hydrolysis of benzo (α) pyrene 7,8-diol 9, 10-epoxides. *J., Am. Chem. Soc.* **101**, 5086–5088.
Wigglesworth V. B. (1969) Juvenile hormone assays in insects with special reference to metabolic breakdown. *Gen. Comp. Endocrinol.* **13**, 541.
Yamoto S. & Block K. (1970) Studies on squalene epoxidase of rat liver. *J. Biol. Chem.* **245**, 1670–1674.
Yang S. K. & Gelboin H. V. (1976) Nonezymatic reduction of benzo (α) pyrene diol-epoxides to trihydroxypentahydrobenzo (α) pyrenes by reduced nicotinamide adenine dinucleotide phosphate. *Cancer Res.* **36**, 4185–4189.

20.11 DISCUSSION

Djerassi: Are you aware of any experiments done to confirm carcinogenicity of your products?

Ellis-Pratt: We have not done such experiments ourselves. But I have seen some work in the literature describing observation of carcinogenicity with products related to ours.

Bowers: I have had no information from anybody who has specifically done these tests. However, I am aware of very successful screening experiments looking for antipathogenic activity against insects.

Ellis-Pratt: Have you not obtained some results with precocene which implicate carcinogenicity?

Bowers: No. We have not. However, we have seen cases where bollworms react to high doses of precocene, but I am not convinced these are cases of carcinogenicity. Rather I believe the bollworms have an efficient means of detoxifying precocene which involves conversions of precocene to its reactive epoxides. This detoxification process allows these worms to survive high doses of this product.

Djerassi: Have you tested for carcinogenicity in malaria fever vectors?

Bowers: I have had no experience with mosquitoes.

Ellis-Pratt: We have done some unsuccessful tests with precocenes. As yet we have no clear indications of carcinogenicity in insects we use in our tests.

Bowers: May I caution you that precocene is a very reactive compound which does in fact react with water. This feature is likely to mask the actual effect of the compound in any biological system.

Ellis-Pratt: We have some kinetic studies with precocene epoxides at very low concentrations. These studies have established that the reaction is third order with respect to water. This seems to suggest that the epoxides are protected hydrophobically in the cell and thus can move around a lot.

Bowers: In addition to that, we have found that the rate of epoxidation in some insects, we regard as insensitive to precocene, is not high as in some of our sensitive species. These tests are not yet complete and we are now doing cross-metabolism-transfer mutations to get further insight into the phenomenon.

Djerassi: Do you think that it is possible to develop a better bioassay for antijuvenile activity in the corpora allata of insects? Do some of the *in vitro* incorporations you are talking about have any potential for further studies to see whether or not juvenile hormone is inhibited by the insects' metabolic process?

Bowers: Yes, I think it is possible and a lot of work is in progress on these lines. I am not, however, aware of new discoveries yet.

Marini Bettolo: I should like to make a comment about epoxides of precocenes. In plants, epoxides are rather widespread in many families. In Compositae, a great number of sesquiterpene epoxides were found and proved to have antitumoural activity. Epoxides have also shown co-carcinogen activity in some cases. Anyhow, they are substances with a rather good reactivity and thus may react with genetic material of the cell. In this respect, they may be considered *cytotoxic* in general. The mechanism of this reaction is not clear; it may be influenced by other groups in the molecule and the conditions of the medium. The fact is that in plants, epoxides are found together with the corresponding diols.

Ellis-Pratt: What actually is the reactivity of these epoxides and what are the substituent groups? What are the effects of these groups?

Marini Bettolo: There is no doubt that epoxides are very reactive compounds. Their reactivity depends on the structure and on the environment in which they exist.

Casida: I would like to add that from our experience with the pyrethroids, we have observed formation of epoxides of pyrethrin molecules by insects. In this way the insects detoxify the pyrethrins.

Unnithan: Would you explain some of the effects of precocene in insects in terms of cytotoxicity?

Bowers: I have no idea about what precocenes do in terms of general toxicity.

Saxena: An activity related to what you are talking about has been reported with extracts of *Bartinia*, a local plant in India. It is not, however, clear whether or not the plant antagonizes the gland that produces juvenile hormones in the insect.

Whitehead: Did you say that corpora allata are the only tissues where epoxidation takes place?

Ellis-Pratt: No. I said that corpora allata are the tissues we looked at which do not contain epoxidehydrase.

Feeny: May I ask if you have done a literature survey to see whether the insect which feeds on *Ageratum* is in any way peculiar?

Bowers: No. I have not.

CHAPTER 21

Effects of Precocenes on the Sorghum Shootfly, Atherigona soccata (Diptera, Muscidae)

G. C. UNNITHAN

The International Centre of Insect Physiology and Ecology (ICIPE), P.O. Box 30772, Nairobi, Kenya

CONTENTS

21.1 Introduction   357
21.2 Materials and Methods   358
21.3 Results   359
    21.3.1 Behavioural and lethal effects   359
    21.3.2 Effects on egg maturation   360
    21.3.3 Effects of fumigation of females on hatching of eggs   361
    21.3.4 Fumigation of eggs   361
    21.3.5 Effect of topical application on pupae   362
21.4 Conclusion   363
21.5 Summary   364
21.6 Acknowledgement   364
21.7 References   364
21.8 Discussion   365

21.1 INTRODUCTION

Precocene I (P I) [(7-methoxy-2,2-dimethyl chromene) (Fig. 21.1(a)] and precocene II (P II) [(6,7-dimethoxy-2,2-dimethyl chromene) (Fig. 21.1(b)] are compounds with antijuvenile hormone (antiJH) activity isolated from the plant *Ageratum houstoniaum* (Bowers, 1976; Bowers, Ohta, Cleere & Marsella, 1976). These compounds induced precocious metamorphosis and sterility in a variety of parametabolous insects including the milkweed bug *Oncopeltus fasciatus* (Bowers *et al.*, 1976; Bowers, 1977; Nemec, Chen & Wyatt, 1978; Pedersen, 1978; Pener, Orshan & De Wilde, 1978; Brooks, Hamnett, Jennings,

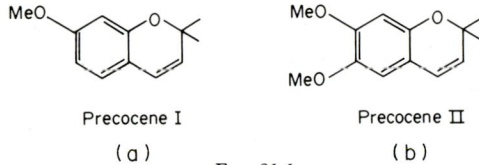

Meo   Meo
                             Meo

Precocene I   Precocene II
(a)   (b)
Fɪɢ. 21.1

Ottridge & Pratt, 1979a,b; Mackauer, Nair & Unnithan, 1979; Unnithan, Nair & Syed, 1980). Morphogenetic and antigonadotropic effects of precocene are caused by the inhibition or selective degeneration and atrophy of the corpora allata (CA) (Bowers, 1977; Bowers & Martinez-Pardo, 1977; Unnithan, Nair & Bowers, 1977; Pener *et al.*, 1978; Masner, Bowers, Kalin & Muhle, 1979; Muler, Masner, Kalin & Bowers, 1979; Schooneveld, 1979a,b, Unnithan & Nair, 1979; Bowers & Aldrich, 1980; Unnithan *et al.*, 1980). Precocene II also inhibited juvenile hormone biosynthesis *it vitro* by CA of *Periplaneta americana* (Pratt & Bowers, 1977). Only active CA of *O. fasciatus* are susceptible to P II (Unnithan & Nair, 1979; Masner *et al.*, 1979). It is suggested that the cytotoxicity of precocene on the CA is caused by its action on metabolism within the gland (Brooks *et al.*, 1979b; Pratt, Jennings, Hamnett & Brooks, 1980).

Among the holometabolous insects, antiJH activity of precocene is reported only in the apple maggot, Colorado potato beetle and in the Mexican bean beetle (Bowers *et al.*, 1976). In the mosquito *Aedes aegypti* neither Cupp, Lok & Bowers (1977) nor Kelly & Fuchs (1978) could demonstrate a specific antiJH action. However, Feldlaufer (personal communication) found that P II treatment of *Aedes* pupae inhibited development of the male accessory gland and the resulting adults gave very poor reproductive performance. Since the male accessory gland has been shown to require JH for development (Ramalingam & Craig, 1977) a specific antiJH action in the male mosquito is revealed.

Substances with antiJH activity are viewed as potential sources of safer and selective "fourth generation" insecticides (Bowers *et al.*, 1976). The objectives of the present investigation are to determine the effects of P I and P II on eggs, pupae and adults of the sorghum shootfly, *Atherigona soccata* Rondani. This muscid fly, about 5 mm in size, is the most important pest of sorghum seedlings in Africa and Asia (Pont, 1972). The egg is laid on the underside of the leaves of sorghum seedlings. A small maggot hatches within 48 h and produces the "dead heart" which destroys the seedling. Larval growth and development is completed within the shoot of the infested seedling.

21.2   MATERIALS AND METHODS

Late third instar larvae of sorghum shootflies, collected from infested sorghum seedlings in the field, were allowed to pupate in an artificial diet. The pupae collected from this diet were kept for adult emergence in an environmental chamber. Eggs were obtained by placing sorghum seedlings in perspex rearing cages containing gravid females and males; the females laid many eggs on leaves in cages.

Treatment at intervals, of females from 4 h to 2-days-old was performed either topically, with 1–10 μg each of P I or P II (Calbiochem, La Jolla, California), in 1 μl acetone, or by fumigation (contact) method. In the latter method 10–20 flies were confined to 5 × 1.8 cm petri dishes with P I or P II residue at a concentration of 1 μg–10 μg cm^{-2} for 0.5–3 h. Similarly, strips of sorghum leaves with 8–12 h-old eggs were fumigated for 16–20 h; the dose ranged from 0.5 μg–10 μg cm^{-2}. The eggs were never in contact with precocene. Pupae, 24–36 h-old, were treated topically with 10 μg each of P I or P II. As control for topical application, adults and pupae were treated with 1 μl acetone, whereas adults and eggs were confined to petri dishes, rinsed in acetone and dried, for appropriate duration to simulate fumigation conditions. Flies were anaesthetized with CO_2 before treatment.

After treatment, the flies were transferred to rearing cages, fed on water and 1:1 brewer's yeast and glucose in the greenhouse, maintained at 27.5 ± 2.5°C, 60 ± 15% RH with the natural photoperiod, or in an environmental chamber maintained at 30°C, 60 ± 10% RH and 12 h photophase. In experiments to determine the effect of fumigation of female shootflies on hatching of eggs, the insects were fed on water and 1:1 baker's yeast and sugar, and kept in the environmental chamber; females were kept with an equal number of untreated males. To determine the effects of precocene on egg maturation, 1-day-old females were treated with P I or P II. Three days later the number of surviving flies with gravid ovaries, number of mature eggs, and number of resorbing oocytes were determined. The data were analysed by using the χ^2 test.

21.3 RESULTS

21.3.1 Behavioural and lethal effects

Experimental and control insects revive from the effect of CO_2 anaesthesia in less than 1 min. The immediate response to precocene treatment is a form of hyperactivity. Flies treated topically with P I and P II fly around rapidly for short distances for a few minutes; soon this activity is reduced to rapid wing beating and spinning. Within a few minutes, the flies are knocked down and appear to be in a state of paralysis. During fumigation, depending on the dose, within 10–50 min after beginning of the treatment, the insects become inactive

TABLE 21.1 MORTALITY OF SORGHUM SHOOTFLY FEMALES WITHIN 3
DAYS AFTER TREATMENT WITH PRECOCENE I AND PRECOCENE II (P I
& P II)

Compound	Method of application	Dose and expo-sure time (h)	No. insects	Per cent mortality (cumulative) within 1 day, 2 days or 3 days		
P I	Topical	10 µg	38	68	89	92
P I	Topical	5 µg	34	41	67	70
P II	Topical	10 µg	13	84	100	—
P II	Topical	5 µg	72	38	63	73
P II	Topical	2 µg	15	6	6	13
Acetone	Topical	1 µg	45	13	22	28
P I	Fumigation	10 µg cm^{-2}, 0.5	28	100	—	—
P I	Fumigation	5 µg cm^{-2}, 1	17	94	100	—
P I	Fumigation	1 µg cm^{-2}, 1	44	13	34	41
P II	Fumigation	10 µg cm^{-2}, 0.5	26	69	81	88
P II	Fumigation	5 µg cm^{-2}, 1	41	95	95	95
P II	Fumigation	5 µg cm^{-2}, 0.5	22	13	68	86
P II	Fumigation	2 µg cm^{-2}, 2	47	38	76	89
P II	Fumigation	1 µg cm^{-2}, 3	19	89	94	94
P II	Fumigation	1 µg cm^{-2}, 0.5	26	7	46	53
Control	—	—	36	0	0	0

and are knocked down. At higher doses of fumigation (5–10 µg cm^{-2}) before the end of fumigation period (0.5–1.0h) a majority of the flies look as if they are dead. Over a period of several hours most insects treated with P I and P II at lower doses recover from the paralysed state. After knock down the insects regurgitate profusely; this is more common after P II treatment. Occasionally, some flies treated with P I ingest, food/water until their abdomens are swollen and the intersegmental membranes stretch tightly. These insects are unable to move and die within a few hours. It appears that P I may upset the control of feeding and excretion. No behavioural changes are observed in control insects.

Both P I and P II are highly toxic to sorghum shootflies and induce very high mortality. Mortality of female shootflies treated with P I and P II are shown in Table 21.1. At higher doses, most insects died before recovering from the initial paralysis. The lethal dose (data not shown) is much lower for male shootflies.

21.3.2 Effects on egg maturation

Normal shootflies mature the first batch of eggs within 3 days after emergence, when fed on brewer's yeast and glucose (Unnithan, unpublished). One-day-old females were treated topically with P I, P II or acetone and

TABLE 21.2 EFFECTS OF TOPICAL APPLICATION OF PRECOCENE I (PI)
AND PRECOCENE II (P II) ON OOCYTE MATURATION AND
RESORPTION IN THE SORGHUM SHOOTFLY

Compound and dose	No. of insects	Age (days)	No. of insects with mature eggs (%)	No. of ovarioles with resorbed oocyte (%)
1. P I 5 µg	10	4	90	30 a
2. P II 2 µg	12	4	75	29 a
3. P II 5 µg	14	4	71	42 b
4. Acetone - 1 µl	17	4	100	31 a

Figures followed by the same letter are not significantly different from each other. Significant level: 1 vs. 2 P <0.01; 2 vs. 3 and 3 vs. 4 P <0.001.

3 days later the number of surviving insects with gravid ovaries, number of mature oocytes and number of ovarioles with resorbing oocytes were determined. From the data shown in Table 21.2, only 75% and 71% of the flies treated with 2 µg and 5 µg P II, respectively, have gravid ovaries, compared to 90% for flies treated with 5 µg P I and 100% for control insects. Frequency of oocyte resorption is significantly higher (P < 0.001) in females treated with 5 µg P II. There is no significant difference in resorption rate among flies treated with 5 µg P I, 2 µg P II and 1 µl acetone each.

21.3.3 Effects of fumigation of females on hatching of eggs

One- and two-day-old females were fumigated with P I or P II to determine whether exposure of gravid females will affect hatching of the eggs. These flies, fed on baker's yeast and sugar, mature eggs within 2 days after emergence (Unnithan, unpublished). All the eggs laid by surviving precocene-treatment and control (CO_2-anaesthetized) insects within 5 days after treatment were collected and the number of eggs hatched were determined. Insects which died within 2 days after treatment did not lay any eggs although they were gravid at the time of treatment. Oviposition was delayed by at least one additional day in gravid females exposed to P I or P II. Also there was an apparent decrease in fecundity in precocene-treated females. There is considerable decrease in hatching of eggs when 2-day-old flies are fumigated with 1 µg cm^{-2} P I for 1 h, 1 µg cm^{-2} P II for 0.5 h and 2 µg cm^{-2} P II for 1 h (Table 21.3). Exposure of 2-day-old flies to 1 µg cm^{-2} P I for 0.5 h and 1-day-old flies to 1 µg cm^{-2} P II for 1 h is without any significant effect on hatching. Examination of unhatched eggs revealed that embryogenesis was also inhibited in these eggs.

21.3.4 Fumigation of eggs

Eight to twelve hours old eggs were fumigated with P I and P II to study the

TABLE 21.3 EFFECTS OF FUMIGATION OF FEMALE SORGHUM
SHOOTFLIES WITH PRECOCENE I AND II (P I & P II) ON HATCHING OF
EGGS LAID WITHIN 5 DAYS AFTER TREATMENT

Compound and dose	Duration of treatment (h)	Age of insects (days)	No. of eggs	Percent eggs hatched
1. P I 1 μg cm^{-2}	0.5	2	164	92 a
2. P I 1 μg cm^{-2}	1	2	220	77 b
3. P II 1 μg cm^{-2}	1	1	151	86 a
4. P II 1 μg cm^{-2}	0.5	2	244	60 c
5. P II 2 μg cm^{-2}	0.5	2	124	59 c
6. Control	—	2	216	90 a

Figures followed by the same letter are not significantly different at 5% level. Differences are significant at 0.1% level except between 2 and 3 where it is only at 5% level.

effects on embryogenesis and hatching. The results are shown in Table 21.4. Fumigation with P I at a dose of 10 μg cm^{-2} for 16 h results in nearly complete inhibition of hatching; only 3% eggs hatch. Inhibition is much less at lower doses of P I. But fumigation with P II, 10 μg cm^{-2} shows no inhibition of hatching. Most of the P I-fumigated unhatched eggs contained fully differentiated larvae, indicating that embryogenesis was apparently unaffected by P I treatment.

TABLE 21.4 EFFECT OF FUMIGATION OF SORGHUM SHOOTFLY EGGS
WITH PRECOCENE I AND II (P I & P II) ON HATCHING

Compound and dose (μg cm^{-2})	Duration of treatment (h)	No. of eggs fumigated	No. of eggs hatched (%)
P I 10	16	100	3
P I 1	20	100	21
P I 0.5	20	100	80
P II 10	20	94	90
Control	—	100	90

21.3.5 Effect of topical application on pupae

Twenty-four to thirty-six-h-old pupae were treated topically with 10 μg each of P I or P II. The results show no significant difference in the number of adults emerged among the precocene-treated and control pupae (Table 21.5). Duration of pupal period in P II-treated pupae is slightly less than in P I-treated pupae (P < 0.01) and in control pupae (P < 0.02).

TABLE 21.5 EFFECT OF TOPICAL APPLICATION OF PUPAE WITH
PRECOCENE I & II (P I & P II) ON ADULT EMERGENCE AND DURATION
OF PUPAL PERIOD

Compound and dose	Number of pupae	Per cent adult emergency	Pupal period (days, $\bar{x} \pm$ SD)
P I 10 µg	35	77 a	6.7 ± 0.6 a
P II 10 µg	40	90 a	6.2 ± 0.9 b
Acetone 1 µl	20	90 a	6.6 ± 0.4 a

Figures followed by the same letter are not significantly different from each other at 5% level.

21.4   CONCLUSION

Exposure of adult sorghum shootflies to relatively low doses of P I and P II leads to high mortality. Some of the behavioural effects of precocene resemble insecticidal effects. Precocene II also showed lethal effects in *Aedes aegypti* (Kelly & Fuchs, 1978), *Oncopeltus fasciatus* (Bowers, 1976; Unnithan & Nair, 1979; Masner *et al.*, 1979). Unlike in hemipterous insects or in the apple maggot (Bowers *et al.*, 1976), precocene does not have any antigonadotropic effect on the sorghum shootfly. However, it is not known whether juvenile hormone, and hence the CA, have any influence on egg maturation in the sorghum shootfly.

Non-specific effects of precocene may account for the delay in ovarian maturation, high rate of oocyte resorption and the apparent reduction in fecundity after P I or P II treatment of adults. Protein deficiency leads to high incidence of oocyte resorption in the sorghum shootfly (Unnithan, unpublished). Precocene II has been shown to inhibit trypsin synthesis in *Ae. aegypti* (Kelly & Fuchs, 1978) and synthetic activity of neurosecretory A cells in *O. fasciatus* (Unnithan, Nair & Kooman, 1978). Precocene II showed antiJH activity in adult males of *Ae. aegypti* when pupal stages were treated (Feldlaufer, personal communication). However, treatment of adult females did not show any antigonadotropic activity, but only a retardation of oocyte maturation (Kelly & Fuchs, 1978).

Both P I and P II have ovicidal activity on sorghum shootfly eggs, although P II is effective only when adults are fumigated. The differential sensitivity of shootfly eggs to P I may be due to the relatively higher volatility of the compound. Precocenes are ovicidal by fumigant action on the eggs of *O. fasciatus* and *Epilachna varivestis* (Bowers *et al.*, 1976); here also embryogenesis does not seem to be affected.

Pupal stages of sorghum shootfly do not seem to be sensitive to topical application of P I and P II. It is not known whether this inactivity of P I and P II on the pupae is due to precocene not reaching the internal tissues. In

Ae. aegypti, Cupp *et al*. (1978) have shown that P II inhibited pupation when applied to fourth instar larvae, and prevented adult emergence when newly hatched larvae were exposed to it; however, pupal stages were not sensitive to P II.

In conclusion, both P I and P II do not seem to have any antiJH activity on the sorghum shootfly. This may be either because of the lack of sensitivity of CA to precocene or due to metabolic breakdown of precocene in some other non-target tissues, like fatbody (Burt, Kuhr & Bowers, 1978), before reaching the CA. Any growth-regulatory activity of precocene is perhaps limited to the ovicidal action. Further research is required to assess the potential of exploiting the lethal and ovicidal effects of precocenes for possible control of the sorghum shootfly.

21.5 SUMMARY

Effects of antijuvenile hormone compounds, precocene I (P I) and prococene II (P II) on eggs, pupae and adults of the sorghum shootfly were investigated. Neither compound appears to have a direct antigonadotropic action, however there is delay in egg maturation and high incidence of oocyte resorption after treatment of adult females with P II. Exposure of gravid females to P I and P II resulted in inhibition of embryogenesis and hatching of the eggs laid after treatment. Fumigation of eggs with P I inhibited hatching, but P II did not inhibit embryogenesis or hatching. Both P I and P II have no discernible effects on pupae when applied topically. There is no indication of P I and P II having any definite antijuvenile hormone activity on sorghum shootflies, however, both compounds show very high toxicity and ovicidal effects.

21.6 ACKNOWLEDGEMENT

I am thankful to Professor W. S. Bowers for critically reviewing the manuscript.

21.7 REFERENCES

Bowers W. S. (1976) Discovery of insect antiallatotropins. *The Juvenile Hormones* (ed. Gilbert L. I.), pp. 394–408. Plenum Press, N.Y.

Bowers W. S. (1977) Antijuvenile hormone from plants; chemistry and biological activity. *Pontificae Academiae Scientarum Scripta varia* **41**, 129–156.

Bowers W. S. & Aldrich J. R. (1980) *In vivo* inactivation of denervated corpora allata by precocene II in the bug, *Oncopeltus fasciatus. Experientia* **36**, 362–364.

Bowers W. S. & Martinez Pardo R. (1977) Antiallatotropins: Inhibition of corpus allatum development. *Science* **197**, 1369–1371.

Bowers W. S., Ohta T., Cleere J. S. & Marsella P. A. (1976) Discovery of insect antijuvenile hormones in plants. *Science* **193**, 542–547.

Brooks G. T., Hamnett A. F., Pratt G. E., Jennings R. C. & Ottridge A. P. (1979a) The action of precocene in milkweed bugs (*Oncopeltus fasciatus*) and locusts (*Locusta migratoria*). *Nature* **281**, 570–572.

Brooks G. T., Hamnett A. F., Jennings R. C., Ottridge A. P. & Pratt G. E. (1979b) Aspects of mode of action of precocenes on milkweed bugs. *Proc. 1979 British Crop Protection Conference — Pests and Disease*, Vol. 1, 273–297.

Burt M. E., Kuhr J. R. & Bowers W. S. (1978) Metabolism of precocene II in the cabbage looper and European corn borer. *Pesticide Biochem. Physiol.* **9**, 300–303.

Cupp E. W., Lok J. B. & Bowers W. S. (1977) The developmental effects of 6,7-dimethoxy-2,2-dimethyl chromene on the preimaginal stages of *Aedes aegypti* (Diptera, Culicidae). *Ent. exp. appl.* **22**, 23–28.

Kelly J. J. & Fuchs M. S. (1978). Precocene is not a specific antigonadotropic agent in adult female *Aedes aegypti. Physiol. Entl.* **3**, 297–301.

Mackauer M., Nair K. K. & Unnithan G. C. (1979) Effects of precocene II on alate production in the pea aphid *Acrythosiphon pisum. Can. J. Zool* **57**, 856–859.

Masner P., Bowers W. S., Kalin M. & Muhle T. (1979) Effect of precocene II on the endocrine regulation of development and reproduction in the bug, *Oncopeltus fasciatus. Gen. Comp. Endocr.* **37**, 156–166.

Muller P. J., Masner P., Kalin M. & Bowers W. S. (1979) *In vitro* inactivation of corpora allata of the bug *Oncopeltus fasciatus* by precocene II. *Experientia* **35**, 704–705.

Nemec V., Chen T. T. & Wyatt G. R. (1978) Precocious adult locust, *Locusta migratoria migratorioides* induced by precocene. *Acta entomologica bohemoslovaka* **75**, 285–286.

Pedersen L. E. K. (1978) Effects of antijuvenile hormone (Precocene I) on the development of *Locusta migratoria* L. *Gen. Comp. Endocr.* **36**, 502–509.

Pener M. P., Orshan L. & De Wilde J. (1978) Precocene II causes atrophy of corpora allata in *Locusta migratoria. Nature* **272**, 350–353.

Pont A. C. (1972) A review of the Oriental species of *Atherigona* Rondani (Diptera, Muscidae) of economic importance. *Control of Sorghum Shootfly* (ed. Jotwani M. G. & Young W. R.), pp. 27–104. Oxford & IBH Publishing Co., New Delhi.

Pratt G. E. & Bowers W. S. (1977) Precocene II inhibits juvenile hormone biosynthesis by cockroach corpora allata. *Nature* **265**, 548–550.

Pratt G. E., Jennings R. C., Hamnett A. F. & Brooks G. T. (1980) Lethal metabolism of precocene I to a reactive epoxide by locust corpora allata. *Nature* **284**, 320–323.

Ramalingam S. & Craig G. B. (1977) The effects of a JH mimic and cauterization of the corpus allatum complex on the male accessory glands of *Aedes aegypti* (Diptera: Culicidae). *Can. Ent.* **109**, 897–906.

Schooneveld H. (1979a) Precocene-induced collapse and resorption of corpora allata in nymphs of *Locusta migratoria. Experientia* **35**, 363–364.

Schooneveld H. (1979b) Precocene-induced necrosis and haemocyte-mediated breakdown of corpora allata in nymphs of the locust *Locusta migratoria. Cell Tissue Res.* **203**, 25–33.

Unnithan G. C. & Nair K. K. (1979) The influence of corpus allatum activity on the susceptibility of *Oncopeltus fasciatus* to precocene. *Ann. Ent. Soc. Am.* **72**, 38–40.

Unnithan G. C., Nair K. K. & Bowers W. S. (1977) Precocene-induced degeneration of the corpus allatum of adult females of the bug *Oncopeltus fasciatus. J. Insect Physiol.* **23**, 1081–1094.

Unnithan G. C., Nair K. K. & Kooman C. J. (1978) Effects of precocene II and Juvenile hormone III on the activity of neurosecretory A cells in *Oncopeltus fasciatus. Experientia* **34**, 411.

Unnithan G. C., Nair K. K. & Syed A. (1980) Precocene-induced metamorphosis in the desert locust *Schistocerca gregaria. Experientia* **36**, 135–136.

21.8 DISCUSSION

Bowers: I am very impressed with the effects of fumigation, especially the mortality figures on flies. It seems fumigation with very small doses is more effective than your topical treatment of flies. How do you explain that?

Unnithan: Yes, you are right. With fumigation, the flies are exposed to the chemical sometimes for only half an hour. This is probably a question of penetration. Precocene I is more volatile, as Dr. Pratt pointed out earlier, and this high volatility makes it more effective when flies are fumigated.

Bowers: Would you imagine some of this getting directly into the trachea and to the tissue?

Unnithan: Yes, I think probably that is what happens. Prococene I probably reaches the tissue faster through the tracheal system than through the cuticle.

Bowers: Would you like to go back and talk about the fact that some years ago you found that *Schistocerca gregaria* was affected by precocene through contact and by fumigation in petridish and less by topical application?

Unnithan: Yes, I could not get any significant effect by topical application of very high doses of precocene on *S. gregaria*, whereas I achieved almost 100% precocious metamorphosis by exposing the third and fourth instar nymphs to an optimum dose of about 15 μg cm^{-2} precocene. This was probably because precocene was getting into the corpus allatum directly through the tracheal system instead of through the circulatory system where precocene may be degraded and detoxified.

Djerassi: Have you tried any field experiment, for example in a small plot of sorghum fumigated or sprayed with precocene solution?

Unnithan: I have not tried that yet. I have observed that strips of sorghum leaves with attached shootfly eggs when fumigated with P I retained the smell of precocene for several days, indicating that activity of precocene may last for several days.

Djerassi: Take a look at what is happening to precocene in the field where it will be exposed to rain and sunshine?

Unnithan: At this stage we do not even know the toxic or other effects of precocene on mammals, or even other non-target animals.

Djerassi: There is no question that we ought to have done some toxicology studies by now. A serious economic pest could be controlled with precocene if it is cleared as being non-toxic, after suitable formulation. However, I would also like to ask if it is known whether fumigation is a really feasible protection, in the practical context, for crops infested with shootfly?

Bowers: I would only say that in the field one of the beneficial effects of gallapol, an acceptable compound today, is its fumigant activity. By the way, this was discovered not in the laboratory, but while doing field application studies.

Djerassi: While talking about fumigation I wanted to make some comments on your paper. When you talk about the use of JH and analogues you indicated relatively limited use against flies and mosquitoes. There is one additional use which is probably a very significant one now approved by the Environmental Protection Agency late last year and made full use of, this year. This is for protection of stored tobacco against tobacco beetles by fumigation. It is very effective and protects stored tobacco for the period of two years which it takes to mature. So there are other applications of this type which are not only against flies. Hopefully, additional application of insect JH analogues, will be found.

Bowers: You were fumigating tobacco with methoprene, were you?

Djerassi: Yes, it was methoprene.

Graham-Bryce: I just wanted to make a comment about the fumigation question. I wonder if the term "fumigation" in itself is a little misleading, in that one tends to think of fumigation as a fairly massive vapour action. If we consider the short-range vapour effects, surely these are fairly widespread in conferring activity to chemicals? Activity of many systemic chemicals, I suspect, ultimately depends on short-range vapour effects outside the leaves.

Djerassi: I am curious. Considering all the experiments done on precocene, how stable is precocene photochemically?

Graham-Bryce: Can we also ask in this connection if there is a figure for its volatility? We have heard it is very volatile but what is its vapour pressure?

Ellis-Pratt: We ought to know the vapour pressure but I don't.

Djerassi: Surely you know how stable it is to sunlight during your experiments?

Ellis-Pratt: Yes, 80% of μg quantities of P I is lost when the solvent evaporates from the petri dish you are coating. The rest must be protected from light and UV. Precocenes run on silica plates rapidly turn yellow in UV light.

Bowers: I agree. It is not very photostable.

Raina: In response to your query about the utilization of precocene for practical pest control, particularly for shootfly, I would like to add that sorghum is a crop grown by subsistence farmers. We are trying to find a method for them which will be economical to use, like plant resistance or even insecticides, which are cheap. This might be a limitation in the case of precocene. Another thing is that its toxicity has still to be tested.

Casida: It seems you call this a highly toxic compound, but if you were to calculate the weight of your doses in micrograms and relate this to the potency of most insecticides you are more than a hundred-fold off in sensitivity. In other words, relative to most insecticides it is not highly potent.

Unnithan: I agree with you, the toxicity of precocene may not be comparable to that of most insecticides.

Jotwani: In the case of shootfly, none of the insecticide applications are foliar applications. None of the insecticides we have on the market have given satisfactory control. One of the reasons is the quick growth of sorghum. Even within 3 days of application there is again infestation and damage. This is why we are not able to succeed with any of the potent insecticides available on the market. Even systemic insecticides, applied as foliar spray, are not successful; only as seed dressing or when applied in the soil was this successful. So I don't think even precocene will be useful as far as shootfly control is concerned.

Djerassi: Is it also true that shootflies are always underneath the lowerside of the leaf when they are not flying?

Jotwani: Yes.

Saxena: I want to comment on your table on egg maturation and resorption. The figures shown on resorption in control and precocene-treated insects are probably not significantly different. Did you have enough replicates?

Unnithan: The difference is significant only in the case of insects treated with 5 μg P II. Oocyte resorption is very high in these. This could be due to some non-specific effects of precocene. There is a report, for instance, that precocene inhibits trypsin synthesis in *Aedes*. The data shown in the table were based on several replicates; but because of the high mortality after P II precocene treatment, it was difficult to find enough surviving insects in one or two treatments.

CHAPTER 22

Structure, Chemistry and Actions of the Piperaceae Amides: New Insecticidal Constituents Isolated from the Pepper Plant

MASAKAZU MIYAKADO, ISAMU NAKAYAMA, NOBUO OHNO and HIROSUKE YOSHIOKA

*Pesticides Research Dept., Institute for Biological Sciences, Sumitomo Chemical Co. Ltd.
4-2-1, Takatsukasa, Takarazuka Hyogo, 665, Japan*

CONTENTS

22.1 Introduction   369
22.2 Materials and Methods   371
    22.2.1 Oleoresin of black pepper fruits   371
    22.2.2 Chemicals   371
    22.2.3 Insects   372
22.3 Results and Discussion   372
    22.3.1 Isolation of insecticidal principles   372
    22.3.2 Structure determinations of the three components of
            band-a   372
    22.3.3 Synthesis of pipercide (1), dihydropipercide (2) and
            guineensine (3)   375
    22.3.4 Joint insecticidal action of pipercide (1) and co-occurring
            compounds isolated from black pepper plants   377
22.4 Conclusions   379
22.5 Summary   379
22.6 Acknowledgements   380
22.7 References   380
22.8 Discussion   381

22.1 INTRODUCTION

The fruits of Piperaceae plants have been known to contain physiologically active principles, and several investigations of the constituents of the fruits

369

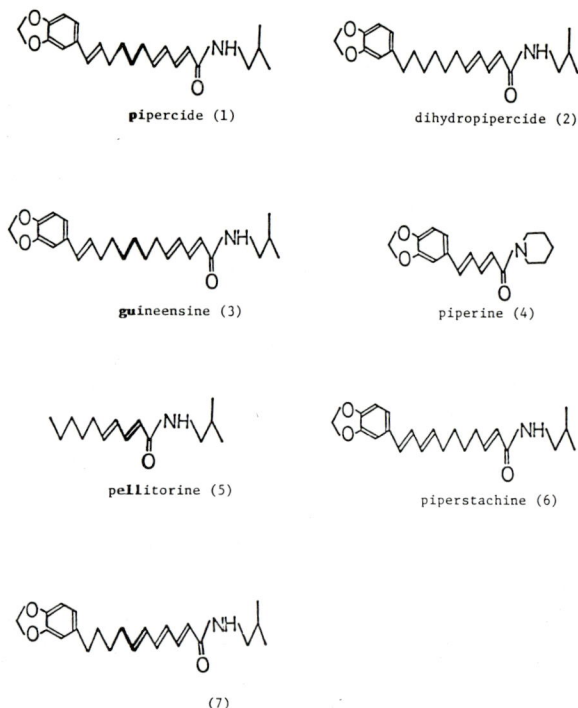

pipercide (1)

dihydropipercide (2)

guineensine (3)

piperine (4)

pellitorine (5)

piperstachine (6)

(7)

FIG. 22.1. Piperaceae amides.

have been published (Grewe, Feist, Neumann & Kersten, 1970; Loder, Moorhouse & Russel, 1966; Joshi, Viswanathan, Gawad & Von Philipsborn, 1975). Among these constituents, a group of unsaturated amides seem to be a major group of secondary metabolites [Fig. 22.1(2)]. Harvill, Hartzell & Arthur (1943) reported that piperine (4), a constituent of the dried fruits of black pepper (*Piper nigrum* L.), was more toxic than pyrethrum (powdered flower heads of *Chrysanthemum cineraliaefolium*) against houseflies. Jacobson (1953) reported that pellitorine (5), one of the constituents of black pepper showed a paralysing effect on houseflies and its lethal activity was one-half of that of pyrethrins. Recently, Su (1977) and Scott & McKibben (1978) reported that crude and purified extracts of black pepper caused high mortality in both rice weevils and boll weevils on topical application. In addition, there are many other studies on the amides of Piperaceae plants. Some of the typical examples for these amides are given in Fig. 22.2. However, almost all of these amides had not been described in any connection with insecticidal activity, so the insecticidal nature of these Piperaceae amides remained obscure up to date. In this paper, the isolation, structure determination, syntheses and insecticidal activity of the new amides which we

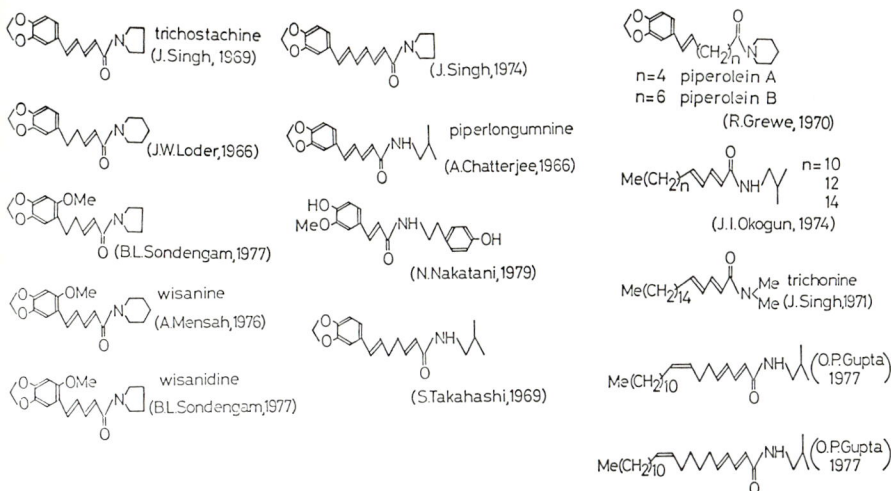

FIG. 22.2. Miscellaneous Piperaceae amides.

named pipercide (1) and dihydropipercide (2) as well as identification of a known amide, guineensine (3) from the fruits of *P. nigrum* L., are summarized.

22.2 MATERIALS AND METHODS

22.2.1 Oleoresin of black pepper fruits

The fruits of black pepper (*Piper nigrum* L.) collected in Sumatra, Indonesia, in 1977 were ground with a Wiley mill. The ground pepper (1.8 kg) was extracted with CH_2Cl_2 for 3 h at 35°C with stirring. After filtration, the filtrate was concentrated *in vacuo* to leave 130 g of yellowish oleoresin.

22.2.2 Chemicals

Synthetic pipercide (1), dihydropipercide (2) and guineensine (3) were used for bioassays (Miyakado & Yoshioka, 1979b; Miyakado, Makayama & Yoshioka, 1980). Pellitorine (5) was synthesized from 1-hexanal and diethyl 3-methoxycarbonyl-2-propenyl-phosphonate (11) by Wittig-Horner reaction. Piperine (4) was purchased from Aldrich Chemical Co. Commercially available extract (20% in n-hexane) was used for the reference compound.

22.2.3 Insects

The pest insects were adzuki bean weevil (*Callosobruchus chinensis* L.), rice weevil (*Sitophilus oryzae*), mosquito larva (*Culex pipiens pallens*) and housefly [*Musca domestica*, WHO/IN-strain and pyrethroid-resistant strain (228-e2b)].

22.3 RESULTS AND DISCUSSION

22.3.1 Isolation of insecticidal principles

The oleoresin from black pepper showed notable insecticidal activity against adzuki bean weevils and mosquito larvae. Throughout the isolation of the insecticidal principles of the fruits of black pepper, adzuki bean weevil had been used as test insects. Isolations of insecticidal principles are schematically described in Fig. 22.3. From band-E on TLC (B), piperine (4), one of the major constituents of pepper was isolated. Although piperine had been reported to be more toxic than pyrethrum (Harvill *et al.*, 1943), we found that piperine was not lethal against houseflies as well as adzuki bean weevils, that is inconsistent with the report of Harvill (Table 22.1). By TLC (C) analysis, band-a showed the highest activity (100% mortality of both male and female) among the four (a–d), at 5 µg/insect applications.

The insecticidal activity of band-a (TLC/C) and related compounds is shown in Table 22.2. The LD_{50} values of piperine (4) was greater than 10 µg. Pellitorine (5) showed moderate activity. On the other hand, band-a showed the highest activity which was a comparable order to that of pyrethrins. However, GLC analysis of band-a indicated that band-a consisted of three components in a ratio of 75:5:20 as shown in Fig. 22.4. Since the contribution for the activity of band-a of each component was not clear, each compound had to be identified and evaluated for insecticidal activities.

22.3.2 Structure determinations of the three components of band-a

22.3.2.1 *Pipercide (1)*
Several recrystallizations of the eluent from band-a from EtOAc gave fine needles which showed a single peak corresponding to peak (A) (Fig. 22.4) on

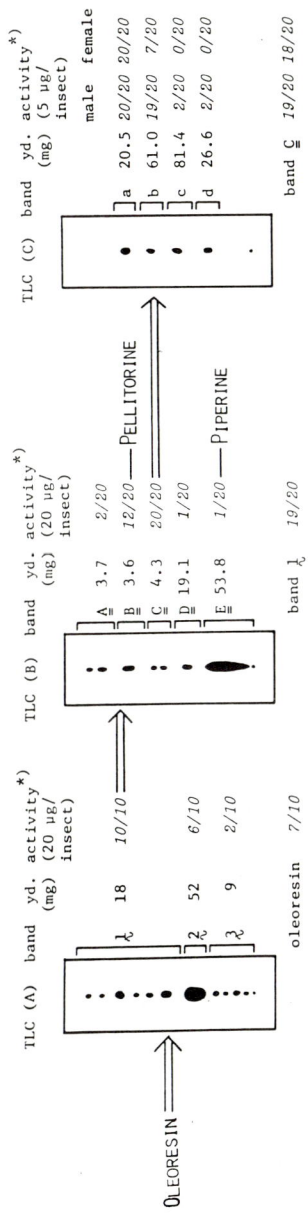

TLC (A) band yd. (mg) activity*) (20 µg/insect)

OLEORESIN

1: 18, 10/10
2: 52, 6/10
3: 9, 2/10

oleoresin 7/10

TLC (B) band yd. (mg) activity*) (20 µg/insect)

A: 3.7, 2/20
B: 3.6, 12/20 —— PELLITORINE
C: 4.3, 20/20
D: 19.1, 1/20
E: 53.8, 1/20 —— PIPERINE

band 1 19/20

TLC (C) band yd. (mg) activity*) (5 µg/insect)

		male	female
a	20.5	20/20	20/20
b	61.0	19/20	7/20
c	81.4	2/20	0/20
d	26.6	2/20	0/20

band C 19/20 18/20

Fig. 22.3. Isolation of insecticidal principles.

TLC (A), applied oleoresin 100 mg, solvent n-hexane-acetone, 1:1 (v/v), spots with UV_{254} and I_2.
TLC (B), applied band 1 100 mg, solvent n-hexane-acetone, 3:1.
TLC (C), applied band C 260 mg, solvent (i) benzene-EtOAc, 2:1, (ii) benzene-EtOAc, 3:1.

*) against adzuki bean weevil, topical application.

TABLE 22.1. INSECTICIDAL ACTIVITY OF PIPERINE (4) AND
PYRETHRINS AGAINST HOUSEFLIES (*MUSCA DOMESTICA* L.,
WHO/IN-STRAINS, CONTACT METHOD, AFTER 24 H)

Material	Concentration (ppm)	Mortality (%)
Piperine (4)	400	0
Pyrethrins	400	100
	200	100
	100	80
	50	30

GLC analysis. This is a new compound which we named pipercide (1) ($C_{22}H_{29}NO_3$, m.p., 114–115°C). The structure (1) was proposed on the basis of spectroscopic data to be (2E, 4E, 10E)-N-isobutyl-11-(3,4-methylene-dioxy-phenyl)-2,4,10-undecatrienamide (Miyakado, Nakayama, Yoshioka & Nakatani, 1979a) and the structure was confirmed by total synthesis (Miyakado *et al.*, 1979b).

TABLE 22.2. INSECTICIDAL ACTIVITY OF PIPERINE (4), PELLITORINE
(5) AND BAND-A APPLIED TOPICALLY TO ADULT ADZUKI BEAN
WEEVILS

| Compound | LD$_{50}$ value μg/insect (48 h) | |
	Male	Female
Piperine (4)	>10.0	>10.0
Pellitorine (5)	2.0	7.0
Band-a	0.15	0.25
Pyrethrins	0.15	0.20

22.3.2.2 Dihydropipercide (2) and guineensine (3)

As the components (B) and (C) were hardly separable from each other by TLC or column chromatography, the structures for components (B) and (C) were proposed on the basis of GLC-MS analysis. The structure of component (B) was supposed to be (2) by analysis of the fragmentation pattern of its mass spectrum. This is also a new amide which we named dihydropipercide ($C_{22}H_{31}NO_3$). The structure for dihydropipercide was confirmed by total synthesis.

The component (C) was identical with guineensine (3) (2E, 4E, 12E)-N-isobutyl-13-(3,4 methylenedioxyphenyl)-2,4,12-tridecatrienamide) isolated by Ologun & Ekong (1974) from the fruits of *Piper guineense* Schum. and Thonn. They had reported the structure determination on spectroscopy basis and insecticidal nature of (3) had not been discussed. The structure for guineensine now has been confirmed by direct comparison with a synthetic specimen.

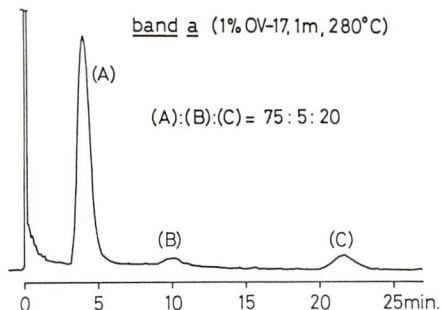

band a (1% OV-17, 1m, 280°C)

(A)

(A):(B):(C) = 75 : 5 : 20

(B)   (C)

0   5   10   15   20   25min.

FIG. 22.4. GLC analysis of band-a.

22.3.3 Syntheses of pipercide (1), dihydropipercide (2) and guineensine (3)

22.3.3.1 Pipercide (1)

The major constituents of Piperaceae plants, piperine (4) and pellitorine (5) have been synthesized by Normant & Feugeas (1964) and Jacobson (1953), respectively. As a more closely related compound to pipercide, Joshi *et al.* (1975) isolated piperstachine from *Piper trichostanchyon* C. DC. and deduced its structure to be (2E, 8E, 10E)-*N*-isobutyl-11-(3,4-metaylene-dioxyphenyl)-2,-8,10-undecatrienamide (6). Later, Viswanathan, Balakrishnan, Joshi & Von Philipsborn (1975) synthesized piperstachine (6) and its 2E, 4E, 6E-isomer (7) and confirmed the proposed structure for piperstachine to be as reported. Pipercide (1) was synthesized via synthetic route shown in Fig. 22.5 (Miyakado *et al.*, 1979b).

The present synthesis was based on a piperonyl + C_6 + C_4 principle employing piperonyl alcohol (15), 1,6-hexanediol (8) and methyl crotonate (10) which are readily available as starting materials. In the usual manner (Vig, Vig, Mann & Gupta, 1975), 6-(tetrahydro-2'-pyranyloxy)-1-hexane-diol was prepared from 1,6-hexanediol (8), which was oxidized with pyridinium chlorochromate (Corey & Suggs, 1975), affording the corresponding aldehyde (9) with a 95% yield. Methyl crotonate (10) was brominated with NBS to give methyl 4-bromocrotonate (Djerassi, 1948, 92% yield), which was converted with triethyl phosphite to diethyl 3-methoxycarbonyl-2-propenylphosphonate (11) according to the method of Davis, Jackman, Siddons & Weedon (1966).

The 2,4-dienoate (12) was prepared by condensation of the aldehyde (9) with the phosphate (11) in the presence of sodium methoxide, whose configurations were predominantly *trans* at C-2 and exclusively *trans* at C-4 (Davis *et al.*, 1966; Pattenden & Weedon, 1968). The protecting group of the 2,4-dienoate (12) was removed by p-TsOH in MeOH to give (2E,4E)-methyl 10-hydroxy-2,4-decadienoate (13) which was oxidised with pyridinium chlorochlorochromate to (2E,4E)-methyl 10-oxo-2,4-decadienoate (14). The phosphonium

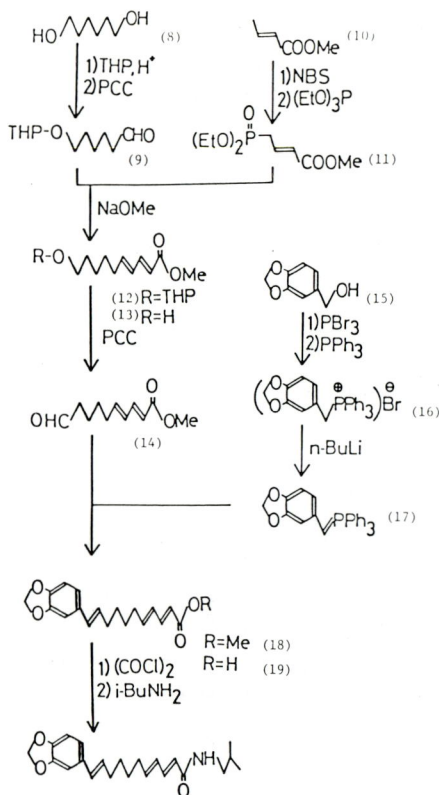

FIG. 22.5. Synthesis of pipercide (1).

bromide (16) was prepared according to the procedure of Grewe *et al.* (1970). The phosphonium bromide (16) in benzene was converted with an equimolar amount of n-BuLi to the Wittig reagent (17), which was coupled *in situ* with the 2,4-decadienoate (14) to give the condensation product (18) in a 70% yield. The ratio of the (E)-isomer to the (Z)-isomer at the C_{10} double bond was about 8:2 as determined by GLC and NMR (House, Jones & Frank, 1964). The E and Z ester mixture (18) was hydrolysed (KOH-MeOH) to give the acid (19). The crude acid was recrystallized twice from benzene to give (2E, 4E, 10E)-11-(3,4-methylenedioxyphenyl)-2,4,10-un-decatrienoic acid (19) as fine needles (60% yield), which was chlorinated with oxalyl chloride. The resultant acid chloride was finally condensed with isobutylsamine yielding (2E, 4E, 10E)-N-isobutyl-11-(3,4-methylene-dioxyphenyl-2,4,10-undecatrienamide (1). The synthetic product was identical in all respects (mixed mp., NMR, IR, MS and UV) with natural piper-cide (1). Thus the structure of (1) was confirmed unambiguously.

22.3.3.2 Dihydropipercide (2) and guineensine (3)

Dihydropipercide (2) and guineensine (3) were also synthesized by essentially the same strategies with that for pipercide synthesis (Miyakado *et al.*, 1980).

22.3.4 Joint insecticidal action of pipercide (1) and co-occurring compounds isolated from black pepper plants

The insecticidal activities of pipercide (1) dihydropipercide (2) and guineensine (3) against azduki bean weevil are summarized in Table 22.3. Among these compounds, dihydropipercide (2) was most potent (relative toxicity, 0.56) and guineensine (3) was still more active than pipercide (relative toxicity, 0.35). Furthermore, a mixture of pipicide and two co-occurring amides showed remarkable joint action as compiled in Table 22.3. In particular, a mixture of the three compounds (1), (2) and (3) in 1 : 1 : 1 ratio demonstrated on available data the highest joint action (relative toxicity, 1.15).

TABLE 22.3 INSECTICIDAL ACTIVITY OF (1), (2), (3), (4) AND (5) AGAINST MALE ADULTS OF ADZUKI BEAN WEEVILS (*CALLOSOBRUCHUS CHINENSIS* L.)

Material	Ratio	Observed relative toxicity[a] (pyrethrins = 1.00)
Pipercide (1)	—	0.23
Dihydropipercide (2)	—	0.56
Guineensine (3)	—	0.35
(1) + (2) + (3)	1:1:1	1.15
(1) + (2)	1:1	0.63
(1) + (3)	1:1	0.70
(2) + (3)	1:1	0.89
Synthetic mixture	b	0.81
Band-a	b	0.80
Pellitorine (5)	—	0.02
Piperine (4)	—	<0.001
(5) + (4)	1:1	0.23
Oleoresin	—	0.01

[a] By dipping method, the activity of pyrethrins (LC_{50}) was about 100 ppm (20 µl/25) after 48 h.
[b] (1):(2):(3) = 75:5:20 (w/w) composition.

We further confirmed that a formulated mixture consisting of the synthetic specimen in identical composition with that of a natural band-a

$$(1) : (2) : (3) = 75 : 5 : 20 \ (w/w)$$

and the natural specimen from band-a were equally active (relative toxicity: the simulated synthetic mixture 0.81, the band-a .80). This also indicates that the joint action is obvious when pipercide (1) is mixed with relatively small

Masakazu Miyakado

amounts of dihydropipercide (2) and/or guineensine (3).

Other different combinations of the Piperaceae amides also revealed the distinct joint action against adzuki bean weevils. Pellitorine (5) exhibited only moderate activity, however a mixture of pellitorine (5) with piperine (4) in 1 : 1 ratio also had notably increased activity. The relatively high insecticidal activity of the crude extracts of *P. nigrum* L. (oleoresin) may be attributed to the joint action owing to the whole Piperaceae amides. The three amides (1), (2) and (3) revealed notable knock-down activity as well as the lethal activity (Table 22.4).

TABLE 22.4. KNOCK-DOWN TIMES OF (1), (2) AND (3) AGAINST MALE
ADULTS OF ADZUKI BEAN WEEVILS (TOPICAL APPLICATION)

Material	Dosage ($\mu g/\circlearrowright$ and KT_{50} (min))			
	0.05	0.1	0.5	1.0
Pipercide (1)	—	11.8	9.2	5.5
Dihydropipercide (2)	—	30.0	13.0	7.9
Guineensine (3)	—	20.5	8.5	6.3
(1):(2):(3) =1:1:1	17.6	6.0	3.6	3.5
Pyrethrins	1.0	—	—	—

Table 22.4 indicates the KT_{50} of these Piperaceae amides when topically applied to male adults of the adzuki bean weevils. When the application dosage was 0.1 μg/insect, pipercide (1) showed the highest activity among the three. The secondary active compound was guineensine (3). In this case also, a mixture of the three amides in 1 : 1 : 1 ratio demonstrated a notably enhanced activity. For the reference compound, the knock-down activity of pyrethrins was also examined. The KT_{50} value of pyrethrins at the dosage of 0.05 μg/insect was only 1.0 min. The unsaturated isobutylamides demonstrated knock-down activities in comparison to those of pyrethrins. Finally, the three amides (1), (2) and (3) showed notable activities against a number of agricultural and household pests in either single or mixed composition (Table 22.5).

Against rice weevils, the three amides were not active up to 4000 ppm. In this case, a mixed application of the three amides in 1 : 1 : 1 ratio also showed a notable enhancement in activity which was about 1/10 of that for pyrethrins. Against mosquito larvae, pipercide (1) is most effective among the three and guineensine (3) is far less active. Finally, the three amides were evaluated for their insecticidal activities against a pyrethrin-resistant strain of houseflies which was selected in our laboratory. Pyrethrins showed no activity up to 400 ppm as expected for the resistant strain. However, guineensine (3) and the mixture exhibited remarkable activity at about 100 ppm. It is noteworthy that

TABLE 22.5. INSECTICIDAL ACTIVITY OF (1), (2) AND (3) AGAINST OTHER INSECTS

Material	Rice weevil[a]	LC_{50}(ppm) Mosquito[b] larva	Housefly[c] (pyrethrin-R)
Pipercide (1)	>4000	0.30	400
Dihydropipercide (2)	>4000	0.92	400
Guineensine (3)	>4000	7.00	120
(1):(2):(3)=1:1:1	3000	0.45	110
Pyrethrins	250	0.010	>400[d]

[a]*Sitophilus oryzae* (dipping, 24 h).
[b]*Culex pipiens pallens* (immersion, 24 h).
[c]*Musca domestica* L. (pyrethrin-resistant strain, 228-e2b, contact, 48 h).
[d]No mortality was observed at 400ppm.

guineensine (3) and the mixture exhibited high mortality against the pyrethroid-resistant houseflies.

22.4 CONCLUSIONS

1. From the fruits of black pepper plants, three insecticidal amides, pipercide (1), dihydropipercide (2) and guineensine (3) were isolated. (1) and (2) are new compounds.
2. The structure of (1), (2) and (3) were confirmed by syntheses.
3. (1), (2) and (3) revealed notable lethal and knock-down activity against adzuki bean weevils. The activity was enhanced when (1), (2) and (3) were administered together.
4. Piperine (4), which had been reported to be insecticidally active, was inactive against adzuki bean weevils and houseflies.

22.5 SUMMARY

Two new amides, pipercide (1) and dihydropipercide (2), were isolated from the fruits of *Piper nigrum* L. as well as a known amide, guineensine (3) as new insecticidal principles. The three insecticidal amides showed high activity against the adzuki bean weevil (*Callosobruchus chinensis* L.). In this connection, we found that piperine (4), which had been previously reported to be insecticidally active, was almost inactive against the adzuki bean weevil.

The structure of these amides was determined on the basis of spectral data and direct comparison upon syntheses.

The insecticidal potencies of these synthetic amides (1), (2) and (3) against the adzuki bean weevil were about one-half to one-third the activity of pyrethrins. However, the activity was 3–4 times enhanced when (1) was administered together with the co-occurring amides (2) and/or (3), indicating an obvious joint action. Moreover, the amides showed notable knock-down activity against the adzuki bean weevil. The insecticidal potencies against a number of agricultural and household pest insects in either single or mixed compositions were also studied.

22.6 ACKNOWLEDGEMENT

We wish to thank Professor K. Munakata, Faculty of Agriculture, Nagoya University, for his kind suggestions throughout the work. We wish to thank Professor K. Mori, Department of Agriculture Chemistry, The University of Tokyo, for his valuable advice for the synthetic work. We are also grateful to Dr. N. Nakatani, Faculty of Science of Living, Osaka City University, for his encouragements.

22.7 REFERENCES

Chatterjee A. & Dutta C. P. (1966) Alkaloids of *Piper longum* L. 1. Structure and synthesis of piperlongumine and piperlonguminine. *Tetrahedron* **23**, 1769–1781.
Corey E. J. & Suggs J. W. (1975) Pyridinium chlorochromate. An efficient reagent for oxidation of primary and secondary alcohols to carbonyl compounds. *Tetrahedron Letters* 2647–2650.
Davis J. B., Jackman L. M., Siddons P. T. & Weedon B. C. L. (1966) Carotenoids and related compounds. Part IX. The structure and synthesis of phytoene, phytofluene, ζ-carotene, and neurosporene. *J. Chem. Soc.* (C), 2154–2165.
Djerassi C. (1948) Brominations with N-bromosuccinimide and related compounds. *Chem. Revs.* **43**, 271–317.
Grewe R., Freist W., Neumann H. & Kersten S. (1970) Über die Inhaltsstoffe des schwarzen Pfeffers. *Chem. Ber.* **103**, 3752–3770.
Gupta O. P., Gupta S. C., Dhar K. L. & Atal C. K. (1977) A new amide from *Piper officinarum*. *Phytochemistry* **16**, 1436–1437.
Harvill E. K., Hartzell A. & Arthur J. M. (1943) Toxicity of piperine solutions to houseflies. *Contrib. Boyce Thompson Inst.* **13**, 87–92. The Merck Index (9th ed.) (1976) 7266, piperine.
House H. O., Jones V. K. & Frank G. A. (1964) The chemistry of carbanions VI. Stereochemistry of the Wittig reaction with stabilized ylids. *J. Org. Chem.* **29**, 3327–3333.
Jacobson M. (1953) Pellitorine isomers II. The synthesis of N-isobutyl-*trans*-2,*trans*-4-decadiena-mide. *J. Am. Chem. Soc.* **75**, 2584–2586.
Joshi B. S., Viswanathan N., Gawad D. H. & Von Philipsborn W. (1975) Piperaceae alkaloids: Part I. Structure of piperstachine; ^{13}C-and ^{1}H-NMR studies. *Helv. Chim. Acta* **58**, 1551–1559.
Loder J. W., Moorhouse A. & Russell G. B. (1966) Tumour inhibitory plants. Amides of *Piper novae-hollandiae* (Piperaceae). *Aust. J. Chem.* **22**, 1531–1538.
Mensah I. A. & Torto F. G. (1976) Wisanine, a novel alkaloid from the roots of *Piper guineense*. *Tetrahedron Letters* 3049–3050.
Miyakado M., Nakayama I. & Yoshioka H. (1980) The *Piperaceae* amides III: Insecticidal joint

action of pipercide and co-occurring compounds isolated from *Piper nigrum* L. *Agric. Biol. Chem. Japan* **44**, (in press).

Miyakado M., Nakayama I., Yoshioka H. & Nakatani N. (1979a) The piperaceae amides I: Structure of pipercide, a new insecticidal amide from *Piper nigrum* L. *Agric. Biol. Chem. Japan* **43**, 1609–1611.

Miyakado M. & Yoshioka H. (1979b) The *piperaceae* amides II: Synthesis of pipercide, a new insecticidal amide from *Piper nigrum* L. *Agric. Biol. Chem. Japan* **43**, 2413–2415.

Nakatani N., Inatani R. & Fuwa E. (1979) Abstracts papers, annual meeting of the Agricultural Chemical Society of Japan, Tokyo, April, p. 387.

Normant H. & Feugeas C. (1964) Synthèse totale de la pipérine. *Compt. Rend. Acad. Sci. Paris* **258**, 2846–2848.

Okogun J. I. & Ekong D. E. U. (1974) Extracts from the fruits of *Piper guineense* Schum. and Thonn. *J. Chem. Soc.* (C) 2195–2198.

Pattenden G. & Weedon B. C. L. (1968) Carotenoids and related compounds. Part XVIII. Synthesis of *cis-* and di-*cis*-polyenes by reactions of the Wittig type. *J. Chem. Soc.* (C) 1984–1997.

Scott W. P. & McKibben G. H. (1978) Toxicity of black pepper extract to boll weevils, *J. Econ. Ent.* **71**, 343–344.

Singh J., Dhar K. L. & Atal C. K. (1969) Studies on the genus *Piper*-IX. Structure of trichostachine, an alkaloid from *Piper trichostachyon*. *Tetrahedron Letters* 4975–4978.

Singh J., Dhar K. L. & Atal C. K. (1971) Studies on the genus *Piper*-Part XII. Structure of trichonine, a new *N*-pyrrolidinyl eicosa-*trans*-2-*trans*-4-dienamide. *Tetrahedron Letters* 2119–2120.

Singh J., Potdar M. A., Atal C. K. & Dhar K. L. (1974) Structure of a new pyrrolidine alkaloid from *Piper trichostachyon*. *Phytochemistry* **13**, 677–678.

Sondengam B. L., Kimbu S. F. & Connolly J. D. (1977) A new piperine-type amide from *Piper guineense*. *Phytochemistry* **16**, 1121–1122.

Su H. C. F. (1977) Insecticidal properties of black pepper to rice weevils and cowpea weevils. *J. Econ. Ent.* **70**, 18–21.

Takahashi S., Kurabayashi M., Ogiso A. & Mishima H. (1969) Structure of futoamide, a constituent of *Piper futokadzura*. *Chem. Pharm. Bull. Japan* **17**, 1225–1228.

Vig O. P., Vig A. K., Mann J. S. & Gupta K. C. (1975) A new synthesis of royal jelly acid. *J. Indian Chem. Soc.* **52**, 538–540.

Viswanathan N., Balakrishnan V., Joshi B. S. & von Philipsborn W. (1975) *Piperaceae* alkaloids: Part III. Synthesis of *N*-isobutyl-11-)3,4-methylene-dioxyphenyl)-undeca-2,4,6-*trans,trans,trans*-trienoic amide and *N*-isobutyl-11-(3.4-methylene-dioxyphenyl)-undeca-2,8,10-*trans,trans,trans*,trienoic amide (piperstachine). *Helv. Chim. Acta* **58**, 2026–2035.

22.8 DISCUSSION

Casida: Some of your biological data seem to suggest that there is a case of negatively correlated cross resistance. I was wondering whether your pyrethroid-resistant strains are more sensitive than the normal strains.

Miyakado: I do not think the resistant strains used in our tests were more sensitive than the normal strains.

Jacobson: Have you tested your compounds for synergistic action?

Miyakado: Yes, we have observed that inclusion of our products in pyrethrum formulation enhances the activity of the pyrethrins.

Djerassi: Have you examined the toxicity of your amides?

Miyakado: Not yet. We plan to do so.

Pryce: Starting from a plant that had been worked with before, I think the excellent results of Dr. Miyakado clearly demonstrate that plant chemists should not be put off by plant materials which had been worked with before.

Djerassi: This is exactly what I said earlier on when Dr. Otieno presented a paper on a new method for separating and quantitating the pyrethrins. I was amazed that despite the long history of the pyrethrins, no practical method for their separation had been developed.

Bowers: My fatherly advice is that do not be put off by what is already known about a plant. Use the information as a stimulant to establish more facts about the same plant.

PART II

The Control of Tropical
Pests and Vectors of
Economic Importance

CHAPTER 23

Review of Vectors of Major Tropical Diseases and their Control

F. L. LAMBRECHT and A. CHALLIER

International Centre of Insect Physiology and Ecology (ICIPE), P.O. Box 30772, Nairobi, Kenya

CONTENTS

23.1	Introduction	388
23.2	Disease Vectors and Their Control	389
	23.2.1 Mosquitoes	389
	23.2.1.1 Anopheline mosquitoes	390
	23.2.1.2 *Culex* mosquitoes	391
	23.2.1.3 *Aedes* mosquitoes	392
	23.2.1.4 *Mansonia* mosquitoes	393
	23.2.2 Flies	393
	23.2.2.1 *Glossina* (tsetse)	393
	23.2.2.2 *Simulium* (blackflies)	395
	23.2.2.3 *Phlebotomus* (sandflies)	396
	23.2.2.4 *Culicoides* (midges)	396
	23.2.2.5 *Tabanids*	396
	23.2.3 Bugs	397
	23.2.3.1 Triatomid bugs (kissing bugs)	397
	23.2.3.2 Bedbugs (*Cimex* spp.)	397
	23.2.4 Ticks and mites	397
	23.2.4.1 Endemic relapsing fever	398
	23.2.4.2 Rickettsiosis	398
	23.2.4.3 Scrub typhus	398
	23.2.5 Lice	399
	23.2.6 Fleas	399
	23.2.7 Snails (Molluscs)	400
	23.2.8 "Mechanical" vectors	403
	23.2.9 Rodents	404
23.3	Summary	405
23.4	Discussion	405

23.1   INTRODUCTION

The majority of important human diseases in the tropics are transmitted through invertebrate vectors, that is an intermediate host which makes it possible for the causative parasite to maintain itself and disperse in subsequent vertebrate hosts, often obligatory to the completion of the parasite's life-cycle. The interruption of the transmission cycle: parasite → vector → host, or the elimination of one of these elements, will achieve eradication of the disease.

Parasitic organisms can be removed from the cycle by the use of effective chemotherapy in the vertebrate host. Removal of the vertebrate host can be achieved by his relocation outside the ecosystem of the vector and/or the parasite. The third method, that of the removal of the vector, is the one here considered during the brief description of some of the more important vector-borne diseases in Africa.

Vector insects belong to families very different in their biology, behaviour and ecology as well as their part played in the disease transmission. To the variety of epidemiological situations and the specific characteristics of vector species correspond a variety of strategies that answer to the following simple questions: where to fight the vector, when and how should control measures have to be applied.

Diseases listed according to the group of vectors by which they are transmitted:

Mosquitoes
(a)   Malaria (*Anopheles*)
(b)   Filariasis (*Anopheles, Culex, Aedes, Mansonia*)
(c)   Yellow fever and other viruses (*Aedes, Culex*)

Flies
(a)   Trypanosomiasis (*Glossina*)
(b)   Onchocerciasis (*Simulium*)
(c)   Leishmaniasis (*Phlebotomus*)
(d)   Filariasis (*Culicoides*, Tabanids)

Bugs
(a)   Chagas' disease (Triatomids)

Ticks and Mites
(a)   Endemic relapsing fever (*Ornithodorus* spp.)
(b)   Rickettsiosis (Ixodid, Argasid)
(c)   Scrub typhus (*Leptotrombiculum* spp.)

Lice
(a)   Epidemic relapsing fever (*Pthirus, Pediculus*)

(b) Typhus (*Pediculus*)

Fleas
    Plague (*Xenopsylla, Pulex*)

Snails
    Schistosomiasis (*Biomphalaria, Bulinus, Oncomelania*)

"Mechanical" insect vectors
Several non-biting insects can carry pathogens of: amoebiasis, conjunctivitis, bacterial and fungi infections. Other potential dangerous insects are venomous arthropods.

Rodents
Rodents are important intermediate hosts (reservoirs) of a large number of disease agents; plague, leptospirosis, toxoplasmosis, typhoid fevers, viruses.

As the purpose of vector control is to disrupt the transmission of pathogens between vectors and man or animals, all these elements have to be taken into account to choose a control strategy. In other words, the vector biology and its epidemiological role as well as the epidemiological pattern of diseases must be known. The technique most effective in the control of these vectors will be different not only for each group of vectors but often for each of the diseases. This is because the circumstances under which each disease occurs differ while the kind of vector, even when belonging to the same group, may have quite different behaviour patterns and breeding sites.

Broadly speaking, the transmission of vector-borne parasitic diseases can be stopped:

1. by preventing vectors to establish and multiply,
2. by eliminating the vector(s),
3. by continuous surveys to prevent circumstances suitable for transmission to develop.

In Africa, some insect-borne diseases are widespread (malaria, onchocerciasis) and require large control campaigns, others, despite their gravity, have more or less a local or regional importance (human trypanosomiasis, plague, leishmaniasis, yellow fever) and require control measures when an outbreak occurs in a focus. Besides the vectors of these important diseases, other insects that generally cause nuisance may become dangerous in certain circumstances (e.g. domestic flies in cholera epidemics).

When known, the distribution area of vector species enables the potential area of diseases to be delimited. Within these areas various existing situations in ecology (habitats) and epidemiology (man or animal vector contact) have to be investigated.

The precise knowledge of species habitats allows control areas to be treated by discriminative treatments. In general, vector insects may be arranged in four main categories:

(a) body insects: fleas, lice, etc.;
(b) domestic insects: flies, mosquitoes, sandflies, bugs, etc.;
(c) peridomestic insects: flies, stable flies, mosquitoes, etc.;
(d) wild insects: tsetse, blackflies, horse flies, etc.

Within their habitats, species spend the major part of their life at rest; resting sites are very important for a selective application of control measures because insects may make contact with the insecticides sprayed on resting substrates. Other important components of habitat are breeding sites where females lay their progeny in the soil, water, refuse, or manure. When readily attainable, breeding sites constitute important places where control measures are applied, particularly when larvae live in the water. Insects are distinguished by their life cycle that consists of different successive stages: egg, larval, pupal and adult stages. These stages fundamentally differ in their habitat, behaviour and nutritional requirements. Appropriate control measures have to be applied: antilarval ones in breeding sites and antiadult ones in resting and feeding sites. The choice between both major strategies (antilarval or antiadult) depends on the ease with which stages can be reached. For example:

(a) Control measures in tsetse are aimed at adult flies only because pupae live buried in the soil. It is easy to reach adults in their resting sites.
(b) In *Simulium*, larvae and pupae are attacked because it is easier to find breeding sites along rivers and streams where females lay their eggs at places with rapid flowing water than to search for resting sites, so far unknown.
(c) In mosquitoes, antilarval and antiadult measures are used because both larva and adult stages live in places inside and outside houses or vicinity.

The objective of vector control is to prevent man or animal-vector contact by preventive or suppressive measures.

It would be possible, under certain circumstances, to apply insecticides over a large area to control a vector, and this in fact may be indicated to stop an epidemic. But to minimize cost and pollution of the environment, chemical applications will be most effective and are done at a far lower cost by residual insecticides applied into precise locations essential to the survival and multiplication of the vector. For all practical purposes, control measures of disease vectors will fall into the following categories:

1. mechanical means,
2. use of residual or non-residual insecticides,

3. use of genetic control (sterile male release technique),
4. biological control.

Prevention measures are limited and generally insure only an individual protection (mosquito netting and meshing, domestic hygiene measures, wearing clothes preventing bitings, use of repellents, etc.). As mentioned above, it is often possible to eliminate vector breeding sites, and this applies to certain mosquitoes, by mechanical means. Environmental sanitation should be considered in the first instance as it may achieve control relatively cheaply and, more important, would result in the permanent elimination of potential vectors. Chemical control is practised by means of natural or synthetic substances. Chemical industry companies discover new compounds every year, but only few of them are eventual candidates for current use in vector control. Potential insecticides must possess a certain number of qualities:

1. to be more toxic to insects than to the rest of the fauna;
2. to have a residual effect, not too excessive; at present, a certain biodegradability is required;
3. not to be phytotoxic;
4. to be stable in stocking conditions.

Resistance to insecticides is a problem that has to be kept in mind in all control campaigns. The World Health Organization (WHO) has standardized methods to detect resistance in vector insects of medical importance. When resistance occurs, palliative measures may be taken: increasing insecticide concentrations, simultaneous use of two insecticides, combination of insecticides with synergists, replacing an insecticide by another of a different chemical family.

23.2 DISEASE VECTORS AND THEIR CONTROL

The use of insecticides is best shown through examples of vector control in Africa.

23.2.1 Mosquitoes

Mosquitoes can be controlled at two stages of their life cycle: during the aquatic stages of the eggs, larvae and pupae; and during the adult stage. In the first system, breeding places have to be searched for and then dealt with by mechanical means, by insecticides or by biological methods.

Mechanical means consist of the draining or filling of swamp and low-lying areas, improvement of irrigation channels and ditches, and other means to stop and prevent breeding in water collections. Insecticides, here called larvicides, are applied directly to the breeding sites or by means of slow-release formulations. Biological control includes the release in the breeding sites of predators, competing invertebrates or certain larvae-eating fish such as the *Gambusia* sp.

Against the adults, insecticides spraying is the main practical means of control; genetic control through sterile male release is still very much in the experimental stage. A "blanket spraying" of non-persistent insecticides may be considered in a confined area in the case of a mosquito-borne epidemic, for instance in a yellow fever epidemic. But otherwise, the most used technique would be the application of residual insecticides to resting and breeding sites of the mosquito vector. To be effective, these sites have to be known precisely so that insecticide applications can be carried out accordingly. For instance, whereas an application of residual insecticides on the inside walls of dwellings may be effective to control certain anophelines, this method will be far less effective for *Culex p. fatigans* (the Bancroftian filariasis vector) because this mosquito rests mainly on curtains, draperies, clothes, the insides of cupboards, and so on.

Many mosquito-vectors bite outside in the open and their control is therefore extremely difficult, except through "blanket-spraying" and the treatment of their breeding sites. Some individual protection may be achieved by the use of repellents. Risks of transmission of malaria and filariasis can be seriously reduced by sleeping under a mosquito net.

23.2.1.1 Anopheline mosquitoes

All forms of malaria are transmitted by mosquitoes belonging to the Tribe Anopheline. Control of anopheline mosquitoes by the application of sprays of residual insecticides to the interior surfaces of dwellings is the principal measure employed in antimalaria programmes. The selection of the insecticide, along with its dosage and spray cycles, is dependent upon numerous factors which vary with the vector and geographical area concerned.

The duration of effectiveness of the insecticides should be determined by suitable methods of biological assessment. These include regular measurements of mosquito densities in sprayed and unsprayed dwellings and the bio-assay of insecticide deposits on treated surfaces. Where transmission period is less than 3 months, a single annual application of residual insecticide may be sufficient in areas where the spray coverage can be completed in a short time; alternatively, lower dosages, repeated several times a year, may be used.

Experiments show that there may be a comparatively rapid loss of insecticidal action on certain surfaces, such as laterite muds of high colloid content. Insecticidal activity is more persistent on surfaces such as thatch, wood or paper. Alkaline surfaces, such as lime and cement, do not destroy chlorinated hydrocarbon insecticides, but they hydrolyse certain organophos-phorus compounds. Removal of deposits by washing or rubbing, or their obliteration by replastering, papering, liming, smoke or soot, may result in losses of active surface as high as 25% in 6 months. In addition, new constructions and temporary shelters introduce untreated surfaces.

Space sprays for mosquito control range from the use of aerosol dispensers in homes, to the use of ground equipment for dispersing mists, fogs or dusts, to aircraft application of sprays and fogs. All treatments are effective in reducing the numbers of mosquitoes, but the reduction is generally only temporary since the sprays have no residual action and kill only mosquitoes upon contact. Re-infestation from outside sources generally occurs within a short time. The use of space sprays for adult control has been favoured in some areas since the selection for resistance may not be as great as that obtained with other methods.

The control of adult mosquitoes outdoors or over large areas can be effectively be done rapidly and at relatively low cost with various dosages of malathion or pyrethroids up to ultra-low-volume sprays. With all space sprays, weather conditions have a pronounced effect on the efficacy of the treatment. Because of the wide range and variety of larval and pupae habitats, larval control of anopheline mosquitoes has only limited application at the present. "Mechanical control" by source reduction of breeding sites is often more efficient, less costly and more permanent. Effective insecticides, applied from the ground or from the air, are: abate and chlorpyriphos. Pellet preparations of other insecticides have also been used with success.

23.2.1.2 *Culex mosquitoes*

Many culicines are disease vectors and many others cause great annoyance because they occur in great numbers and are ferocious biters. The most effective means of control are aimed at reduction of breeding places and the application of larvicides.

Many culicines are domestic and peridomestic breeders with larvae many times found in very small containers, broken bottles, empty tins, discarded tyres, and so on. A large number of insecticides can be used as larvicides against culicines, several have become ineffective because of resistance in certain areas. Larvicides may be applied by hand or power-sprayers, or sprayed from an aircraft as emulsions, solutions in oil, wettable powder, or in the form of granules or pellets.

Among the culicines, *Culex pipiens fatigans (Culex quiquefasciatus)* is one of the most obnoxious because of its worldwide distribution, its very wide range of breeding sites, including brakish and heavy polluted waters (sewers, latrine pits, cess pools, etc.), and its great vectorial capacity for transmitting *Wuchereria bancrofti* (Bancroftian filariasis, or elephantiasis). In heavy polluted breeding sites, the amount of effective insecticides may have to be two to three times as high as the normal dosage.

23.2.1.3 Aedes mosquitoes

Mosquitoes of the genus *Aedes* are vectors of yellow fever and other virus diseases. Among the potential thirteen vector species in Africa, the most important ones are: *Aedes (Stegomyia) aegypti, Ae.* (S.) *simpsoni, Ae.* (S.) *africanus,* and *Ae.* (S) *luteocephalus.*

The natural focus of yellow fever is widespread throughout Africa in forests and gallery forests. Monkeys constitute the main reservoir in East Africa whereas, in West Africa, insect vectors are considered as reservoirs. The passage of the amaril virus from the natural focus to man generally occurs in two steps: monkeys are infected in the canopy by an insect vector which transmits the virus to men frequenting the natural focus. Men infected in the natural focus introduce the virus into towns and villages. The epidemization of the disease is carried out by *Ae. aegypti* in West Africa and by *Ae. simpsoni* in East Africa.

Vaccination constitutes the main permanent measure against yellow fever but, in addition, vector control allows the epidemics to be prevented and jugulated. It is not conceivable to control vectors in the enzootic focus due to the very dispersed distribution of these insects. The vector species of yellow fever differ by their distribution, ecology and anthropophyly. *Aedes (S.) aegypti* is a pantropical species. Biting behaviour differs in relation to light, the habitat requirements and bioclimatic zones. Yellow fever vector control is practiced in two ways:

1. the interepidemic periods when measures are taken to keep the vector population at low levels;
2. the epidemic periods when intensive measures are taken to control all potential vectors.

Aedes breeding sites include the following natural and man-made larval habitats:

1. *natural:* tree holes, leaf-axils, large leaves and other plant parts on the ground, rock holes, etc.;
2. *domestic:* earthenware jars and other containers used in water-storage;
3. *peridomestic:* in the vicinity of houses: barrels, earthenware jars, calabashes, tins, used tyres, broken bottles, pieces of machinery and all other cavities that can collect rainwater.

All these different breeding sites are important to the epidemiology of yellow fever; domestic and peridomestic breeding places make man–vector contact permanent and intensive. Control measures are mainly directed to these sites because it ensures rapid break of contact between man and the vector, while natural sites are far more difficult to reach because of their very nature and their dispersion over very large areas. The majority of domestic and peridomestic *Aedes* breeding places being man-made can, therefore, also be avoided through proper sanitary measures and this would avoid to a large extent the risks of epidemic upflares. Chemical control can be achieved by the use of abate in water-storage vessels (1 ppm); fogging with fenitrothion twice a week. The vegetation bordering the home can be sprayed with malathion.

When an epidemic occurs, the speed of control is essential. This can be achieved by spraying non-toxic insecticides over towns and villages in order to knock down and kill all adult vectors. Malathion or fenithrothion are used for space spraying from the ground by means of knapsacks or vehicle-mounted sprayers, or from the air by means of fixed-wing aircraft or by helicopter. Two treatments at 10 to 14 days interval will achieve a drastic decrease of adult vectors, sufficient to halt transmission of the disease.

23.2.1.4 *Mansonia mosquitoes*

Mansonia mosquitoes are, in certain parts of the tropics, the vectors of *Brugia malayi*, a filariasis disease similar to bancroftian. The larvae and pupae live attached to the stems and roots of aquatic plants. One of the most effective means of control is through the use of herbicides to kill the host-plants. Against the adults, the same methods as those used for anopheline control are effective. Many *Mansonia* species are exophilic, e.g. they do not bite or rest inside houses so that residual or space spraying may not be very effective for those species.

23.2.2 Flies

23.2.2.1 *Glossina* (tsetse)

Tsetse are widely distributed in forest and savannah areas in Africa south of the Sahara. They are responsible for the transmission of *Trypanosoma* which cause a fatal disease in man, called sleeping sickness, and a debilitating and often fatal disease in livestock, called "nagana".

Trypanosoma rhodesiense, causing an acute type of sleeping sickness, and the disease in domestic animals, mainly caused by *T. vivax* and *T. congolense*, are carried with no apparent ill-effect in wild animals. Treatment of the human patient or sick cattle is, therefore, not adequate to prevent further transmission. Gambian sleeping sickness, however, caused by *T. gambiense*, is essentially a man-to-man transmitted disease.

Glossina vectors involved in the human and livestock trypanosomiasis belong mainly to two major groups: (1) the so-called *palpalis* group found in forested and riverine vegetation and the main vector of the man-to-man transmitted Gambian sleeping sickness, and (2) the *morsitans* group in various types of woodland savannahs, containing species responsible for Rhodesian sleeping sickness in man, and for livestock trypanosome infections. Tsetse are larviparous insects which lay their larvae about every tenth day. The third instar larva pupariates in the soil in shady and humid places under vegetation. The duration of pupal stage is about 30 days at 25°C. In some parts of Africa, seasonal changes in temperature and humidity are important; the tsetse life-cycle varies greatly (27–60 days, for pupal stage). These variations have to be taken into account for insecticide treatment timing. During the dry season, species concentrate in restricted suitable areas in certain parts of Africa.

The resting sites of tsetse are more and more studied because their knowledge permits selective use of insecticides in control campaigns. At night, tsetse rest on leaves whereas during day they choose the woody parts of plants (tree trunks, the underside of horizontal and inclined branches). An important point is the space occupied by resting flies. Along river banks, species rest in the parts of gallery forests close to the riverbed. Resting sites are usually found near the ground. For savannah species, ecotones are important concentration places.

Mechanical means of control of tsetse is possible. *Glossina* are fairly well confined to their tree habitats and any drastic modification of the tree-cover will result in their disappearance. However, except in restricted zones, the removal of tree-cover is very expensive, undesirable if applied to large areas and regrowth has to be checked to prevent the return of the tree-cover. The wild animal reservoir of trypanosomiasis can be reduced or eliminated by game control. Again this is an undesirable method which has now been almost completely abandoned. It is finally the control of the *Glossina* vector by insecticidal applications that prevails as the main practical method. Two major systems are used:

1. ground spraying of residual insecticides in the resting places of the insect;
2. aerial application of the infested areas.

The first method needs a very precise knowledge of the resting sites and seasonal dispersion of the fly. The aerial spraying can be applied more quickly and therefore is specially indicated in case of epidemics or to reduce the fly population quickly over large areas.

The ecology, population structure, dispersion and resting sites are quite different between *Glossina* of the *palpalis* group and those of the *morsitans* group. Strategies for their control, whether by ground spray or by aerial spray, have to be adjusted to these differences and local circumstances. Riverine vegetation is treated by spraying the parts particularly frequented by species. In *G. morsitans* habitats discriminative spraying of vegetation is confined to forest island, clump thickets and woody riverine vegetation (Sudan zone) and to *doka* woodland (Guinea zone). Where habitats are wide, strips about 80 m are treated with about 100-m intervals of unsprayed vegetation. Selective spraying is applied to tree trunks over about 25 cm in diameter up to a height of about 1–2 m from the ground (Sudan zone) and to the underside of all tree branches, between 1.5 and 4 m from the ground.

During the dry season insecticides are applied once only. Large areas of tsetse-infested country can be treated with insecticides applied from a fixed-wing aircraft or from a helicopter. The fixed-wing technique, developed in East Africa, has been used for the application of persistent insecticides at high dosages, now it is used for the application of low-dosage sprays (ULV). Applications are done along parallel paths at intervals of 200 to 300 m. Four or five applications are needed at 12 to 21 days interval. The advantage of spraying by helicopter is that it does not require landing fields. Also applications can be made in confined and well-defined areas (discriminative application). Helicopters can fly at 1 to 3 m above the vegetation and follow all the meanders of gallery forests at relatively low speed (24 to 32 km/h) and also fly over light woodland at 40 to 60 km/h.

The large-scale application of genetic control of tsetse by the sterile male technique is still largely experimental. One of the main problems is that of mass rearing. Biological control through the introduction or mass release of predators, parasites or pathogens also has been purely experimental.

23.2.2.2 *Simulium* (blackflies)

Blood-sucking flies of the family Simuliidae are vectors of onchocerciasis (river blindness) in Africa, Mexico and Central and South America. In other areas where this disease does not occur they constitute a serious pest problem. In Africa, *Simulium damnosum* is responsible for large depopulations in many parts of the Volta River Basin. Riverine villages have been abandoned in cyclic fashion with a retreat when more than half the population became blind, and the return to the fertile river banks when the population was vigorous enough to risk the consequences.

The aquatic stages of *Simulium* are found in fairly fast running water with a certain oxygen content, the larvae and pupae attached to slightly submerged rocks or vegetation. Abate applied to infested water-courses is the method of choice for control of blackflies and the reduction of onchocerciasis transmission. The insecticide can be dispensed by the drip method or by aerial spray, as an emulsion, a suspension or an oil solution. For local control, complicated equipment is unnecessary. Any metal container in which holes have been punched can be used. For long stretches of infested river courses, such as the WHO Volta River Basin Project, aerial applications are made. Insecticides can also be applied onto the water surface from motorized sprayers.mounted on boats. In addition to Abate, DDT, chlorphoxim, chlorpyrifos, Methylchloropyrofos and phoxim can be used. Applications are made at 7 to 10 days intervals for a month, repeated if necessary over longer periods.

The major problem with *Simulium* control is that the insect has a tremendous dispersion potential, in some cases covering more than 100 km in a few days. Re-invasion of cleared areas is therefore a major concern of the vast WHO Volta River Basin Project, aimed at the eradication of onchocerciasis by the year 2000. The twenty years control operation and surveillance is dictated by the fact that the adult *Onchocerca* filaria living in the tissues of the human host has an estimated life-span of 20 years during which infective microfilaria are produced. Besides *S. damnosum*, other *O. volvulus* vectors are known from scattered foci. *S. neavei* is a vector of the disease in forested areas. The larvae and pupae are found attached to a fresh-water crab so that control measures have to be adjusted to these circumstances.

23.2.2.3 Phlebotomus *(sandflies)*

Several species of sandflies are vectors of leishmaniasis. The disease is known in Africa under two forms:

1. cutaneous, caused by *L. tropica* and
2. visceral, caused by *L. donovani*.

Cutaneous leishmaniasis is mainly a man-to-man transmitted disease although animal reservoirs, mainly domestic dogs (and cats), have been proven to carry the parasite in certain areas.

Visceral leishmaniasis is carried by a large number of animals such as dogs and other canines, cats and wild rodents. Man is occasionally infected from these sources through the phlebotomine vector.

The animal reservoir and breeding of the Phleobotomine vector may be close to human dwelling — the fly has a very short flight-range and the breeding sites may be only at a short distance from an infected dwelling. Sandflies often breed in rubbish heaps, cracks in the soil and walls, and in animal lairs. Some of these can be eliminated by simple mechanical measures.

The systematic spraying by residual insecticides of breeding places and the interior and exterior of doorways, windows and other openings is usually very effective in the control of the vector.

Sandflies are often especially active at dusk. Repellents may be a means of reducing risks if exposure to sandflies is unavoidable.

23.2.2.4 Culicoides *(midges)*

In North America and Australia the name "sandfly" has been applied to species of the genera *Culicoides*, *Leptoconops*, and *Styloconops* (family Ceratopogonidae). Many species of these breed in moist soil, some in fresh-water locations, others in salt-marshes washed by daily tides. The *Culicoides* of *Acanthecheilonema (Depetelonema)* filaria in tropical Africa breeds in rotting banana stumps. Sandflies are of great economic importance regarding the tourist trade because of their vicious biting habits on many beaches of the world.

23.2.2.5 Tabanids

The large and cosmopolitan family of Tabanidae includes the blood-sucking flies, commonly known as host-flies, deer-flies, green-heads, breeze-flies and mango-flies. These flies can severely annoy animals and humans. Two species of mangoflies, *Chrysops silacea* and *C. dimidiata*, are vectors of *Loa loa* (filaria) worms in parts of West and Central Africa. Tabanid flies may also be mechanical vectors of certain important diseases of animals; trypanosomiasis anthrax, tularaemia and surra. Control of tabanids is very difficult. Little is known of their breeding sites or adult resting sites.

23.2.3 Bugs

23.2.3.1 Triatomid bugs (kissing bugs)

Species of *Triatoma*, *Panstrongylus* and *Rhodnius*, also known as cone-nose bugs, assassin bugs, china bedbug or kissing bugs, are vectors of Chagas' disease in Mexico, Central and South America. The bugs feed on blood of man and animals to whom they can transmit *Trypanosoma cruzi*, causing an infection called Chagas' disease. These bugs spend most of their time in cracks, fissures and other hiding places in walls and ceilings of human dwellings and in animal niches and nests. Where housing is primitive, conditions may be ideal for triatomid infestation and for the transmission of *T. cruzi*. The parasite is not confined to humans but is also found in a very wide range of animals. Chagas' disease is of great public health importance in Central and South American countries, estimated to infect more than 7 million people.

The method of choice for the permanent control of *Triatoma* bugs is the elimination of breeding and hiding places, including provision for improved dwellings and the clean-up or removal of animals nests around and under dwellings. Insecticides should be regarded as supplementary to this procedure. The use of Dieldrin or BHC has been proven the most satisfactory.

23.2.3.2 Bedbugs (Cimex spp.)

Bedbugs feed on blood, principally that of man. They are spread by clothing, baggage, second-hand bedding and furniture, and by laundry. Bedbugs may also be serious pests in animal and poultry houses and in laboratory animal quarters. Bedbugs are usually found about the tufts, seams and folds of mattresses and bed-covers; later they spread to crevices in bedsteads, behind basement boards, window and door casings, picture frames and furniture, loosened wall paper, cracks in plaster and partitions.

Direct application of insecticides into the hiding places will control bedbugs. The effectiveness of control depends upon the thoroughness with which the spray is applied. Bedbugs have developed resistance to DDT in many areas. Alternative insecticides are lindane, diazinon and malathion.

23.2.4 Ticks and Mites

Ticks can transmit viruses, bacteria, rickettsioses and various important diseases affecting domestic animals. Many of the tick-borne diseases are transmitted from generation to generation through the eggs or larval stages. In addition, ticks may feed on different animals during the completion of their long life-cycle and, if infected, may spread the infection to several animal species.

Generally, ticks are very difficult to control in most circumstances. Area control of ticks can be obtained with applications of any of several chlorinated hydrocarbon insecticides. A thorough knowledge of the habits of target species often allows control measures to be applied to limited areas rather than to the entire countryside. To control ticks in cattle herds, the "dip" method is used, that is, the animals are forced to wade through a trough filled with a mixture of anti-tick chemicals. With regard to human diseases, the following are important public health tick-borne problems:

23.2.4.1 Endemic relapsing fever

This disease is caused by *Borrelia recurrentis* and transmitted by several species of *Ornithodorus* ticks, in Africa principally by *O. moubata*. The presence and behaviour, including feeding on man, of this tick is closely related to the environment and the type of dwelling of the inhabitants. The parasite can be maintained for a long time in a single dwelling because of the very long lifespan of the vector (several years, even without a feed) and the fact that it can be transmitted to the next generation of ticks through transovarian contamination. *Ornithodorus* ticks hide during the day in cracks in the floor and walls and feed, usually at night, rapidly engorge, and promptly leave the host to hide again, for several weeks or months.

Control of this vector can be achieved by mechanical means by sealing hiding places, avoiding contact by improving sleeping quarters and furniture. Spraying the inside of the houses and especially in the potential hiding places with a residual insecticide, such as BHC, is usually effective.

23.2.4.2 Rickettsiosis

A large number of tick genera and species are involved in the transmission of various Rickettsioses: tick-borne typhus, caused by *R. rickettsia*, Kenya tick typhus, caused by *R. conori*; Q. fever, caused by *R. (Coxiella) burnetti*. Many Rickettsia are maintained in a variety of animals. However, as in relapsing fever, transovarian transmission is common. Measures designed to reduce tick populations are generally impractical. In selected areas, direct application of selected insecticides such as chlordane, dieldrin, lindane, diazinon, BHC give excellent control of some vector-ticks.

23.2.4.3 Scrub typhus

Scrub typhus is caused by the *Rickettsia tsutsugamushi*, transmitted by *Leptotrombiculum akamushi* and *Trombicula deliensis*. The disease is prevalent in Asia and the South Pacific. Although not recorded as disease vectors in other areas, related species are known for their problems to man. Area treatment with insecticides and personal protection with repellents are the chief means of dealing with "chigger" infestation. The animal reservoirs of this disease are several species of field mice.

23.2.5 Lice

The sucking lice include three human species: *Pediculus humanus* (body lice); *P. h. capitis* (head lice); and *Phthirus pubis* (crab lice). Since the body louse is the vector of epidemic typhus and epidemic relapsing fever, effective control measures of this species have received most attention. Epidemic relapsing fever is one of the three important diseases of man that are associated with lice. Before adequate individual sanitation and hygiene and the advent of DDT, great epidemics have occurred as recently as the period of World War I and its aftermath in Russia, Central Europe and Africa. Together with typhus, it is a disease typical of population movements and gatherings in war and refugee-camps.

Relapsing fever is caused by *Borelia recurrentis* which develops in the louse after a single infectious feed on a human carrier. Man acquires the parasite by crushing the infected louse, usually in the act of scratching. Louse-borne epidemic typhus is a disease of ancient origin and widely distributed in Europe, North Africa, Central and South America. The causative organism, *R. prowazeki* spreads easily whenever human beings are concentrated in close quarters, such as in times of war, famine or large religious gatherings. Mortality may range from 10 to 100%; it increases with age. Infections are acquired through crushing the infected louse and from the faeces.

The epidemic of Naples in 1943 threatened the whole population during World War II, with mortality as high as 81% until the epidemic was stopped by the effective delousing campaign organized by the Allied armies and based upon the use of DDT. Epidemic typhus is a poverty-associated disease. Susceptibility and pathogenicity increase with malnutrition.

Trench fever is a sort of milder form of typhus first noticed during World War I, hence its name. It is caused by *R. quintana*. Man acquires the infection through crushing the infested louse and from faeces.

Body lice can be controlled by treatment of bedding and clothes with 10% DDT, 1% Lindane, 5% Carbaryl, 1–3% Malathion.

23.2.6 Fleas

Plague, the Black Death, that decimated whole populations during the Middle Ages in Europe, is caused by a bacillus *Yersinia (Pasteurella) pestis* and is transmitted by various flea species. Wild rodents are the natural reservoir of plague; infections in man may start directly from "wild" infections through their co-habitation with man. Within a few days of feeding infested blood, bacteria develop rapidly in the oesophagus and the proventriculus of the flea which becomes blocked by the great mass of organisms. This obstruction prevents blood from entering the mid-gut so that the flea begins to starve. Frantically trying to feed, it will try several hosts and at each probe regurgitate large numbers of bacilli into the bite-wound, thereby inoculating the organisms into the host.

The principal vectors of plague are the flea, *Xenophysylla* species, and the human flea, *Pulex irritans*. The human plague occurs in three forms:

1. bubonic plague, the most common, transmitted by the fleas from rodent to man or man to man;
2. pneumonic plague is spread by the airborne route through inhalation of exhaled droplets from other carriers;
3. septicaemic plague may result when bacilli are injected directly by the fleas into the capillaries or develop from bubonic or pneumonic plague. The latter form usually kills the victim in a matter of a few days.

The disease was of worldwide distribution but has now been reduced to seasonal outbreaks in a number of regions, including parts of Africa. Methods of control are essentially preventive through decent sanitation of human habitation. Early diagnosis and specific treatment by antibiotics is highly effective. The most effective control measure is the control of rats on ships, docks and warehouses by rat-proofing or periodic fumigation. Strict international regulations describe measures applicable to ships, aircraft and land transport arriving from plague areas. All ships should be free of rodents or periodically de-ratted. Rat proofing of buildings of seaports and airports requires the application of appropriate insecticide with residual effect every 6 months, and de-ratting with effective rodenticide.

Xenopsylla cheopsis, the principal transmitter of plague and of endemic or murine typhus, had been readily controlled by the application of 5% or 10% DDT dust to rat runs and harbourage areas. The greater toxicity of the 10% dust to other ectoparasites normally less susceptible to DDT, such as the cat flea, makes it the formulation of choice. In areas of resistance to DDT, malathion has been used in 4 to 5% dust formulations. To avoid the spread of plague from contaminated areas, dichlorvos evaporators (one evaporator to 4 m³) have been put into cargo containers. Aircraft are treated with micronized powder of chlorpyrifos (40%), fenthion (20%) propoxur (64%) or resmethrin (25%).

23.2.7   Snails (Molluscs)

Several species of snails are the intermediate invertebrate hosts of human

Schistosomiasis (Bilharzia). Two major human species infest man in Africa, *Schistosoma mansoni* and *S. haematobium* (+*S. intercalatum* in parts of western Africa). The parasite, a Trematode helminth, causes serious damage to tissues of the liver, intestines and kidneys, fatal in heavy infections and debilitating in the chronic state.

The life-cycle of the parasite is as follows: eggs produced by the fertilized female and expulsed in the faeces (in case of *S. mansoni*) or in the urine (in case of *S. haematobium*), arrive in water-courses, ponds and so on. From the eggs, larvae (miracidia) escape and find their way into a fresh-water snail of the right species. Inside the snail, the miracidium changes form and multiplies, and escapes from the snail in the form of free living "cercariae". These burrow under the skin of a human being during bathing, washing or wading in a stream. Inside the human host, the parasite develops into mature male and female worms with production of eggs by the female.

Schistosomiasis is becoming a major disease in the tropics, mainly as a result of large agricultural development programmes using dam and irrigation systems, all of which will become eventually ideal breeding places for fresh-water snails. Compared to many other infectious diseases, little overall progress has been achieved in the control of bilharziasis. Again, the transmission could be achieved through preventive methods as schistoso-miasis is a typical man-made disease, and this in various ways. Firstly, the tremendous expansion of the disease in many tropical areas of the world is the result of the enormous increase of snail habitats created by irrigated crops and the formation of artificial lakes. Secondly, and equally important, it is man who contaminates the waterways where the snails breed by the unsanitary deposit of human faeces. Thirdly, man infects himself by coming into contact with these contaminated waters, mainly as a result of the absence of clean public water supply forcing him to drink, wash and bathe in natural waters.

Taking into account that chemotherapeutic treatment of the human host presents many logistic difficulties and that the prevention of contamination by eggs of the waters through health education faces even greater prob-lems, the control of Schistosomiasis is, at present, largely achieved through the control of the snail vector. In most endemic areas the control of the intermediate snail host populations offers the best opportunity for the reduction of bilharzia transmission. It is generally most effective when combined with one or more of the other methods. All methods of control have both advantages and disadvantages. The method or methods selected in a given area will depend upon local conditions. In most areas, snail control will be the most effective single measure, but often a combination of methods will be advantageous. The available methods of snail control are: (1) environ-mental control; (2) chemical control; and (3) biological control.

Environmental control can be achieved through elimination of breeding sites of the snails or the modification of their habitat by such techniques as weed clearance, and a good ditch and stream management. The method of

biological snail control has long had a wide appeal. It has been suggested that snails might be controlled by predators such as ducks, fish, insect larvae and leeches. A competitor snail, such as *Marisa cornuarientis*, proved effective, in Puerto Rico, to render some habitats unsuitable for the schistosome vector *Biomphalaria (=Australorbis)*.

Chemical control of the intermediate hosts of bilharziasis is predicated on the assumptions that they are the weakest link in the life-cycle of the parasites and that molluscicidal compounds can be used to produce rapid and devastating reductions in snail populations. In practice, molluscicides have not always lived up to their promise. This has been due largely to the diversity and extent of the habitats, the physiological differences, the habits and the ecology of the various snail species involved. Lack of consideration of one or more of these factors is apt to lead to the selection of the wrong compound or formulation, the use of inadequate equipment and inefficient methods of application. It should be obvious that control will not be satisfactory unless an effective molluscicide comes into contact with practically all of the snails in a given habitat. There are both advantages and disadvantages to the employment of molluscicides. The most effective molluscicides will destroy schistosome cercariae and kill all, or nearly all, the snail hosts and snail eggs. Thus transmission of schistosome infection is stopped almost immediately. Unfortunately, repopulation of bodies of water usually occurs, requiring the repeated application of the chemical to achieve permanent control. However, it has been possible in certain instances to eradicate the snail hosts by application of molluscicides. These instances have involved relatively simple physical and ecological situations in circumscribed areas. Large-scale eradication has not been achieved by chemical methods. However, the development of more effective molluscicides and more efficient methods of application gives hope that this method will become more effective as time goes on. Nearly all effective molluscicides are toxic in varying degrees to other organisms as well. Some must be handled with particular care. It is essential that the application of the chemical to bodies of water will not result in harm to the human or animal population. Most molluscicides are toxic to much of the aquatic fauna and some to the flora. If used in improper concentrations, they may harm cultivated crops. Thus far, there is no evidence of accumulation of any such chemicals to phytotoxic levels in the soil, although the possibility should be kept in mind.

It is not possible to describe an ideal or typical molluscicide. Rather, it should be emphasized that diversity and versatility are to be sought. However, there are certain general characteristics that are desirable in a molluscicide: (1) it should have low toxicity for man, domestic animals, game and fish; the chemical should kill not only the snails but their eggs; (2) it should be active at very low concentration since this reduces transport costs significantly; (3) it should have a low cost so that it can be employed economically; (4) it should be reasonably safe in the hands of trained but relatively unskilled people; (5) it should be usable with simple, durable equipment; (6) it should be stable in storage and in the habitat after use; and (7) tests should be available to measure the concentrations used in the field. No practical molluscicide is likely to have all these characteristics, and it is quite possible for an excellent molluscicide to lack more than one of them.

Some herbicides are also molluscicides, and usually kill snails at concentrations below the herbicidal level. In some endemic situations herbicides can be used to control both weeds and snails. Not only does this achieve two objectives simultaneously, but the removal of weeds results in a less favourable habitat for the snails which use the vegetation for food, for deposition of eggs and for protection.

23.2.8 "Mechanical" vectors

Many vertebrates and invertebrates have been proven or suspected of being the route by which certain pathogens are dispersed by simple means of transportation without the involvement of biological development. More often the pathogens are simply picked up by the legs or feet of the "mechanical" vector from contaminated food or other surfaces and deposited directly onto the human host or onto his food.

No doubt that the ordinary house fly, *Musca domestica*, is a major intermediate mechanical host of several bacterial pathogens and can be, under particularly unsanitary conditions, the principal source of epidemics of many bacterial infections.

The spread of amoebiasis by the housefly has been suggested, other intestinal and diarrhoeal disorders, especially the various typhoid fevers, have been associated with housefly pollution. Trachoma, a serious eye infection especially prevalent in drier areas of the tropics, is essentially spread by flies; *M. sorbens* is said to be the main agent of spread.

Another insect high on the list of "mechanical" vectors are cockroaches. They are, probably, a major source of food contamination and have been even, at one time, accused of the spread of the leprosy bacillus. Although venomous arthropods such as scorpions, wasps and spiders are not disease vectors, they are of concern to those engaged in the control of insects of public health importance. The risk of scorpion stings can be greatly reduced by eliminating their hiding places beneath loose stones or bark or fallen trees, boards, piles of lumber, accumulation of debris, stored household effects in basements or places adjacent to houses. In an area where scorpions are abundant, clothing should be shaken out and shoes inspected before dressing. Control can be obtained with spray solutions of malathion, lindane, or dieldrin.

Most wasps kill destructive insects and are, therefore, beneficial. But when wasps build nests too close to the house or in shrubbery where children play, they should be destroyed. Wasps can be controlled by applying an insecticidal spray or dust to their nest. Treatments should be made in the evening or early morning when the wasps are less active and most of them are in the nest. Chlordane (5%), DDT (10%) or dieldrin (1%) dust formulations can be applied to the opening of the nest. Sprays include oil solutions, emulsions and wettable powders of chlordane, DDT, methoxychlor, malathion and pyrethrins.

Most spiders are beneficial because they feed on harmful insects, but some are poisonous to man. Trash, piles of lumber or brick, weeds, woodsheds and similar hiding places are favoured by spiders. If near the house, many spiders may enter. The elimination or cleaning of their breeding places is a first and important step in control. Sprays from emulsion concentrates, wettable powder, or solutions can be used: chlordane, lindane, dieldrin and malathion.

23.2.9   Rodents

Rodents are not only sources of spoilage of food and of the "mechanical" spread of pathogens but are also hosts and carriers of parasites which develop in them, such as plague, leptospirosis, rickettsioses, viruses and helminths. Sanitation is absolutely essential to the permanent control of commensal rats and mice; the use of rodenticides should be regarded as supplementary to sanitation. A programme of control should include: (1) rat-proofing of building, (2) elimination of harbourages, (3) elimination of food and water-supplies, and (4) destruction of the rodents themselves. Rat-proofing can be done at the time of construction with little extra cost, whereas rat-proofing later on is usually expensive. The proper storage of food and the proper disposal of refuse and garbage are essential.

Rodenticides are usually used with bait. Yellow cornmeal is a readily acceptable and inexpensive bait for several days in selected places or containers before introducing the poisoned bait. Anticoagulants are widely used as rodenticides, such as warfarin, pival, diphacinone and fumarin. The effect of the anticoagulants is cumulative and their rodenticidal action depends upon small amounts of poison being consumed every day. The chemical should be available to the rats for a period of at least 1 or 2 weeks. Thallium sulphate as a 0.5–1.5% bait and zinc phosphide as a 1–3% bait are also used as rodenticides. Thallium sulphate has no distinctive odour or taste and is highly toxic, making it more dangerous to use. As it is absorbed through unbroken skin, rubber gloves have to be worn when handling or mixing.

In all cases, unconsumed baits should be collected and safely disposed of by burying deep in the ground and at least 10 m from water wells.

All pesticides are toxic to some degree, and care in handling all types should be routine practice. Precautions have been described in *Toxic Hazards of Pesticides to Man* (Wld. Hlth. Org. techn. Rep. Ser.).

23.3   SUMMARY

The control of arthropod-borne diseases presents the originality of being carried out by tackling another link of the epidemiological chain than man or domestic animals: the vector.

Vector insects belong to families very different in their biology, behaviour and ecology as well as their part played in the disease transmission. To the variety of epidemiological situations and the specific characteristics of vector species corresponds a variety of strategies that answer to the following simple questions: where to fight the vector, when and how control measures have to be applied.

In this paper we propose to show different strategies used against the main vectors of diseases, taking some examples in Africa.

23.4   DISCUSSION

Menn: In your catalogue of vectors, etc., you made no mention of anthrax.

Lambrecht: This occurs of course but is usually due to bad procedure and hygiene. The disease is transmitted to man by handling of diseased hides and carcasses. Tabanid flies may be involved in transmitting the disease between animals.

Bowers: What about the control of schistosomiasis?

Challier: This is a completely man-made disease which could be controlled by proper environmental management based on a proper understanding of the natural history of the snails involved, and the use of molluscicides.

Menn: As I recall a potent molluscicide was isolated from an Ethiopian plant and appeared to be very promising. Why has its development not been followed up?

Lambrecht: The plant you are referring to is the so-called "soap berry", *Phytolacca dodecandra*. It was largely advertised in several scientific papers and by the WHO and I do not understand the reasons why there has been no follow up of this very promising plant.

Nayudama: Often such things are not followed up because no one will profit commercially.

Pinchin: In fact it has been found in Brazil that the juices from sisal have molluscicidal activity although the active ingredient is not known.

However, it has not been exploited mainly due to transportation problems and it goes to waste.

Szabo: It appears that tsetse control is achieved by chemical means. Are there any alternatives?

Challier: Other than insecticides no other method is employed but application is now done at very low dosages. In small areas it is possible to use screens but for large areas insecticides are used.

Menn: I believe in Volta that sterile males of blackfly have been used in the ratio of ten sterile males to one wild. Is this strategy effective?

Challier: Yes they are competitive and have some effect in reducing the population.

Young: Considering the ban on DDT in many of the developed countries, how does this apply in Africa?

Lambrecht: Legislation here in Kenya allows the use of DDT at present and it is used in large areas.

Young: Is DDT still the insecticide of choice and is it still available here?

Lambrecht: Yes for the tsetse fly it is very effective in dry conditions and it is still available here at present, I believe for blackfly one of the organophosphates is very useful. In fact for tsetse fly there has been no significant development of resistance to DDT so here it is the insecticide of choice.

Bowers: I feel that DDT was a good insecticide for agricultural use and it is still endorsed by some Governments. However, its supply in the USA is threatened, although I can cite a case where it has been used recently in the USA. When bubonic plague carried by prairie dogs was threatening, it was used.

Kumar: I would like to draw your attention to an excellent paper by Professor Tahori, on the evolution and use of DDT in which he detailed the uses of DDT with emphasis on its indoor use and not its agronomic use. However, I feel that each country should have the right to make its own decisions regarding its use — of course it is mostly produced in the developed countries and many manufacturers have stopped producing. Therefore it is becoming more difficult and expensive for the developing countries to obtain.

Bowers: I was wondering why so many of these pests are world-wide because of dissemination by ships, planes and so forth, and yet tsetse is still confined to Africa? That's very fortunate of course for other countries but I was wondering why? Has it ever been threatening outside Africa?

Lambrecht: Well, I think that's related to the very peculiar reproduction of the tsetse. They have to deposit the living larvae in a suitable place. So if you take a fly which is ready to deposit the larvae outside Africa, first of all every 8 days a larva is produced, so within 8 days the larva has to be deposited in the right place. The chance of doing that outside Africa carried by 'plane, for instance, is very small. I'm sure that in certain places, e.g. the southern USA, there is vegetation that is suitable for the fly. I'm not sure from the climatic point of view. Below 16°C there is no tsetse reproduction, no development of the larvae. So in areas where the mean temperature is below 16°C for any substantial length of time during the year, e.g. the winter months, the fly will not survive. The distribution in Africa is in fact a prehistoric event. Fossil tsetse have been found in Colorado, USA, and date from the oligocene/miocene period, 20–25 million years ago. Continental drift separated the North American fly belts from the African ones. Tsetse became extinct in the North American continent but survived in tropical Africa.

Arunga: There was a record of one species in the south Arabian peninsula, but whether it still exists today I don't know.

Lambrecht: *Glossina austeni* was reported a long time ago in the extreme south-western tip of Arabia but has since then disappeared, whether by control measures or by natural causes we don't know.

CHAPTER 24

Agricultural Pests of Crucial Economic Importance in the Tropics and their Control

R. KUMAR

Department of Zoology, University of Ghana, Legon, Ghana

CONTENTS

24.1	Introduction	410
24.2	Some Features of Tropical Agriculture	410
	24.2.1 Diversity of crops	410
	24.2.2 Reasons for diversity in tropical crops	411
24.3	Variations in the Status of Pests of Tropical Crops	412
24.4	Ecological Factors	413
	24.4.1 Monoculture	413
	24.4.2 Quality and quantity of food	416
	24.4.3 Host/natural enemy relationship	416
	24.4.4 Introduction to new environments	416
	24.4.5 Intra- and interspecific compact	416
24.5	Economic Factors	417
	24.5.1 Economic change	417
24.6	Most Important Pests of Tropical Crops	417
	24.6.1 Drug plants (coffee, tea and cocoa)	420
	24.6.2 Fibre crops (cotton and jute)	422
	24.6.3 Cereals and sugarcane	424
	24.6.4 Vegetable oil crops (coconut, oil palm and pyrethrum)	424
	24.6.5 Grain legumes (cowpeas and groundnuts)	425
	24.6.6 Root and tuber crops (cassava and yams)	425
24.7	Cosmopolitan and Pantropical Pests	426
24.8	Current Methods of Pest Control on Tropical Crops	426
	24.8.1 Drug plants (coffee, tea and cocoa)	426
	24.8.2 Fibre crops (cotton and jute)	427
	24.8.3 Cereals and sugarcane	427
	24.8.3.1 Parasites	427

        24.8.3.2 Pathogens   428
      24.8.4 Controlled use of pesticides   428
      24.8.5 Crop resistance   428
      24.8.6 Cultural practices   429
              24.8.6.1 Vegetable oil crops (coconut, oil palm and
                     pyrethrum)   429
              24.8.6.2 Grain legumes (cowpeas and groundnuts)   430
24.9   Control Strategies for Tropical Crop Pests   430
      24.9.1 Pests   431
      24.9.2 Management and economic aspects   431
      24.9.3 Environment   431
24.10 Needs of Developments of Relevant Technology for the Control of
      Tropical Pests   432
24.11 Summary   433
24.12 References   434
24.13 Discussion   435

24.1 INTRODUCTION

Occurrence of 48 major tropical crops have been listed, 407 insect pests of major and 778 pests of minor importance (Hill, 1975). The figures do not include numerous other species recorded on these crops as well as the stored product pests. A review of pests of crucial importance among these is not an easy or even a realistic assignment. According to Whitney (1973) yield losses on major groups of crops in tropical Africa range from 5–56% in cereals; 5–100% in grain legumes, and 20–80% in roots and tubers. In Asia about 100 insect species occur on paddy rice and of these twenty are major pests (Pathak, 1977). However, as noted by Walker (1975), estimates of crop losses due to pests are few and very scattered in literature. Further, the status and type of pests varies according to diverse factors (see below) and for an understanding of the basic problems of pest management, an appreciation of the background to their occurrence in the tropical agroecosystem is essential. For this reason, I first wish to consider some features of tropical agriculture, followed by variation in pests of crops in the tropics. This will provide the basis on which the most important pests of some tropical crops can be discussed more fully, and their management considered meaningfully.

24.2 SOME FEATURES OF TROPICAL AGRICULTURE

24.2.1 Diversity of crops

Tropical agriculture is characterized by the diversity of it crops. In the temperate regions, the principal crops belong to a small number of species in a few families. There are a few cereals (mainly wheat, corn, soya beans and barley), some grain legumes, crops of family Cruciferae, root crops of genus *Beta*, tubers (potato, *Solanum tuberosum*) and a few fruits such as apples, plums and raspberries. In the tropics, on the other hand, a variety of cereals (rice, maize, sorghum and some seven types of millets), grain legumes (groundnut, different types of beans including soya beans, cowpea, gram, etc., and lentils), roots and tubers (yams, cassava, sweet potato, dasheen, tannia, arrowroot and sago) are grown. Further tropical crops provide commodities which are important to world commerce. These include drug plants (coffee, kola, tea, cocoa, tobacco, opium, etc.), fibres (cotton, jute, kenaf, sunn, hemp, sisal, etc.), vegetable oils and fats (coconut oil, oil-palm, castor oil, cocoa butter, pyrethrum, etc.), rubber, spices (tumeric, nutmeg, cinnamon, cardamon, etc.) and essential oils (citronella, gingergrass oil, lemon grass oil, basil oil, etc.).

24.2.2 Reasons for diversity in tropical crops

The diversity of crops in the tropics must be attributed to particular environmental conditions prevailing in different parts of the tropical world. Thus factors such as solar radiation, temperature, daylength and rainfall interact in different ways in the monsoon, dry, subhumid and humid regions. These factors have a mutual influence on one another and operating together, they give an area its ecological identity. The elements mentioned consitute the climate of a place, though a particular element may be of special significance in a particular area. Thus, in West Africa the most important climatic element is rainfall and it exerts its ecological influence by its distribution. It will be clear from Table 24.1 that in the West, the rainfall distribution largely determines the climax vegetation which in turn decides the suitability of a particular area for various agricultural pursuits.

Further, inadequate resources and the generally poor education of the tropical farmer greatly influence the practice of agriculture. Farmers tend to depend on nature to provide agricultural inputs and are unable to manipulate the soil fertility to their advantage. Unlike their counterparts in temperate regions they are usually unable to combat pests. However, while our small, independent, illiterate farmers are highly adaptable and responsive to their changing circumstances, little support has been given to labour-intensive peasant agriculture by agricultural administrators. Thus improved crop varieties or improved methods of pest control seldom reach the farmers. The problem also lies in our lack of knowledge of the potentialities of our resources. For example, how much land do we have that is suitable for the growth of such and such crops?

TABLE 24.1. VEGETATION ZONES IN WEST AFRICA AND THEIR
AGRICULTURAL USE

Rainfall (in. approx)	Climax vegetation	Agricultural use
<5	Desert	Nil
5–10	Desert scrub	Sparsely populated with nomadic pastoralists
10–25	Desert grass (Sahel) savannah	Pastoral country of low carrying capacity
25–35	Grass (Sudan) savannah	Marginal for crop use but intensively cultivated for millet, guinea corn and yams; important pastoral area for livestock farming
35–45	Low tree (Guinea) savannah	Suited to a very wide range of annuals, some perennials grow in higher rainfall areas; game ranching possible, livestock farming not important due to tsetse
45–80	Mixed deciduous forest	Suited to a wide range of annuals and many perennials (e.g. banana, cocoa, tea and coffee); livestock farming not important
80–120	Tropical forest, evergreen	Wide range of annual crops grown; especially favourable for perennials requiring high rainfall (e.g. rubber, oil palm, cocoa, banana); livestock farming not important

24.3   VARIATION IN THE STATUS OF PESTS OF TROPICAL CROPS

From the foregoing account it would be realized that pests of crops and their status in the tropics are likely to vary according to the local environmental conditions. Thus, the pests of cocoa are quite different in West Africa, South America and South-east Asia. Even within West Africa the pest complex varies from country to country. The production of cocoa, which is largely determined by its pest fauna (mainly capsids and mealybugs in West Africa), has in the recent years shown a steady increase in production in the Ivory Coast while in Ghana, a neighbouring country with similar climatic, edaphic and demographic conditions, and land tenure system as in the Ivory Coast, production has declined drastically. Similarly, excepting the pantropical green scale (*Coccus vividis* (Green)), major pests of coffee are seldom the same where coffee is grown in the tropics. The same diversity of pests is found on tea, cotton, sugarcane and rice, coconuts and other palms. There are, no doubt, exceptions to this statement: for example, the sorghum midge, *Contarinia sorghicola* (Coquillett) attacks sorghum wherever the crop is grown. The green stink bug, *Nezara viridula* (L.), is generally agreed to be a cosmopolitan pest on cotton, tomato and castor; *Heliothis armigera* (Hub.) is cosmopolitan in the Old World on tobacco, sorghum, cotton, groundnuts, okra and pulses. Table 24.2 lists forty-three cosmopolitan and pantropical pests of crops in the tropics.

Where the same pest occurs widely, its pest status is not the same. Damage inflicted varies according to country, crop and season as well as the many other factors, including farming practices. Some insect pests have increased in importance in the recent years. For example, the brown planthopper, *Nilaparvata lugens* (Stål), a minor pest in the past, has become a major pest of rice throughout tropical Asia. Losses due to this pest have been estimated at about US $300 million in eleven countries (Dyck, 1977). The whorl maggot, *Hydrellia* sp., once a pest of little significance, has become a major pest in the Philippines. In Kenya, the emergence of leaf miners, *Leucoptera* spp. and yellow-headed borer, *Dirphya nigricornis* (Olivier), as major pests of coffee has been quite perplexing. In some West African countries such as Ghana, a shield-bug, *Bathycoelia thalassina* (h.-S.), has recently become a major pest. The basic causes of changes in pest status of these and other insects are really not well understood. A complex of many factors influence the nature of pest, the damage caused by it, and its effect on yield. Some of these factors are ecological, others are economic.

24.4   ECOLOGICAL FACTORS

24.4.1   Monoculture

Uvarov (1967) has summarized the effects of monoculture on tropical and subtropical crops. He has shown how some well-known pests of crops such as sorghum, maize and sugarcane exist in the wild as feeders on wild grasses and are too few to be noticed. However, when the host crop is grown on a large scale, they infest to pest proportions. A recent West African example is the variegated grasshopper, *Zonocerus variegatus*, which has emerged as a destructive pest on cassava monocultures in Nigeria but is of little importance in Ghana. Many of the modern pests of stored products exist at insignificant levels in the field but large-scale storage of food enables them to reach pest proportions. However, Way (1971) has argued that in the long run crop monocultures "may be a favourable characteristic merely through a simple diluting effect on colonizing pests". For further discussion on diversity and stability concepts in relation to tropical insect pest management, reference should be made to the article on the subject by Way (1976).

TABLE 24.2 A LIST OF MAJOR COSMOPOLITAN AND PANTROPICAL
PESTS OF TROPICAL CROPS (COMPILED FROM HILL, 1975)

Name of pest	Distribution	Crop attacked
HOMOPTERA		
Aphididae		
Aphis gossypii Glover	Cosmopolitan	Capsicums and cotton
Aphis fabae Scopoli	widespread	Soybean
Myzus persicae (Sulz.)	widespread	Capsicums, peach and cotton
Toxoptera aurantii (B. de F.)	Pantropical	Citrus
Toxoptera aurantii (B. de F.)	Cosmopolitan	Cocoa
Rhopalosiphum maidis (Fitch)	Cosmopolitan	Maize and sorghum
Diaspididae		
Ischnaspis longirostris (Sign.)	Pantropical	Coconut and date palm
Aonidiella aurantii (Maskell)	Cosmopolitan	Citrus
Aspidiotus destructor (Sign.)	Pantropical	Coconut
Chrysomphalus aonidum (L.)	Cosmopolitan	Fig, citrus, coconut, date, palm and mango
Pseudococcidae		
Ferrisia virgata (Ckll.)	Pantropical	Coffee and guava
Planococcus citri (Risso)	Pantropical	Citrus
Planococcus citri (Risso)	Cosmopolitan	Cocoa
Planococcus citri (Risso)	Widespread	Papaya
Saccharicoccus sacchari (Ckll.)	Pantropical	Sugarcane
Pseudococcus abonidum (L.)	Pantropical	Oil palm
Coccidae		
Coccus viridis (Green)	Pantropical	Citrus and cocoa
Saisselia oleae (Bern.)	Pantropical	Citrus and olive
Magarodidae		
Icerya purchasi Mask.	Pantropical	Citrus and guava
Aleyrodidae		
Bemisia tabaci (Genn.)	Widespread	Cassava
Bemisia tabaci (Genn.)	Cosmopolitan	Sweet potato, tobacco, tomato and cotton
HETEROPTERA		
Pentatomidae		
Nezara viridula (L.)	Cosmopolitan	Cotton, castor, vegetables, tomato
Thripidae		
Selenothrips rubrocinctus (Glard)	Pantropical	Guava and mango
Thrips tabaci Lind.	Cosmopolitan	Onion, tobacco and tomato
Heliothrips haemorrhoidalis (Bouche)	Cosmopolitan	Tea

Name of pest	Distribution	Crop attacked
-do-	Pantropical	Citrus
-do-	Widespread	Cocoa
LEPIDOPTERA		
Noctuidae		
Heliothrips armigera (Hub.)	Cosmopolitan in the Old World	Tobacco, sunflower, sorghum, cotton, groundnuts, okra and pulses
Agrotis ipsilon (Hfn.)	Cosmopolitan	Tobacco and seedlings
Pyrallidae		
Ephestia cautella (Hb.)	Widespread	Date, palm and fig
Marcuca testulalis (Geyer)	Pantropical	Pulses and soybean
Gele chiidae		
Phthorimaeae operculella (Zeller)	Pantropical	Potato
DIPTERA		
Anythonyiidae		
Hylemya platura (Meigh.)	Cosmopolitan	Onions, pulses and seedlings
Cecidomyiidae		
Contarinia sorghicola (Coq.)	Pantropical	Sorghum
COLEOPTERA		
Cuculionida		
Cosmopolites sordidus (Germ)	Pantropical	Banana
Cylas formicarius (F.)	Pantropical	Sweet potato
Sitophilus zeamais (Motsch)	Cosmopolitan	Maize
Bruchidae		
Callosobruchus spp.	Pantropical	Soybean
Callosobruchus spp	Cosmopolitan	Pulses
Acanthoscelides obtectus (Say)	Cosmopolitan	Pulses
Lasioderma serricorne (F.)	Pantropical	Cocoa beans, tobacco, groundnut, peas and beans, flour and foodstuffs
Tenebrionidae		
Tribolium castaneum (Herbst)	Cosmopolitan in warmer parts	Maize, wheat and other stored grains
Silvanide		
Oryzaephilus sp.	Cosmopolitan in warmer parts	Stored grain in other plant and animal stored products

24.4.2   Quality and quantity of food

Modern crops are selected for their high yield of large seeds and fruits which provide more nutrition than in their wild progenitors. Such crops, when grown over large areas, provide virtually unlimited food to a phytophagous insect which is able to multiply rapidly to virtually an unlimited extent.

24.4.3   Host/natural enemy relationship

The effects of parasites, predators and disease in keeping their hosts under control are now well recognized. However, modern agricultural practices, e.g. the use of pesticides, tend to upset this balance. There is further evidence that some pest enemies, such as predators, due to their slow rate of development, are often unable to keep pace with the increasing numbers of their fast-breeding hosts which become pests.

24.4.4   Introduction to new environments

We are all aware of examples of the introduction of pests from their original homes to new environments. According to Wharton (1966), the majority of Australian pests are exotic and have been introduced in that country over the past 190 years. Within a natural ecosystem there is a well-established system of enemies but a new environment may lack such checks on the rapid growth of a pest population. Similarly, when crops are introduced into new environments, local insects may find them more delectable and soon become pests on them.

24.4.5   Intra- and interspecific competition

It is generally recognized that the greater the diversity of the biota, the greater the chances of stability (Southwood & Way, 1970). However, if there are few species and plenty of food available to them, with reduced competition, they can multiply and specialize in their new habitats, thereby becoming pests. Uvarov (1967) has drawn attention to the work of Russian entomologists who carried out quantitative studies of the insect populations of virgin grass steppe and of new wheat fields in Kazakhstan. One of their important findings was that there was a greater specific variety in the insect populations of virgin steppe which has a better numerical balance between its species. In the wheat fields, on the other hand, there were a few insect species which very soon became the worst pests of wheat.

24.5 ECONOMIC FACTORS

24.5.1 Economic change

In terms of overall costs and returns, small crop losses are tolerable when investments are low but they would become unbearable when expenditures are high. This means there is a tendency to set lower economic thresholds, and insect damage which would otherwise be ignored, is likely to become important and attributable to the presence of pests in the crops. In spite of the current rapid increase in human populations, food production hardly keeps apace and this leads to unrealistic emphasis on eradicating any animal or plant that might lessen the yield.

It will thus be clear that in the vast and varied areas of the tropical world, ecological and economic factors favouring the build up of pests vary greatly and in turn affect the pest status. So it is very difficult to generalize on the status of the pests throughout the tropics.

24.6 MOST IMPORTANT PESTS OF TROPICAL CROPS

It was stated earlier that a very large number of insects are associated with tropical crops. Fortunately, most of them are unimportant. Table 24.3 lists major tropical crops and the number of major and minor pests associated with each crop. Lamb (1974) listed most important pests of coffee, tea, cotton, cocoa, sugarcane, rice and coconuts (see Table 24.5 for summary). Hill (1975) lists major and minor pests of these crops. A comparison of the view of the two authors as to the status of these pests is as follows:

1. *Coffee* (Table 24.4). Out of fourteen most important pests of coffee mentioned by Lamb (1974), Hill (1975) considers four of these as minor pests and does not even include two of these in his list. Further, a number of important pests of this crop from Africa are not included.

2. *Tea* (Table 24.5). Lamb (1974) lists six most important pests of tea but four of these are considered as minor pests by Hill (1975) and two are not even mentioned by this author.

3. *Cotton* (Table 24.6). Of the sixteen most important pests of this crop mentioned by Lamb (1974), three are considered minor by Hill (1975) and two do not figure in his listings. A number of major pests from Africa and Central America are not at all listed.

TABLE 24.3 MAJOR TROPICAL CROPS AND THEIR PESTS (COMPILED
FROM HILL, 1975)

Crop	No. of major pest	No. of minor pest
Apple	3	10
Avocado	3	7
Banana	5	19
Brassicas	8	10
Brinjal	3	7
Capsicums	5	15
Cashew	4	4
Cassava	3	9
Castor	5	14
Citrus	30	32
Cocoa	7	24
Coconut	10	25
Coffee	35	39
Cotton	27	34
Cucurbits	4	6
Date Palm	4	11
Fig	3	21
Grass	4	4
Groundnut	14	35
Guava	5	18
Mango	3	26
Maize	25	25
Millet (Bulrush & Finger)	13	??
Oil Palm	7	6
Okra	6	8
Olive	2	15
Onions	2	2
Papaya	1	12
Passion fruit	3	4
Peach	7	10
Pineapple	1	10
Potato	5	11
Pulses	22	32
Pyrethrum	1	6
Rice	41	30
Rubber	1	16
Simsim	2	16
Sisal	2	4
Sorghum	18	18
Soybean	7	20
Sugarcane	13	32
Sunflower	5	12
Sweet potato	8	29
Tea	8	24
Tobacco	7	27
Tomato	6	15
Wheat (barley & oats)	7	10
Yam	2	14

TABLE 24.4. MOST IMPORTANT PESTS OF COFFEE IN THE TROPICS
AND THEIR GEOGRAPHICAL DISTRIBUTION (AFTER LAMB, 1974)

Pest	Family	Method of attack	Distribution
Coccus viridis (Green)	Coccidae (Homoptera)	Leaves, shoots, berries	Pantropical
Habrochila spp.	Tingidae (Heteroptera)	Defoliating leaves	Africa
Antestiopsis spp.	Pentatomidae (Heteroptera)	Green berries, buds	Africa, India, Malaysia
Zeuzera coffee Niet[b]	Cossidae (Lepidoptera)	Stem borer	India, Sri Lanka Vietnam, Indonesia, New Guinea
Leucoptera spp.	Lyonetiidae (Lepidoptera)	Leaf miners	Central and South America, Africa
Dichocrocis crocodera (Meyr)[a]	Pyralidae (Lepidoptera)	Leaf roller	Africa
Cephanodes hylas (L.)[b]	Spingidae (Lepidoptera)	Leaves	Africa, Vietnam, Malaysia
Epicampoptera spp.	Drepanictae (Lepidoptera)	Defoliation of leaves	Africa
Apate monachus F.	Bostrychidae (Coleoptera)	Stem borer	Africa, Central and S. America
Xylotrechus spp.[a]	Cerambycidae (Coleoptera)	Trunk borer	South-east Asia
Anthores leuconotus Pasc.	-do-	-do-	Africa
Bixadus sierricola (White)[b]	-do-	Ring barking	Africa
Hypothenemus hampei (Ferr.)	Scolytidae (Coleoptera)	Berry borer	Pantropical
Xylosandrus compactus (Eich)[b]	-do-	Twig borer	Africa, South-east Asia, Pacific

[a]Not mentioned by Hill (1975)
[b]Considered as a minor pest by Hill (1975).

4. *Cocoa* (Table 24.7). Of the five most important pests of cocoa mentioned by Lamb (1974), only one is considered as of major importance by Hill (1975), one is of minor importance and the remaining three are not even mentioned. No major cocoa pests are listed from South America.

5. *Sugarcane* (Table 24.8). Of the five most important pests of Lamb (1974), only one is listed as a major pest by Hill (1975), three are recognized as minor and one is not even considered as a minor pest. Some of the major pests of this crop in Africa, for example, *Eldana saccharina* Wlk., are not even mentioned.

6. *Rice* (Table 24.9). The six most important pests listed by Lamb (1974) are also listed as major pests by Hill (1975). Again some of the major pests of this crop in Asia (e.g. *Nilaparvata lugens*) and Africa (e.g. *Maliarpha separatella* Rag.) are not even mentioned.

7. *Coconuts and other palms* (Tables 24.10 and 24.11). The three most important pests listed by Lamb (1974) are also listed as major pests by Hill (1975).

420   R. Kumar

TABLE 24.5. MOST IMPORTANT PESTS OF TEA AND THEIR
GEOGRAPHICAL DISTRIBUTION (AFTER LAMB, 1974)

Pest	Family	Method of attack	Distribution
Scirotothrips bispinosus (Bagn.)[a]	Thripidae	Leaves	South India
Helopeltis theivora Waterh[b]	Miridae	Buds, young leaves	India, Sri Lanka
Homona cofearia (Niet)[b]	Tortricidae (Lepidoptera)	Tea tortrix leaf roller	India, Indonesia, Sri Lanka
Cydia leucostoma(Meyr)[b]	-do-	Flushworm, leaf roller, buds and upper leaves	India, Indonesia, Sri Lanka, Taiwan
Xylosandrus formicatus Eichn.[b]	Scolytidae	Tea shot hole borer	Sri Lanka India
Xyleborus compactus (Eich.)[a]			Japan

[a]Not mentioned by Hill (1975).
[b]Considered as a minor pest by Hill (1975).

There may be many reasons for the disagreements on the status of the tropical pests of the crops mentioned above. It is likely that the two authors have assigned an importance to the pests which reflects their own experiences in particular areas of the tropics. On the other hand, little is known about the causes of outbreaks of many tropical pests or why they differ in importance locally. With these reservations, and the factors mentioned earlier, it seems that the undermentioned pests warrant serious study in the tropics:

24.6.1   Drug plants (coffee, tea and cocoa)

Coffee bugs, *Antestiopsis* spp. (Heteroptera), on arabica coffee require detailed investigations. Although these bugs occur in West Africa they are not as yet significant pests of the crop in that region, while they are a major pest in East Africa (Greathead, 1966; Le Pelley, 1973).

Coffee berry borer, *Hypothenemus hampei* (Ferr.) (Coleoptera), is indigenous to tropical Africa but has been carried to other continents, and is now a pest of coffee in Brazil and Indonesia and requires attention. This pest can continue to breed in stored coffee. The control of pantropical scale *Coccus viridis* (Green) requires a thorough understanding of the attendant ants and its incidence varies from country to country. Currently, leaf miners, *Leucoptera* spp. (Lepidoptera), and yellow-headed borer, *Dirphya nigricornis* (Oliver) (Cerambycidae), have become important pests on coffee in Kenya.

Bathycoelia thalassina (H.-S. Heteroptera) is fast becoming a major pest of cocoa in West Africa and has been reported as a serious pest of cocoa in

TABLE 24.6. MOST IMPORTANT PESTS OF COTTON AND THEIR
GEOGRAPHICAL DISTRIBUTION (AFTER LAMB, 1974)

Pest	Family	Method of attack	Distribution
Empoasca lybica (de B.)	Cicodellidae	Leaf scorch	Africa, Spain, Israel
E. fascialis (Jac.)	-do-	-do-	Africa
E. devastans Dist.[a]	-do-	-do-	India
Aphis gossypii Glover	Aphididae	Attacks leaves boll shedding and vector of cotton antho-cyanosis virus in Brazil	Cosmopolitan
Bemisia tabaci (Genn.)	Aleyrodidae	Vector of cotton leaf curl virus	Africa
Helopeltis schoutedeni Reut.	Miridae	Shoots, bolls	Africa
Dysdercus superstitiosus (F.)	Pyrrhocoridae	Cotton stainer	Africa
Caliothrips impurus (Pr.)	Thripidae	Dark cotton leaf thrips	Africa
Sphenoptera spp.[b]	Buprestidae	Stem borers	India, Africa
Podagrica puncticollis Weise[a]	Chrysomolidae	Cotton flea beetle	Africa
P. pallida (Jac.)[b]	-do-	Hambuk flea beetle	Africa
Anthonomus grandis Boh.	Curculionidae	Larvae and adults attack flowers and fruits	North and Central America, West Indies
Pectinophora gossypiella (Saund.)	Gelechidae	Bolls	Africa[c]
Diparopsis spp.	Noctuidae	Bolls	Africa
Heliothis armigera (Hb.)	Noctuidae	Bolls	Cosmopolitan
Alabama argillacea (Hb.)[a]	-do-	Leaf worm	Americas

[a]Considered as a minor pest by Hill (1975).
[b]Not mentioned by Hill (1975).
[c]Actually pantropical.

Uganda; in West Africa the species of this genus are rarely serious pests.
Several species of *Helopeltis* occur on tea wherever the crop is grown. *Helopeltis
theivora* (Waterhouse) is a major pest in India and Sri Lanka. Individual pests
are of importance locally, for example, *Distantiella theobroma* on cocoa in
Ghana, *Sahlbergella singularis* on the same crop in Nigeria and *Planococcoides
njalensis*, vectors of swollen shoot disease throughout West Africa, should
continue to receive attention on a regional basis. Their importance lies in the
fact that countries such as Ghana have until recently been the world's largest
producers and exporters of Cocoa beans.

TABLE 24.7. MOST IMPORTANT PESTS OF COCOA AND THEIR
GEOGRAPHICAL DISTRIBUTION (AFTER LAMB, 1974)

Pest	Family	Method of attack	Distribution
Planococcoides njalensis (Laing)[a]	Pseudococcidae	Vector of swollen shoot virus	West Africa
Sahlbergella sigularis Hg.	Miridae	Shoot and pod feeder	West Africa
Distantiella theobroma (Dist.)[b]	Miridae	Shoot feeder	West Africa
Pantorhytes spp.[a]	Curculionidae	Larval stem borer	New Guinea
Xyleborus morstatti Hag.[a]	Curculionidae	Larval stem borer	Africa

[a]Not mentioned by Hill (1975).
[b]Considered as a minor pest by Hill (1975).

TABLE 24.8. MOST IMPORTANT PESTS OF SUGARCANE AND THEIR
GEOGRAPHICAL DISTRIBUTION (AFTER LAMB, 1974)

Pest	Family	Method of attack	Distribution
Perkinsiella saccharicida Kirk	Delphacidae	Vector of the persistent sugarcane Fiji disease virus	Australia, South East Asia, Hawaii, South America, Africa, Pacific, Madagascar
Aeneolamia varia saccharina Dist.[a]	Cercopidae		West Indies Central & South America
Chilo auricilius Dug.[b]	Pyralidae	Sugarcane stem borer	India, Java, Taiwan
Diatraea sacchralis (Fab.)[a]	Pyralidae	Oriental sugarcane borer	America
Rhabdoscelus obscurus (Boisd.)[a]	Curculionidae	Cane borer	New Guinea, Celebes, Pacific, Taiwan, East & South America

[a]Considered as a minor pest by Hill (1975).
[b]Not mentioned by Hill (1975).

24.6.2   Fibre crops (cotton and jute)

Principal pests of cotton are those insects that attack the bolls or the flower buds which precede them. For this reason, in Africa, the cotton bollworms (Lepidoptera), mainly *Pectinophora gossypiella* (Saund.) (Gelechildae), *Heliothis armigera*, *Earias* spp., *Diparopsis* spp. and *Spodoptera* spp. require priority attention. The key insect pests of cotton in the tropical Central America are

TABLE 24.9. MOST IMPORTANT PESTS OF RICE AND THEIR
GEOGRAPHICAL DISTRIBUTION (AFTER LAMB, 1974)

Pest	Family	Method of attack	Distribution
Leptocorisa acuta (Tuub)	Coreidae	Feed heads	Asia, New Guinea Australia, Pacific
Chilo suppressalis (Wlk.)	Pyralidae	Larval stem borer	Asia, Widespread, Australia
Nymphula depunctalis (Guen.)	Pyralidae	Paddy case borer	India, South-east Asia, Australia, South America
Tryporyza incertulas (Wlk.)	Pyralidae	Yellow stem	India, South-east Asia, Philippines
T. innotata (Wlk.)	Pyralidae	White stem borer	Indonesia, New Guinea, Australia, Philippine, Malaysia
Sesamia inferens (Wlk.)	Noctuidae	—	India, South-east Asia, New Guinea, Solomon Island

TABLE 24.10. MOST IMPORTANT PESTS OF COCONUTS AND OTHER
PALMS AND THEIR GEOGRAPHIC DISTRIBUTION (AFTER LAMB, 1974)

Pest	Family	Method of attack	Distribution
Oryctes rhinoceros (L.)	Scarabaeidae	Stem borer of palm	India, South-east Asia, Pacific New Guinea
Rhynchophorus ferrugineus (Oliv.)	Curculionidae	Stem borer of coconut, date, sago, and other palms	India, Sri Lanka, Asia, South-east Asia, New Guinea, Pacific
R. palmarum (L.)	Curculionidae	Coconuts and other palms and sugarcane	Central and South America

the boll weevil, *Athonomus grandis* (Boh.), and the cotton leaf worm, *Alabama argillacea* (Hb.) (Lepidoptaera). Other important pests in this area include the bollworm, *Heliothis zea* Boddie, several species of armyworm *Spodoptera* spp., loopers such as *Trichoplusia ni* (Hubner) and the whitefly, *Bemisia tabaci* (Genn.). In the United States a great deal of attention has been devoted to the tobacco budworm *(Heliothis virescens)*, pink bollworm *(Pectinophora gossypiella)*, cabbage loopers *(Trichoplusia ni)* and the beet armyworm (*Spodoptera exigua* (Hb.)). The cotton bollworms and pests such as *Dysdercus* spp. need to be studied on a regional basis to solve the peculiarly local problems. *Empoasca* species indigenous to Africa have been recorded from twenty-four countries on this continent, and are responsible for severe local attacks from time to time. Detailed investigations of this pest are required.

TABLE 24.11. MOST IMPORTANT PEST OF PYRETHRUM

Pest	Family	Method of attack	Distribution
Thrips nigropilosus Uzel	Thripidae	Eggs inserted into stems and leaves; adults feed on foliage	Kenya, Tanzania, Europe, Hawaii, Canada, USA

Jute is an important fibre crop extensively cultivated in Southern Asia. Semilooper, *Anomis sabulifera* (Gn.), stem weevil, *Apion corchori* Mshl., and yellow mite, *Polyphagotarsonemus latus* (Banks) are considered to be the most destructive pests of jute.

24.6.3   Cereals and sugarcane

Most important pests of rice, maize and sugarcane are the stem-borers in Asia, Africa and America. Species of the genus *Chilo* are important on sugarcane in Asia and Africa while *Diatraea* is important on this crop in the Americas. Several species of *Sesamia* are pests of sugarcane in Asia, Africa, Mediterranean, New Guinea and Solomon Islands. *Eldana saccharina*, now fast assuming notoriety as a destructive pest of sugarcane, is confined to Africa and requires detailed attention. The species of borers attacking sugarcane also tend to attack maize and rice as well. However, *Maliarpha separatella* tends to attack only rice and is widely distributed in Africa and is also found in India, Sri Lanka, Burma, China, Papua and New Guinea. It is a serious rice pest in Madagascar especially where there is continuous cropping of rice. *Buseola fusca* (Fuller) attacks maize and sorghum in most countries in Africa south of the Sahara, north of which it is replaced by *Sesamia cretica*, *S. nonagrioides*, and the pyralids, *Ostrinia nubilalis* and *Chilo* species (Walker & Hodson, 1976). *Diopsis* spp. (Diopsidae: Diptera) are also restricted to Africa and can be serious pests. *Tryporyza* species are important pests of paddy in Asia.

Sorghum midge *(Contarinia sorghicola)* is a major pest which attacks sorghum wherever it is grown. Maturing grains may be attacked by the larvae of the midge and damage may also occur close to the harvest (Harris, 1976).

24.6.4   Vegetable oil crops (coconut, oil palm and pyrethrum)

Rhinoceros beetle, *Oryctes rhinoceros* (L.) (Scarabaeidae), is the most destructive pest of coconuts in India, South-east Asia, Pacific, New Guinea. *O. monocoros* (Oliv.) attacks palms in West Africa. The latter prefers coconut but will under certain conditions attack oil palm as well. The red or Asiatic palm weevil, *Rhynchophorus ferrugineus* (Oliv.) (Curculionidae), South American palm weevil, *R. palmarum* (L.), and African palm weevil, *R. phoenicis* (F.), are among the most destructive insect pests of oil palm and require intensive studies. The leaf mining hispid, *Coelaenomenodera elaeidis* Mlk. is at present considered to be the most important oil palm pest in West Africa (Mariau, 1976) and requires detailed study. *Aspidiotus destructor* (Sign.) is one of the most serious pests wherever palms are cultivated. Large populations are associated with poor agronomy and unfavourable weather conditions.

Besides their function in human nutrition, coconut and palm oil also provide important industrial products. The acreage of almost all the producer countries under these crops has shown a tendency to increase. With this increase, the economic importance of these pests is likely to increase as well.

Pyrethrum thrip, *Thrips nigropilosus* (Uzel), is an important pest of *Chrysanthemum cinerariifolium* which provides the valuable insecticide, pyrethrum.

24.6.5   Grain legumes (cowpeas and groundnuts)

Grain legumes are a cheap source of protein in tropical countries where malnutrition is a problem. The area under their cultivation has shown a steady increase. As monocultures have been established, the pest problems have multiplied. Rachie & Roberts (1974) indicated that in Africa there are 15 major and 100 minor pests of cowpea alone. They estimated that effective control measures often result in 10–30 times increases in yield. The pyralid moth, *Maruca testulalis* (Gey.), has an extremely wide distribution and is able to infest, at least in some major grain legumes, nearly all the aerial parts of the plants. Destruction of the major stages of growth, flowering and seed production constitutes a major limitation to the productivity of grain legumes wherever *M. testulalis* has established itself (Raheja, 1973). Apart from its major damage to cowpea, groundnuts and various species of beans (*Phaseolus* spp.), it attacks thirty-five species of plants, most of which are Papilionaceous (Akinfenwa, 1975). Other important pests of cowpeas and groundnuts requiring attention are: Brown leaf beetle, *Ootheca mutabilis* Sahlb. (Chrysomelidae: Coleoptera); Coreids (*Acanthomia* spp. and *Mirperus jaculus* Thunb.) are important on cowpeas only. *Aphis craccivora* (Koch.) transmits ground rosette virus which causes the plants to become stunted (see Anonymous, 1973 for pests of groundnuts).

24.6.6   Roots and tuber crops (cassava and yams)

The scarabeids (yam beetles) *Heteroligus indes* (Billb.) and *Prionoryctes* caniculus (Arr.) are important pests of yams in Africa. Whiteflies, *Bemisia tabaci* (Genn.) (Aleyrodidae), are responsible for transmission of cassava mosaic. The mealy bug, *Phenacoccus manihoti* (M. Ferrero.), has recently established itself in Zaire, Congo and Angola and requires serious investigation. A number of other insects occur on these plants but for the most part they are unimportant (see Anonymous, 1978 for pests of tropical root crops).

24.7 COSMOPOLITAN AND PANTROPICAL PESTS

Of the thirty-six species of these pests listed in Table 24.2 such as *Heliothis armigera*, *Nezara vividula*, *Chrysomphalus aonidum*, *Thrips tabaci*, *Heliothrips haemorrhoidalis*, *Ceratitis capitata* and *Spodoptera exempta* require serious study as they infest a number of important crops. However, their importance varies from place to place. On the spot biological data are required to assess the suitability of available control measures.

24.8 CURRENT METHODS OF PEST CONTROL ON TROPICAL CROPS

24.8.1 Drug plants (coffee, tea and cocoa)

Extensive use of cultural practices along with controlled use of pesticides is known to give the best results in controlling pests of these crops. For example, pruning to keep coffee bush open greatly reduces the incidence of coffee bug on this crop because the bugs prefer dense foliage. The collection and burning of unpicked and fallen berries at least fortnightly, during fruiting peaks and at least monthly otherwise, checks the incidence of the berry borer on Robusta coffee. Parasite introduction against the pest has been unsuccessful in East Africa.

In the Indian Tea Gardens, red spider, *Oligonychus coffeae* (Niet.), can be an important pest and persists on a few old leaves as well as the scale leaves at the base of shoots during cold weather. Removal of these leaves during the lean period greatly reduces the incidence of the mite on the crop (Das, 1960).

The pest problems of cocoa and their control are well documented and studied in West Africa (Kumar, 1979), thanks largely to the substantial research carried out at Cocoa Research Institute, Tafo (Ghana). While accelerated and continued research is necessary into biological and economic aspects of cocoa problems (Kumar, 1975), enough information is available to decide how West African countries can maintain their positions as leading exporters of cocoa beans. However, due to administrative shortcomings, the cocoa industry in countries such as Ghana is steadily declining.

24.8.2   Fibre crops (cotton and jute)

In Africa, cotton entomology is especially well developed in the Sudan where, at times, pests and disease hazards have brought cotton production to the brink of disaster. But local research work in entomology, plant pathology and plant breeding has been able to deal effectively with all problems posed by pests and diseases (Ripper & George, 1965).

It is true that in most parts of the world, cotton pest control relies almost exclusively on the use of insecticides applied on a regular schedule. However, the extensive use of pesticides has led to severe environmental problems and economic crisis for the industry during the 1965–70 period in Central America. To overcome these problems, integrated pest-control programmes had to be developed (see Bottrell & Adkisson, 1977 for a review of cotton insect pest management).

In East Africa, legislation has been passed to ensure that there is a close season for cotton-growing in order to prevent population build up of pink bollworm which is monophagous on the cotton plant. The legislation stresses that all cotton plants should be uprooted and destroyed (or burned) by a certain date and no seedlings should be planted until the following rains arrive. But the legislation is seldom enforced.

For the control of pests of jute, only insecticides have been used so far. There is urgent need of research leading to the development of integrated control measures.

24.8.3   Cereals and sugarcane

Little success at control of stem borers by biological and chemical methods has been achieved so far, but cultural practices and development of resistant varieties have shown good promise.

24.8.3.1 Parasites

Actual manipulation of egg-parasites for biological control of rice stem borers and rice leaf folder has been attempted in the People's Republic of China. About twelve species of *Trichogramma* have been identified. Mass-rearing and field release of *T. australicum* is widely practiced. The parasites are reared at commune and brigade levels.

In India, three egg parasites — *Trichogramma japonicum, Telenomus beneficiencis* and *T. schoenobii* — are being tried for stem borer control. If the level of egg-parasitization approximates 70% then insecticidal sprays are considered unnecessary.

24.8.3.2 Pathogens

In East Africa Waiyaki (1972a,b) has found *Bacillus thuringiensis* effective against borers and armyworm. However, this bacterium has little effect on *Busseola* in field trials. Among pathogens, *B. thuringiensis* has received major attention for control of several rice insects in China. More than fifty strains have been isolated and some being mass produced at various communes and production brigades and used for control of yellow stem borer, leaf folder, rice skipper armyworms, etc.

Viruses have been used for *Heliothis* control in the Americas and several commercial formulations are available. (More attention to earlier detection of outbreak, more resistant to light formulations should make this method attractive.)

24.8.4 Controlled use of pesticides

In the past the use of pesticides for stem-borer control generally has seldom taken timing of application into consideration. Pesticide efficiency can be improved by better timing of application, based on pest surveillance and forecasting, and by application only at predetermined pest thresholds. One needs to know how long the pest is at a stage which can be controlled, the duration of insecticidal effect, and the amount of attack which can be economically allowed to develop. For example, at Kitale in Kenya, trials on *Busseola* indicated (Walker, personal communication) that control of larvae is only necessary between 10 and 40 days after egg laying starts. Endosulfan granules will control larvae for more than 40 days, so if this insecticide is used, only one application rather than two of DDT at present is necessary soon after egg laying starts.

24.8.5 Crop resistance

This appears to be a most promising way of reducing pest attack. Even moderate resistance is useful as it can be combined with other control practices. Resistance of maize to stem borer and leaf aphid is of increasing importance in the USA. Some resistance to *Busseola fusca* has been found in Nigeria as well as to *Sesamia calamistis*. These lines are being tested in an international programme.

ICIPE Scientists at IRRI in the Philippines have developed varieties which are resistant against brown planthopper. They have screened several thousand varieties and breed resistance into improved rice varieties. Several such varieties are currently being utilized in insect pest-management programmes in various Asian countries resulting in savings of millions of dollars.

24.8.6 Cultural practices

Numerous cultural practices are available for the control of stem borers in the tropics. Here the main emphasis is to prevent insect damage rather than destroy an existing infestation. For this purpose a knowledge of the pest life-cycle and of the population of the different stages is required. Once this information is known, trials can be carried out on the effect on pest and crop loss of changed time and method of sowing, cultivation, fertilization, duration of maturation, time and method of harvest and crop rotation. The following are some of the available methods:

(a) *Trap crops* — in the Philippines growing of a trap crop 20 days ahead of the main rice crop attract more brown planthopper on the trap crop. The control of the pest is easier, cheaper and safer for the natural enemies and yields are higher in fields with trap crops.

(b) *Rotation* — In parts of India and Pakistan, legume crops are often grown after rice which starts to rot and become unsuitable for stem borers. Rotating of rice with another annual crop has been recommended. This helps to break the life-cycle of the pest (see Adjei-Twum, 1971 for rotation of vegetable crops in the tropics).

(c) *Pest free periods* — planting of maize in Ghana during pest-free periods is widely practised. In China this method is widely used against rice stem borers as well.

(d) *Burning of stubs* — this method is employed to control aestivating stem-borer larvae and pupae. But studies in Australia on white stem borer *Tryporyza innotata* showed that burning killed only 10% of the larvae that were in the stems above the ground but it destroyed most of the pupae of the parasite, *Bracon* sp.

Many other methods of control employing pheromones, physiological and genetic measures, systems analysis, etc., are currently being developed to control pests of cereal crops. However, as stated earlier nowhere is the need greater than in the tropics to develop methods to suit particular local conditions.

Harris (1976) has carried out extensive work for the control of sorghum midge. He believed that development of resistant varieties and effective use of cultural control measures should facilitate integration of chemical and non-chemical methods for the control of this pest. Much work remains to be done to achieve these objectives.

24.8.6.1 Vegetable oil crops (coconut, oil palm and pyrethrum)

There is urgent need for extensive investigation on the pests of these crops to develop suitable control measures. However, control of certain pests such as the oil palm hispid by biological means such as the introduction of parasites seems promising.

24.8.6.2 Grain legumes (cowpeas and groundnuts)
Research carried out in West Africa indicates that the most effective methods of preventing insect damage in grain legumes are early planting, mixed cropping with cereal crops, and clean weeding. In West Africa traditionally the insect-susceptible grain legumes such as soybean, bambarra nut and groundnut are planted as the sole crop. For the control of groundnut aphid, cultural control by early planting, close spacing, etc., is advised. Seed dressing gives protection against the pest for about 5 weeks.

24.9 CONTROL STRATEGIES FOR TROPICAL CROP PESTS

In most parts of the tropics, the small farmer is at present and is likely in the future to be responsible for the greater part of agricultural production. Though it must be realized that large monocultures of wheat and rice are successfully working in many countries of South and South-east Asia and, nearer home, Sudan has converted the barren desert and grass savanna into a vast cotton-growing area. But these changes have been brought about by widespread changes in agricultural practices including multiple cropping, irrigation and increasing use of fertilizers and pesticides. In most countries of the Third World, however, pest-management strategies must in their broadest sense relate to the needs of the small farmer (see Farrington, 1977 for some aspects of research-based recommendations versus farmers' practices). Ten to twelve million small farmers live in West Africa 8 million in Nigeria alone. The need for efficient extension services to reach these farmers cannot be over-emphasized. The value of the treatments may be lost unless applied at the right rate and the proper time. Extension services are often the most neglected aspect of crop-protection programmes of developing countries in spite of being the vital element providing the feedback through the various links to policy-makers. To ensure that the farmer receives maximum benefit from research, he must have easy access to the extension service of the Ministry of Agriculture or by specialist organizations. This requires a major effort by each government to train a large body of suitable manpower to reach the small farmer. Further pest management must itself be seen in the wider perspective of agricultural improvement — including farming system, new varieties, fertilizers and farmers' education. With diversity of crops and pests mentioned earlier, the immense potential for manipulating the environment in diverse parts of the tropics to keep pests in check remains unrealized and must be regarded as an example of tropical underdevelopment. To this must also be added the fact that many research results are available but are seldom applied.

It appears to me that to tackle the problem of tropical underdevelopment so far as pests are concerned, the following information on pests in the tropics is required:

24.9.1 Pests

(a) *Taxonomy and ecology:* completion of our knowledge of the taxonomy and ecology of the pests.
(b) *Life tables:* construction of life tables of the insect pest and determination of the mortality factors by simple regression analysis. This would provide an idea of the role of natural enemies in the regulation of the pest population as well as a rough evaluation of the time and places of their occurrence.

24.9.2 Management and economic aspects

(a) *Crop loss:* data on crop loss in relation to pest intensity and precise definition of economic injury levels. This must be done at both local and regional levels.
(b) *Harvesting regimes:* extensive studies on harvesting regimes in relation to pest and disease incidence on food crops.
(c) *Pesticides:* to obtain information on the impact of pesticides on natural enemies with a view to avoiding spraying during the periods of their abundance.

24.9.3 Environment

Agro-climatology: to determine the relation between pest build-up and environmental factors so as to be able to manipulate these factors in a plan of pest management (with a view to preventing the pest from causing economic damage) use of prediction methodologies on the outbreak of pests (with a view to develop controlled use of insecticides).

The above information (see Fig. 24.1 for interaction of pests, crop management and environment) can only be obtained by intensive and detailed investigations locally and there is here an urgent need for more research. Therefore, for the developing world, the days of effective control of pests by purchasing insecticides and application technology from the advanced world are fast running out.

Fɪɢ. 24.1. Diagrams depicting interaction of pest, crop and environment.

24.10   NEED FOR DEVELOPMENT OF RELEVANT TECHNOLOGY FOR THE CONTROL OF TROPICAL PESTS

(a) *Research* — there is urgent need for far more research; intensive and detailed investigations on pests, management aspects and environment are required on the spot in all parts of the tropics.

(b) *Scientific manpower* — development of high and intermediate level of local scientific manpower and background for the establishment of long-term sound pest-management programmes. It should be emphasized that it is not sufficient to have a scientist. He must be of good calibre and interested in the specific job and thus the administrators must have the option to pick and choose. The present position where too few scientists are working on vast and complex problems in most parts of the tropical world is highly unsatisfactory. For example, in Kenya only two entomologists are currently engaged on pest problems of coffee grown over 100,000 acres of land; in Ghana five entomologists are working on pest problems of cocoa spread over an area of 3.69 million acres; and in Nigeria barely four entomologists are employed on cocoa planted over 2 million acres of land.

(c) *Crop protection techniques* — emphasis must be on simple crop-protection techniques relevant to small farm holders. For example, use of simple cultural practices such as regular pruning and removal of diseased material can bring considerable increase in yield of cocoa and coffee with little of such inputs as pesticides. However, where insecticides are to be used, there is the need for development of suitable equipment for the tropical farmer. Lisinger, Price & Herrera (1978) in surveys conducted in the Philippines found that farmers were applying an extremely low rate of the recommended insecticides as foliar sprays against rice pests. It was discovered that although the farmers mixed the correct amount of insecticide, due to difficulties of carrying the recommended amount through paddy field, they could spray only about one-tenth of the required amount. Recent innovations aimed at increasing farm productivity without recourse to heavy machinery include the use of solar energy in powering light crop sprayers (Wijewardene, 1979). The use of cattle in operating simple farm machinery needs to be developed further.

(d) *Resistant varieties* — efforts should be intensified in the development of varieties of crops resistant to pests. These efforts have consistently paid handsome rewards.

(e) *Organization of workshops* — there is an urgent need for the organization of workshops on the following pest management strategies:

(i) *Forecasting of pest attacks*
Accurate forecasting of pest attacks before they actually take place would enable control measures to be planned with maximum efficiency. There is the need for determination of economic thresholds on a routine basis as well as for standardizing the techniques of assessing losses due to pests.

(ii) *Cultural practices*
Information is required on what is known and what remains to be done in developing down-to-earth cultural practices for the tropical farmer. Cultural practices are labour intensive. In advanced countries, there is a shortage of labour but in developing countries with enough surplus manpower, labour-intensive practices should be encouraged. With increasing costs of fertilizers due to rises in petroleum prices, use of farm manure as the fertilizer needs to be investigated in detail. This would include the use of green-manure, compost, mud manure and nightsoil, etc.

(iii) *Extension services*
Blueprints for organization of extension services to benefit small-scale farmers need to be developed. Most developing countries are particularly deficient in this area. The added costs for pest-management advisory services are only a small fraction of the money saved by minimizing the use of chemical pesticides.

24.11   SUMMARY

Tropical agriculture is characterized by a diversity of its crop which must be attributed to particular environmental conditions prevailing in different parts of the tropical world. Further, the inadequate resources and poor education of the tropical farmer influence agriculture greatly. The pests of a crop and their status also varies according to the local environmental conditions. Damage varies according to the country, crop and to the season as well as many other factors, such as monoculture, quality and quantity of food supply, host/natural enemy relationship, intra- and interspecific competition, and economic factors, etc. It is very difficult to generalize on the status of particular pests when we consider the tropics as a whole.

In making a comparison of the listings of major pests of some tropical crops, differences of opinion regarding the status of the pests were noted. Little is known about the causes of outbreaks of many tropical pests or why their importance differs locally. Important pests of some stimulants, fibre crops, cereals and sugarcane, vegetable oil crops, grain legumes, root and tuber crops and pests of stored products are outlined. Control strategies for tropical crop pests are briefly discussed.

24.12 REFERENCES

Adjei-Twum D. C. (1971) Rotation of vegetable crops in the tropics. *World Crops*, pp. 10–15.

Akinfenwa S. (1975) Bioecological study of *Maruca testulalis* (Geyer) (Lepidoptera, Pyralidae) in the Zaria area of Northern Nigeria. M.Sc. Thesis, Ahmadu Bello University, Zaria (unpublished).

Anonymous (1973) Pest control in groundnuts. PANS Manual No. 2, COPR, London. 197 pp.

Anonymous (1978) Pest control in tropical root crops. PANS Manual No. 4, COPR, London. 235 pp.

Bottrell D. G. & Adkisson P. L. (1977) Cotton insect pest management. *Ann. Rev. Ent.* **22**, 451–81.

Das G. M. (1960) Occurrence of the red spider, *Oligonychis coffeae* (Nietner), on tea in North-east India in relation to pruning and defoliation. *Bull. ent. Res.* **51**, 415–426.

Dyck V. A. (1977) Paper presented in the Brown Planthopper Symposium, April 18–22, 1977. *Int. Rice Res. Inst., Los Banos, Philippines*. Mimeographed.

Farrington J. (1977) Research-based recommendations versus farmers' Practices: some lessons from cotton-spraying in Malawi. *Expl. Agric.* **13**, 9–15.

Greathead D. J. (1966) A taxonomic study of the species of *Antestiopsis* (hemiptera, Pentatomidae) associated with *Coffea arabica* in Africa. *Bull. ent. Res.* **56**, 515–554.

Harris K. M. (1976) The sorghum midge. *Ann. Appl. Biol.* **84**, 114–118.

Hill D. (1975) *Agricultural Insect Pests of the Tropics and their Control*. Cambridge University Press, London. 516 pp.

Kumar R. (1975) (ed.) *Proceedings of 4th Conference of West African Cocoa Entomologists*. Zoology Department, University of Ghana, Legon. 202 pp.

Kumar R. (1979) *Cocoa Industry in Ghana*. Popular Technology London.

Lamb K. P. (1974) *Economic Entomology in the Tropics*. Academic Press, London. 195 pp.

Le Pelley R. H. (1973) Coffee insects. *Ann. Rev. Ent.* **18**, 121–142.

Litsinger J. A., Price E. C. & Herrera R. J. (1978) Paper presented at 9th Annual Conference of Pest Control Council of the Philippines. May 3–6, Manila.

Mariau D. (1976) In *Oil Palm Research* (eds. Corley R. H. V., Hardon J. J. & Wood B. J.), pp. 369–383. Elsevier, Amsterdam.

Pathak M. D. (1977) Defence of the rice crop against insect pests, *Ann. N.Y. Acad. Sci.* **287**, 287–295.

Rachie K. O. & Roberts L. M. (1976) Grain legumes of the lowland tropics. *Advances in Agronomy* **26**, 1–132.

Raheja A. K. (1973) Insect pest complex of grain legumes in northern Nigeria. *Proc. IITA Grain Legume Improvement Workshop*, pp. 295–299.

Ripper W. E. & George L. (1965) *Cotton Pests of the Sudan*. Blackwell Scientific Publications, Oxford. 345 pp.

Southwood T. R. E. & Way M. J. (1970) Ecological background to pests management. In *Concepts of Pest Management* (eds. Rabb R. L. & Guthrie N. E.), pp. 6–28. North Carolina State Univ. Raleigh, USA.

Uvarov B. (1967) Problems of insect ecology in developing countries. *PANS* **13**, 202–213.

Waiyaki J. N. (1972a) Efficiency of *Bacillus thuringiensis* against larvae of *Busseola fusca* and *Chilo partellus*. Miscellaneous report 785, Tropical Pesticides Research Institute, Arusha, Tanzania. pp. 3.

Waiyaki J. N. (1972b) Notes on the effect of *B. thuringiensis* on the African armyworm, *Spodoptera exempta*. Misc. report 792 TPRI, Arusha, Tanzania.

Walker P. T. (1975) Pest control problems (pre-harvest) causing major losses in world food supplies. *FAO Plant Protection Bulletin* **23**, 34, 70–77.

Walker P. T. & Hodson M. J. (1976) Developments in maize stem borer control in East Africa, including the use of insecticide granules. *Ann. Appl. Biol.* **84**, 111–114.

Way M. J. (1971) A prospect of pest control. Inaugural Lecture Imperial College of Science and Technology. London, pp. 127–162.

Way M. J. (1976) Diversity and stability concepts in relation to tropical insect pest management. Entomological Society of Nigeria *Occasional Publication* No. **18**, 68–93.

Wharton R. H. (1966) Exotic insects and insect borne diseases. Introduction. *News Bull. Ent. Soc. Od.* No. 28: 2.

Whitney W. K. (1973) *Plant Protection of Tropical Food Crops in Africa*. Ibadan: IITA.

Wijewardene R. (1979) Solar energy powers light crop sprayer. *West African Technical Review*, April 1979: p. 87.

24.13 DISCUSSION

Wandiga: I would like to ask about the neglect of storage pests such as in grain. I believe in Kenya the percentage of food lost in storage is of the order of 20–50%. Traditional practices seem to be neglected.

Kumar: Storage pests are of course a problem all over the world, many of these problems, however, can be prevented by following simple routine practices. For example, a certain maize pest, *Scitophylus zea mays*, can only fly about ½ mile, therefore do not have your stores within 1 mile of your crop, otherwise it will start in your store, fly to your crop and from there is carried back to store at harvest. For storage many hygienic measures are known. In Ghana, for example, we use ethylene dibromide. We supply the farmers with the sacks and the ethylene dibromide to keep the pests out, but then the farmers often store the grain not properly dried and once fungus sets in nobody will buy the maize.

For each of these things there are usually appropriate control measures. For example, before we export cocoa to the UK we fumigate it with pyrethrins at 4.00 p.m. every day before shipment. They will not buy it if it is infested. So mostly there are solutions to these problems. But you have to find out what the insect's life-cycle is, when should I spray, or should I spray at all, what is the economic threshold, etc.? For stored products the technology is already available. Some good experiments were carried out in Nigeria. They tried to organize the people into cooperatives and got good results. But I notice in Kenya you have much better extension services than in other parts of Africa. You have almost 1 extension worker to 1000 farmers.

Szabo: You said that the production of cocoa is increasing in the Ivory Coast and decreasing in Ghana and you attributed this to pest factors. In saying this you seem to be overlooking a very important fact in comparing the two countries. Mainly that the Ivory Coast has the best control systems in West Africa including chemical pest control which is not quite the case in Ghana. So I agree that you need a campaign for reinforcement of extension services. This is at the base of the problem and is perhaps contributing greatly to the shifting pattern of yields, etc.

My second point is on the pyrethroid resistance problem and this is a problem that very directly relates to the economy of the developing countries. It is a known fact that alternating the chemicals used in control applications reduces or eliminates the development of resistance. Who is doing this? Why is it not being done? The large international companies do not have an interest to sell, say, a carbamate for this purpose. That is, to suggest the use of a carbamate only every second or third year and to use a competitor's product the other years. So obviously this would be the job of the local research people and local governments in the developing countries to advise the farmer on these practices.

Kumar: Of course we have problems of resistance. After BHC we have switched to carbamates, and already there are signs of resistance, then there'll probably be a switch to pyrethroids. After that I don't know what? But what I'm saying is that if all the routine practices are followed then the money you spend on insecticides will be reduced. Actually it is difficult to think of a time when it will not be sensible to use insecticides for pest control, but we should be able to use them in an improved controlled way. That should be the correct technology for our farmers.

You have made a very important point here. If you trace the history of pest management, that is what happened with the advent of DDT. With high-pressure salesmanship from the chemical companies, people concentrated on spraying only DDT. With aircraft spraying, and high cost, high-return technology the entomologists were considered redundant — we don't need them. Some scientists warned against this but were ignored. That was the greatest mistake that was made. People were so carried away that they overlooked these problems. They were considered as academic niceties for which research funds were not normally available. Then resistance developed and the entomologists were needed again.

In the developed countries the departments of agriculture expected the universities to do it. But here the extension worker is highly important and he should be backed by local research to find out when to apply, etc. But on the whole this work is not being done. In Ghana, for example, political upheaval destroyed a good system.

Szabo: If I may respectfully disagree with your approach to the stored products protection. As you said technologies are available. But what sort of technologies? You mentioned chemicals that can be used for sterilization of containers, etc. But are these technologies adaptable to small farmers and their conditions? Are these technologies really practical at this level? I fully appreciate your approach, e.g. choosing your storage space a mile from your crops, etc., but what about neighbours' fields intermixed with yours such that you simply cannot do this? So, there may have to be adaptations to these technologies. Have we really examined this? Made the necessary adaptations? In very few countries do we have any direction or advice and education. This is why I guess the World Food Conference in 1975 made a mandate to especially emphasize this as a global problem. Because, although the technologies exist, the greatest losses still occur mostly in the developing countries and during the storage period. This loss goes up to an incredible 50%.

Kumar: If we want to discuss stored products we need to go into much more detail. The FAO and the University of Ghana are trying to set up a "post-harvest technology centre" where some of these problems will be dealt with at the farmer level.

Taylor: I have a brief point to make. We know the technology that is correct. But forgetting ethylene dibromide which is in fact totally unsuitable for farm storage, the simplest method for the farmer is the use of a dilute dust insecticide, and I have personally carried out a survey questionnaire world-wide recently. We have found that in very few countries are suitable insecticidal dusts available for the farmer. You might find these dusts available at an urban centre and then they are often in large containers. A small pack suitable for the farmer is not marketed because it's too costly. So generally speaking, although the technology is known for stored grains, the chemicals are just not available. Although Kenya is an exception. In 1968–70 we had a scheme here for farmers, through the Freedom from Hunger campaign, actually involved in getting these chemicals. But in most countries they are just not available.

Kumar: I would like to comment here that in W. Africa, the Government said okay you have harvested your cocoa we will store it, handle large-scale fumigation at various strategic points. They have been able to develop these hermetically sealed silos (as in Australia) although the cost is higher. But often we have the problem that the farmer will keep a few bags for family use and these get infested. So we have the problem of keeping these clean.

Menn: You seem to emphasize surveys of insect pests and the cost, but it seems to me what may be very important is to determine threshold levels of these populations in order to drastically reduce the use of chemicals and also to introduce methods of better timing. You didn't mention such things at all.

Kumar: Well, just to give an example, if we get two capsids for every 100 cocoa trees, we will spray — or if 5% of a crop in Kenya is attacked by stem borer, spraying is recommended. So this sort of economic threshold should be developed and I think I mentioned that we should develop accurate, quantitative, reproducible methods for economic thresholds, especially in Africa. These should be standardized. Then economic threshold will be comparable, loss assessments will be comparable. This sort of standardization has never taken place. I consider this assessment and forecasting to be so very important, that in Ghana, I wish to have monitoring stations in strategic places, to give simple down-to-earth advice and assistance; you want light traps, we'll provide the bulbs, etc. This is done in Sudan. In West Africa we also have problems getting competent technicians. For example, one of mine recorded 32 days in one month so all that data went out the window. This sort of thing has to be taken into consideration.

CHAPTER 25

Contribution of ICRISAT to Studies on Plant Resistance to Insect Attack

W. REED, K. V. SESHU REDDY, S. S. LATEEF, P. W. AMIN and J. C. DAVIES

*International Crops Research Institute for the Semi-Arid Tropics (ICRISAT),
Patancheru P.O. 502 324, Andhra Pradesh, India*

CONTENTS

25.1	Introduction	439
25.2	Sorghum	440
	25.2.1 Shootfly	440
	25.2.2 Stem borers	441
	25.2.3 Midge	442
	25.2.4 Earhead bug	442
25.3	International Sorghum Pest Resistance Testing Programme (ISPRTP)	443
25.4	Pearl Millet	444
25.5	Pigeonpea	444
25.6	Chickpea	445
25.7	Groundnut	446
25.8	Summary	448
25.9	References	448

25.1 INTRODUCTION

The International Crops Research Institute for the Semi-Arid Tropics (ICRISAT) has a mandate to serve, in particular, the small farmers of the semi-arid tropics. The target crops are: sorghum (*Sorghum bicolor* (L.) Moench), pearl millet (*Pennisetum americanum* (L.) Leeke), pigeonpea (*Cajanus cajan* (L.) Millsp.), chickpea (*Cicer arietinum* L.) and groundnuts (*Arachis hypogaea* L.). All these crops suffer major losses caused by insect pests, yet all, except groundnuts, are generally grown without pesticide protection in the target farmers' fields. One of the most attractive options for pest management in such circumstances is the provision of cultivars that are more resistant to insect pests and that yield more in the low input systems than the currently available materials. A brief summary of our progress in this area to date is presented for each of the target crops.

25.2   SORGHUM

At the ICRISAT Centre, three sorghum pests of major international importance are found in our fields and so have been selected for study. These are the shootfly (*Atherigona soccata* Rond.), a stem borer (*Chilo partellus* Swin.), and the midge (*Contarinia sorghicola* Coq.). A fourth pest, the earhead bug (*Calocoris angustatus* Leth.), is important in certain parts of India, and related species may have damaging effects in certain situations in Africa, so we have added it to our management studies.

The major objectives of our programme include the development of screening techniques for the major pests of sorghum, identification of lines from the available germplasm having resistance characteristics, and assisting breeders with screening of materials emanating from their programmes.

25.2.1   Shootfly

Sorghum shootfly, *Atherigona soccata* Rond., occurs over much of Africa and Asia, attacking seedlings and young plants. It is responsible for severe crop losses, especially in late-sown sorghum and in areas where the dry season is not prolonged (Doggett, 1972; Rao, Rana & Jotwani, 1978; Seshu Reddy & Davies, 1978). The use of resistant varieties of sorghum is likely to be the most useful method of minimizing damage by this pest since it is recognized that varieties of sorghum differ in the extent of damage by shootfly (Ponnaiya, 1951; Pradhan, 1971; Soto 1974; Jotwani, 1978).

A large-scale screening operation was started in the rainy season, 1974, on a newly opened block of pesticide-free land at the ICRISAT. To ensure that high, repeatable, and uniform levels of shootfly are available at the susceptible stages of sorghum growth, a fish meal-spreading technique (Starks, 1970) has been modified and adopted. Spreader rows of a known susceptible sorghum hybrid (CSH-1) are sown about 3 weeks after the break of rains and shootflies are attracted by using fish meal. The test material is sown 3 weeks later and flies emerging from the spreader rows are attracted to the rows of test seedlings, again using fish meal. Oviposition is observed on 80–100% of the plants under test. Known susceptible checks are included as controls.

A large number of lines that had been screened for resistance to shootfly earlier in India and elsewhere, and some untested ones, have been exposed to heavy shootfly attack over several seasons. Resistance is assessed by counts of eggs and of "dead hearts" produced on the material under test. At harvest, a count of the number of harvestable heads is taken.

More than 300 lines have been identified as having significant levels of resistance/tolerance to shootfly. This resistance has been found to be mainly due to non-preference for egg laying and secondary or recovery resistance. There appears to be a definite link between non-preference for oviposition and the presence of trichomes (minute hairs) on the under-surface of leaves (Maiti, Bidinger, Seshu Reddy & Davies, 1980). In many lines, the presence of trichomes appears to be associated with a glossy leaf surface.

It is now possible to obtain differential levels of shootfly attack by utilizing a combination of sowing dates, spreader rows and fish meal on the breeders' material (Seshu Reddy & Davies, 1978).

25.2.2   Stem borers

The stem borer, *Chilo partellus* Swinhoe, is one of the more serious pests of sorghum in India (Jotwani & Young, 1972) and in the lowland areas of East Africa (Ingram, 1978). It is also present and is potentially important in other areas of the semi-arid tropics.

At the ICRISAT, initial attempts to screen the germplasm were in fields in which stalks from the previous crop had been allowed to stand between seasons and subsequently irrigated, to encourage the pupation of resting larvae. This resulted in some infestation on the test material but it was not great enough for screening and selection.

Consequently, it was decided that we would have to produce large numbers of larvae in our laboratories for introduction into the field plots in order to produce infestations that would allow intensive and repeatable screenings. A suitable diet and a technique that allowed us to produce very large numbers of these larvae were developed (Seshu Reddy & Davies, 1978a) and we can now infest large areas of sorghum. The newly hatched larvae are mixed with finger millet seed and introduced to the whorls of 25-to 28-day-old sorghum plants with the aid of a dispenser that is a modification of the one developed at CIMMYT. Each plant receives two applications, with about six larvae per plant in each. The subsequent leaf damage and "dead heart" incidence is recorded. This method has been found to give a rapid and accurate screening and is enabling us to identify tolerant lines from among our large World collection of sorghum germplasm.

Little information is availabale on the mechanism of resistance to this pest. It appears that non-preference for egg laying is not a factor (Roome, unpublished). The importance of leaf texture cannot, however, be ruled out and it has been suggested that thin stems are not favoured. A preliminary study by scientists of the Centre for Overseas Pest Research, London, of the HCN and phenolic acid levels in sorghum cultivars at the ICRISAT Centre (Woodhead *et al.*, unpublished) has indicated that there are no distinct or repeatable differences in the levels of these chemicals between the resistant and susceptible cultivars. It was also recorded that the phenolic acid levels varied over time and with different agronomic situations. Further studies are planned on these chemicals.

25.2.3   Midge

Sorghum midge, *Contarinia sorghicola* Coq., is an extremely important pest and a potential source of crop loss in almost all areas where sorghum is grown. The populations of this pest can be greatly increased by poor agronomic practices, and the pest appears to be on the increase particularly in areas where early-maturing and improved cultivars are grown in the same areas as later-maturing local cultivars.

Observations over 6 years have shown that the levels of this pest on the ICRISAT Centre are low. Efforts were made to develop methods of ensuring high midge challenge at another location, using spreader rows of mixed-age cultivars, but were not effective.

In co-operation with the Maharashtra Department of Agriculture, an attempt was made to screen materials that had already been identified in other programmes as midge resistant, in an area where midge was known to be endemic and serious. However, these trials were only partially successful. We are now in the initial stages of developing a technique that, we hope, will allow us to usefully screen for resistance to this pest at Dharwar, in Karnataka State.

Work elsewhere has shown that non-preference is involved in the resistance so far discovered and tolerance does not appear to be important (Teetes, Wuenache & Johnson, 1979).

25.2.4   Earhead bug

The head bug, *Calocoris angustatus* Leth., is very damaging in parts of India, seasonally. Little is known of the biology, carryover or even loss levels caused by this insect, or by head bugs in general. The number of Heteroptera in sorghum heads are often very high, but their significance in economic terms is unknown. The quality of grain can be severely affected.

Screening techniques are still being worked out, but the results to date show that this work will be extremely complex and difficult. Sowing cultivars of mixed maturities immediately after the break of monsoons as spreader rows and sowing the test material protected against shootfly 3 weeks later has shown promise, but further observations are necessary. In general, the compact-head types have been found to be more susceptible than the open-headed types. The detection of resistance is made difficult by the fact that test cultivars mature at different dates which can result in a line appearing to be resistant when it is merely an "escape" since the peak attack by *C. angustatus* is avoided. Attempts are being made to sow trials within which all cultivars have a similar maturity. Careful observations on bug numbers and flowering dates are a necessary prerequisite for satisfactory identification of resistance.

25.3 INTERNATIONAL SORGHUM PEST RESISTANCE TESTING PROGRAMME (ISPRTP)

The prime objective of our sorghum programme is to increase the production of sorghum in the SAT. To achieve this, it will be necessary to provide national programmes with genotypes having not only higher yield potential, but also greater yield stability than those currently grown by the farmers. In order to succeed in those objectives, it is important that the sorghum germplasm be screened across a range of environments and pest situations, to ensure that broad-based resistance to pests are located.

Since 1976 we have been sending pest-resistance nurseries to several countries to evaluate resistance in different environments. The nurseries include lines that have been found to be resistant at the ICRISAT and those identified by our colleagues in the All-India Programme, with whom we enjoy close cooperation. The number of such nurseries that have been despatched are presented in Table 25.1.

TABLE 25.1. ICRISAT PEST-RESISTANCE NURSERIES SENT FOR MULTILOCATIONAL TESTING

Year	Combined			Shootfly			Stem borer			Midge		
	E	C	L	E	C	L	E	C	L	E	C	L
1976	24	9	11	—	—	—	—	—	—	—	—	—
1977	22	3	3	22	5	7	22	6	7	16	3	3
1978	—	—	—	20	8	13	20	7	12	15	6	8
1979	—	—	—	20	8	18	20	9	14	15	10	15

E, entries; C, countries; and L, locations.

The number of successful returns has been disappointing but, at locations where proper observations had been taken, it was clear that the entries selected by our programme showed higher degrees of resistance than local check entries. Some of the shootfly and midge-resistant lines were promising in Kenya, Mali and Upper Volta. There is an increasing demand for identified resistant lines from within India. Currently, several of the identified lines are also being used by our ICRISAT breeders for crossing to agronomically superior material.

25.4   PEARL MILLET

It is generally believed that insect pests do not pose any serious threat in the cultivation of pearl millet. However, this belief is partially contradicted by the long list of more than seventy insect pest species recorded on this crop at ICRISAT (Seshu Reddy & Davies, 1979). The truth is that the actual economic importance of these pests has not been fully worked out, owing to their periodic and sporadic occurrence in farmers' fields. In recent years there appears to have been a considerable change in the status of some of the minor and sporadic pests, which have now assumed the status of serious pests in some of the millet-growing areas in India.

At the ICRISAT Centre, observations to date on pearl millet cultivars showed that the pest incidence was generally low with the exception of sporadic attacks by *Mythimna separata* Walk. and *Heliothis armigera* Hub. However, observations on millet pests will be continued. Most of our entomological work on pearl millet will have to be done in West Africa where pests appear to cause substantial losses to this crop.

25.5   PIGEONPEA

There are several insect species which can reduce the yields of pigeonpea, with variation in the composition of the pest complex between and within countries and from year to year. In India, where over 90% of the world's recorded production of this crop is grown, and in several other countries, the major loss is caused by lepidopteran borers that attack the flowers and pods. This complex is usually dominated by *Heliothis virescens* (F.) and *H. zea* Boddie causing major losses in the Americas. In addition the podfly, *Melanagromyza obtusa* Mall. causes considerable losses in most Asian countries, but particularly in the important production areas of northern and central India.

Work on the screening of the available germplasm and other materials for reduced susceptibility to the pod borers and podfly was initiated at the ICRISAT in 1975 (Lateef, 1977; Davies & Lateef, 1978). A methodology for open-field screening, with augmentation of the natural pest attacks where needed, was developed. To date, nearly 10,000 lines have been screened. Initial evaluation is of small unreplicated plots of the test materials which are compared with check cultivars of comparable maturities in pesticide-free conditions. All those entries that yield less and also suffer greater pest damage than the checks are rejected at this stage. Any lines of interest are tested in trials of a narrow maturity range with increasing replication in each year, each trial containing relevant check cultivars and lines that are of known high susceptibility. Advanced-stage testing is in balanced lattice square design trials with sixteen or nine entries in each. This design has been found to give useful increases in efficiency which is very welcome in this work where spatial and temporal variations in pest attack result in a high coefficient of variation.

From the beginning it was appreciated that the search for reduced susceptibility to the pests, particularly to *H. armigera*, would be difficult and our experience has confirmed this view. The large plant size, lengthy growing season, compensatory habit, and high incidence of outcrossing have all posed problems and slowed our progress. Nevertheless, we now have several lines that differ markedly in their susceptibility to both *H. armigera* and podfly. We have been looking for and have found, lines that differ not only in their susceptibility to attack by pests, but also those that tolerate attacks and yield a reasonable crop under heavy pest threats in the unsprayed situation. Of particular importance is the ability to compensate for early losses and there appear to be considerable differences among pigeonpea cultivars in this important character.

We found that some *Atylosia* spp., which are close relatives of pigeonpea, have considerable resistance to both pod borers and podfly. We have found a pod wall barrier and antibiosis. In cooperation with our breeders we have been screening derivatives of the intergeneric crosses between *Atylosia* spp. and pigeonpea in an attempt to transfer the resistance. To date, this approach does not appear to have been as productive as our germplasm screening but the cost of this work is low and it is probably worth persisting with for at least two more seasons.

We are continuing our open-field screening work with individual plant and population selection. Laboratory feeding tests and other experiments are under way in an attempt to distinguish the mechanism of the differences between more and less susceptible cultivars. We already have some useful leads on the possible causes of susceptibility differences to podfly oviposition. Seeds of our selections have been given to our ICRISAT breeders for crossing and to entomologists in India for testing in other environments.

25.6 CHICKPEA

Chickpea is the major pulse crop of India and is of considerable importance in several other countries, particularly those in the Mediterranean region. It has a very restricted range of pests but *Heliothis* spp. attack this crop from the seedling to the maturity stage and can be locally devastating. At present our plant-resistance studies on this crop at ICRISAT are confined to *Heliothis armigera* but we are also helping with work at ICARDA in Syria where the leaf miner, *Liriomyza cicerina* (Rondani), can be very damaging.

Chickpea is much easier to handle than pigeonpea, with a relatively short season, small plants, and little or no outcrossing. Here, however, the small plant size does give us the problem of individual plant escape, so individual plant selection in open-field screening has not been very productive so far.

As with pigeonpea, we have found considerable and consistent differences in the susceptibility of different lines to *H. armigera* attack in the 12,000 germplasm accessions and breeders' materials that we have so far screened. Here again we rejected many lines on the basis of results from unreplicated small-plot screening, then embarked on trials of narrow maturity ranges with increasing replication of the more promising lines. In this crop again, the ability to compensate for early losses is very marked and we are selecting not only for resistance *per se* but also for tolerance and compensation which are expressed in the yields under our pesticide-free conditions.

All the green parts of chickpea have a dense cover of glandular hairs which exude a very acidic liquid. This exudate, with a pH of approximately 1.3 and a high content of malic acid, is thought to be a factor in limiting the range of pests that attack this crop. There appear to be differences between cultivars in the amount and concentration of the exudate and this may be associated with differences in observed susceptibility. We hope to develop further co-operation with the Max Planck Institute for Biochemistry at Munich in the chemical studies of this exudate and of other aspects of susceptibility differences among the pulses.

Our breeders have made crosses of the more and less susceptible cultivars that we have identified and we are now screening the F_2 populations of these. We have begun multilocation testing of the selections and will supply seed, where available, to any interested workers. Further details of our screening, and other entomological studies, on both chickpea and pigeonpea, are recorded in our Pulse Entomology Departmental Progress Reports which are available on request.

25.7 GROUNDNUT

Insect pests have probably been generally underestimated as yield-reducing factors in groundnuts. On this crop, insects are of particular importance as vectors of viral and other diseases, with *Aphis craccivora* Koch spreading rosette disease in Africa and peanut mottle in several countries. Jassids spread phyllody or witches' broom, and desap the foliage by feeding, thus causing poor pod filling. Thrips can also cause severe damage to plants by feeding as well as by acting as vectors of bud necrosis, a disease now known to be caused by the tomato spotted wilt virus. Other insects including wireworms, termites, and earwigs are of particular importance, for they act as agents in allowing entry of the mycotoxin-producing fungi into the pods which they have bored or scarified, while damage from pests such as the seed beetle (*Caryedon serratus* Ol.) can start before harvest and continue in the store.

Of all the ICRISAT crops, groundnut has probably the greatest potential for the economic use of pesticides, but our groundnut entomology programme is utilizing/employing much of its resources to the search for resistance to insect pests. At this stage the screening work is restricted to the four major pests that are common in the groundnut fields at ICRISAT Center, these being, *Frankliniella schultzei* (Trybom.) EM the major vector of bud necrosis disease, *Empoasca kerri* Pruthi, *A. craccivora*, and different termite species. Although this programme is relatively young, we have already identified lines that show promise for resistance to pests.

Over 1500 groundnut lines have so far been screened in our fields for thrips damage and we have recorded obvious differences in susceptibility. Five lines have been found to have high levels of resistance as confirmed by feeding tests in the laboratory; on these lines the fecundity of the thrips was substantially reduced. In addition, wild species relatives of the groundnut have been field-and laboratory-screened; *Arachis chacoense* Krap. et Greg. was found to be highly resistant while other species showed a range of resistance to thrips.

A jassid, *E. kerri*, is particularly abundant from late August to early September at the ICRISAT. The attacked plants show typical tip-yellowing and in the laboratory, one insect per plant can cause wilting in young seedlings. In our field tests we have so far screened 250 lines by counting the nymphs on the three youngest leaves on each of the ten plants, 60 to 80 days after sowing, and comparing these counts with those from adjacent plots of standard cultivars. We have already recorded high resistance in two lines. Caged seedlings of one of these lines were able to withstand very large populations of jassids without wilting, while plants of a standard cultivar (M-13) wilted and died when similarly exposed. The resistant cultivars were found to have thick cuticles and high tannin deposits. In addition one line had pubescent leaves, with tannin-filled hairs.

Over 1000 lines were screened for resistance to aphids in the seedlings stage in screenhouses but no noticeable level of resistance was observed. In a modified screening procedure, month-old plants of 200 lines were screened but again with little or no success. Of the several wild species tested, however, *Arachis villosa* Benth. and *A. chacoense* were immune, while other wild species showed a high to moderate degree of resistance. Additionally, some interspecific hybrids were found to be relatively resistant, so future work will be concentrated upon these.

448   W. Reed, K. V. Seshu Reddy, S. S. Lateef, P. W. Amin and J. C. Davies

Termites are known to be serious pests of groundnuts in Africa and India; Johnson (1978) estimated up to 10% losses to these pests in northern Nigeria. Pods left in the ground after maturity suffer scarification and we have used this factor in our preliminary screening efforts. We leave pods of many cultivars in the ground well after maturity in plots that are known to be infested with termites, then score the late harvested pods for scarification. Here again we have found consistent differences in susceptibility over 2 years of testing, with two lines being particularly promising. Groundnut lines that have been found resistant to the various pests are already being supplied to national programmes. Our thrips-resistant lines have been sent to Brazil where *Enneothrips flavens* Moulten is particularly damaging. Promising lines are also being supplied to a programme in the USA where resistance to both jassids and thrips is required. We expect to supply the lines showing reduced termite susceptibility to co-operators in Africa and other areas of India.

25.8 SUMMARY

This paper presents a summary of the progress of ICRISAT's studies on plant resistance to insect attack on its target crops. On *Sorghum bicolor*, progress in screening for resistance to shootfly (*Atherigona soccata*), a stem borer *(Chilo partellus)*, midge (*Contarinia sorghicola*), and a head-bug (*Calocoris angustatus*) is described. Pearl millet (*Pennisetum americanum*) suffers relatively little damage by insect pests at the ICRISAT Centre, so a screening programme has not been developed but it is probable that work will be concentrated in our co-operative programme in Africa, where insect pests appear to be of greater importance to this crop. On pigeonpea (*Cajanus cajan*), progress is reported on the search for reduced susceptibility of *Heliothis armigera*, and to podfly (*Melanagromyza obtusa*) on the former crop. On groundnut (*Arachis hypogaea*), progress in screening for resistance to *Frankliniella schultzei*, *Empoasca kerri*, *Aphis craccivora* and termites is presented.

25.9 REFERENCES

Davies J. C. & Lateef S. S. (1978) Recent trends in grain legume pest research in India. In Singh S. R. *et al.*, *Pests of Grain Legumes: Ecology and Control*, pp. 25–31. London, Academic Press.
Doggett H. (1972) Breeding for resistance to sorghum shootfly in Uganda. In *Control of Sorghum Shootfly* (eds. M. G. Jotwani and W. R. Young), pp. 192–201. New Delhi: Oxford and IBH Publishing Co.
Ingram W. R. (1958) The lepidopterous stalk borers associated with Gramineae in Uganda. *Bull. Ent. Res.* **49**, 367–383.
Johnson R. A. (1978) Soil pests of groundnuts. National Seminar on Groundnut Production, Kano, Nigeria, 1978.

Jotwani, M. G. (1978) Investigations on insect pests of sorghum and millets with special reference to host plant resistance. *Research Bulletin of the Division of Entomology, IARI, New Delhi* (New Series No. 2), Nov. 1978.

Jotwani M. G. & Young W. R. (1972) Recent development on chemical control of insect pests of sorghum. *Sorghum in Seventies* (eds. N. G. P. Rao & L. R. House), pp. 377–398. New Delhi: Oxford and IBH Publishing Co.

Lateef S. S. (1977) Pest control strategy in the semi-arid tropics with specific reference to pigeonpea (*Cajanus cajan*) and chickpea (*Cicer arietium*). *Chemicalisation of Plant Production in the Tropics and Subtropics*, Vol. 3, pp. 106–116. Karl Marx University, Leipzig.

Maiti R. K., Bidinger F. R., Seshu Reddy K. V. & Davis J. C. (1980) Nature and occurrence of trichomes in sorghum lines with resistance to sorghum shootfly. *ICRISAT Departmental Progress Report-3, Sorghum Physiology.* ICRISAT.

Ponnaiya B. W. X. (1951) Studies on the genus sorghum II. The cause of resistance in sorghum to the insect pest, *Atherigona indica* M. *Madras University J.* **21**, 203–217.

Pradhan S. (1971) Investigations on insect pests of sorghum and millets. *Final Technical Report 1965–70. IARI, New Delhi.*

Rao N. G. P., Rana B. S. & Jotwani M. G. (1978) Host plant resistance to major insect pests of sorghum. *Proceedings of an advisory group meeting on the use of induced mutations for resistance of crop plants to insects*, pp. 63–78. International Atomic Energy Agency, Vienna.

Seshu Reddy K. V. & Davis J. C. (1978) The role of the entomology program with reference to the breeding of pest-resistant cultivars of sorghum at ICRISAT. Presented at the Symposium on Strategies for Insect Pest Control through Integrated Methods, 16–17 Aug. 1978, IARI, New Delhi.

Seshu Reddy K. V. & Davis J. C. (1978a) A new medium for mass rearing of the sorghum stem borer, *Chilo partellus* Swinhoe (Lepidoptera: Pyralidae) and its use in resistance screening. *Indian J. Plant Protection* **6**, 48–55.

Seshu Reddy K. V. & Davis J. C. (1979) Pests of sorghum and pearl millet, and their parasites predators, recorded at ICRISAT Centre, India, up to August 1979. *Departmental Progress Report-2, Cereals Entomology.* 23 pp.

Soto P. E. (1974) Ovipositional preference and antibiosis in relation to resistance to a sorghum shootfly. *J. Econ. Ent.* **67**, 265–267.

Starks K. J. (1970) Increasing infestations of the sorghum shootfly in experimental plots. *J. Ent.* **63**, 1715–1716.

Teetes G. L., Wuenache A. L. & Johnson J. W. (1979) Mechanisms of sorghum midge-resistant sorghums. *IX International Congress of Plant Protection and 71st Annual Meeting of the American Phytopathological Society. 5–11 Aug. 1979, Washington D.C.*, Abstract No. 445.

CHAPTER 26

Botanicals and their Derivatives in Vector Control

G. C. LABREQUE

Joint FAO/IAEA Division of Isotope and Radiation Applications of Atomic Energy
for Food and Agricultural Development, Vienna, Austria

CONTENTS

26.1 Introduction   451
26.2 Insecticides   452
26.3 Repellents   455
26.4 Chemosterilants   456
26.5 Conclusion   457
26.6 References   458

26.1 INTRODUCTION

From earliest times, insects have threatened man's health and survival. In his agrarian or urban environment, they found ecosystems favourable to high reproduction rates and extended longevity. However, concurrent with the evolvement of man's society was the development of measures to control pests affecting his health and that of his domesticated animals. For over a thousand years he has availed himself of various materials of inorganic and organic origin to accomplish these ends. It is well known that many plants manufacture chemicals that protect them against insect depredation and the extract from these plants affects metabolism and response in species other than those attacking the plant from which the chemical was derived. The Romans used hellebore as an insecticide (Fletcher, 1974) and that the use of *Derris* has been common practice among the Chinese (Fukami & Nakajima, 1971). Sabadilla and nicotine were recognized as effective insect control agents as early as the seventeenth century (Schmeltz, 1971; Crosby, 1966). However, it was not until the early 1800s when Jumtikof discovered the insecticidal effectiveness of ground pyrethrum flowers, did a reliable substance of botanic origin become available to control insect pest problems (Fletcher, 1974; Matsui & Yamamoto, 1971). From that time, evidence can be found on the use of other botanicals for vector control, but in most cases, usage

451

was extremely limited since the trend to pest control with chemicals changed to inorganic and later to organic formulations. Although DDT was synthesized in 1874, its efficiency as an active pesticidal agent was not discovered until 1939 (Fletcher, 1974). From then on and until the mid-1950s, the marked efficacy of the newly developed organic insecticides resulted in a new attitude of euphoria among investigators that, at last, means were at hand to eliminate disease vectors. It is truly unfortunate that this goal could not be realized; the development of resistance, and in many cases lack of understanding and of technical skill, led to their profligate overuse with the resultant awareness by concerned parties of their detrimental effect on fragile ecosystems.

26.2   INSECTICIDES

Why during the interval of 100 years (from the time pyrethrum became known until the advent of organic pesticides) pyrethrum and other botanicals did not receive some of the recognition and use later awarded to the organo-chlorines and organo-phosphates is still conjecture. It could be attributed to production costs or limited availability. Moreover, the greater awareness to the needs of human, animal and crop protection from pest depredation; and the rapid development of more highly developed socio-economic and agri-economic systems that burgeoned at the time DDT came on the scene, spurred the utilization of this readily available and inexpensive synthetic organic pesticide. Investigations for new insecticidal compounds of botanical origins did not abate despite the increased research devoted to the discovery of new organo-chlorines and organo-phosphorus pesticides, and numerous derivatives of *DERRIS, quassia, sabadilla, hellebore, RYANIA, Lycopersicon, Haplophyton, Annona, Tripterygium* and Mammein were isolated with many showing strong activity against various vectors (Anon., 1967; King, 1954; Crosby, 1971; Table 26.1). In addition to pyrethrum, nicotine and rotenone, camomile, anabasine, quassia, oil of eucalyptus, larkspur and sabadilla are still recognized as efficient insecticides (Frear, 1969). Screening programmes at the USDA Insects Affecting Man and Animals Research Laboratory formerly in Orlando and now in Gainesville, Florida, evaluated over 550 plant extracts and pesticidal activity was observed in many of these (Anon., 1967; King, 1954; Table 26.2).

Unfortunately, for the most part, the investigations in Florida and elsewhere were never pursued beyond field studies. Most of these studies lacking the funds necessary to pursue investigations of a promising compound beyond this point since the cost of obtaining a label for a pesticide is now in the vicinity of US $15–30 million (Goring, 1977; Watson & Brown 1977; Table 26.3). The private sector is reluctant to pursue investigations in the area

TABLE 26.1. BOTANICAL DERIVATIVES OF VARIOUS PLANTS

Derivatives of *Derris*:
  rotenone, deguelin, elliptone, sumatrol, toxicarol, malaccol, munduserone,
  pachyrrihizone, dolineone, erosone and amorphin

Derivatives of quassia and related genera:
  quassin, norquassin, pseudoquassinolic acid, alloquassin and isoquassin
  neoquassin, *O*-acetyl-neoquassin, x-*O*-methyl-neoquassin, -*O*-methyl-enoquassin and
  norneoquassin

Lycopersicon hirsutum — Tridecanone

Ester alkaloids of sabadilla (*Veratrum sabadilla*):

  cevacine, cevadine, veratridine, vanilloylveracevine and sabadine (sabatine)

Alkaloids of hellebore (*Veratrum album*) and related species:
  germitetrine, germitrine, neogermitrine, germanitrine, germerine, germidine,
  germbudine, neogrembudine, protoveratridine, protoveratrine, zygacine,
  methylbutyrylzygadenine, veratroylzygadenine and vanilloylzygadenine

Derivatives of *Ryania speciosa*:
  ryanodine, anhydroryandonine, triacetyl-, ryanodol and anhydroryanodol

The alkaloids of *Haplophyton cimicidum* (Mexican cockroach plant):
  haplophytine, eburnamine, isoeburnamine, *O*-methyleburnamine, cimicidine, cimicine,
  haplocine and haplocidine

Alkaloids of *Tripterygium wilfordii* (Thundergod vine):
  wilforine, wilfordine, wilforgine, wilfortrine, wilforzine, wilfordic acid and hydroxy
  wilfordic acid

Derivatives of Mammein (Mamey):
  Mammein, *O,O*-dimethyl-mammein, *O,O*-diacelyl-mammein, dihydro-mammein,
  O,O-diacetyl-dihydro-mammein and ISO-mammein

of botanicals since the various patents necessary to insure the return on the costs of producing a pesticide are often unobtainable. Moreover, one must bear in mind that one of industry's prime objectives is to derive a nominal return on funds expended. The demand for specific pesticides to protect major crops such as cotton, rice, tobacco, or wheat will receive first consideration since the volume of the pesticide used will insure a greater income and a more rapid recovery of the original investment than that obtained from a pesticide synthesized for medical or veterinary use where the volume is limited. Botanicals (except for pyrethrum) have been unable to compete not only because of the patent problem but also because in most cases production costs were excessive, the materials were less effective than those synthesized from petroleum products and primarily the sources frequently were unable to supply the demand. With the tenfold increase of fossil fuel bases for pesticides, the realization that reliance on pesticide is no longer a solution to pest control, the knowledge that a pesticide having a lower order of toxicity may actually function more effectively in an integrated pest-management programme and the need for developing countries to develop an agricultural programme such

TABLE 26.2. EFFECTIVENESS OF VARIOUS PLANT EXTRACTS
EVALUATED AS PESTICIDES AGAINST VARIOUS PEST SPECIES (ANON,
1967; KING, 1954)

| | | | Vector | |
| | | | Mosquito | Houseflies |
Toxicant	Louse	Tick		
Rotenone		×	×	
Caffeine	×			
Nicotine	×	×	×	×
Sabadilla	×			
Citronellal oxime	×			
Heliopsis longipes	×	×	×	
Trypterigui wilfordii	×		×	
Tillandsia usneoides			×	
Strophantin	×			
Echinacea pallida	×		×	
Dryopteris marginalis			×	
Hydnocarpus antihelminthica	×			
Melia azedarach			×	
Oils[a]	×			

[a]Oils of almond, anise, cajeput, matico, nutmeg, parsley seed and rosemary are also good as lousicides.

as the raising of plants from which botanical pesticides could be developed to provide foreign currency, these factors could well cause renewed interest in pesticides of botanical origin.

TABLE 26.3. 1975 ESTIMATED COSTS IN $M FOR DISCOVERING AND
DEVELOPING A NEW PESTICIDE (GORING, 1977)

Synthesis, screening, field research and development	5–8
Metabolism, environmental, residual and toxicological studies	2.25–3
Formulation, pilot plant and registration	2.75–4
Total	10–15

A case in point is pyrethrum. Although its derivatives have a higher order of toxicity than other existing botanical insecticides, these products are now competitive in price with most pesticides of organic synthesis origin and modern technology has increased its effectiveness especially in the area of residual activity. These compounds could have a greater role not only in vector control but in the control of agricultural pests if the supply could meet with the demand. In vector biology, their effectiveness is nearly unrivalled. They function not only as toxicants, larvicides, adulticides, ovicides, but also as smoke and insect repellents, flushing agents to determine population density, and in some cases as chemosterilants. A representative list of vectors against which it is effective

TABLE 26.4. VECTOR CONTROL WITH PYRETHRUM

Vector	Pyrethrum
Lice	dusts and powders
Bedbugs	dusts and sprays
Mosquitoes	adulticide, larvicide, repellent
Cockroaches	dust, sprays, flushing agents
House flies	dust-sprays, adulticide, larvicide, and chemosterilants
Tsetse	adulticide-repellent
Sand flies	adulticide-screen treatment
Horn flies	adulticide, repellent
Stable flies	adulticide, repellent
Horse flies	adulticide, repellent
Mites	adulticide
Deer flies	adulticide, repellent
Fleas	adulticide

and some of its uses are presented in Table 26.4 (Anon., 1978, 1967; Feinstein, 1953; Herms & James, 1966; King, 1954; LaBrecque, Wilson & Gahan, 1958; Smith, 1973).

Another interesting aspect is that these derivatives have been highly instrumental in leading to the discovery of the pyrethroid insecticides. Although not derivatives, these analogues of pyrethrum have been effective substitutes to organochlorine and organo-phosphates insecticides where insect resistance to these chemicals has been encountered (Roberts & Zimmerman, 1980; Roberts, 1980; Roberts, Baldwin, Pinson, Walker & Meish, 1980; Van der Vloedt, Baldry, Everts, Pak, Roman, Kuzoe & Cullen, 1980). With the development of resistance to organochlorine and organo-phosphorus insecticides it is hoped that research for new pesticides will continue to follow this trend indicated with the pyrethrum derivatives and pyrethroids since their mode of action is completely different and cross-resistance need not be a concern. Since most botanicals have dissimilar modes of toxic activity, it can be conceived that rotenone, nicotine and other botanicals could be the next compounds to be investigated as sources of new pesticide derivatives or bases for organic synthesis (Yamamoto, 1970).

26.3 REPELLENTS

Extracts from various plants have been used from time to time in repelling and preventing attacks by blood feeding disease vectors. Their use and distribution were primarily limited to an area where the plant was indigenous and readily available. Until 1938, oil of citronella, an extract of *Andropogon* and *Cymbopogon vardus*, was the most widely and commercially available mosquito repellent (Fletcher, 1974; Herms & James, 1966). In the late 1930s and early

1940s, German armed forces used pyrethrum as an insect repellent in their North African and European campaigns but because of its limited protection time, it was not widely accepted and other countries opted for various formulations of organic chemicals (Herms & James, 1966). At periodic intervals since the time, pyrethrum alone or in combination with piperonyl butoxide suspended or dissolved in oil, water, alcohol, dusts and other carriers have been evaluated but none provided more than 4 to 5 hours of protection time (Smith, 1973). Recently, Galun found that laboratory animals could be given full protection from bites of the tsetse fly, *Glossina morsitans*, and from the tick, *Ornithodoros tholozani*, for periods of 10 and 7 days respectively with various formulations of microencapsulated pyrethrum applied to the skin of laboratory animals at the rate of 0.6 mg cm^{-2} (Galun, Ben-Eliahu & Ben Jamar) This protection time far exceeds that of any formulation derived either of plant or organic synthesis and bodes well for the use of this plant derivative in the protection of humans and livestock. Once again, the pyrethroids have also indicated an activity similar to that of their plant-derived predecessors.

Tests with the pyrethroid permethrin (3-phenoxyphenyl) methyl *cis, trans* (+)-3-(2,2-dichlorethyl)-2,2-dimethyl) cyclopropane carboxylate) when applied to clothing rivals the results in protection time being effective in repelling over twelve disease vectors and was effective against them at a dose of 1 mg cm^{-2}. Protection from the bites of *Aedes aegypti*, *Anopheles quadrimaculatus*, *Stomoxys calcitrans* and *Amblyomma americanus* was obtained after protective clothing had been exposed to 42 days of weathering and exposure to approximately 8.0 cm of rainfall. The chemical is not a repellent *per se* and its specific mode of action has not yet been definitely identified; but be it irritation, repellency, or toxicity, it achieves its objective in preventing insect bites (Schreck, Posey & Smith, 1978).

26.4 CHEMOSTERILANTS

In the light of the status of many of the present insecticides, high production costs, development of resistance, and deleterious effects to non-target organisms, it is necessary for those involved in vector control to not only develop new pesticides, or to reassess earlier chemicals that previously did not meet the high criteria required to compete favourably with existing pesticides, but to re-evaluate and formulate other approaches to vector control in protecting man and animals. The reliance on one mechanism is no longer applicable in most cases and resort to multiple control approaches in a co-ordinated programme have high priority. The use of the Sterile Insect Technique (SIT) is an ideal mechanism that can readily incorporate with various control measures.

Interest in the possibility of using sterility control pests is widespread and research continues to expand. The potential for control with chemicals that can produce sterility in both sexes far exceeds that of insecticides alone, and that of a sterilant affecting females only, equals for the most part those results obtained with pesticides. Theoretically, a pesticide alone applied to produce a 90% selection pressure against an insect having a five-fold rate of increase/generation will necessitate at least twenty generations to achieve zero. However, a sterilant bait affecting both sexes applied under the same conditions will produce the same results in four generations (LaBrecque & Smith, 1968). A female sterilant bait alone will produce a control as effective as a pesticide; however, when included with the toxicant, will achieve the same degree of control within five generations (Meifert & LaBrecque, 1971; Meifert, LaBrecque & Rye, 1969). This potentiation results not only in reduced costs for chemicals and labour, but also the presence of two diverse mechanisms affecting reproductive physiology and survival will reduce the selection for resistance to both insecticide and chemosterilant. It has been observed that, in some cases, irradiation used to produce sterility can at times produce some loss in insect viability, thereby reducing the effectiveness of the sterile-insect technique as an approach to control. At other times, the availability of an irradiation source is inadvisable or economically impractical to conduct feasibility studies in the field. In these cases, chemosterilants can often replace irradiation as a sterility-induction method.

Many sterility-inducing chemicals are available but their mutagenicity often precludes their use. However, many sterilants of botanical derivation have been synthesized and have proven effective (Table 26.5) against the house fly. Reserpine, colchicine and extracts of *Aristolochia indica* have sterilized both sexes, whereas *Sterculia foetida* extracts and seretonin affect only females. Females of the face fly have been also sterilized by *Sterculia* and heliotrine as well and the screwworm female with colchicine and emetine. Some of the better botanical insecticides — pyrethrum, rotenone, nicotine — have been evaluated with some sterility derived; however, the sterility effect is presumed to be lack of oviposition due to toxicity rather than on the effect upon gonadal development. The need to produce and refine additional control techniques for future use in the development of the new concepts in disease vector management is of high priority and the sterile-insect technique ranks uppermost as a mechanism to be further assessed, expanded and utilized.

26.5 CONCLUSION

It appears that cost still remains the primary deterrent to the development and commercialization of new pesticides. The demands for new pesticides for vector control are still too small in most cases to warrant expenditure of over

TABLE 26.5. BOTANICALS AND DERIVATIVES EFFECTIVE AS INSECT
CHEMOSTERILANTS

Reserpine	Mexican fruit fly (Benschoter, 1966; La Brecque & Smith, 1968)
	Olive fly (Fytizas & Bacoyannis, 1968)
	Housefly (Hays, 1965; Hays & Amerson 1967; Hays, Hays & Mims, 1969; Wicht & Jays, 1967; Guerra *et al.*, 1972; Guerra
	Tobacco budworm (Guerra, 1972; Guerra *et al.*, 1971; Masner *et al.*, 1970)
	Tribolium (Huot & Carrivault, 1967; Jays, 1967)
Sterculia foetida	Housefly (Beroza & LaBrecque, 1967)
	Face fly (Lang & Treece, 1971)
Colchicine	Screwworm (Chamberlain & Hopkins, 1960)
	Fruit fly (*Drosophila*) (Jacob, 1958)
	Housefly (LaBrecque & Smith, 1968)
Heliotrine	Fruit fly (*Drosophila*) (Clark, 1959)
	Face fly (Zapanta & Wingo, 1968)
Cafeine	Fruit fly (*Drosophila*) (Clark & Clark, 1968)
Serotonin	Housefly (Hays, Hays & Mims, 1969)
Emetine	Screwworm (Crystal, 1964)
Pyrethrum	Cigarette beetle (Tenhet, 1947)
	Housefly (Fye & LaBrecque, 1976)
Aristolochia indica	Housefly (Mathur & Sharma, 1980)
Rotenone	Housefly (Fye & LaBrecque, 1976)
Nicotine	Housefly (Fye & LaBrecque, 1976)

US $30 million per pesticide. The need for excess production to regain the funds expended makes reliance for the development of new chemicals the province of agricultural development rather than specifically for medical and veterinary entomology.

The favourable aspect to this rather sombre outlook is that because of their nature, botanicals for the derivation of pesticides are ideally fitted for regional development and can provide monetary supplement to many areas where agricultural production cannot compete with the highly developed agro-industrial advances of developed countries. Coupled with the desirability for a more selective pesticide showing lessened effect on non-target organisms for inclusion in integrated pest-management programmes, there is a need to reassess not only promising botanicals for sources of toxic materials but also those phytochemicals which at one time were rejected as being non-competitive compared with the highly effective but non-selective pesticides of the past decades.

26.6 REFERENCES

Anon. (1967) Materials evaluated as insecticides, repellents and chemosterilants at Orlando and Gainesville, Florida. 1952–1964. *USDA-ARS Agric. Handbook* No. 340, p. 423.
Anon. (1978) *Pyronone Insecticide and Related Base Chemicals.* Fairfield American Co.

Benschoter C. A. (1966) Reserpine as a sterilant for the Mexican fruit fly. *J. Econ. Ent.* **59**, 333–334.

Beroza M. & LaBrecque G. C. (1967) Chemosterilant activity of oils, especially oil of *Sterculia foetida*, in the housefly. *J. Econ. Ent.* **60**, 196–199.

Chamberlain U. F. & Hopkins D. E. (1960) Effect of colchicine on Screwworms. *J. Econ. Ent.* **53**, 1133–1134.

Clark A. M. (1959) Mutagenic activity of the alkaloid heliotrine in *Drosophila*. *Nature (Lond.)* 163–731.

Clark A. M. & Clark, E. G. (1968) The genetic effects of caffeine in *Drosophila melanogaster*. *Mutat. Res.* **6**, 227–234.

Crosby D. G. (1966) Natural pest control agents. In *Adv. Chem. Series* **53**, 1–15.

Crosby D. G. (1971) Minor insecticides of plant origin. In *Naturally Occurring Insecticides* (eds. Jacobson M. & Crosby D. G.), pp. 177–242. Marcel Decker Inc., N.Y. 584 pp.

Crystal M. M. (1964) Antifertility effects of antihelminthics in insects. *J. Econ. Ent.* **57**, 606–607.

Feinstein (1953) *Yearbook of Agriculture*, p. 520. US Dept. Agriculture.

Fletcher W. W. (1974) *The Pest War*, p. 201. Basil Blackwell, Oxford Press.

Frear (1969) *Pesticides Index*, p. 399. College Science Publishers, State College PA.

Fukami H. & Nakajima M. (1971) Rotenone and rotenoids. In *Naturally Occurring Insecticides* (eds. Jacobson M. & Crosby D. G.), pp. 71–98. Marcel Decker, Inc., N.Y. 584 pp.

Fye R. L. & LaBrecque G. C. (1976) *Bibliography of Arthropod chemosterilants*, ARS-S-93, p. 54.

Fytizas E. & Bacoyannis A. (1968) Action of reserpine on *Dacus eleae* Gmel. (Fr.). *Ann. Epiphyt.* **19**, 623–628.

Galun R., Ben-Eliahu M. N. & Ben Jamar, D. (1979) Long term protection of animals from tsetse bites through controlled-release repellents 207–218. *Proc. IAEA/FAO Symp., Vienna, 9 May 1979*, p. 468.

Goring C. A. I. (1977) The cost of commercializing insecticides. In *Pesticide Management and Insecticide Resistance* (eds. Watson D. L. & Brown A. W. A.), p. 638. Academic Press, N.Y.

Guerra A. A. (1972) Sterility induced in tobacco budworms by combination of reserpine and gamma irradiation affected by age and sex of pupae. *J. Econ. Ent.* **65**, 1281–1283.

Guerra A. A., Wolfenbarger D. A., Hendricks D. E., Garcia R. D. & Raulston, J. R. (1972) Competitiveness and behaviour of tobacco budworms sterilized with reserpine and gamma irradiation, *J. Econ. Ent.* **65**, 966–969.

Guerra A. A., Wolfenbarger D. A. & Lukefahr M. J. (1971) Effects of substerilizing doses of reserpine and gamma irradiation on reproduction of the tobacco budworm. *J. Econ. Ent.* **64**, 804–806.

Hays S. B. (1965) Some effects of reserpine, a tranquilizer on the housefly. *J. Econ. Ent.* **58**, 782–783.

Hays S. B. & Amerson G. M. (1967) Reproductive control in the housefly with reserpine. *J. Econ. Ent.* **60**, 781–783.

Hays B., Hays R. L. & Mims I. S. (1969) Comparative effects of reserpine and serotonin creatine sulfate on oviposition in the housefly. *An. Ent. Soc. Amer.* **62**, 663–664.

Herms W. B. & James M. T. (1966) *Medical Entomology*. p. 616. The Macmillan Co., N.Y.

Huot L. & Corrivault G. W. (1967) Les substances neuroleptiques et le comportement des insectes. V. Etude comparative de l'action de la reserpine et de quelques-uns de ses derives sur le *Tribolium confusum* Duval. *Arch. Int. Physiol. Biochem.* **75**, 745–753.

Jacob J. (1958). A study of colchicine induced sterility in the female fruit fly, *Drosophila melanogaster*. *Growth* **22**, 17.

King W. V. (1954) Chemicals evaluated as insecticide and repellents in Orlando, Florida. *USDA-ARS, Agric. Hand.* No. 69, p. 397.

LaBrecque G. C. & Smith C. N. (1968) *Principles of Insect Chemosterilization*, p. 346. Appleton Century Crofts, N.Y.

LaBrecque G. C., Wilson H. G. & Gahan J. B. (1958) Synergised pyrethrums and allethrin baits for the control of resistant house flies. *J. Econ. Ent.* **51**, 798–800.

Lang J. T. & Treece R. E. (1971) Sterility and longevity effects of *Sterculia foetida* oil on the face fly. *J. Econ. Ent.* **64**, 455–457.

Martin H. (1970) The chemistry of insecticides. *Ann. Rev. Ent.* pp. 149–161.

Masner P., Huot L., Corrivault G. W. & Prudhomme J. C. (1970) Effect of reserpine on the function of the gonads and its neuro-endocrine regulation in Tene-brionid beetle. *J. Insect Physiol.* **16**, 2327–2344.

Matsui M. & Yamamoto I. (1971) Pyrethroids. In *Naturally Occurring Insecticides* (eds. Jacobson M. & Crosby D. G.), pp. 3–70. Marcel Decker Inc., N.Y. 584 pp.

Meifert D. W. & LaBrecque G. C. (1971) Integrated control for suppression of a population of houseflies, *Musca domestica*, L. *J. Med. Ent.* **8**, 43–45.

Meifert D. W., LaBrecque G. C. & Rye J. R. (1969) Housefly, *Musca domestica*, control with chemosterilants and insecticides. *Fla. Ent.* **52**, 55–60.

Roberts R. H. (1980) Effectiveness of ULV ground aerosols of phenothrin against mosquitoes, houseflies, and stable flies. *Mosq. News.* **41**, 251–253.

Roberts R. H., Baldwin K. F., & Pinson J. L., (1980) Effectiveness of nine pyrethroids against *Anopheles quadrinaculatus* Say and *Psorophora columbiae* (Dyae and Knab) in Arkansas *Mosq. News*, **40**, 43–46.

Roberts R. H. & Zimmerman J. H. (1980) Chigger mites, efficacy of control with two pyrethroids. *J. Econ. Ent.* **73**, 811–812.

Schmeltz I. (1971) Nicotine and other tobacco alkaloids 99–136. In *Naturally Occurring Insecticides* (eds. Jacobson M. & Crosby D. G.), pp. 99–136. Marcel Decker Inc., N.Y. 584 pp.

Schreck C. E., Posey K. & Smith D. (1978) Durability of permethrin as a potential clothing treatment to protect against blood-feeding arthropods. *J. Econ. Ent.* **71**, 397–400.

Smith C. N. (1973) Pyrethrums for control of insects affecting man and animals. In *Pyrethrum* (ed. Casida J. E.), pp. 226–239. Academic Press, London. 323 pp.

Tenhet J. N. (1947) Effect of pyrethrum on oviposition of the cigarette beetle. *J. Econ. Ent.* **40**, 910.

Watson D. L. & Brown A. W. A. (1977) *Pesticide Management and Insecticide Resistance*, p. 638. Academic Press, N.Y.

Wicht M. C., Jr. & Jays S. B. (1967) Effect of reserpine on reproduction of the housefly. *J. Econ. Ent.* **60**, 36–38.

Yamamoto I. (1970) Mode of action of pyrethroids, nicotinoids, & rotenoids. *Ann. Rev. Ent.*, pp. 257–272.

Zapanta H. M. & Wingo C. W. (1968) Preliminary evaluation of heliotrine as a sterility agent for face flies. *J. Econ. Ent.* **61**, 330–331.

PART III

Requirements for Pesticide Production and Utilization

CHAPTER 27

Formulation and Application of Biologically Active Chemicals in Relation to Efficacy and Side Effects

I. J. GRAHAM-BRYCE

East Malling Research Station, East Malling, Maidstone, Kent, UK

CONTENTS

27.1 Introduction 463
27.2 Efficiency of Utilization 464
27.3 Spray Characteristics 465
27.4 Spray Application 466
    27.4.1 Fate of residual deposits 467
    27.4.2 Controlled release 468
27.5 Behaviour-controlling Chemicals 470
27.6. Summary 471
27.7 References 471
27.8 Discussion 472

27.1 INTRODUCTION

Developments in analytical techniques in recent years have led to impressive progress in the isolation of biologically active chemicals and the identification of their structures. In contrast, knowledge of how to use such chemicals to best advantage for pest and disease control has advanced much less. Since the scope for influencing biological response is substantial, this neglect is unfortunate. The case for devoting more attention to methods of use is strengthened by the rapidly rising costs of introducing new products which have now reached the order of $20 million per marketable product. The proportion of these costs accounted for by developing the active compound into the successful product is much greater than that attributable to the initial discovery of the biological activity (see, for example, Green, Hartley & West, 1977).

In this context, the objectives of formulation and application methods can be defined as the achievement of safe, economical and convenient handling, together with efficient delivery to the target organism and minimum exposure of non-target organisms, thereby promoting both efficacy and selectivity. It is the second part of this definition, the influence on biological effects, which is considered in this paper.

27.2   EFFICIENCY OF UTILIZATION

Efficiency of delivery cannot be assessed simply, particularly for preventative treatments, or compounds acting other than by direct kill. Nevertheless one useful index of efficiency for toxicants is the fraction of the applied dose received by, or required to kill damaging levels of the pest or pathogen concerned. Estimates for a wide variety of different pests and types of treatment indicate that this fraction exceeds a small percentage only in very specialized circumstances and is frequently less in order of magnitude (Graham-Bryce, 1977a,b). Any such estimates must be subject to some uncertainty, but they indicate conclusively that an overwhelming proportion of the pesticide applied for pest and disease control is distributed in the environment without contributing usefully to the intended objective. While it is unrealistic to expect highly efficient utilization, and indeed this may be undesirable if the benefits of arresting an infestation at an early stage are to be attained, such uncontrolled and unintended wastage must be unsatisfactory. At the very least the figures suggest that techniques for delivering pest-and disease-control chemicals require careful examination and that attempts to devise improved methods of dosage transfer should be rewarding.

Approaches to obtaining these improvements should comprise three inter-related components: a thorough understanding of pesticide behaviour in the environment; knowledge of how receiving organisms respond to different patterns of chemical distribution; and an imaginative approach for formulation and application which will match the spatial and temporal distribution of the active ingredient to the requirements for selective control. These principles are relevant for any type of chemical employed against pests and diseases and to any type of action, but in applying them to naturally occurring and related chemicals, the particular characteristics of these classes must be taken into account. Thus, for example, some compounds in these categories, such as the pyrethroid insecticides (Elliott, Janes & Potter, 1978), are outstandingly active and thus applied at very low rates; this also applies to many behaviour-controlling chemicals which in addition may be very expensive so justifying relatively elaborate formulations. Other naturally derived substances may be photo unstable and there may be more specific considerations such as when employing compounds which act by inducing a response from within the plant.

The subject is very large, as is the scope for ingenuity. Comprehensive treatment is not possible within the space of this paper but an attempt is made to illustrate the basic principles with a few selected examples which suggest some new approaches and demonstrate that detailed analysis can reveal relationships which are not obvious at first sight.

27.3 SPRAY CHARACTERISTICS

Whatever techniques are developed in future, spray application will remain of major importance as a means of achieving the essential dispersal of the control agent over the extensive area in which control is usually required. Requirements for spray application will be illustrated by reference to insect control, although the principles apply with variations in detail to other classes of damaging organism.

With spray application, the operator has under his control a range of factors which can be adjusted to obtain optimum results. These include drop size, composition and density, volume rate, choice of carrier and of additives to modify drop production, retention and absorption. In seeking efficiency of utilization, a very important consideration is the dose needed to kill the insect concerned, particularly in the case of direct interception which is often of considerable significance even for compounds which can act systemically. Thus, if a single drop of, say, 100 μm diameter contains a just lethal dose, well over 90% of the content of a 300 μm drop will be wasted even if it hits the insect. Consideration of the statistical probability of contact (Graham-Bryce, 1977a) indicates that even smaller drops would be more efficient. If it is assumed that all insects in an infestation have an equal chance of intercepting the drops in a spray, it can be shown that both the proportion of insects which escape a lethal dose and the proportion which receive a wasteful dose exceeding that required for kill decrease as the size of content of the drops is decreased by adjusting the spraying technique so that progressively more drops are required to give the lethal dose.

Various studies (e.g. Reay & Ford, 1977) give general support to this theoretical prediction that performance improves with smaller drop size and increasing frequency of contact, although many other factors must be taken into account as made clear in the analysis by Ford, Reay & Watts (1977). It is instructive to examine the implications of this principle for naturally-occurring and related compounds, for example, the very potent synthetic pyrethroids such as NRDC 161. Table 27.1 gives characteristics of spray drops of different sizes for this insecticide, calculated for the idealized case of a spray comprising drops of uniform size. For illustration it is assumed that the

TABLE 27.1. CHARACTERISTICS OF SPRAY DROPS FOR NRDC 161
(APPLICATION RATE 50 g ha^{-1} IN 10 l ha^{-1} CARRIER)

Diameter (μm)	No. of drops per ha	Amount per drop, ng
50	1.5×10^{11}	0.33
100	1.9×10^{10}	2.63
300	7.1×10^{8}	70.6

Compare LD$_{50}$ values: *Phaedon cochloeariae* 0.19
    (ng per insect)   *Anopheles stephensi* 0.04
                 Plutella xylostella 1.13

compound is applied at 50 g ha^{-1} at a volume rate of 10 l ha^{-1}, which is reasonably representative of commercial practice.

It will be seen that, except for *Plutella xylostella* at the smallest drop size considered (50 μm), the amount present per drop greatly exceeds the median lethal dose. These conditions would therefore be far from the objective identified above of dividing the lethal dose between as many drops as possible for maximum efficiency. This could be achieved only by increasing the volume rate, which would be very undesirable in view of the logistic and economic advantages of low volumes, or by decreasing the drop size. Table 27.1 indicates, however, that this decrease would need to be substantial and would give drops unrealistically small. In these terms, it could therefore be suggested that modern insecticides might be too active to be used efficiently by direct interception and that their performance might be even better if appropriate application methods could be devised.

27.4   SPRAY APPLICATION

These arguments presuppose that it is possible to produce the drop sizes required. Until relatively recently, this was far from the case. Conventional hydraulic sprayers give a wide range of drop sizes, between 1 and 1000 μm diameter. A convenient measure of the range of drop sizes is the ratio of the volume mean diameter (VMD) to the number mean diameter (NMD) which exceeds unity unless all the drops are the same size. Typical values for hydraulic sprays are 10–20. Great improvements have been obtained by the development of controlled droplet application (CDA) based on rotary atomizers which give VMD/NMD ratios of the order of 1.5, very close to the target value of 1. This represents a substantial advance in that if optimum drop characteristics can be specified, much can now be done to produce them.

A major difficulty remains. The requirements for efficient utilization dictate the use of small drops — in some cases as shown above, very small drops indeed. The smaller the drop, however, the greater the risk of drift. In this situation, therefore, the quest for efficacy and concern for the environment are in direct conflict. Fortunately a possible solution to this dilemma has recently emerged. This is the use of electrostatic charging. The basic concept is well established; it has been appreciated for many years that the pattern of initial retention of small particles can be very significantly altered by giving them a high electrostatic charge. In the past, attempts to exploit this potential in practice for improved drop capture have been frustrated because the necessary equipment was cumbersome and expensive. Modern techniques offer much better prospects and several promising developments have been described. For example, Arnold (1979) found that deposition of water-based sprays of permethrin on field beans (*Vicia faba*) could be increased approximately three-fold when electrostatic charging was applied to a rotary atomizer spray at a volume rate of $36 \, \mathrm{l} \, \mathrm{ha}^{-1}$. The electrodyn system (Coffee, 1979) is restricted to oil-based formulations but has the very great advantages that the formation and projection of the drops is achieved electrostatically, in addition to their charging, in a system containing no moving parts. The equipment is simple, has a low power consumption and gives good control of drop size which can be altered by electrical control. Again, deposition is considerably increased compared with comparable uncharged systems.

The ability to provide depositions to the required specification has thus markedly improved. This should provide an even greater stimulus to attempt to define what these specifications should be; at present there is very little systematic information about the optimum drop sizes, compositions and densities for different crop-pest problems. Any such consideration of spray specifications should include consideration of selectivity. Selectivity has many aspects; in the case of insecticides selectivity between harmful and beneficial insects, both pollinators and natural enemies, is of particular concern but appears to have been almost totally neglected in considerations of CDA. Differences in selectivity between different types of application may be considerable: theoretical calculations by Graham-Bryce (1975) indicated that a ten-fold difference between, for example, vapour uptake from a residual deposit and direct interception of spray could be expected.

27.4.1 Fate of residual deposits

With any of the treatments considered so far, but particularly for residual deposits, an important reason for apparent low efficiency of utilization is that much of the applied dose is lost from the area of application by decomposition or by physical processes of removal, so that enough material must be applied to compensate. With regard to evaporation, pesticides generally have relatively low volatility compared with many conventional laboratory chemicals, but this can still be very significant for compounds applied at rates equivalent to a layer only 1 μm thick over the surface of a soil or crop canopy subjected to a substantial evaporative potential. This was emphasized in studies by Phillips (1974) who found that up to 95% of the applied dose of dieldrin could be lost in only 24 h from cotton leaves in controlled environment rooms at 40°C and a windspeed of 60 m min^{-1}. Potential losses in the field may be estimated approximately by relating evaporation of the biologically active chemical to that of water which has been measured experimentally in the field for a wide variety of experimental conditions, using the relationship (Hartley, 1969) that relative evaporation rates are proportional to the product pM where p is the vapour pressure and M the molecular weight. Table 27.2 gives estimates of the amounts of some representative pesticides, including two pyrethroids, which could evaporate under typical British summer conditions, calculated from this relationship. It will be seen that the evaporative potential is considerable compared with normal rates of application, particularly when it is appreciated that many agricultural climates are much more extreme. Such figures also imply that although the compounds concerned have relatively low vapour pressures these may be quite sufficient to allow short-range vapour effects.

TABLE 27.2. POTENTIAL EVAPORATIVE LOSSES FROM 1 HA INERT
SURFACE

	Vapour pressure (mm Hg)	Molecular weight	Estimated loss
Permethrin	3.4×10^{-7}	391	0.34 kg year^{-1}
NRDC 161	1.5×10^{-8}	505	0.05 kg year^{-1}
DDT	1.9×10^{-7}	355	0.49 kg year^{-1}
Parathion	3.8×10^{-5}	291	1.22 kg week^{-1}
Dimethoate	8.5×10^{6}	229	1.47 kg month^{-1}

27.4.2 Controlled release

The concept of controlled release follows logically from the considerations in the preceding section, the obvious objective being to extend persistence in the intended location and decrease the amount dissipated into the environment to no good purpose. It is, however, important to be clear about the concentration–time–distance patterns required and attainable; there is usually some sort of inverse relationship between biological availability and persistence. Figure 27.1 summarizes some of the principle involved and demonstrates the argument for economy of material. For illustration it is assumed that the requirement for control is to maintain a level of 1 unit of concentration (shown in arbitrary units on the vertical axis) from time 10 to time 20 (shown in arbitrary units on the horizontal axis). This is indicated by

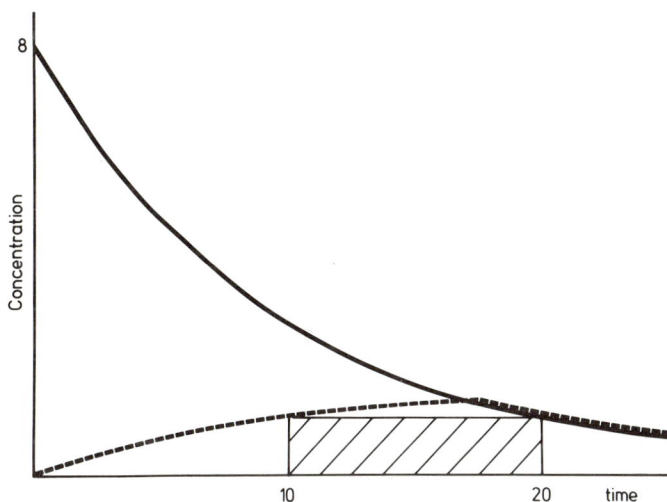

Fɪɢ. 27.1. Principles of delayed release.

the cross-hatched area. Ideally, for maximum efficiency we should have a source which introduced 1 concentration unit at $t = 10$ that was maintained by a steady supply until time 20 and then withdrawn. For the conditions assumed in Fig. 27.1, which include exponential decay and no consumption in the toxic process, the amount of chemical consumed in this idealized case would be 1.04 units which would be the absolute minimum required. Withdrawal is clearly quite impracticable so the most economical situation which could be conceived in practice would be to allow decay (at the assumed exponential rate) following cessation of the source after time 20. In this case the total consumption would be 2.04 units. If the material could be introduced only by conventional means so that it was all freely available at $t = 0$ and was again lost exponentially, it would be necessary to supply 8 units in order to have the necessary concentration of 1 unit at $t = 20$, and this would therefore be the amount consumed. This is indicated by the solid line in Fig. 27.1. Finally, if a controlled release formulation provided the active ingredient at a constant rate from $t = 0$, the supply would have to continue until $t = 17.3$ when the resulting residue could be left to decay: the consumption would be 2.78 units (broken line).

The potential saving in active ingredient and decrease in environmental contamination is therefore substantial. While it is most unlikely that this full theoretical potential will ever be achieved, formulation techniques can now go a long way towards it. For example, significant control over release rates can be achieved by microencapsulation (Phillips, 1974; Simkin, 1980). Microencapsulation has other potential advantages: it can increase rainfastness, protect active ingredients which are decomposed photochemically, mask

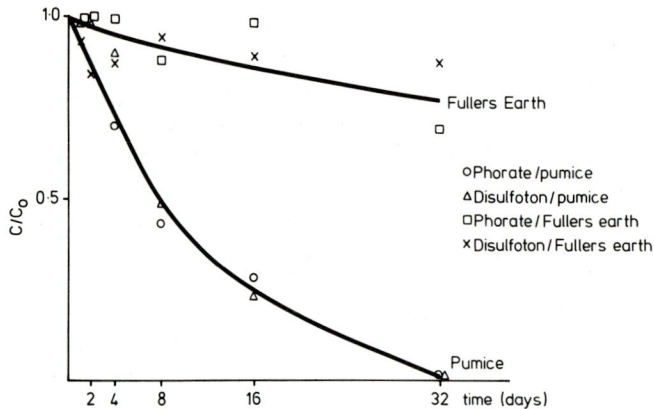

Fig. 27.2. Losses by evaporation from contrasting granular carriers expressed as a fraction of original content (C/C$_0$).

repellent properties and decrease contact action without affecting stomach-poison action, thus improving selectivity against leaf-chewing insects. An interesting example of the advantages of these properties is in the use of microencapsulation in baits for leaf-cutting ant control to delay the rapid action of pyrethroids (Phillips & Etheridge, 1977). Various other sophisticated dealayed-release devices have been developed, including plastic microtubes and multiple-layer plastic strips. However, it should not be overlooked that several longer-established and cheaper formulations including stickers such as amine stearates (Phillips & Gillham, 1968) and granules can give significant control over rate of release. For example,Fig. 27.2 shows results obtained by Graham-Bryce, Stevenson & Etheridge (1972), investigating evaporation of disulfoton and phorate, from different carrier materials in controlled environment rooms. Rates of loss from Fuller's earth granules were greatly retarded compared with those from pumice which has a more open structure. It is clear that, as with application methods, formulation offers many possibilities for controlling pesticide delivery, emphasizing again the need to define the biological requirements so that this potential for control can be fully exploited.

27.5   BEHAVIOUR-CONTROLLING CHEMICALS

The arguments for a better understanding of dosage transfer, advanced for pesticides in preceding sections, apply even more forcibly for chemicals acting other than by direct kill. The well-established concept of preventing pest damage by manipulating their responses to behaviour-controlling chemicals holds much promise as a potentially very powerful and selective method of control. This promise remains unfulfilled and although the use of pheromones and other attractants in traps for monitoring pest populations is well established, there has been disappointingly little practical success in using such substances for direct control. One reason for this may be that there has been insufficient recognition of the importance of choice. Organisms in an environment effectively permeated with attractant or repellent substances from multiple sources are unlikely to distinguish and respond to any particular stimulus. Such considerations suggest that combining attractants and repellents should enhance the activity of each. Analysis of the patterns of spread from pheromone sources and experimental studies with model systems (Graham-Bryce and Hartley, 1979) give support to this conclusion, indicating that vapour gradients to which insects respond are very sensitive to local sinks. It is clear from these and other studies that the arguments for a better understanding of transfer processes in the environment and of the behaviour and receiving characteristics of target and non-target organisms advanced for pesticides in the preceding sections apply even more forcibly for chemicals acting other than by direct kill.

27.6 SUMMARY

In relation to biological effects, the objective of formulation and application methods for chemicals used against pests and diseases should be to ensure effective delivery to target organisms and minimum exposure of non-target organisms, thereby promoting both efficacy and side effects. Assessment of existing practices suggests much potential for improvement. The design of formulations and application techniques for optimum biological effects requires knowledge of the physico-chemical properties and environmental behaviour of the active compound in relation to the locations and responses of target and non-target organisms. Special considerations in the case of naturally-occurring plant products may include their stability, potency and the nature of their biological effect. Examples illustrating these considerations for both toxic agents and behaviour-controlling chemicals are discussed and appropriate formulation and application methods examined.

27.7 REFERENCES

Arnold A. J. (1979) Field trials comparing charged and uncharged spray systems. *Proc. 1979 Brit. Crop Prot. Conf.* **1**, 289–294.
Coffee R. A. (1979) Electrodynamic energy — a new approach to pesticide application. *Proc. 1979 Briut. Crop Prot. Conf.* **3**, 777–790.
Elliot M., Janes N. F. & Potter C. (1978) The future of pyrethroids in insect control. *Ann. Rev. Ent.* **23**, 443–469.

Ford M. G., Reay R. C. & Watts W. S. (1977) Laboratory evaluation of the activity of synthetic pyrethroids at ULV against the cotton leafworm, *Spodoptera littoralis*, Boisd. In *Crop Protection Agents, their Biological Evaluation* (ed. McFarlane N. R.). Academic Press Ltd. (London).

Graham-Bryce I. J. (1975) Selective insecticidal action. *C.R. 5e Symp. Lutte integree en vergers.* OILB/SROP, pp. 315–326.

Graham-Bryce I. J. (1977a) Crop protection; a consideration of the effectiveness and disadvantages of current methods and of the scope for improvement. *Phil. Trans. R. Soc. Lond.* B, **281**, 163–179.

Graham-Bryce I. J. (1977b) Crop protection chemicals: a framework for future development. *Med. Fac. Landbouww. Rijksuniv. Gent.* **42**, 819–829.

Graham-Bryce I. J., Stevenson J. H. & Etheridge P. (1972) Factors effecting the performance of granular insecticides applied to field beans. *Pestic. Sci.* **3**, 781–97.

Graham-Bryce I. J. & Hartley G. S. (1979) The scope for improving pesticidal efficiency through formulation. *Advances in Pesticide Science*, Part 3 (ed. Geissbuhler H.). Pergamon Press, Oxford.

Green M. B., Hartley G. S. & West T. F. (1977) *Chemicals for Crop Protection and Pest Control.* Pergamon Press, Oxford.

Hartley G. S. (1969) Evaporation of pesticides. In *Pesticidal Formulations Research: Physical and Colloidal Chemical Aspects.* ACS Advances in Chemistry Series No. 86, pp. 115–134.

Phillips F. T. (1974) Some aspects of volatilisation of organochlorine insecticides. *Chemy. Ind.* pp. 193–197.

Phillips F. T. & Etheridge P. (1977) *Rep. Rothamsted Exp. Stn. for 1976*, p. 173.

Phillips F. T. & Gillham E. M. (1968) Effect of formulation on persistence of deposits of DDT on foliage. *Rep. Rothamsted Exp. Stn. for 1967*, pp. 170–171.

Reay R. C. & Ford M. G. (1977) Toxicity of pyrethroids to larvae of the Egyptian Cotton Leafworm, *Spodoptera littoralis* (Boisd): II. Factors determining the effectiveness of permethrin at ultra-low volume. *Pestic. Sci.* **8**, 243–253.

Simkin J. (1980) Microencapsulated natural pyrethrum, an improved insect. repellent, this volume, pp. 151–163.

27.7   DISCUSSION

Djerassi: You have described an interesting dispensing device. Does it have potential for wide use in industry?

Graham-Bryce: The device is being made available in limited quantities for evaluation. Already, it has been tested in a number of trials with very encouraging results. After these tests, the device will be made available in large quantities. I think it has great potential for wide application.

Casida: The device compares very well with the effect of microencapsulation of pyrethroids particularly in the control of social insects like the bees and ants.

Graham-Bryce: Yes, it is a useful device for delivering a toxicant which acts through odour and contact.

Bernays: This is where the key difference between the two techniques lies. Microencapsulated pyrethrins are effected when the toxicant contacts the target. No odour is involved. The device described by Dr. Graham-Bryce would be effective in cases where odour is involved as well. However, to protect more than a few plants, you really need to have an area to spray in which the insects cannot make a choice.

Graham-Bryce: I wonder whether any definitive experiments in support of what you've said have been done?

Bernays: Carrying out these experiments is difficult because we do not know the attractive odour for many insects. Furthermore, it is extremely difficult to provide the insects with something really attractive.

CHAPTER 28

Research, Development and Commercialization of Triatomine Control Agents

RICHARD PINCHIN

Nucleo de Pesquisas de Produtos Naturais, Centro de Ciências da Saude,
Universidade Federal do Rio de Janeiro, Rio de Janeiro, Brazil

CONTENTS

28.1	Introduction	475
28.2	Juvenoids	476
	28.2.1 Field trials	476
28.3	Organo-Phosphates	477
	28.3.1 Field trials	478
28.4	Phytochemicals	480
28.5	Commercialization of the Pesticides	481
28.6	Summary	481
28.7	Acknowledgements	481
28.8	References	482
28.9	Discussion	482

28.1 INTRODUCTION

American trypanosomiasis, Chaga's disease, remains one of the major public health problems in Brazil. There is at present little prospect of either immunization of a curative or prophylactic drug becoming available for large-scale use, and the principal method of combating continues to be the control of the vector-triatomine bugs.

BHC (30% gamma isomer) is the insecticide currently used in Brazil. Its use in intensive control campaigns has brought good results in the past, though it must be stressed that this has only occurred in the more economically favoured regions of the country, such as Sao Paulo state and near the larger cities (Rocha e Silva, Dias Junior & Guarita, 1969). It is certain that socio-economic development, with improved housing conditions and better education, are vital factors in limiting the incidence of disease transmission. However, this is a long-term solution and the need for effective control measures continues.

BHC is only really effective if applied two or more times a year, and is followed up by continual vigilance and repeated application in residual foci. Although various administrative and economic difficulties have hampered past national control campaigns (Freitas, 1974/5), the underlying cause is the sheer magnitude of the problem: the disease is endemic in a large area of Brazil with many millions of people at risk to infection. Consequently, many areas have not been included in past spray campaigns, and others are sprayed very infrequently, thus the disease remains unchecked.

28.2 JUVENOIDS

Conscious of the increasing relative importance of this problem, and also of the decline in interest of many of the major pharmaceutical companies in research into drugs for a disease which affects many, but poor per capita people, the Brazilian National Research Council has prompted many Chagas's disease research projects during the last 10 years. Our own interest began with a study of the action of juvenile hormone analogues (JHAs) on triatomine bugs. To date, over 100 compounds have been synthesized and screened, and some structure-activity relationships established (Pinchin, Oliveira Filho, Figueiredo, Muller, Gilbert, Szumlewicz & Benson, 1978a). Some of the analogues have been sufficiently active (Fig. 28.1) to warrant further development. However, problems arising from the comparatively short JH-sensitive phase of the bugs' life-cycle and the lack of stability of the active JHAs had to be overcome before a field trial could be attempted. The bugs are most sensitive to JHAs during the first half of the instar, and the development times from egg to adult can be longer than 6 months. Thus, to have any effect on a natural population of bugs infesting a house, the JHA must have exceptional residual activity.

We consequently turned our attention to the development of slow-release formulations for these JHAs (see also Pinchin, Oliveira Filho, Figueiredo, Muller & Gilbert, 1978b), and of the many types tested, two, a rubber-based paint and a polymer-based powder, were outstanding. The paint, which contained merely 100 mg m^{-2} of JHA, was active for 7 years, whereas the powder retained activity for 10 months.

28.2.1 Field trials

Our first field trials with these materials, although a number of juvenilized

Compound	Av. score Dose µg / nymph 1.0	0.1	Compound	Av. score Dose µg / nymph 1.0	0.1
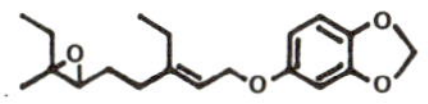 (structure)	4.0	1.4	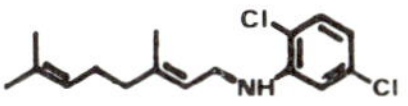 (structure, Cl, Cl)	4.0	1.4
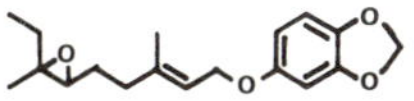 (structure)	4.0	1.1	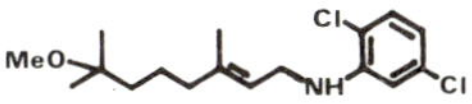 MeO- (structure, Cl, Cl)	2.8	–
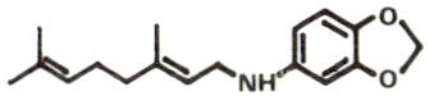 (structure)	3.4	–	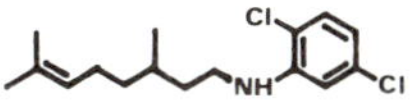 (structure, Cl, Cl)	3.1	0.6
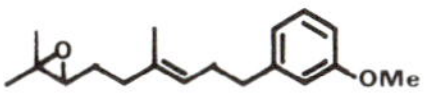 (structure, OMe)	4.0	1.3	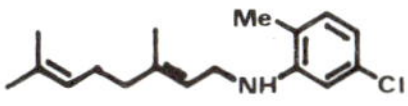 Me- (structure, Cl)	3.9	3.3

Fig. 28.1. Juvenile hormone analogues active on fifth instar *Panstrongylus megistus*. Average juvenilization scores (4 = completely juvenilized supernumary nymphs, 0 = inactive, and scores 1–3 represent partially juvenilized adultoids) are given for a single topical application of 1.0 and 0.1 µg/nymph (see Pinchin *et al.*, 1978c).

bugs were produced, were not very successful because of difficulties in physically maintaining the JHA formulations in the house for any length of time. Cardboard strips, bearing the paint on the reverse side, were nailed to the walls in order to provide an artificial JHA-containing hiding place. These were soon removed by the house owner. The powder, applied as a dust, was lost by mechanical erosion.

It is important to note that JHAs could be used to advantage in the same formulation as conventional insecticides: the most insecticide-tolerant stage, the fifth instar, is also the most JH-susceptible stage. Much slow-release formulation development has since been carried out, with insecticides as the active ingredients. This has considerably speeded up the process of information feedback from the biological tests to the formulation laboratory, and several of the better formulations have already entered field trials.

28.3 ORGANO-PHOSPHATE

The results obtained in the laboratory for two organo-phosphorus insecticides, chlorpyrifos and malathion, were particularly encouraging (Fig. 28.2). There was little loss in activity for the formulations based on PVC (polyvinylchloride) after ageing for 2 years, compared to those based on the synthetic elastomer ABS (copolymer of acrylonitrile, butadiene and styrene) (Pinchin, Oliveira Filho, Muller, Figueiredo, Perlowagora-Szumlewicz &

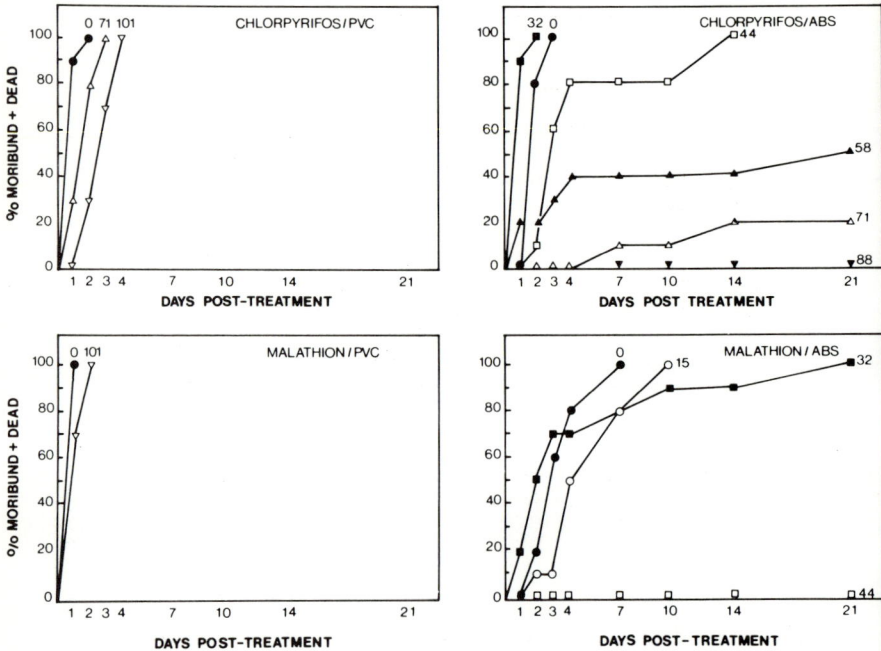

Fig. 28.2. Persistence of activity of chlorpyrifos and malathion, formulated at 5% in ABS and PVC, and aged for up to 101 weeks. Bioassays performed after 0 (—●—), 15 (—○—), 32 (—■—), 44 (—□—), 58 (—▲—), 71 (-△—), 88 (—▼—), and 101 (—▽—) weeks (from Pinchin *et al.*, 1978c).

Gilbert, 1978c). These formulations were originally designed to be applied as water-dispersable powders, thus avoiding any change in the spraying technique and equipment currently in use in Brazil. The active ingredient, unlike emulsion or ultra-low-volume formulations, is not rapidly absorbed by the substrate, typically mud or adobe, but is retained within the particles. Furthermore, an important and relevant behavioural characteristic of triatomine bugs is the camouflage phenomenon (Zeledon, Valeiro & Valeiro, 1973), in which the bugs habitually cover themselves with dust from their surroundings. If this dust contains the pesticide, then the insects may help to dose themselves. For slow-release formulations this is important, as diffusion of the insecticide from the organic polymer matrix to the wax of the cuticle may well take some time. A short contact time with a formulation which does not permit the bug to remove particulate matter is not nearly as effective.

28.3.1 Field trials

Many problems, particularly with particle aggregation and suspensability,

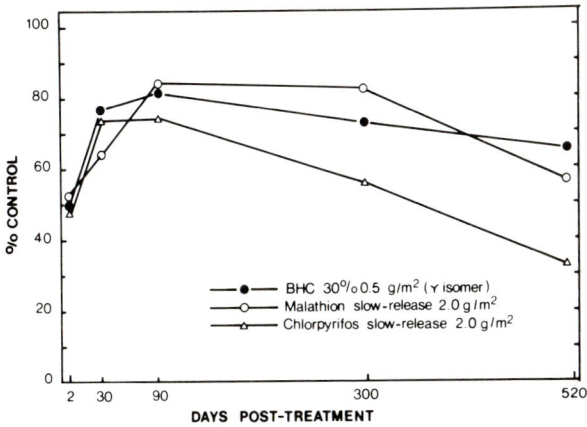

Fig. 28.3. Control of *Triatoma infestans* with slow-release formulations (based on PVC) of chlorpyrifos and malathion.

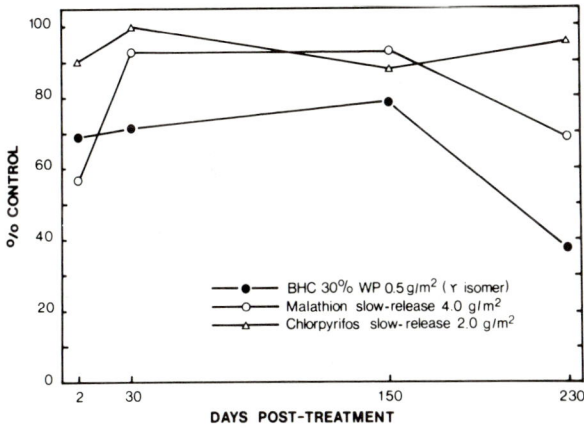

Fig. 28.4. Control of *Triatoma infestans* with improved slow-release formulations (based on PVC) of chlorpyrifos and malathion.

had to be overcome before field performance (Pinchin, Oliveira Filho, Ayala & Gilbert, 1978d) reflected the laboratory results (Figs. 28.3 and 28.4). More recently several commercially available insecticides were also being tested in the field, with particularly good results for the persistent synthetic pyrethroids, permethrin and decamethrin.

Our goal for all these persistent insecticide formulations is to maintain acceptable levels of control for long periods, up to a year at least. This will permit the spray teams to treat much larger areas at each spraying, and with a better result than currently obtained for BHC treatment.

An alternative approach, designed to eliminate the cost of a national spray programme, is to supply local authorities with insecticidal materials which can be applied in infested houses by unskilled labour without the need for spray equipment of any kind. One such material, a malathion-based white emulsion-type paint, is already under test in Brazil. It is safe to the user, oral LD_{50} (rats) > 5000 mg kg^{-1} and dermal LF_{50} (rats) > 5000 mg kg^{-1}, and a fairly high dosage of 10 g m^{-2} (active ingredient) can be used without any observable reduction in the cholinesterase level of the inhabitants. Although there is the advantage of the house being painted white, a financial incentive will probably also be necessary to induce house owners to paint their property. Much more information, both on the residual activity and cost-effectiveness, has to be accumulated for this method of vector control. As the distribution material and campaign organization will be carried out by local health authorities, it is hoped that this technique will circumvent the administrational difficulties often blamed for the failings of the national Chagas's disease campaigns.

28.4   PHYTOCHEMICALS

Other lines of basic research on triatomine control methods include studies on the action of the precocenes and repellents. Precocene II is active on the second through fourth instars of *Panstrongylus megistus* and the precocious adultoids formed have vector potential and are unable to reproduce. However, a much more active analogue will have to be found before this class of compound can be considered for practical use. Similarly, the use of repellents for preventing disease transmission does not seem to be very promising. A starved bug will sometimes feed on skin treated with diethyl toluamide at 10 mg cm^{-2}. Pyrethrin, and some of the synthetic pyrethroids, were more effective in preventing feeding, probably by the irritant action on the bugs (also used for flushing-out during triatomine surveys) rather than by actual repellency.

28.5   COMMERCIALIZATION OF THE PESTICIDES

The industrialization of the promising products has been considered. As mentioned earlier, one of the reasons often given for the non-execution of a national triatomine control campaign is the lack of funds. As all of these products will certainly be more expensive than BHC, though they may prove to be cost-effective in comparison, it is to be expected that they will also be subject to the inconsistencies of governmental purchasing policies. From the point of view of a manufacturer, it will be more attractive to supply materials direct to local authorities rather than to a central governmental agency.

It is obviously commercially important for a new product to reach several different markets, and the possibilities of using the formulations originally developed for triatomine control for other purposes, particularly for cockroach and mosquito (larvicide) control, is currently under investigation.

Our formulation development policy has concentrated, wherever possible, on using cheap, locally produced raw materials, with the exception of most of the insecticides which are imported into Brazil for agricultural and domestic use. Necessary information, such as toxicology, product shelf-life, and analytical methods are being acquired and some patents have been applied for. As yet, more biological tests, especially designed to demonstrate the long-term cost-effectiveness, are needed before such products can compete with established pesticides.

28.6 SUMMARY

American trypanosomiasis, Chaga's disease, remains one of the major public health problems in Brazil. There is at present little prospect of either immunization or a curative or prophylactic drug becoming available for large-scale use, and the principal method of combating continues to be the control of the vector-triatomine bugs.

28.7 ACKNOWLEDGEMENTS

This work was supported by the Brazilian National Research Council (CNPq), the Ministry of Planning (FINEP), and the Research Council of this University (CEPG/UFRJ). The author is indebted to Dr. B. Gilbert, A. M. de Oliveira Filho, M. J. Figueiredo, C. A. Muller, J. A. da Silva Netto and C. A. C. Ayala for their contributions to this project.

28.8 REFERENCES

Freitas C. A. De (1974/5) Alguns aspectos da epide miologia e profilaxia da doenca de Chagas no Brasil. *Rev. Bras. Malar. Doencas Trop*, **26/27**, 61–92.
Pinchin R., Oliveira Filho A. M. De, Figueiredo M. J., Muller C. A., Gilbert B., Szumlewicz A. P. & Benson W. W. (1978a) Screening and structure-activity relationships of synthetic

juvenile hormone analogues for *Pans trongylus megistus*, a primary vector of Chaga's disease in Brazil. *J. Econ. Ent.* **71**, 950–955.

Pinchin R., Oliveira Filho A. M. De, Figueiredo M. J., Muller C. A. & Gilbert B. (1978b) Slow-release juvenile hormone formulations for triatomine control. *Trans. Roy. Soc. Trop. Med. Hyg.* **72**, 322–323.

Pinchin R., Oliveira Filho A. M. De, Muller C. A., Figueiredo M. J., Perlowagora-Szumlewicz A. & Gilbert B. (1978c) Slow-release insecticides for triatomine control: activity and persistence. *Rev. Bras. Malar. Doencas Trop.* (in press).

Pinchin R., Oliveira Filho A. M. De, Ayala C. A. C. & Gilbert B. (1978d) Slow-release insecticides for triatomine control: preliminary field trials. *Rev. Bras. Malar. Doencas Trop.* (in press).

Rocha E., Silva E. O. Da, Dias Junior J. & Guarita O. F. (1969) Suspensao do rociado no combate ao *Triatoma infestans* em areas do Estado de Sao Paulo, Brasil. *Rev. Saude Pub.*, Sao Paulo **3**, 173–181.

Zeledon R., Valerio C. E. & Valerio J. E. (1973) The camouflage phenomenon in several species of Triatominae (Hemiptera, Reduviidae). *J. Med. Ent.* **10**, 209–211.

28.9   DISCUSSION

Onuorah: You mentioned the possibility of putting control agents in the paint used for decorating new houses. Is DDT effective in paint?

Pinchin: No. Not really since these bugs are not much affected by DDT anyway. Difficulties arise with finding suitable formulation. The most successful control agents we found were permethrin and BHC.

Johnson: Have there been any adverse effects on the environment or the population as a result of these studies?

Pinchin: To my knowledge only a few cats have died. No adverse effects on humans have been recorded.

Johnson: Are there any pheromones or attractants in these bugs?

Pinchin: Some little work has been carried out on this aspect and so far an aggregation pheromone is suspected in the fecal material. But there is no sign of any sex pheromones.

A large part of the problem is in the attitude of the people themselves. Basically they couldn't care less. The majority of them are of a very low educational standard. In fact, it is considered unlucky if you do not have the bugs in your house. So what do you do?

Bowers: I would suggest using a mixture of a juvenile hormone and a toxicant as a control method. Perhaps it might even be possible to design a compound which has both insecticidal and juvenile hormone activity.

Pinchin: Indeed that might be a useful approach.

Kahumbura: When did you actually carry out these studies? Was it during the dry season and what was the peak biting period?

Pinchin: These studies were carried out entirely in the dry season. Conditions during the wet season would make it next to impossible. During the dry season hatching is reduced and adults and later nymphs prevail. But during the wet season hatching increases dramatically and there is a population burst of the younger nymphs. As for the peak biting period, we don't know for sure. They seem to bite at almost any time of the day, but we suspect that it is greatest during the early evening.

CHAPTER 29

The Production of Pesticides in Developing Countries including Natural Products

THE SECRETARIAT OF UNIDO

CONTENTS

29.1	Introduction	486
29.2	Crop Losses	486
29.3	Prevention of Crop Damage	486
	29.3.1 Use of chemicals	487
29.4	Pesticide Production in the Third World	488
	29.4.1 Advantages and difficulties in pesticide establishment	489
	29.4.2 The role of agriculture	490
	29.4.3 Possible co-operation between countries	491
29.5	Natural Products	492
	29.5.1 Pyrethrum extract	492
	29.5.1.1 Properties of the extract	493
	29.5.1.2 Production practices	494
	29.5.1.3 Economic importance	495
	29.5.2 Nicotine	495
	29.5.3 Rotenone	496
	29.5.4 Ryanodine	496
	29.5.5 Sabadilla	497
	29.5.6 Other useful natural products	497
29.6	New Approaches to Pest Control	498
29.7	Formulation of Pesticides	499
	29.7.1 Procedures involved	500
29.8	Economic Benefits from Pesticides	502
29.9	Regulation of the Industry	504
29.10	Conclusions	505

29.1 INTRODUCTION

Protection of crops from pests is part of the total task to produce two ears of corn where only one grew before. According to Wortman, crop-production equation is made up of a number of multipliers (factors) such as fertilizer, plant variety, soil condition, water, climate, pest and weed control if any one of these inputs is deficient or missing, the yields start converging on zero. Thus, no water, no weed or pest control, the yield approaches zero. In other words, the successful agriculture is the product of several favourable conditions which may be provided by nature or carved out with hard work by man.

From time immemorial, man has been in a continual fight with pests (insects, plant diseases, rodents, weeds, etc.) that compete for his food supply, damage his forests and livestock. National frontiers are meaningless to these voracious enemies of mankind, demanding all nations' co-operation in self-defence.

29.2 CROP LOSSES

It is hard to estimate the global crop damage caused by pests. A plausible figure comes from Cramer's book entitled *Plant Protection and World Crop Production*, indicating that food losses due to insects and plant diseases are of the order of one-third of the total production. Losses caused by insects and rodents to crops during storage and to livestock are not included, nor those losses caused by competing weeds. Thus, the agricultural production foregoes some 75 billion dollars or approximately 35% in potential income due to losses caused by pests. Apart from economics, this is an enormously alarming figure in light of the demographic forecasts which indicate a doubling of the world's population in about 40 years.

There must be a concerted effort on the part of mankind to keep up with the increasing demand for food, both qualitatively and quantitatively, by improving the "physical input factors", e.g. land, climate, seeds, water, fertilizers, pesticides, animals, tools and fuel. From Wortman's Yield Equation, it is evident that the pesticide use has to grow at a rate commensurate with the growth of other inputs affecting agricultural production. When increasing the use of fertilizers, the extra nitrogen seems to make some crops more susceptible to numerous diseases and insect damage and makes the weeds grow more vigorously and compete more fiercely with the crops. Mechanization of agricultural production often requires such prerequisites as clean terrain, healthy crops, etc.

29.3 PREVENTION OF CROP DAMAGE

Pesticides are an important component of the physical inputs required for realizing the best crop yields and avoiding losses caused by various pests (Table 29.1). One could metaphorically also say that the function of pesticides

TABLE 29.1. AREAS AND NATIONS IN ORDER OF PESTICIDE USAGE
HA^{-1} AND IN ORDER OF YIELDS OF MAJOR CROPS

Area or nation	Pesticide use		Crop yields	
	g/ha	Rank	kg/ha	Rank
Japan	10,790	1	5480	1
Europe	1870	2	3430	2
United States	1490	3	2600	3
Latin America	220	4	1970	4
Oceania	198	5	1570	5
Africa	127	6	1210	6

in the complex mechanism of the effective and modern agricultural production is maintenance, and we all know that no complex equipment can properly function without adequate maintenance and care. Unfortunately, experience teaches that this consideration is often ignored. There is no sense in building dams and irrigation canals or using large quantities of fertilizers in order to increase agricultural production if the safety of the expected higher yields is not secured. This important interrelation of the physical input requirements is unfortunately often overlooked by the planners particularly in developing countries.

29.3.1 Use of chemicals

The worldwide usage of chemical pesticides has been rapidly increasing as plant production became more intensive. Tropical regions were in a rather unenviable position some years ago according to statistics in this respect, as shown by data taken from *The Production Yearbook of the Food and Agriculture Organization* (of 1963).

Today the world market of pesticides is growing faster than the US and other developed country markets. *Farm Chemicals* predicts US pesticide sales of 2.7 billion dollars in 1980 and 3.3 billion dollars in 1984, an increase of 19% during the 5-year period (*Farm Chemicals*, Sept. 1977). During the same time period, the publication predicts a 24% increase in world sales, from 8.1 billion dollars in 1980 to 10.0 billion dollars in 1984 (see Table 29.2). These rates of growth reflect the maturing of the pesticides market, especially the US and other developed countries where most acreage and livestock, which would benefit from pesticide application, have already come under treatment. They also reflect the fact that the base upon which the rate of growth is calculated has become considerably larger. For example, the projected two billion dollar increase from 1980 to 1984 in the world pesticide market would represent more than a 55% increase over the 1971 base of 3.6 billion dollars.

TABLE 29.2 US AND WORLD PESTICIDE MARKETS, 1971–1984[a] USER'S
LEVEL

	United States market		World market	
	$ billion	% change	$ billion	% change
1971	1.4	—	3.6	—
1974	2.1	49	6.3	77
1980	2.7[b]	28	8.1[b]	27
1984	3.3[b]	19	10.0[b]	24

[a]1977 price level.
[b]Estimated by *Farm Chemicals*.
Source: 1971 and 1974: "World pesticide markets", *Farm Chemicals*, Sept. 1975, p. 45; adjusted
        to 1977 price level by EPA.
        1980 and 1984: "A look at world pesticide markets", *Farm Chemicals*, Sept. 1977, p.38.

The fastest growing market is in Africa, with sales projected to increase by
182% between 1980 and 1984 (see Table 29.3). Other fast-growing markets
are Central and South America (32%), Asia and the Far East (28%) and the
Middle East (26%).

29.4   PESTICIDE PRODUCTION IN THE THIRD WORLD

One may ask why this lengthy introduction is required before dealing with
specific aspects of establishing national pesticide industries in developing
countries. It is because the importance of the use of pesticides in the "yield
equation" is still not generally and fully recognized. For example, the fourth
annual meeting of the Inter-American Economic and Social Council held in
Buenos Aires in March 1966 recommended a number of agricultural
measures, including the acceleration of agrarian reform, the establishment of a
regional system of agricultural insurance and the formulation of specific
projects for promoting the use of fertilizers on a regional basis in order to
accelerate the development of an efficient agricultural production. The need
for financial insurance was properly recognized by the Council; however, the
need for technical insurance in other words, the use of pesticides to preserve
the crops through harvesting, was surprisingly ignored.

It is sometimes argued that substantial differences exist between fertilizers
and pesticides in respect to availability of know-how, relative importance and
volumes, toxicity, cost and product turn-over. Even if some of these points
used for justifying the preferential promotion of fertilizer industries are valid,
it should be kept in mind that according to forecasts, the combined global
production value of pesticides will surpass the value of fertilizer production
within a couple of years. This trend is another strong indication of the relative
importance of the use of pesticides in modern agriculture.

TABLE 29.3. WORLD PESTICIDE MARKET GROWTH, 1974–1984 USER'S LEVEL

Area	$ million			Projected increase from 1980 to 1984 (%)
	1974	1980	1984	
Western Europe & British Isles	1301	1728	1946	13
Eastern Europe including USSR	527	815	921	13
Middle East (Egypt, Syria, Greece, Turkey, Israel, Lebanon, Sudan)	150	305	386	26
Africa	92	210	593	182
Asia (Iran, Pakistan, Afghanistan, India)	157	244	313	28
Far East including People's Republic of China	480	969	1245	28
Australasia	96	145	174	20
Central and South America	410	825	1092	32
North America	1977	2812	3291	17
	5138	8053	9961	24

Source: "A look at world pesticide markets", *Farm Chemicals*. Sept. 1977, p. 42.

29.4.1 Advantages and difficulties in pesticide establishment

After establishing the importance of using pesticides in agricultural production let us discuss and evaluate the advantages and difficulties in establishing pesticides industries in developing countries.

Industrialization is one of the chief objectives of every developing country. A typical difference between developed and developing countries is the relative position of agriculture and industry in their economies. As development goes forward, the relative importance of industry is gaining, frequently resulting in a tendency to devote insufficient resources to the agricultural sector. However, in a number of countries, experience has demonstrated that a lagging agriculture can jeopardize industrialization and the growth of the economy as a whole. Industries based on agricultural raw materials play an important role in the industrialization of the developed countries and also had a beneficial feedback effect on agricultural production itself. Such industries are estimated to account for about half of the total value added and nearly two-thirds of the employment in manufacturing industries in developing countries. Obviously, the industries serving agriculture (fertilizer, pesticides, agricultural machinery) make a substantial contribution to a country's agricultural and, consequently, industrial development.

Under the conditions of the developing countries, which are in effect trying to achieve an agricultural and industrial revolution simultaneously, the building of an "agro-oriented" industry seems to be the first and most important logical step. Such an industry concentrating on consumer goods and inputs needed by the agricultural population and production has a strong stimulating effect on the balanced development of agriculture and industry.

In attempting to set up a local pesticide industry, one should utilize all available techniques used in modern industrial planning to ascertain the economical feasibility and viability of the project under consideration. This approach utilizes a number of exploratory steps common for all industries, such as analysis of current and projected national and regional markets, availability of raw materials, technical know-how and capital investment, adequacy of the energy supply and network of transportation, estimate of investment-turnover rates, profitability and survey of human resources. The evaluation of all these aspects has to be generally interrelated to the current and expected world practice and progress in technology. All these are general rules more or less applicable to the pesticide industry. However, there are aspects in which establishing a pesticide industry may come under special and different consideration.

29.4.2   The role of agriculture

It was earlier mentioned that there is an interdependence between agriculture and industry, which is as a rule strongest in the economies of developing countries. It is a complex inter-relationship. Agriculture supplies food for the industrial labour force and raw materials for the industry. Besides this basic function, agricultural exports in developing countries must provide the bulk of the foreign exchange earnings for the importation of the capital goods required for industrialization unless the country is exceptionally rich in mineral resources. Agriculture lends labour and finance to industry. The agricultural population provides a market for industrial products, primarily for equipments and materials, such as pesticides and fertilizers, used in agricultural production. It is of vital interest for each developing country to bring its foreign exchange earning sectors of the agricultural production to a maximum efficiency. This can be achieved only by utilizing advanced techniques, including the use of pesticides and fertilizers. However, if these materials are to be imported using foreign exchange of limited availability at a given time, for which the capital goods required for industrialization are also competing, the inevitable result will be a curtailment of these vital supplies. From all this it is obvious that establishing a local pesticide manufacturing industry in those developing countries where agricultural productivity is exploitable for earning badly needed foreign exchange, for promoting industrialization, is highly recommendable.

29.4.3   Possible co-operation between countries

Once this conclusion has been reached in principle, the next question to be answered is, What kind of approach is realistic to reap the maximum benefit? Conditions permitting, the integrated production of all important pesticides would seem to be the most attractive. Integrated production, as used here, would include the manufacture of the most important intermediates, technical active materials, and formulated end-products. But conditions (market demand, raw materials, know-how, etc.) are not usually favourable in most developing countries to this approach. One way to overcome this inherent difficulty is to pool markets and resources in regional co-operation. Such co-operation between countries having common crop-protection problems seems desirable in order to unify the fragmented markets and facilitate bulk and profitable production of the most widely used pesticides that would pass for chemical commodities. This category of products including DDT, methyl parathion and 2,4-D to name only a few, are offered on the world market by foreign producers from industrialized countries at keenly competitive prices made possible by operating fully back integrated manufacturing plants of huge capacity. The erection of highly back integrated manufacturing complexes is, of course, of such a high capital investment requirement that few developing countries may be capable to afford it.

Balancing the disadvantage of operating plants of lower productivity in the category by applying tariff protection, price support and other regulatory measures may buttress new industries but only for a limited time and on a limited basis.

However, the problems and possibilities are somewhat different in the category of speciality pesticides:

(a) they are manufactured and marketed as proprietary products under the protection of an effective patent;

(b) high profit margins are incorporated in their selling prices.

The feasibility of locally manufacturing a speciality pesticide may be hampered or enhanced by these factors. If the originating company is holding a valid patent on the composition, manufacture or use of a product which is considered for local production in a developing country, the co-operation of the patent holding company has to be obtained. This co-operation may take the form of a licensing agreement covering the transfer of technology and know-how, a joint venture or a direct and exclusive foreign investment.

However, if the company which is in proprietary position refuses to co-operate, there is little left that could be done.

On the other hand, if there is no valid patent involved, the feasibility of local production is enhanced by the favourable price structure. This could mean that smaller than optimum scale operation still may prove profitable without much back integration and large capital investment, even if the country is lacking a strong raw-material base. Cases can be visualized in which the manufacturing would consist of a single finishing step of the synthesis based on imports of the required intermediates and subsequent formulation.

29.5   NATURAL PRODUCTS

Tropical developing countries have a unique potential in the field of botanical pesticide industries because all important plants providing such pesticides are indigenous in the tropical areas. The pyrethrum industry takes the foremost importance in this group both by production value and number of people to whom it provides a livelihood.

29.5.1   Pyrethrum extract

Pyrethrum is a naturally occurring insecticide obtained from the dried flowers of *Chrysanthemum cinerariaefolium* Vis. (*Pyrethrum cinerariaefolium* Trev.), a member of the family Compositae. The pyrethrum plant is a perennial, having long simple stems that bear large daisy-like flowers, usually with deep rose-coloured petals surrounding the yellow centre or disc. The dried flower heads contain insecticidal principles collectively called "pyrethrins", of which six have been isolated and chemically identified as Pyrethrin I, Pyrethrin II, Cinerin I, Cinerin II, Jasmolin I and Jasmolin II. Although there is a small pyrethrin content in the stem and leaves of the plants, the main concentration is in the flowers, particularly in the discs or achenes, and in commercial practice only the flowers are harvested.

Though the plant is a perennial, the yield of dried flowers ha^{-1} declines over time, thus necessitating periodic replanting. The yield, which may vary considerably depending on the strain of the species, the nature of the soil and the altitude, ranges from 550 to 1350 kg of dried flowers (10% w/w moisture) ha^{-1}. The production of pyrethrins has been found to depend very largely on the amount of chilling or cold weather which the plant receives. Consequently, the plant can be successfully grown in the tropics only at altitudes which provide a cool or cold period. The pyrethrin content of the dried flowers varies from 0.7% to about 1.6%.

Since the pyrethrin content of the flowers reaches a maximum at the time of full bloom (about 4 months after planting), the flowers are picked at this time by hand, dried and were previously packed for export to many countries. However, at present, the export of pyrethrum as dry whole flowers is rare. Instead, they are ground (there is still some use for powdered pyrethrum), the pyrethrum is extracted with an organic solvent such as kerosene, and the solution is dewaxed and concentrated. More than 80% of the total output of pyrethrum is now extracted in this way and comes into the market as "pyrethrum concentrate" which contains 20–25% pyrethrins.

As mentioned above, the highest yields of pyrethrins result when the flowers are hand-picked in their full bloom. Therefore this industry can hardly develop or thrive in an industrially developed country with high wages.

The original home of the pyrethrum flower is said to have been China, and the Middle and Near East. From these areas, in the nineteenth century, it spread into Europe, the United States, and then Japan, Africa and South America. At the beginning of the twentieth century, Yugoslavia and Japan became the principle producing countries; by 1941 Japan was the major producer, but after World War II pyrethrum output declined sharply; at present, Kenya stands first followed by Tanzania, Rwanda, Uganda, Congo, Ecuador and Papua New Guinea.

Precise production figures are not available, but the world's dry flower production has been estimated to vary between 15,000 to 25,000 tons per annum.

29.5.1.1 *Properties of the extract*

As insecticides, pyrethrins have several desirable properties. First, they are extremely toxic to insects and have a remarkably wide scope of effectiveness. They can be used against most common household pests such as mosquitoes, houseflies, cloth moths, bed bugs, ants, cockroachs, lice and silverfish. They can also be used with success against many pests which affect domestic and farm animals, and also many stored product insects such as grain beetles and the warehouse moth. They are also effective against a number of forest and agricultural pests, but their potential in this field could not be exploited because of their light sensitivity. Their toxicity is characterized by the unusually rapid paralytic effect — "knock-down" — on flying insects. Secondly, they have a low toxicity on mammals. In fact, pyrethrins have a remarkable record of safety towards humans, other mammals and plants. Thirdly, though minor isolated examples of immunity have been recorded, insect resistance to pyrethrins has not been a practical problem. Fourthly, they are a powerful insect repellent. Finally, pyrethrins are rapidly degraded by the combination of sunlight and air and therefore represent little of the environmental hazards which are usually associated with certain other classes of persistent insecticides.

NPIPM–Q

All these properties have enabled pyrethrins to compete successfully in the insecticide market. They are the active ingredients in many household insecticidal preparations, livestock and pet sprays, and are used as dusts or sprays to get rid of insect pests in food-processing plants, dairies and restaurants. Before DDT, pyrethrins were also sold for use against cabbage loopers, cabbage moths, diamond back moths, and some other agricultural pests. But their high cost (about \$130 per kg) and the fact that they break down rapidly in sunlight have hitherto limited their use in agriculture and forestry. Recently, however, it has been shown that very small quantities (about 100 g ha^{-1}) of pyrethrins are required for the control of some field pests. This coupled with improved formulation and application techniques, e.g. the ULV or ultra-low volume method, could possibly reduce the cost and extend the use of pyrethrins to the control of pests in agriculture and forestry.

At present, in spite of the progress made in the manufacture of synthetic analogues, the market for natural pyrethrins is believed to be strong. The safety of the product and its non-persistence make it a strong contender as an "environmentally safe" insecticide.

29.5.1.2 *Production practices*

Basically pyrethrum extraction involves dissolving out the active insecticidal ingredients, the pyrethrins, from the dried flowers with a solvent and then evaporating the solution to recover the solvent and to obtain a concentrated crude extract, which is then either packed in drums for export or refined into a pale extract which can be formulated into the final product (usually aerosol sprays).

For extraction, the flowers must first be ground. Hammers attached to a high-speed rotor, spin around and beat the flowers against the cylindrical casing of the mill where they break down into powder. This operation has to be carefully controlled because if the ground flowers become too coarse, the achenes containing the pyrethrins are not broken open, and if too fine, a dense mass is produced which the solvent cannot penetrate.

After being milled, the ground flowers, known as grist, are fed into the extraction vessels where extraction is effected in a counter-current operation. A somewhat oversimplified picture of the extraction process can be obtained by imagining a vertical extractor where grist is fed in at the top and passes slowly down the extractor while fresh solvent is pumped in at the bottom and rises to the top. As the solvent rises in counter-current flow to the grist, it first washes the exhausted material at the bottom, then continues upwards, dissolving what soluble matter remains. Finally, before leaving the top of the extractor, it comes in contact with and removes the first flush of extractives from the fresh grist. In this way, completely exhausted material, known as "marc", leaves the bottom of the machine and a solvent rich in pyrethrins called "miscella" leaves the top.

The solvent used (isohexane) is like petrol, volatile and flammable, so that great care has to be taken in its use. Moreover, since it is also expensive one has to recover it for re-use. At the end of the extraction the exhausted flowers, called the "marc", is first drained of the solvent. The steam is passed through it to evaporate any solvent remaining in it. The resultant mixture of steam and solvent vapour is condensed, the condensed water separated and discarded, and the liquid solvent recovered and returned to the process.

The last stage of the process is distillation. The isohexane "miscella", rich in pyrethrins, is pumped into stills where it is concentrated by applying moderate heat and high vacuum. When all the solvent has been driven off in this way, there remains a "concentrate" which normally contains between 30 and 35% pyrethrins, the remainder consisting of oleo-resins. The solvent driven off is condensed and re-used.

The production of dewaxed and decolorized pale extract is usually done by the manufacturing companies who buy the crude pyrethrum extract from the producing countries. However, the producing countries are now moving towards doing their own refining. In keeping with this trend, Kenya has already established their own pyrethrum-extract refinery while Rwanda is in the process of installing a refinery unit with the assistance of UNIDO.

29.5.1.3 *Economic importance*

The greatest economic impact of the pyrethrum industry is in the pyrethrum-growing areas themselves. Tens of thousands of farmer families are making a living through growing the pyrethrum flowers enabling them to break away from subsistence farming and to reap the benefits of growing a cash crop.

The second most important feature of this industry is that of hard-currency earning, since the large majority of the end product sells in industrially developed countries. In addition, as an environmentally safe insecticide, pyrethrum has an important role to play. Its use is expanding and the demand outstrips the supply. It is already a major crop in Kenya, Tanzania and Rwanda and to a lesser degree in New Guinea and Ecuador.

Other botanical insecticides that gained wide market acceptance at one time or another are nicotine, rotenone, ryania and sabadilla in that order of importance.

29.5.2   Nicotine

Nicotine, as alkaloid, is prepared from waste tobacco, *Nicotiana tabacum*, or from *N. rustica* either by steam distillation in the presence of alkali or by solvent extraction (trichloroethylene) in the presence of alkali and re-extraction from the solvent by dilute sulphuric acid. The alkaloid is a non-persistent, non-systemic contact insecticide and an effective fumigating agent with some ovicidal properties. Usually it is marketed as the 95% pure alkaloid or as nicotine sulphate (40% alkaloid). The addition of soap or alkali is required to aqueous dilutions of the latter to liberate the nicotine. It is also formulated as a 3 to 5% dust. For fumigations, nicotine is applied to a heated metal surface or nicotine "shreds" are burnt.

The great merit of this natural insecticide is that its production uses an agricultural by-product which has little or no other commercial value. While demand still exists for this uniquely effective insecticide in the industrially developed countries, its production was abandoned long ago and current supplies come from a few more advanced developing countries. Current global production is estimated to be less than 200 tons of 95% nicotine, and supplies are tight.

29.5.3 Rotenone

Rotenone is the trivial name of the insecticidal component of the roots of certain *Derris* and *Lonchocarpus* spp. The roots of *Derris selliptica* are known as derris-root, tuba-root or aker-tuba; the roots of *Lonchocarpus utilis*, *L. urucu* and *L. nicou* are known as barbasco or cube in Spanish-speaking countries of South America, as haiari in British Guiana, as nekoe in Dutch Guiana and as timbo in Brazil.

The insecticidal properties of derris root, also long used as a fish poison by the natives of China and East Africa were known for centuries, but its commercial exploitation started only in the second decade of this century, when its extract was patented in England as an insecticidal spray. It is a selective non-systemic insecticide with some acaricidal properties, environmentally friendly, because it is of low persistence and non-phytotoxicity.

Historically Zanzibar has produced significant quantities of derris root, however, current production is much reduced because of competition by synthetics. No recent statistics are available on the present status of the global derris production but data in Zanzibar going back to 1962 indicate that the average yield was 1290 lb acre^{-1} representing a production value of 2600 shillings or 1500 shillings net income after the deduction of production cost. It would seem that a natural product with the above properties and economy should be able to continue to have a role in providing adequate supplies of safe and non-residual insecticides.

29.5.4 Ryanodine

Ryania is the insecticidal component of the ground stemwood of *Ryania speciosa*, a shrub native to Trinidad and the Amazon basin. Ryanodine, the insecticidal component, is water soluble thus water extracts are easily obtainable. This product, introduced in 1945, is a selective stomach insecticide causing cessation of feeding and a slow death and has been effectively used against lepidopterous larvae, including the European cornborer, sugarcane borer and codling moth. For use, the powdered root, stem and leaves are mixed with an innert carrier, a formulation claimed to be more stable to UV light than extracts.

29.5.5   Sabadilla

Sabadilla, the extract of the seed of the plant *Schoenocaulon officinale* Gray, is a selective contact insecticide effective against domestic pests and houseflies. While relatively non-toxic to warm-blooded animals, it is an irritant to mucous membranes that curtails its use.

In addition to the above described most known botanical pesticides there is a string of natural products which hardly moved out, if at all, of the initial discovery stage. These products are the victims of the large-scale production and commercialization of synthetic products which are nologies suitable to large-scale operations. Perhaps it is worth while to mention a few of the products which encountered this fate:

29.5.6   Other useful natural products

The saponin of *Sapindus saponaria* (Mharita) is an excellent snail control agent (molluscicide) which could perhaps be used in the control of schistosomiasis, presently based on the control of the snail by synthetic chemicals.

The extract of quassia (*Quassia amara*) has been known to be a potent aphicide, but its use is very limited and virtually abandoned now.

Mammea americana, Mandulea sericea, Pachyrrhizus erosus are other plants which contain insecticidal products.

Altogether thousands of higher plant species have been examined for insecticidal properties and more than 2000 have been shown to have some activity. In many instances, this activity was in the range of commercial botanical compounds such as rotenone and often these plants were used as folk remedies against insects. For instance, *Nicandra physaloides*, also known as the "Peruvian ground cherry" or the "shootfly plant", was used as an insecticide in India which also displayed a marked insect-repellent activity. The plant distributed in a room repels flies, in a greenhouse it makes whitefly disappear or a few hundred plants around a stable, protect the animals from being bothered by flies. Although these properties have been known and use has been made thereof for decades, its active ingredient called "nicandrenon" was isolated and identified only lately. This is just one example in the field of natural pesticides where highly valuable products have completely escaped the attention of scientists.

Aside from higher plants, ferns, fungi and bacteria may have insecticidal components. It is perhaps enough to refer to the well known case of the *Bacillus thuringiensis*. This micro-organism which can be produced in a fermentation process contains spores with a crystalline endotoxine which is harmless to humans, animals and useful insects and safe for the environment. It has specific activity against some lepidopterous larvae with high gut pH and is being used on tobacco, cotton, soybeans, fruit trees and forested areas. The production of this natural insecticide would seem to particularly fit developing countries because of its valuable properties and relatively simple production technology.

The aforementioned natural products all possess insecticidal activities. However, there are others which are known to be toxic to other types of pests: for the sake of illustration a few are listed in Table 29.4.

TABLE 29.4. NATURAL PLANT PRODUCTS TOXIC TO DIFFERENT PESTS OTHER THAN INSECTS

Plant	Active ingredient	Activity
Red quill (*Urginea maritima*)	Scilliroside	Rat poisoning
Streptomyces, Penicillium and other related fungi	Antibiotics (Actidion, streptomycin griseofulvin, terramycin, etc.)	Antimicrobial
Cymbopogon (nardus grass)	Citronella oil	Insect repellent
Conifers	Pinosylvin and thujaplicins	Fungicidal
Garden pea	Pisatin	Fungicidal
Black walnut (*Juglans nigra*)	5-Hydroxy-1,4-naphtoquinone	Herbicidal
Geranium spp.	Extract	Virus inhibitor
Brassica spp.	Glucobrassicin	Plant growth regulator

29.6   NEW APPROACHES TO PEST CONTROL

Recently, much research work goes into the investigation of biological and chemical control agents of insect growth and behaviour. Juvenile hormones, regulating the development of insect larvae, and chemosterilants interfering with insect-reproduction processes are in the centre of the work on the insect growth-control agents. While these efforts may result in new techniques of

practical value for developing countries, at present, there are many questions
to be answered before they could be applied on a large scale.

Chemical control agents of insect behaviour, such as pheromones, are closer
to gaining practical application because environmentally they are benign and
application in form of baits require small quantities of products. New bait
techniques could lead to a substantial reduction of the insecticide requirements of developing countries thus the developments in this field are to be
followed closely.

Among the biological pest control agents *Bacillus thuringiensis*, discussed
under Section 29.5.6, is the most important product by the sheer fact of its
well-tested activity, broad application and environmental neutrality. The
discussion of pesticides would not be complete without mentioning a number
of new biological (non-chemical) agents under development, as shown on
Table 29.5.

TABLE 29.5. BIOLOGICAL AGENTS AND THEIR PROPOSED USAGE

Biological agent	Proposed use
Virus of *H. zea* (NPV)	*Heliothis* spp. (cotton)
Virus of tussock moth	Forests
Virus of gypsy moth	Hard-wood forests
Virus of *Autographa californica*	Lepidopteran larvae
Sawfly virus	Forest
Bacillus popillea	Japanese beetle, lawn
Bacillus sphaericus	Mosquitoes (larvae)
Nosema locusteae (protozoan)	Locust, rangeland
Colleotrichum gloeosporioides (mold)	Weed control (rice)
Hirsutella thompsonii (mold)	Citrus mites
Phytophtora citropthora (mold)	Citrus, milkweed vine

Some of these agents, once cleared for practical application, may become
attractive production targets for developing countries because of one or more
of the following reasons: available raw materials, ease of production
technology, low capital investment requirement and special interest in the
pest control of large domestic crops (e.g. rice, citrus, etc.).

29.7 FORMULATION OF PESTICIDES

Some countries, for one reason or another, may have to shrink away from
ambitious plans of manufacturing technical active materials. For them, the
road is still open to local formulation of pesticides.

29.7.1   Procedures involved

Formulation of pesticides is an essential manufacturing operation in which the basic pesticide (toxicant) is converted to a more or less diluted physical form in which it is more suitable for use in pest control. The operation includes physical admixture with one or more ingredients from the classes of solid and liquid diluents, emulsifying agents, wetting agents, dispersants, deactivators, etc. The net result is a formulated product which can be measured and applied accurately and has high effectiveness against pests it has been designed to control.

A large part of the pesticide formulations frequently consists of diluents, in extreme cases amounting to as much as 98% or 99% of the finished product. Most countries have local sources of part or all of the inert diluents, even though they may not manufacture basic pesticides or surfactants. The potential for saving in foreign exchange through using local carriers and diluents is evident. Additional savings are possible through use of local labour, reduction in transportation costs through strategically located plants and associated advantages.

Investment in formulating plants is not excessive. Further, a single suitably designed plant can formulate a wide variety of basic pesticides. Thus, there is little danger of obsolescence of the formulating plant as contrasted with such hazards in construction of a plant for a single basic pesticide.

The general objective of formulation is to assure the highest degree of efficiency with the lowest possible dosage of the toxicant and best economy. To achieve this complex goal, several types of formulation had to be designed to accommodate a large variety of requirements and conditions as determined by the physical properties of the toxicant and the nature of the pests. The best known types of formulation are shown in Table 29.6.

For best results, the type of formulations has to be selected with great circumspect, taking into account also such factors as available equipment for application, availability of water for diluting the concentrates and wettable powders, mammalian toxicity of the active material in reference to the available safety equipments, ecological and environmental conditions, etc.

Wettable powders, capsules and all the liquid formulations are further diluted with water before application. They must disperse spontaneously in water of all hardness and, with the aid of gentle agitation, remain uniformly dispersed throughout the spraying period. Dusts, granules, baits and pellets are usually used without further dilution or modification.

From Table 29.6 it can be seen that fillers, solvents and diluents are present in different types of formulations in a widely varying proportion, thus determining the magnitude of the potential local material contribution. This can be exploited to some degree to ensure maximum utilization of local resources, but one must remember that the formulation of pesticides is

TABLE 29.6. PERCENTAGE AND RANGE OF ACTIVE MATERIAL CON-
TENTS OF WET AND DRY FORMULATIONS

A. Dry formulations	Range of active material contents, %
1. Dusts	0.5–5 (10)
2. Wettable powders	25–80
3. Granules	1–20(40)
4. Pellets	1–10 (25)
5. Baits	0.1–10
6. Capsules	0–100
B. Liquid formulations	
1. Aqueous concentrates	10–50
2. Emulsifiable concentrates	5–80
3. Flowable concentrates	25–60

primarily activity and property-oriented technology in which interest is
centred in the chemical and physical compatibility of pesticides with
formulation ingredients and the environment. The physical and chemical
properties of the toxicant frequently define not only the most suitable
formulation and mode of application but also the limitation of the
concentration latitudes.

The most important dry carriers and diluents used in pesticide formulations
are inorganic materials, principally of natural origin. They include minerals
such as attapulgites, diatomite, vermiculites, clays, montmorillonites, talcs
and pyrophyllites. These are processed for use in pesticide formulations by
many techniques, ranging from simple drying and pulverizing to washing,
air-floating, calcining, etc. Their properties are imparted by the crystalline
and mineralogical structure as well as by the composition. The properties of
these carriers and diluents are often enhanced by the processing conditions.
The most important characteristics of the solid carriers are particle size,
sorptivity, bulk density, surface acidity and chemical compatability, flowabil-
ity, dustability, abrasiveness and water breakdown (disintegration) in the
case of granular preparations.

The carriers of emulsifiable concentrates are organic solvents. Because most
pesticidal chemicals are insoluble in water, it is also necessary to use solvents
in the preparation of liquid concentrates frequently used for the impregnation
of dry formulations. The different types of solvents which are used for pesticide
formulations may be classified by composition, chemical type, structure and
functionality. Hydrocarbons and petroleum distillates are economically the
most important solvents. They are further classified as aliphatic or aromatic
types, a distinction important both economically and functionally. Occasionally polar solvents may also be used. Ketones, esters, glycols and glycol ethers
should be mentioned in this group. Solvents are characterized by the following
functional properties: distillation range and boiling point, specific gravity

(density), Kauri-Butanol value (solvency), aromatics content, flash point, water miscibility, viscosity, toxicity, colour and odour.

From all said, it must be evident that in order to establish the extent to which locally available carriers and solvents can be used in pesticide formulations, a systematic and comprehensive survey and analysis is to be carried out. Difficulties of this kind should not discourage the initiation of local enterprises since technical assistance and consultations from international organization (UNIDO, FAO) are available.

29.8   ECONOMIC BENEFITS FROM PESTICIDES

The potential benefits of a local pesticide formulation may be very substantial for a developing country. The major general benefits can be as follows:

(a) reduction in cost of transportation by locating the formulation plant at a strategically well-selected site;
(b) saving substantial amounts of foreign exchange otherwise used both for the purchase of diluents and its transportation (cif cost);
(c) increase in national labour imput and purchasing power of the workers;
(d) potential reduction in cost to the farmer;
(e) minimizing the risk of an early destructive degradation (ageing) of the formulated products by reducing the shelf-life period;
(f) generating associated industries, such as the mining and processing of mineral fillers; and
(g) reducing dependence on foreign suppliers.

Apart from the general economic benefits, there are several specific objectives that can be attained by establishing local pesticide-formulation industries.

The assumption that foreign suppliers are in a good position to offer a larger variety and selection of pesticides against all common pests is most likely correct. However, it would seem irrational to expect these foreign suppliers to continually conduct special formulation and field research tests to find solutions to the numerous specific problems of an average size developing country unless the market increase deriving from such research promises to be conspicuously sizeable. Engagement in such research for companies with foreign bases would prove to be highly costly, too. It would require continual monitoring of the gradation of pests, ecological and environmental effects of the use of pesticides and quasi-servicing of the total local machinery of plant protection. In the case of a relatively small market, this cannot be a lure or challenge, but rather, an inconvenience and losing proposition for a foreign company.

Problems that may require special local attention may be associated with a unique climate, unusual pests, available means and equipments for the application of pesticides, local human attitudes, environment, etc. In any of these relations, significant differences may exist within a single country. There are tremendous differences between the climates of the "high lands" and lower tropical regions. As a consequence, the insect population and plant disease may widely vary from one region to the other, not only in kind, but also in behaviour and relative importance to the agricultural production. Differences also exist in the relative degree of development of the agricultural techniques and production. The less developed areas may be lacking equipment for the safe application of some toxic pesticide or specialized equipment for a particular formulation that may be available in the more advanced parts of the country. This also must be taken into account in the formulation and distribution of the pesticides in order to facilitate the transition from primitive production techniques to more advanced ones. The toxicity of pesticides may get a different connotation and importance in developing countries from that existing in industrially developed countries. Due to a deficiency and scarcity of safety equipment and protective clothing, acute toxicity may rise to an overriding importance in developing countries. While this is a lesser concern in developed countries, there the use of "hard" pesticides is watched apprehensively and a restrictive trend in the regulations against using them is gradually stiffening. Although the selective use of highly toxic pesticides may not be entirely avoidable, the danger of fatal accidents can be diminished by introducing dry formulations of lower toxicant concentration. The addition of a conspicuous colorant may also be useful in reducing the occurrence of accidental poisonings. Some regions may be so poorly provided with permanent water sources that spraying would impose on the farmers a real hardship. For such area, special dust formulations should be made available to allow the farmers to keep up with the progress and improvements of the modern agricultural production.

The required number of pesticide applications during the growing season is of great concern in every economy. In industrially developed countries the reason for concern is the shortage and cost of labour. The problem may be equally important in developing countries for other reasons such as inadequate capacity in equipment required in the application of pesticides, shortage in water sources, social attitudes, etc. One way to reduce the number of applications is to develop formulations based on the combination of two or more pesticides or pesticides and fertilizers that can be applied simultaneously. The correct choice of such combinations should take into account, besides the compatability of the components, the local geographic and climatic conditions, pest population and ecology. This is also a task which clearly cannot be expected to be solved by foreign supplies.

29.9 REGULATION OF THE INDUSTRY

It would be superficial and incorrect to assume that by establishing a local pesticide manufacturing or formulation industry these problems would be automatically solved and all the specific objectives met. The first and most indispensable step towards meeting the aforesaid objectives is the establishment of a regulatory, advisory and testing body which would define the specific objectives for the pesticide industries, conduct field experiments with the products developed and designed by the industry to satisfy the special local requirements and finally to make recommendations as to the judicious, safe and effective use of such products. Some developing countries already have some sort of organization to tackle this task. Most of them are inadequately staffed, endowed and organized. Others do not have such organizations at all.

The need for adaptive research is indeed basic for planning a successful programme of agricultural development in all developing countries. Given the opportunity, agricultural scientists and technologists can apply existing principles and concepts to find answers to problems of expanding agricultural productivity. This ability to find answers through basic and adaptive research and through technological innovations within a country is distinctly different from already knowing the answer. The adaptation of existing methods and techniques to indigeneous needs is not the same as transplanting unmodified foreign techniques. There is a wrong and persistent public impression that agricultural science already has the answers to the problems of increasing food production in developing countries, giving rise to the "know-how, show-how" fallacy, implying that successful foreign practices can be applied with equal effectiveness in far different climes and cultures.

The importance of organizing these institutions cannot be over-emphasized, even if building a local pesticide manufacturing or formulation industry is not imminently planned. These institutions and regulatory bodies should have an equally important role and responsibility in testing the products of foreign suppliers, setting minimum standard requirements in terms of activity, purity, etc., enforcing the observance of these standards, making recommendations concerning the registration of pesticides, determining the schedule of the recommended applications and advising the farmers thereon, and checking pesticide residues in harvested products destined for foreign export and domestic consumption. Eventually, the services provided by this organization may become invaluable in the surveillance, monitoring and checking of the environmental pollution attributable to the manufacture and use of some pesticides.

Whenever the question of establishing local pesticide industries arises, most of the industrial production experts are quick to point out that, as a rule, the cost of production of pesticides is higher in developing countries than in an industrialized country. Most frequently the following reasons are quoted:

1. lack of raw materials;
2. modest market demand, justifying low production capacities only;
3. lack of back-integration of the production due to the aforecited reasons;
4. overhead cost rendered higher by the necessity to employ two or three times more personnel for accomplishing the same job, because of great local distances and poor communication systems; and
5. high transportation cost because of the remoteness of some agricultural areas from transportation centres.

First, one may note that all these reasonings would also apply to any other industry. Secondly, it is obvious that the cited adverse conditions would apply to a highly varying degree as one compares the complexity of the production of a commodity-type bulk pesticide to simple technologies required for the production of some speciality pesticides and formulation processes. Thirdly, it should be pointed out that the acceptable rate of return from any economic development programme is somewhat arbitrary and is partly based upon a subjective judgement which includes the balancing of several factors: the desire to maximize profits, the desire to maximize social welfare, the existence of alternate opportunities of investment, etc. For instance, a public enterprise might be willing to accept a lower rate of return for compelling social benefits, in this instance, the encouragement of the growth of the agricultural and industrial sectors.

To improve the investment-return balance, regional co-operation can be established between countries having common crop protection interests and problems. This should lead to broader markets and to close co-operation and co-ordination in sharing responsibilities in the overall economic planning and conservation of the environment in respect to the manufacture and use of pesticide.

29.10 CONCLUSIONS

The aspirations and immediate goals of industrial development in developing countries are, to a large degree, determined or influenced by socio-economic considerations and interests. Even in technical terms, the aspirations of a developing nation in a specific area may be vastly different from those of an industrialized nation. For instance, the modern concept of building huge manufacturing plants to optimize the economy of production is generally valid. Still, a developing country could be more attracted to the idea of giving preference to the construction of specially designed multipurpose manufacturing plants that would satisfy a modest market demand in a series of products, rather than building a large capacity single-product plant. Unfortunately, there are no blueprints or designs available for such multipurpose plants. they would have to be individually worked out according to the technology requirements of all product components. Manufacturing industries of industrially developed countries are usually unprepared for designing such multipurpose plants because of their different type of orientation. For these kinds of tasks the assistance of specialized agencies and engineering firms would be required.

UNIDO can provide assistance to the governments of developing countries in the field of pesiticide production, if requested, in such areas as analysis of accruable benefits, by installing formulation facilities, pre-feasibility and feasibility studies of pesticide manufacturing projects, establishment and start-up of pesticide production or formulation plants (including the transfer of the required production technologies), evaluation of technical studies and proposals prepared by technical agencies and consulting firms, promotion of foreign capital investments, designing of multipurpose pesticide manufacturing plants and organization of in-plant training programmes.

CHAPTER 30

Industrial Production of Pesticides: A Note on the Cost/Benefit Problem in Developing Countries

Y. F. O. MASAKHALIA

Ministry of Economic Planning and Community Affairs, Nairobi, Kenya

CONTENTS

30.1	Introduction	507
30.2	The Role of Pesticides in Industrial Development	507
	30.2.1 Industrial raw materials and agricultural exports	507
	30.2.2 Food production	508
30.3	Problems of Pesticide Production in Developing Countries	509
	30.3.1 Use of non-conventional raw materials	510
30.4	The Cost of Using Pesticides	511
30.5	Conclusion	511
30.6	Summary	512
30.7	References	513
30.8	Discussion	513

30.1 INTRODUCTION

The international community has set a target of 25% of world industrial output as the share of developing countries by A.D. 2000 as compared to 7.9% in 1973. The attainment of this target entails the implementation of a comprehensive economic development strategy in which intensive use of national resources and the development of local enterprise for processing of domestic raw materials must be important elements.

30.2 THE ROLE OF PESTICIDES IN INDUSTRIAL DEVELOPMENT

30.2.1 Industrial raw materials and agricultural exports

One of the important groups of industries likely to contribute substantially to the Lima target of developing countries' industrial output are those based on local resources in these countries. Local resources of mineral, animal and agricultural origin are the major export earners of the countries. If they are locally processed, their local value added will enhance the contribution of industries to their national incomes and also increase their export earnings. An increasing domestic supply of agro- and animal-based raw materials such as cotton, jute, sisal, tobacco, oil seeds, rubber, sugarcane, fruits, vegetables and hides and skins is therefore a prerequisite for the envisaged growth in the resource based industries. Similarly an increasing supply of agro-based export commodities such as coffee, tea, cocoa beans, etc., is necessary for Third World countries to earn adequate foreign exchange to import the necessary capital equipment and intermediates including oil to facilitate the rate of industrialization aimed at. In view of the enormous proportion of agricultural raw materials and agro-based export items that are being destroyed both in the fields and storage by pests, insects, rodents and weeds, the need for the production and application of appropriate pesticide to protect agro- and animal-based raw materials and export items becomes apparent in achieving the industrial targets of the Third World countries.

30.2.2   Food Production

Furthermore, the production and judicious use of pesticides are necessary to save the large quantities of food grains that are currently destroyed. These losses result in food shortages which often necessitate the use of scarce foreign exchange for food importation by developing countries which can hardly afford parting with these valuable resources in this manner.

It is estimated that the current world population of 4000 million is increasing at the rate of 70 to 75 million per year. According to the UN population projections the total population of the world is expected to reach 6400 million by the year 2000. Whereas it took two million years for the world population to reach the 1000 million level, recent developmental achievements have brought about a situation where it has taken only 15 years, between 1960 and 1975, to add an additional 1000 million people on to the world's population. By the year 2000, it is estimated that 80% of the population or 5000 million people will be living in the Third World countries. This highlights the basic problem of food supply to the world's poor areas. These regions of the world are already undernourished. To add to this, it is estimated that by 1985 the low-income food deficit countries with a per capita income of less than US $200 p.a. will face a deficit of 48 million tons of cereals per year. As the world demand for food increases, the constraints on efforts to further expand food production becomes conspicuous. It is therefore vitally important that a world-wide effort be made to increase the output of food and protect produced food from destruction on the field and in storage.

A large variety of pest species share with man his food. These pests include viruses, bacteria, fungi, insects, birds and rodents. Weeds reduce food production through competition with plants for nutrients such as water, sunlight and fertilizers. Crop losses from insects, diseases and weeds range from 46% for rice to 24% for wheat. The aggregate is estimated at 35% of potential production of all crops. For the whole world, losses from insect pests, diseases and weeds of food grains alone are estimated to be US $70-90 billion per year (Cramar, 1967). It can, therefore, safely be concluded that the direct benefits of the production and use of pesticides in the developing countries can be enormous.

In the specific case of Africa, "it is estimated that almost 42% of the potential value of crops is lost because of plant diseases and pests prior to harvest. Losses from diseases are estimated at nearly 13% while the loss from weeds is estimated roughly at 16%. The total value of lost output in recent years in Africa have been estimated at $8,000 million" (Cramar, 1967) per annum (at 1965 prices) and severe losses also occur in storage.

30.3   PROBLEMS OF PESTICIDE PRODUCTION IN DEVELOPING COUNTRIES

Most of the pesticides currently in use were developed by transnational chemical companies in the industrialized countries of the West. The basic research and development for the production of new pesticides involves a large number of scientists working for a number of years requiring substantial expenditure in research and development. The synthetic chemists who produce new materials, "screen" their material to test whether they have the desired biological and chemical activity. The chances of a newly synthesized compound being marketable may be one in ten thousand. Most multinational research firms regard research and development as a necessity for maintaining their position in the market as well as to provide new processes and products. In view of the huge expenditures involved in research and development they charge high prices for their products.

The developing countries on the other hand are constrained in respect of technical manpower, financial resources as well as the size of their own markets to undertake research and development. In Kenya, for example, pesticides are imported in all forms from active chemicals to formulated materials. Most multinational pesticides manufacturers are represented in the Kenyan market either through their own branches or through agents. In 1977 Kenya imported 8400 tonnes of various types of pesticides, fungicides and herbicides valued at $24.4 millions., The small size of the Kenya market for individual pesticides is shown by the technical materials needed in organo-phosphorous insecticides imported annually in the country (Table 30.1)

The quantities are obviously low and have, therefore, not provided a favourable basis for establishment of a local production capacity. We are now examining the possibility of obtaining technology for production of some of these items from a sister developing country which has developed an appropriate technology for production of these items.

Table 30.1. ANNUAL INSECTICIDE IMPORTS IN KENYA

Insecticide	Tonnage
Fenitrothion	150
Diazinon	60
Malathion	40
Dimethoate	25
Fenthion	30

Dependence on chemical materials for pesticides made abroad creates another problem for the developing countries. The price charged by the chemical firms are what the market can bear. Furthermore, when the branches of these chemical firms are represented in the developing countries they either directly import pesticides for distribution or import technical material for local formulation of pesticides for distribution. It is difficult for the local firms to compete with the transnational corporations, which through the system of transfer pricing and optimizing profits on a global basis have advantages over purely local firms.

30.3.1 Use of non-conventional raw materials

The basic chemical raw materials required for the manufacture of pesticides are expensive and are becoming more so with the rapidly increasing prices of petroleum-based chemicals. With the available technology it is possible for the developing countries to modify the processes and use less expensive and more readily available raw materials. In Kenya, for example, we have possibilities to utilize molasses-based alcohol as well as chlorine which is a by-product in the manufacture of caustic soda. It is important to note that with the ever-increasing prices of petroleum-based chemicals, the developing countries should be able to utilize agro-based alcohol and its derivatives as substitutes for petroleum products.

Therefore, in any attempt to produce pesticides in developing countries, the size of their markets, the availability of technologies suitable for small-scale production and substitution of expensive imported raw materials with those locally available are important aspects to be taken into account. When markets are smaller than what the smallest pesticides unit can manufacture it is advisable to undertake a co-operative venture between contiguous developing countries on a multinational basis.

30.4 THE COST OF USING PESTICIDES

In the foregoing paragraphs the broad magnitudes of the potential benefits of the use of pesticides on food grains, agricultural and bovine raw materials for industries and agro-based exports have been indicated. But the use of pesticides is not without its costs. As research, development and manufacture of pesticides are undertaken mostly in the industrialized countries, the biological and chemical tests of these pesticides is done in the eco-systems of those countries. When developing countries import these pesticides they are applied to completely different eco-systems whose reactions to pesticides has not been studied. Very often the multinationals dealing in pesticides undertake aggressive sale promotion activities in the developing countries to sell pesticides which are prohibited in the industrialized countries as health hazards.

It is now becoming obvious that some of the well-established and popular methods of pest control are becoming a threat to the human and animal species. The indiscriminate use of DDT and other chlorinated hydrocarbons as pesticides and herbicides is beginning to threaten many species of animal life and even man. It is reported that traces of DDT have been found in the penguin of distant Antarctica as well as in children in the villages of Thailand. At one time it was reported that the concentration of DDT in mothers' milk in the US exceeded the tolerance levels established for foodstuffs by the Food and Drugs Administration of the US.

To make matters even more serious intensive research is not generally undertaken in the developing countries at national levels in the testing of pesticides. Whenever a pesticide is sprayed, it is being applied to a system which is very imperfectly understood. It is therefore imperative to develop and produce pesticides specific to the pests against which it is desired to be applied to maintain the environmental balance.

30.5 CONCLUSION

In conclusion it may be stressed that the manufacture and use of pesticides is important to the developing countries to sustain the supply of good-quality agro-based and bovine industrial raw materials for local processing, to ensure the conservation of food production in order to feed the increasing population of these countries and to minimize wastes of agro-based export commodities. Pesticides, therefore, have an important role to facilitate the developing countries in achieving the aim target of contributing 25% of world industrial production by A.D. 2000. But in view of the concentration of manufacturing activities of pesticides in the industrialized countries, it is necessary that the developing countries adapt manufacturing processes and sizes of operation to suit their own requirements, that they develop and modify processes to use more readily available local resources, particularly agro-based raw materials and that they undertake either individually or in co-operation with others, testing of pesticides in their own eco-systems to maintain the environmental balance.

30.6   SUMMARY

The current world population of 4000 million is increasing at the rate of 70–75 million additional people per year, and is expected to reach 6400 million by 2000. About 80% of this population will be in Third World countries raising the basic problem of food production to feed it. Food production is determined by the use of scientific cultivation methods which includes the application of fertilizers, pesticides, herbicides, etc. This paper is confined to the discussion of costs and benefits of the manufacture of pesticides and their applications in the developing countries.

A large complex of pest species share with man his food. Crop losses from insects, diseases and weeds are enormous and are estimated to be around 35% of potential production of all crops. An estimate by UNIDO places the total loss from insects, weeds and diseases between $70 to 90 billion per year the world over. The total value of lost output due to pests and diseases in Africa is estimated at US $8000 million annually.

But the methods of pest control currently in use are becoming a health hazard and environmental problem. In some cases they are becoming a threat to the human and animal species. Some pesticides have created new pests and have destroyed millions of organisms valuable to man. Unfortunately, the extent of research undertaken in the production and testing of pesticides has not been adequate as in the case of human drugs. Therefore, very few pesticides are specific to the pests against which they are applied.

Indiscriminate use of pesticides is likely to cause serious economic and environmental damage unless care is taken to reduce the application of broad spectrum pesticides specific to the pests against which the pesticides are desired to be applied. From the commercial point of view, however, it is not financially attractive to develop and market specific pesticides or to study new methods of integrated pest control. Very often the multinationals dealing in pesticides undertake agressive sales promotion activities in developing countries to sell pesticides which are prohibited in the industrialized countries. Therefore, it is advisable for governments in the developing countries to sponsor research institutions to develop and evaluate pesticides. It is also necessary to develop alternative techniques of pest control in the place of conventional chemical pesticides. In view of the damage caused by the broad spectrum pesticides it is advisable for the developing countries to:

(a) develop non-persistent chemicals;
(b) research into the use of plants resistant to insect attack;
(c) introduce insect enemies such as predators and parasites;
(d) use bacterial toxins;
(f) release sterile male insect pests to progressively reduce their population.

30.7   REFERENCE

Cramar H. H. (1967) *Plant Protection and World Crop Production*, UNIDO Publication.

30.8   DISCUSSION

Gubler: You said that Kenya imports pesticides worth $25 million annually. Can you tell us the corresponding annual export value of Kenya's agricultural products?

Monteiro: I do not know the annual export value of Kenya's agricultural products. However, against the pesticide import bill of $25 million, revenue earnings in 1978 from coffee and tea, Kenya's principal agricultural crops amounted to about $508 million. Further, pyrethrum, sisal, cotton, spices, horticultural crops, cashewnuts, maize and other agricultural products earned Kenya substantial revenues during the same year. The export sale value of Kenya's agricultural products is therefore quite in excess of the cost of importing pesticides.

Szabo: I would like to comment that the countries whose economies rely heavily on agriculture and therefore have to import fertilizers, farm machinery and other necessary agricultural inputs, should seriously think of setting up plants to manufacture these items locally to avoid the obviously dangerous and expensive dependence on foreign sources of these important items.

Monteiro: That is a very helpful suggestion. In fact the Kenya Government is currently negotiating the erection of such plant in Kenya with a group of experts from India.

CHAPTER 31

Development of Appropriate Technology (Equipment) for Utilization of Pesticides in the Tropical Environment

RAJNI P. PATEL

Faculty of Engineering, University of Nairobi, Nairobi, Kenya

CONTENTS

31.1	Introduction	515
31.2	Application of Pesticides	516
	31.2.1 Static devices	516
	31.2.2 Dynamic devices	517
	31.2.2.1 Experimental results	517
31.3	Conclusions	519
31.4	Summary	521

31.1 INTRODUCTION

Generally pesticides are applied in atomized form. This is because of the fact that the pests to be controlled are of microscopic nature. Of course for control of pests on animals, e.g. ticks, etc., it is not necessary to apply pesticides in atomized form. Indeed the precise nature of pesticide application will depend on the pests to be controlled. It is obvious, therefore, that there is a need for close collaboration between technologists and scientists trying to control pests. The scientist provides basic specifications for the application of pesticide and the technologist then develops an appropriate equipment.

From the designer's point of view the specification should include the following information:

(a) whether the pesticide is to be applied in solid, or liquid or gaseous form
(b) whether the pesticide will corrode the common metals and materials used in fabricating the equipment;

515

(c) specific harmful effects of the pesticide to human and domestic animals; and in general environmental effects; and

(d) any special precaution to be taken.

31.2 APPLICATION OF PESTICIDES

When a pesticide is used in liquid form its application can be achieved in a variety of ways. As, for example, for a small application in a house one uses a simple hand pump or a pressurized container. Basically the devices used to apply liquid pesticides can be classified in two categories, namely, static and dynamic devices. In the former the atomization of liquid takes place due to the static pressure whereas in the latter it is due to the dynamic forces acting on the liquid.

A very common device operating on static pressure is a nozzle. The use of nozzles are very common in both ground and aerial applications of pesticides. Note that in all the static devices the basic elements required are: (a) a container, (b) a pump and (c) nozzles.

For dynamic devices a method has to be found to impart dynamic forces on liquid pesticides. An example of such a device is a rotating disc which is fed at the centre with a liquid pesticide. This is sometimes called a spinning disc atomizer. The basic elements involved here are: (a) a container, (b) a motor and (c) a spinning disc. The liquid flows in a thin layer over the spinning-disc surface and eventually is broken up into droplets at the edge. Indeed the spinning-disc technique in applying pesticides has not been widely used because of its limitation. However, there is an equipment which is used in aerial spraying and it is claimed to be operating on the spinning-disc principle.

This paper gives some experimental results on sprayer for aerial application and its performance is compared with an equipment currently used.

31.2.1 Static devices

As mentioned before, the static devices incorporate some means of pressurizing liquid pesticide and forcing the pesticide through a nozzle. Figure 31.1 shows some of the equipment commonly used for applying pesticides. Note that the important element in each of these is the nozzle. In general a simple nozzle will produce a conical spray of atomized liquid pesticide. However, it is possible to produce a plane spray or rotating conical spray field by appropriate modifications on the nozzles. It should be mentioned that although the nozzle is the simplest equipment for the application of pesticides the whole system may become quite complicated. This is clearly demonstrated in Fig 31.1. As a matter of fact some of these devices are being fabricated in

Kenya. Although it is not generally recognized by the designers that durability, saving of materials and energy input, easy repair and use of recyclable materials are important factors to be considered in developing this equipment.

31.2.2   Dynamic devices

For application of pesticides on a small scale, a spinning-disc device has been developed. However, for large-scale applications, as, for example, spraying large plantations by an aircraft, a device shown in Fig. 31.2 can be used. Basically this equipment operates as follows: Liquid pesticide is fed into the cylinder either under gravity or using a pump. The perforated cylinder is kept stationary and it is enclosed by a concentric screen cage which is rotated by a windmill. Thus the power required to operate this device is extracted from the windmill. The pesticide is ejected radially from the fixed cylinder and it impinges on the rotating-screen cage. At the cage the liquid breaks up into small droplets. Indeed the droplet size thus produced will be dependent, apart from other parameters, on the physical characteristics of the screen and the dynamic forces exerted by the cage. Furthermore, there will be negligible amount of liquid flow along the surface of the cage. The flow phenomena in this equipment is quite different compared to that of a spinning disc.

In Fig. 31.2 is also shown a new device developed by the author. The equipment was meant to be used in place of the above device. It should be mentioned that it was necessary to understand the flow phenomena in the previous device before the new equipment was developed. The equipment consists of a rotating hub carrying a number of blades. The hub is mounted on a hollow stationary shaft through which the liquid can be fed either under gravity or by a pump. To reduce viscous drag on the rotating hub a specially designed chamber is fixed at the end of the shaft. The liquid then passes through the roots of hollow blades and is ejected near the tips.

A quick glance at Fig. 31.2 indicates that the new equipment is quite small. As a matter of fact it is about 16 times lighter. The saving in dead weight itself is of considerable importance when the equipment is supposed to be carried by an aircraft. Apart from this, performance tests were also carried out on these devices.

31.2.2.1 Experimental results

The performance of the two dynamic devices was based on the measurement of droplets produced by them. Also from dimensional analysis it was concluded that:

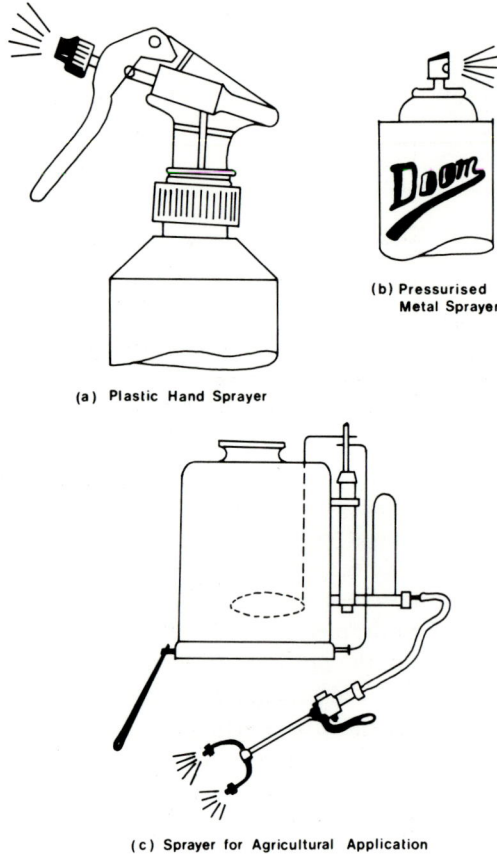

(b) Pressurised
Metal Sprayer

(a) Plastic Hand Sprayer

(c) Sprayer for Agricultural Application

FIG. 31.1. Static devices for applying pesticides: (a) plastic hand sprayer, (b) pressurized metal sprayer, (c) sprayer for agricultural application.

$$\left(\frac{d}{D} = f \frac{m}{\mu \omega^2 D^2} \, , \, \alpha \right)$$

where d is the mean diameter of droplets,

D is the diameter of blade tip or the diameter of the cage,

m is mass flow rate,

μ is the viscosity of liquid,

ω is the angular speed

and α is the blade setting (pitch angle).

Hence the results are presented in Fig. 31.3 using the above non-dimensional parameters.

From Fig. 31.3 it is clear that there is a functional relationship between

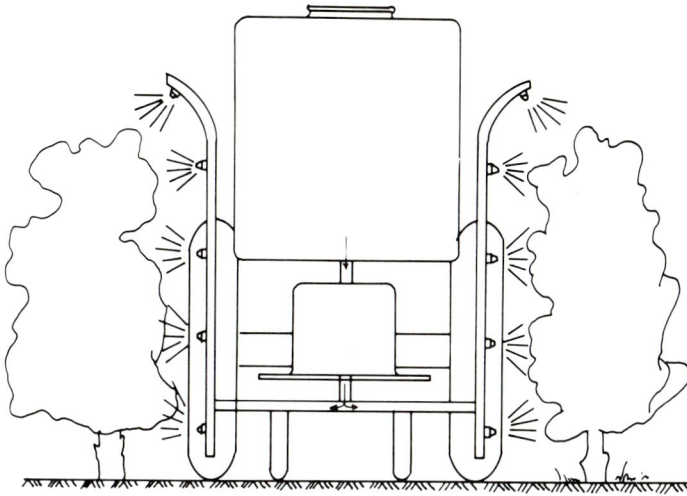

(d) Tractor Driven Sprayer

(e) Cattle Sprayer

FIG. 31.1 (d) tractor-driven sprayer and (e) cattle sprayer.

various parameters as shown above. Furthermore, the performance of the new equipment is better since it produces smaller drop sizes compared to those produced by the other device.

31.3 CONCLUSIONS

In developing equipment for pesticide application it is necessary to lay down clear specifications. These should be based on collaborative efforts of all the people involved with the problems of pest control. Indeed one must then look at the availability of various components required to produce a desired equipment or system rather than start reinventing the components. Often in complicated devices there is a need to understand physical phenomena

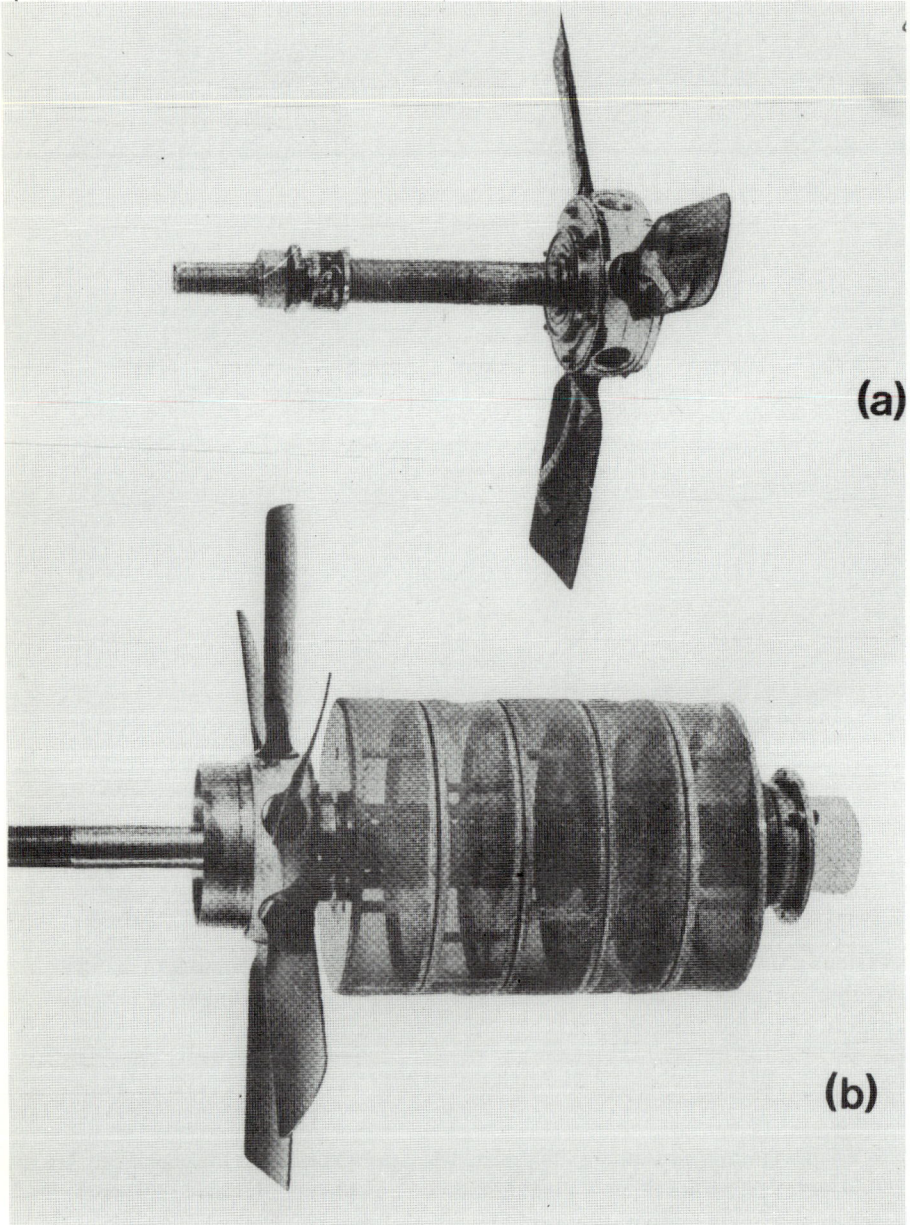

Fig. 31.2. Atomizers: (a) the new atomizer and (b) the micronair atomizer.

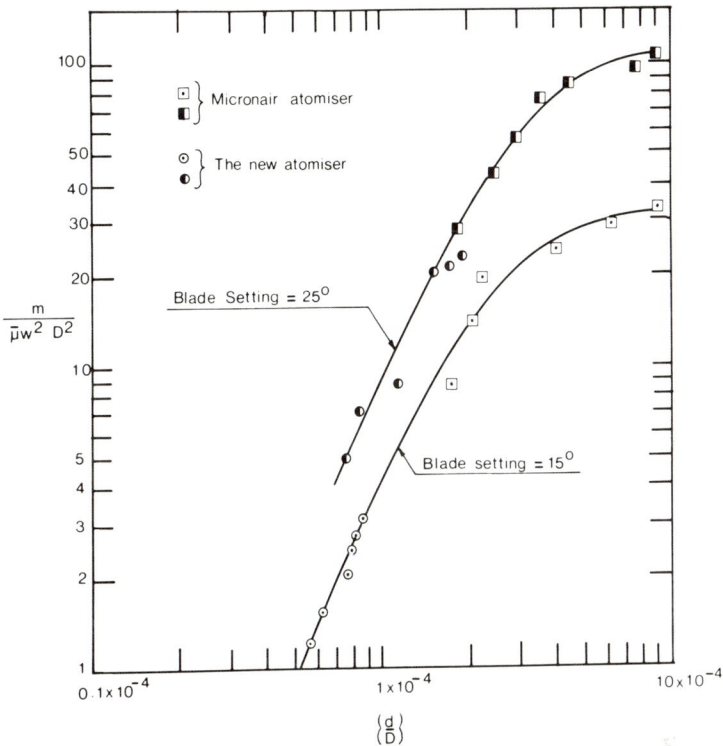

COMPARISON OF THE NEW AND THE MICRONAIR ATOMISERS

FIG. 31.3. Comparison of the new atomizer and the micronair atomizer.

involved. An example of a device for which the physical phenomena was not understood clearly is given here. Using available components a new equipment was designed and it is shown that its performance is better than the existing device.

31.4 SUMMARY

This paper describes the basic requirements for developing equipment for pesticides applications. Equipments which use pesticides in liquid form are described under two categories, namely static devices and dynamic devices. Some experimental results of an atomizer developed by the author for aerial application of pesticides are also presented.

CHAPTER 32

Registration Requirements for Pesticides

JAMES W. YOUNG

Zoecon Corporation, 975 California Avenue, Palo Alto, California 94304, USA

CONTENTS

32.1	Introduction	523
32.2	Regulations on Pesticides	524
	32.2.1 Registration	524
	32.2.2 Tolerance	524
32.3	Data Requirements for Registration	525
	32.3.1 Product chemistry	525
	32.3.2 Environmental chemistry	525
	32.3.2.1 Hazard evaluation — wildlife and aquatic organisms	527
	32.3.2.2 Metabolism	527
	32.3.2.3 Hazard evaluation — humans and domestic animals	528
	32.3.3 Efficacy testing	530
	32.3.4 Label development	530
32.4	Tolerances	531
	32.4.1 Elements of residue testing	533
	32.4.2 Example of evaluation of a proposed tolerance	534
32.5	International Consideration of Pesticides Registration and Tolerances	535
32.6	Conclusion	536
32.7	Summary	537
32.8	Acknowledgement	538
32.9	Discussion	538

32.1 INTRODUCTION

It is appropriate that the regulatory aspects of pesticide development be recognized as critical in any protocol for moving a new pest-control discovery from the research stage to commercialization. In this discussion of government regulation of the use of pesticide chemicals, I will be referring almost exclusively to that process as it applies to the United States. The procedures there can be taken as representative of the registration process in many of the industrialized nations. The premarket testing requirements imposed by most nations reflect an increased worldwide awareness of the need to document the effects of new pesticide chemicals on non-target organisms prior to their widespread commercial introduction. This attitude may for all practical purposes be considered universal.

32.2 REGULATIONS ON PESTICIDES

Two regulatory activities are of primary concern in the development and introduction of a new pesticide chemical.

32.2.1 Registration

A pesticide registration may be defined as the procedure of obtaining approval to market a pesticide product containing one or more active ingredients in a specific formulation.

In most nations it is illegal to sell a pesticide product without first securing government approval. In the United States the Environmental Protection Agency (EPA) is required to review data on product safety and effectiveness prior to granting a registration.

32.2.2 Tolerance

A pesticide tolerance is the maximum residue of a pesticide chemical which is allowed to remain in or on a raw agricultural commodity or a processed food product. If residues exceed this tolerance level, the crop may not be used for consumption by man or by food-producing animals.

For purposes of this discussion, we will assume that we are faced with the task of anticipating the regulatory requirements which will be imposed on a new pesticide product whose practical application includes both food and non-food uses. The product may be used, for example, for control of an insect disease vector and also for protection of one or more food crops. Since use on food crops is involved, planning must account for both registration and the establishment of tolerances.

In the United States pesticide registration and the establishment of a tolerance are separate administrative procedures. Both involve scientific review of safety data, but they have fundamentally different purposes. The mechanism of granting registration involves a detailed review of the proposed label for the product and determination that the proposed use will not pose unreasonable risks to the pesticide user or to the environment.

The function of a pesticide tolerance, on the other hand, is that of protecting the public health. It is considered that exposure of the public to pesticide residues in foods is an involuntary exposure of persons not directly involved in the decision to use the pesticide in the first place. The mechanism for establishment of a tolerance involves consideration of chronic toxicity data (to simulate long-term exposure to residues in the food) and data quantifying and chemically identifying the residues remaining in or on food commodities. After consideration of these data, an acceptable public exposure level is determined and an analytical method suitable for enforcement purposes is made available.

32.3   DATA REQUIREMENTS FOR REGISTRATION

Evaluation of an application for registration of a new pesticide product, whether for food or non-food use, involves consideration of the following several categories of data: Product Chemistry, Environmental Chemistry, Metabolism, Mammalian Toxicology, Fish and Wildlife Toxicology and Efficacy. All of these data are then considered in the context of the proposed label. The label should be written so as to give not only clear instructions for use of the product, but to state clearly any potential hazard to non-target organisms, including the pesticide user. We will consider each of these requirements in some detail.

32.3.1   Product chemistry

The principal purpose in supplying product chemistry data is to completely characterize the active ingredient and other constituents of the formulation. Of particular concern to the regulatory agency is the likelihood that the product may concern impurities of toxicological significance. General Product Chemistry data requirements are indicated in Table 32.1.

32.3.2   Environmental chemistry

The environmental fate of a pesticide is a very major concern. The extent of testing usually required of an agricultural pesticide is illustrated in Table 32.2. The principal objective of these experiments is to determine the likelihood that

TABLE 32.1. PRODUCT CHEMISTRY DATA REQUIREMENTS

Active ingredient	Formulation
Identity	Composition
How purity is determined	Purity of inert ingredients
How manufactured?	How formulation manufactured?
What impurities are possible theoretically?	Storage stability
Limits of detection of impurities	Physical properties
	Chemical properties
Chemical properties	How to determine active ingredient content?
Physical properties	
Storage stability	

the pesticide or any of its breakdown products may tend to accumulate in the food web through mechanisms other than direct contamination of a treated food crop. For example, a water-soluble pesticide may tend to be easily leached through the soil into subterranean potable water supplies. Alternatively, soil residues of a pesticide may be easily taken up by rotated crops in succeeding seasons, and lead to otherwise unexpected residues.

TABLE 32.2. ENVIRONMENTAL CHEMISTRY DATA

Hydrolysis
Photodegradation
Volatility
Mobility in soil
Leaching
Adsorption on soil
Uptake by plants
   — Soil residues
   — Water residues
Effect on soil microorganisms
Effect on activated sludge (waste-digesting microbes)
Dissipation (field stability) studies
   — Water
   — Soil
   — Model Ecosystem

The testing requirements indicated in Table 32.2 illustrate the spectrum of environmental chemistry requirements. The ease of hydrolytic and photochemical degradation must be determined, and if toxicologically significant breakdown products are observed, residue tests must take this fact into account. It is also customary to test for adverse effects on certain vital soil micro-organisms, as well as on microbes important in the treatment of waste.

Most experiments concerning the environmental fate of pesticide residues are usually conducted with radio-labeled material. By standard techniques one can monitor the rate of disappearance of parent compound, the binding of residues to soil, and the tendency of the pesticide or its metabolites to leach from soil. Using a model ecosystem, the tendency of the material to accumulate in the food web can be determined. The overall objective of Environmental Chemistry testing is to characterize the persistence and mobility of the pesticide and its breakdown products. These properties are considered in assessing the environmental hazard which may result from use of the product.

32.3.2.1 Hazard evaluation — wildlife and aquatic organism

Closely connected in the assessment of risk associated with pesticide use is consideration of risk to fish and wildlife. The usual testing requirements for an agricultural pesticide are indicated in Table 32.3.

TABLE 32.3. HAZARD EVALUATION TO WILDLIFE
AND AQUATIC ORGANISMS

Avian testing:
    Acute oral — one species
    Subacute feeding — two species
    Reproduction — two species
    Field test (small pen to full-scale tests)

Aquatic organisms
    Invertebrate LC_{50}
    Fish LC_{50} — two species
    Other tests may be required if exposure is possible

Avian testing usually involves acute and subacute feeding studies and reproduction studies on two representative species — one aquatic and one upland game species (usually Mallard Duck and Bobwhite Quail). If such tests indicate a tendency of the material to be toxic or to affect reproduction, a simulated field test is usually required.

Hazard to aquatic organisms is determined by testing on a freshwater invertebrate (usually *Daphnia magna*) and two species of freshwater fish one representative of cold water habitat and one of warm water (usually rainbow trout and bluegill sunfish). If the pesticide is intended for aquatic use or where long-term contamination of fish-bearing waters is possible due to runoff, it is usually necessary to conduct chronic testing on fish. Any direct or indirect contamination of salt water leads to additional testing requirements on representative estuarine or marine organisms.

32.3.2.2 Metabolism

The series of metabolism tests indicated in Table 32.4 are of major importance in addressing a number of questions:

(a) How is the material metabolized by soil microorganisms? What is the chemical identity of the metabolites?

(b) Does the material tend to accumulate in fish? Does it accumulate as the parent pesticide or as a metabolism product?

TABLE 32.4. METABOLISM STUDIES

Microbes		
— Aerobic	Determine identity of important	
Anaerobic	Soil/Aquatic metabolites	
Fish		
— Static	Tendency to bioaccumulate?	
Dynamic	Chemical identity??	
Mouse		
Rat		
Dog		
Bovine — Required if the crop or by-products are used as cattle feed		
Plant — Compare with mammalian metabolism		
Purpose:		
— Correlate with toxicology studies		
— Guide for residue chemistry studies		

(c) How is the material metabolized by mammals? This question is critical in interpreting toxicology studies. Further, is there significant interspecies divergence in metabolism? If so, one must ask which laboratory test animal is the better model for simulating human toxicology and metabolism. Bovine metabolism studies are required when the crop or any processing by-products are used as cattle feed. Examples of this practice are the feeding of citrus pulp and cottonseed meal. It is possible that if the feed contains residues of the pesticide, residues also may be found in the meat and milk of cattle. Knowledge of the chemical identity and locus of concentration of any residues, whether as parent pesticide or as metabolities, is a prerequisite to meaningful residue studies.

(d) How do target plants metabolize the pesticide? This becomes of paramount importance when one considers that toxicology studies are taken to be representative of human exposure to pesticide residues remaining in food. The conventional means of conducting such studies is to add the pesticide material directly to the diet of laboratory animals. This experimental approach is valid only in situations where important plant metabolities are identical to mammalian metabolism products. If plants produce unique breakdown products not found in mammals, it may be necessary to conduct special feeding studies on these plant metabolites.

The underlying basis for metabolism studies, then, is to aid in the understanding of toxicological observations and to determine whether there is a need to monitor residues of toxicologically significant metabolites.

32.3.2.3 Hazard evaluation — humans and domestic animals

The standard battery of toxicology studies usually conducted on a new pesticide product is indicated in Table 32.5. It is helpful in the design of a meaningful toxicology program to consider the reasoning behind each of the

TABLE 32.5. HAZARD EVALUATION TO HUMANS AND DOMESTIC
ANIMALS (MAMMALIAN TOXICOLOGY)

Active ingredient	Formulated product
Acute oral	Acute oral
Acute dermal	Acute dermal
Acute inhalation	Acute inhalation
Acute delayed neurotoxicity	Eye irritation
Subchronic oral	Dermal irritation
Subchronic dermal	Dermal sensitization
Subchronic inhalation	Dermal sensitization
Subchronic neurotoxicity	
Chronic feeding — two species	
Oncogenicity — two species	
Teratogenicity — two species	
Reproduction	
Mutagenicity	

several toxicology requirements. It can be noted by examination of Table 32.5 that testing is required both on the active ingredient and the end-use formulated product. The testing required for the formulated product is primarily of an acute nature (one time exposure to the test substance). The intent is to determine whether persons involved in the manufacture, transport and application of the formulated pesticide produce will be exposed to acute hazards. Similarly the active ingredient itself is subject to acute testing, primarily to assess the potential of hazard to workers involved in the manufacture or formulation of the active ingredient.

Subchronic toxicology studies, in which test animals are exposed continuously to the test substance for a duration substantially less than a lifetime, are also important in assessing potential hazard to individuals exposed to a pesticide product through moderately long-term activities such as application or manufacturing. Choice of exposure route(s) (oral, dermal, or inhalation) is made based on the likelihood of actual human exposure via one or more of these routes.

Longer term studies usually required are chronic feeding studies in a rodent and a non-rodent, and oncogenicity studies, usually of lifetime duration, in two rodent species. Teratogenicity studies are usually conducted on a rodent and non-rodent. This test calls for administration of the test substance to pregnant females during gestation to assess the potential to cause birth defects. A related long-term test, usually conducted on rats, assesses the potential for general reproductive effects. In this experiment the test substance is administered in the diet of parents prior to mating and to the females during gestation and while nursing. After weaning, the offspring receive the same treated diet until they are mature. These first-generation offspring are then mated and the cycle is repeated with continuous exposure through two or three generations.

Testing to assess mutagenic potential has begun to play an increasingly important part in risk assessment for pesticides. Numerous short-term assays have been developed in microbial and in cultured mammalian cell systems to indicate the potential for direct interaction with, or other effects on genetic material. Not only are these tests capable of detecting substances which may introduce deleterious mutations into the human gene pool, they have also been shown to have significant value in predicting potential for oncogenic hazard. The latter function of these assays is of particular importance because they supply relatively short-term and inexpensive information which is useful in assessing the potential for a substance to be carcinogenic.

We will return to further consideration of chronic toxicity tests in the course of discussing pesticide tolerances.

32.3.3   Efficacy testing

The United States Environmental Protection Agency (EPA) has traditionally required extensive demonstration that a new pesticide product is effective for the uses claimed on the label. Very recently it has been decided that for certain types of products the review of efficacy data may be deemphasized so that more of the EPA's resources can be devoted to hazard evaluation. Despite this change in regulatory emphasis, the demonstration of product effectiveness remains a critical aspect of pesticide development. The principal objectives in efficacy testing are noted in Table 32.6.

TABLE 32.6. EFFICACY TESTING

Define use pattern
   Rate of application
   Frequency of appliction — timing
   Mode of application
   Importance of variety, maturity, cultural
   Practices
   Climatic or geographical limitations

Demonstrate effectiveness
   Target pest(s)
   Host crop

Phytotoxicity — Alone and in combination with other materials

32.3.4   Label development

As noted in the introduction, the culmination of all of the reviews associated with registration of a pesticide product is in development of the label, whose

TABLE 32.7. LABEL

Product identity
Composition
Warning statement
Antidote statements
Disposal instructions
Detailed instructions for use

principal component parts are noted in Table 32.7. The product identity and composition sections of the label are relatively straightforward and are dictated by the product chemistry data filed with the registration application. The warning statement may take the form of a simple "Caution — Keep out of reach of children" for a substance with low acute hazard. At the other extreme, a pesticide with a very high degree of acute toxicity may be required to carry the statement "Danger — Poison" with a skull and crossbones sign. The evaluation of environmental chemistry and fish and wildlife toxicity data dictates the extent of environmental hazard warning statements. It would be expected, for example, that a pesticide with a known high toxicity to aquatic organisms would bear strong warning statements to avoid application to fish-bearing water or to locations where runoff to such waters would be possible.

The label must also include very detailed instructions for use of the product. The instructions delineate target pests and host crops where the product has been shown effective and clearly states any limitations, including varieties of non-target plants known to be adversely affected by the pesticide.

32.4   TOLERANCES

When a pesticide is to be applied to a food crop, the product registration must be supplemented by a tolerance for residues of the pesticide on that specific commodity. It is assumed that the food commodity will contain residues of the pesticide or its breakdown products. For this reason, treated crops are not allowed into the channels of trade unless a tolerance has been established for residues of the pesticide on the food commodity. As noted earlier, the establishment of a tolerance is considered in the United States to be an administratively separate function of the regulatory agency. The elements of a pesticide-tolerance petition, as outlined in Table 32.8, are chronic feeding studies in laboratory animals, and quantitative determination of residues of the pesticide or significant metabolites in the treated crop or in any processed commodities prepared from the treated crop. If the agricultural commodity or a processing product or by-product is fed to

TABLE 32.8. ESTABLISHMENT OF PESTICIDE TOLERANCES BASIC
INFORMATION REQUIREMENTS

Information required	Source
Residue identification	Radiochemical metabolism and degradation experiments
Residue quantitation	Chemical analysis of field-treated agricultural commodities
Toxicological	(a) Laboratory animal feeding experiments
Evaluation	(b) Estimation of hazard to man using — Safety factors — Dietary analysis

livestock for example cattle, it is also necessary to determine residues and establish tolerances in meat and milk.

Analysis of these food residues, together with consideration of typical contribution of each food material to an average person's diet, allows one to calculate the theoretical maximum residue contribution to an individual's diet. This quantity is then compared with feeding levels in laboratory animal studies which are shown to cause no observable affect. These "presumed safe" levels from the animal studies are adjusted by means of a safety factor and an acceptable daily intake (ADI) for man is determined. The regulatory decision on the allowability of the tolerance petition is based on a comparison between the acceptable daily intake derived from animal feeding studies and the theoretical maximal residue contribution to the human diet as projected from the analysis of residue studies.

Not directly involved in the tolerance setting process but also related to pesticide residues are considerations such as the effect of pesticide residues on product quality and acceptability. It is required, for example, to demonstrate that the pesticide does not adversely affect taste, processability, or other qualities of the food product. Consider as an example the case of cottonseed bearing residues of a pesticide. Cottonseed is processed to produce cottonseed oil, a human food. It must be demonstrated that the oil bears no unacceptable favours. Cottonseed meal, a by-product of oil processing, is an important protein-rich animal-feed supplement for cattle and poultry. Not only must the level of residues in the meal be determined but the palatability of the meat to cattle and poultry must be considered. It also is necessary to measure the residues (and propose tolerances) for the pesticide in meat, milk and eggs. Furthermore, egg producers insist on knowing whether such residues will have an adverse effect on the taste or appearance of their product, the thickness of shells, and so on.

The details of product acceptability testing requirements vary considerably from one crop to another. Even when not required by regulatory agencies, this type of testing must be conducted by the pesticide developer as protection against liability claims. One can imagine, for example, the consequences of overlooking the fact that residues of a fungicide on wheat have been known to leave large quantities of flour unsuitable for bread production because of toxicity to yeast!

32.4.1   Elements of residue testing

Measurement of pesticide residues in food commodities constitutes a major aspect of the tolerance-development process. The principal objectives of a well-planned residue program are indicated in Table 32.9. In monitoring direct residues one must demonstrate the fate of pesticides in the food crop with respect to time. If metabolism studies indicate the need to follow the fate of significant metabolities, this also must be factored into the experimental design.

TABLE 32.9. RESIDUE CHEMISTRY

Methodology on each commodity
  (a) Direct; treated commodities
  (b) Indirect; residues resulting from feeding to food-producing animals

Field trials
  (a) Use rate
  (b) Exaggerated rate
  (c) Multiple application

Define rate of residue decline

Determine maximum residue likely to result from application according to label

Define preharvest interval

Field residue trials must be planned and executed with great care if they are to yield high-quality data. Applications of the pesticide are made at the proposed use rate and at exaggerated rates, on a multiple treatment schedule representative of the maximum application frequency allowed by the label. The crop is sampled at a sufficient number of time points to completely characterize the rate of decline of the residues under a full range of environmental conditions. Analysis of these crop samples may lead to the requirement for a "preharvest interval" — a minimum period of time which must elapse after the last treatment to ensure that residues will be below tolerance levels at time of harvest. The preharvest interval, if called for, will constitute an element of instructions for use on the product label, often taking the form "Do not harvest within __ days of treatment."

Of equal importance in defining pesticide residues is analytical methodology. One must recall that measurement of levels significantly below 1 ppm is required. Plant materials, milk and animal tissues contain an abundance of extractable natural products which must be cleanly separated from the pesticide-residue analysis represents the absolute state of the art in microanalytical techniques.

32.4.2 Examples of evaluation of a proposed tolerance

It is instructive to examine a specific example to clearly understand the steps involved in evaluating a new tolerance petition.

Consider that potatoes are treated with an insecticide for control of an insect pest. Residue studies conducted by analyzing potatoes following actual field applications of the insecticide show that the maximum residue likely to result from the use of the product is 1 ppm. This level, 1 ppm, is therefore proposed as a tolerance level representing pesticide residue likely to result from good agricultural practices (1 ppm is equivalent to 1 mg pesticide residue kg^{-1} of potatoes). It is further known that potatoes contribute 7% to the diet of a typical American person. This typical person weighs 60 kg and consumes 1.5 kg of food per day. His theoretical maximal daily intake of residues of the pesticide from potatoes can therefore be expressed as:

1 mg/kg	×	0.07	×	1.5 kg/day	= 0.105 mg/day
residue		per cent		daily	theoretical
level		potatoes		food	maximum
potatoes		daily		intake	residue
		diet			contribution
					(TMRC)

The calculation described above allows one to arrive at an upper limit of pesticide residues to which a representative person will be exposed.

To complete the evaluation of the tolerance it is next necessary to consider what level of residues of the pesticide can be presumed safe in the human diet. Suppose that the no-observable-effect level (NOEL) in a chronic fat-feeding study is 20 parts pesticide per million parts of food, and that the rat has been shown to be the laboratory animal most sensitive to this particular pesticide. For the rat, it is known that 20 ppm in the diet is equivalent to 1 mg of the pesticide kg^{-1} body weight day^{-1}. In calculating an acceptable daily intake (ADI) for man it is conventional to apply 100-fold safety factor to the NOEL in a lifetime animal-feeding study. This safety factor is intended to account for differences in sensitivity both between individuals and between species. Application of the safety factor to the 1 mg kg^{-1} day^{-1} level for the rat allows one to calculate an ADI for man of 0.01 mg kg^{-1} day^{-1}. The maximum permissible intake for a 60-kg person is, therefore, 0.60 mg of the pesticide day^{-1}

TABLE 32.10. MULTICROP RESIDUE INTAKE ANALYSIS

Commodity	% Diet	Daily[a] intake (g)	Estimated tolerance (ppm)	Residue exposure in daily diet (mg)	Daily intake (mg) cumulative
Potatoes	7.0	105.0	1.0	0.105	0.105
Cottonseed oil (margarine, shortening, edible oils)	2.29	34.4	0.5	0.017	0.122
Dairy products	26	390	0.05	0.020	0.142
Meat and poultry	11.47	172	0.05	0.009	0.151
Broccoli, B. sprouts, cabbage, lettuce, corn, cauliflower	3.61	54.15	1.0	0.054	0.205
Eggs	3.00	45	0.05	0.002	0.207
Soybeans, nuts, other dry beans	1.0	15	1.0	0.015	0.222

[a]Assuming a total intake of 1500 g per 60-kg man.

Since potatoes could theoretically contribute $0.105 \text{ mg}^{-1} \text{ day}^{-1}$ to a maximum permissible intake of $0.60 \text{ mg}^{-1} \text{ day}^{-1}$, the proposed tolerance would be acceptable.

If our example is to be a practical one, it must also be considered that the pesticide will probably be used on crops other than potatoes. If this is the case it is necessary to propose tolerances and repeat the tolerance-evaluation process for each crop in which residues are likely to be found. Suppose that the candidate pesticide will also be applied to cotton, to a variety of vegetable row crops (broccoli, cabbage, corn, etc.) and to soybeans and related crops. Since we know that cottonseed meal is also used as feed for cattle and poultry, it will also be necessary to consider the residues in meat, milk and eggs resulting from livestock feeding.

A detailed multicrop residue analysis suggested by the proposed use pattern is shown in Table 32.10. Note that the cumulative daily intake of pesticide residues resulting from all proposed uses for the product is $0.222 \text{ mg day}^{-1}$. Recall that the analysis of toxicology data supports a maximum permissible daily intake of 0.60 mg day^{-1}. It is therefore likely that all of the proposed tolerances would be approved and that the pesticide could be applied to all of these crops once the appropriate registrations are granted.

32.5 INTERNATIONAL CONSIDERATION OF PESTICIDES REGISTRATION AND TOLERANCES

Most of the developing countries are not able to expend the resources needed to conduct a full analysis of the potential risks of pesticides before allowing their use in local situations. They have in the past and will for sometime continue to rely upon safety evaluation conducted in the industrialized nations. That this fact carries with it certain obligations on the part of pesticide suppliers is reflected in new regulations governing the export of pesticides from the United States. Such materials will soon be required to bear cautionary statements (in appropriate languages) equivalent to those for products sold within the US. Furthermore, exported pesticides not registered in the United States will be required to carry a label statement to that effect, and the government of the importing country will be informed of the regulatory status of the product.

Responsible officials in developing countries must be alert to new information concerning the safety of imported pesticides and make decisions after full consideration of risks and benefits.

Since the early 1960s, the difficulties posed by trade in food materials between nations has been the subject of substantial deliberation and action on the part of the United Nations and other international bodies concerned with pesticides. The concerns of these bodies are based on the fact that food is a very major item of international trade, and that different nations can and do establish different food standards. In the case of pesticide tolerances, approved levels may vary widely between different countries. This fact causes concern on two fronts:

(a) Importing nations are concerned with the safety of imported foods to their citizens.
(b) Exporting nations are concerned that Food Standard Regulations may serve as non-tariff barriers to trade.

Since 1962, the Codex Alimentarius Commission, a joint undertaking of the World Health Organization (WHO) and the Food and Agriculture Organization of the United Nations, has been taking steps to provide international guidance on this problem. It has been their objective to coordinate international food standards, including pesticide residue tolerances, with the two-fold guiding principle of protecting public health and promoting fair-trade practices. Although a full discussion of the activities of this group is beyond the scope of this paper, I suggest that their work should be familiar to any person involved in formulating national policy concerning pesticides and international commerce in food.

32.6 CONCLUSION

In concluding this discussion, it is safe to say that the concept of pesticide residues and tolerances is not generally understood. It is important, however, that thoughtful individuals involved in the development of novel pesticides take this very important aspect of the regulatory process into consideration. It is equally important that we recognize that there are many controversial issues surrounding the establishment of pesticide tolerances. A few of these are itemized here without a statement of my personal opinion to stimulate your consideration of this very important problem.

(a) How appropriate is the 100-fold safety factor?

(b) Why is the intake analysis based on maximum residue expectations? Residue data are developed from raw agricultural commodities rather than from food ready for consumption. It is also not considered that not all persons will be exposed to commodities treated with a particular pesticide.

(c) Should we consider the fact that a "National average" typical dietary intake is a poor approximation for certain age groups, ethnic groups and localities?

(d) Which residues in milk constitute 100% of the diet of some infants, are treated in a separate calculation which accounts for this fact? Should special safety factors or toxicology studies be applied to the evaluation of milk tolerances?

(e) What about residues of more than one pesticide material on a particular crop? Are the toxic effects of such residues simply additive or are they likely to be synergistic?

(f) How should regulatory agencies treat food residues of pesticides which have shown the potential to cause cancer in laboratory animals?

(g) How well are pesticide tolerances enforced? What is the likelihood that the public is frequently exposed to residues in excess of tolerances?

32.7 SUMMARY

An overview of the registration process for pesticides in the United States will be presented. The lecture will outline data requirements which address questions of human and environmental hazard. In particular, requirements for metabolism, toxicology and environmental chemistry data will be discussed, as well as data concerning the effects of pesticides on non-target organisms. Product performance (efficacy) testing and product labeling will be discussed briefly. The procedure used to establish tolerances for pesticide residues on food commodities will be described by means of examples. This process involves consideration of residue data, toxicology, food consumption data and certain safety factors.

The effects of registration procedures in the United States upon pesticide developments in other nations will be discussed. The work of the *Codex Alimentarius* Commission to develop guidelines for regulation of pesticide residues in food will be reviewed.

32.8   ACKNOWLEDGEMENTS

I would like to thank my colleagues Terry Burkoth, Norma Jean Galiher and David Schooley for their many contributions to this paper and to my understanding of the scientific elements of the pesticide regulatory process.

32.9   DISCUSSION

Odhiambo: I have a question regarding hazard evaluation. As part of the requirements for registration where would you need to do toxicological testing on humans?

Young: Yes, that is a very important question. The EPA does not have the authority to conduct human testing. The FDA does and requires it in the clinical testing phase of drug evaluation. The EPA is not allowed to ask for human data, so we do no experiments on humans with a few rare exceptions. Say if you do a dog, rat and mouse test and they all come out differently, the question appears, which is more like human metabolism and which do we choose as the best model? In such a case pharmacodynamic and metabolism experiments are really critical. We would ask, what is the rate of metabolism; what is the degree to which the material is sequestered in man; what is the ratio of urinary to fecal excretion, etc.? You compare these with the animal tests and choose the best toxicological model. We rarely do this because of the prohibition on human testing. That can only be done if there are volunteers. In one case at Zoecon Professor Djerassi was all set to be such a volunteer, but it turned out that it wasn't necessary.

Bowers: In the event of a serious difference in drug metabolism between your normal animals would you go to primate testing?

Young: Yes, a primate test would probably be required. However, every pesticide developer hopes that a primate test wouldn't be required because these are very expensive. These are normally sub-chronic tests. A proper chronic test would likely be a 10-year experiment on a primate.

Bowers: A 2-year feeding experiment would be a truly sub-chronic test?

Young: Yes, indeed.

Menn: While the EPA has no authority to require human studies, it would be possible to do *in vitro* metabolism studies with human liver enzyme preparations, or microsomal mixed-function oxidases, to obtain a metabolic pattern for the pesticide. These data can be compared with the animal *in vivo* studies.

Young: That's a great idea. Has this been done?

Menn: Yes, some studies have been carried out at Cornell.

Bowers: In terms of unusual analyses I have an amusing side. It concerns an insecticide applied to grapes which regardless of its efficacy as an insecticide was found to keep birds from the grapes because of its taste. It was finally registered not only as an insecticide but also as a bird repellent. I think this emphasizes the importance of organoleptic evaluation of treated food crops

Young: Another occurrence which is not really organoleptic analysis, is also quite amusing. There was once a fungicide used on grains which went all the way through registration, etc. Then after the first crop was harvested and it was in the storage silos, etc., they tried to make bread from the flour, only it wouldn't rise. The fungicide was in fact killing the yeast. So one has to try to anticipate these sort of problems.

Bowers: Does the Delany clause apply to pesticide residues?

Young: That's for direct food additives. Insecticides are not direct additives and are therefore not covered by the Delany clause.

Odhiambo: How closely do the EPA and FDA link up and collaborate in matters dealing with problems of pesticide tolerances in the USA?

Young: They certainly don't collaborate in matters of registration, but there is some collaboration in tolerance matters. Mostly the EPA does the majority of the work where pesticides are concerned and they have the authority to enforce the pesticide tolerance levels. The FDA does not have a formal oversight but they are involved directly in seeing that toxicological studies are performed according to good laboratory practice as defined by the FDA.

Menn: The FDA has jurisdiction on those pesticides used to protect livestock from insect pests which may be considered also as endoparasites, for example, cattle grubs.

Gebreyesus: It seems that in countries like the USA, Western Europe, etc., the levels of tolerances we are talking about depend on the level of technology, or the level of civilization, whereas in the developing countries we are trying simply to make enough food available. In your opinion are there pesticides that have been developed which could be produced at a fairly cheap price, but because of higher restrictions on tolerances in the developed countries they have not been able to reach the market. And if there are such, would they be made available to the developing countries at a cheap price which would contribute to food production?

Young: In answer to the first part, I simply don't know! But what you say in principle I'm sure is possible. If the intake of pesticide residues is equal to the acceptable daily intake, the EPA may not issue any additional tolerances. What could happen is that if a pesticide has reached its limit of tolerances in the USA but your country asks the manufacturer to establish a tolerance on this crop which is important in your country so that we can assure your government that is safe, the EPA might not issue that new tolerance if it would lead to an intake above the acceptable limit. However, what you say in principle could happen but I'm just not aware of it.

Bowers: We're talking about a situation similar to DDT which is not allowed in the developed countries at all but is used elsewhere.

Gubler: But only in the hygiene area, DDT is not used on food crops.

Menn: That's not strictly true, because in India DDT is widely used to protect stored grains.

Gebreyesus: True! But as the chairman said the other day, perhaps we should not be worried about the DDT in the fat but should be worried about how much fat we have. Here we are talking about simply producing enough food to accumulate any fat in the first place.

CHAPTER 33

Legal Protection of Industrial Property — Patents

DONALD W. ERICKSON and JULIUS J. MENN

Zoecon Corporation, Palo Alto, California 94304, USA

and

KURT GUBLER and ERICH A. HORAK

Ciba-Geigy Limited, CH-4002 Basle, Switzerland

CONTENTS

33.1	Introduction	541
33.2	The Patent System	542
33.3	Towards International Uniformity	543
33.4	Unity and Novelty Conditions	544
	33.4.1 Opposition	545
33.5	Publication	545
33.6	The Application	546
	33.6.1 Specification — title, field and utility	546
	33.6.2 The claim	547
33.7	Filing a Patent	548

33.1 INTRODUCTION

The principal forms of legal protection of industrial property are patents, trade secrets, trademarks, copyrights and unfair competition laws. The most important and most frequently used form of legal protection for the research organization is patents. The second most important form of protection for the research organization is trade secrets which sometimes is referred to as proprietary know-how. Trade secrets or know-how can be a very effective form of protection; however, it is considered to be a high-risk form of protection. It is subject to loss, for example, through inadvertent communications or lax security. Trade secrets offer no protection, obviously, against the independent third party making the same or similar discovery and putting it into practice. Further, in today's atmosphere of more and more disclosure of technical background by reason of increasing requirements of health and environmental government agencies, it has become more difficult to use the trade secret form of protection. Patents therefore take on an even more significant role in protecting the discoveries of the research laboratory.

33.2   THE PATENT SYSTEM

A patent may be viewed as a contract between the inventor and the government. The inventor, on the one hand, grants to the government the disclosure of his invention, with the right of the government to publish the invention, and free use, by all, of the invention after expiration of the patent. The government, on the other hand, grants to the inventor the exclusive right to his invention for a certain term of years. Most usually, the term of exclusivity is from 15 to 20 years, depending upon the particular country. In Kenya, for example, the term of exclusivity is 20 years from the date of filing the patent application, which is true for many countries.

By reason of the foregoing contractual arrangement — the patent system — all parties benefit (the inventor, the government and the people) and science benefits. Without disclosure, i.e. publication, science cannot progress. Scientific intercommunication, exchange of knowledge of discoveries, is essential. The incentive provided by the patent system, the exclusivity for a term of years, promotes scientific intercommunication which is the foundation on which scientific progress depends.

Patents also provide an incentive for additional discoveries. The patent position of one research laboratory may, and often does, prompt a second research laboratory to make additional and new inventions in order to avoid the exclusive position of the first laboratory. Patents furnish a foundation, the exclusivity for a term of years, to support the investment of capital in order to commercialize new scientific discoveries which, in turn, support the research laboratory.

In view of today's vigorous regulatory requirements and scrutiny of the government prior to introducing new products into the market, commercialization of new inventions is a long road and exceedingly costly. The term of years needed to recoup both the research investment and the development investment becomes longer and the risk higher. Patents are essential in providing the exclusivity needed in order to justify and protect the investment. Without the exclusivity for a term of years provided by the patent grant, business would be seriously hampered in the introduction of new products.

Indeed, the number of new products would seriously diminish to the disadvantage of the public.

A patent term of 20 years may seem like a very long time. It is not. One must bear in mind that patents are applied for at the time of the research discovery. It takes many years of further research, testing, development and toxicology for safety and health before the research discovery is refined and reaches the market. In the case of pharmaceuticals and pesticides, the time span is generally a minimum of 6–8 years and an investment of many millions. Therefore, the importance of and need for a strong and effective patent system increases with the research conducted in a particular country. Recently, this has been exemplified by changes which strengthen the patent system in Switzerland, Italy and Japan. It is hoped that additional countries will follow these examples by providing an effective patent system.

33.3 TOWARDS INTERNATIONAL UNIFORMITY

Patents are based on national law. There are no international patent laws. Efforts to harmonize and move toward uniformity relative to the patenting procedure, however, are underway. Examples are the European Patent Convention (EPC)* and the Patent Cooperation Treaty (PCT)†. The EPC and PCT represent very significant steps towards international uniformity of the patenting procedure. Not the least of the immediate advantages of the EPC and PCT are efficiency in the patenting procedure and economy through standardization of procedure, minimization of translations from various languages and elimination of duplicate prior art searching by the examination countries. Upon completion of the patenting procedure under either EPC or PCT, the resulting patent grant is, nevertheless, subject to the national laws of each country in which the patentee may need to enforce the patent grant. The inventor, after filing for patent in the national country, has the option of filing for patent in additional countries on a country-by-country basis or by the newer route of the EPC or PCT or a combination thereof. In view of the relative newness of the EPC and PCT, some research organizations continue to use the old route of country-by-country patent filing as well as the new routes provided by the EPC and PCT.

One of the main characteristics of the patenting procedure is that it is different in almost every country; with the exception of the recently introduced EPC patent and PCT patent procedures. In general, the patenting procedure flows as represented on Fig. 33.1.

*EPC has entered into force on 7 October 1977. Patent applications first accepted on 1 June 1978.

†PCT has entered into force on 24 January 1978. Patent applications first accepted on 1 June 1978.

```
                        Patent Application
                               |
        Registration  <────────────────>  Registration
                    \                          |
                     \                          v
                      \                Official Examination
                       \               |Suitability
                        \              |Unity
                         \             |Novelty
                          \            |Subject Matter,
                           \           |        Patentability
                            \          |Utility
                             \         |Technical Advance
                              \        |Invention Level
                               \       |Non-Obviousness
                                \               |
                                 \              v
                                  \    Opposition Proceeding
                                   \   (Public Examination)
                                    \       /
                                     v     v
                                    Patent
```

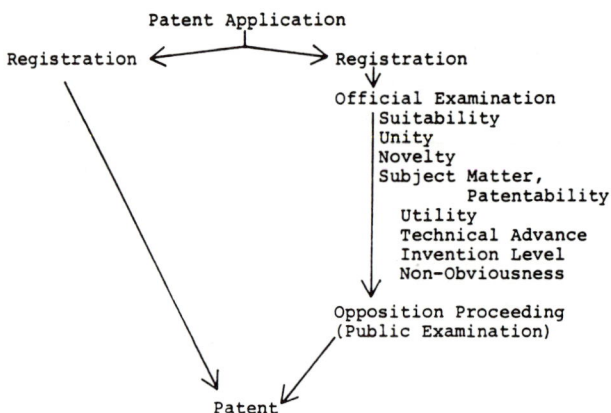

FIG. 33.1. Flow of patenting procedure.

Referring to the foregoing outline, in some countries the patent application filed by the inventor is examined by the Patent Office only for formal matters. In other words, an administrative inspection determines that the correct form, documents and fees have been filed. These are oftentimes referred to as "registration countries", for example, Belgium. The patent application proceeds directly to patent grant without further examination. In most countries, various degrees or levels of official examination are applied. The variations of official examination are not applied uniformly from country to country, and not all categories listed above are applicable.

Some countries examine only for suitability and unity of invention under the law. Suitability is reviewed to determine whether the invention sought to be protected is prohibited. For example, a patent application for a chemical compound in a country such as Mexico, Spain or the Netherlands would be unsuitable because a chemical compound is not entitled by national law to be patented. A current situation on the question of suitability of the invention has been recently accepted by the United States Supreme Court for hearing. The case* involves the issue of whether a living organism, a new strain of bacteria, may be protected by patent in the USA. The US Patent Office has held that a new bacterium cannot be protected by patent under the US patent laws. The British Patent Office, on the other hand, considers under UK patent law a new bacterium a protectable form of invention and granted British patent number 1,436,573 claiming the bacterium itself.

33.4   UNITY AND NOVELTY CONDITIONS

Most countries require unity of invention. In other words, only one invention may be protected in one patent. If the patent office determines that

*In the Chakrabarty, 197USPQ72(CCPA 1978).

two or more inventions are present in one patent application, the applicant (inventor) will be required to file a division application for each invention in excess of one. Many countries include examination for novelty. The standard of novelty is not uniform, however. Generally, novelty is a determination of whether the invention is different from the prior art. A lesser number of countries examine the subject matter or patentability of the invention. The standard of patentability varies with the national laws. All sub-categories of patentability listed above are not necessarily applicable to any single country. A principal feature of the EPC and PCT is to lend some uniformity to the standard of patentability on which patent applications are examined. If achieved, the standard will then be subject to the interpretation of the legal system of the individual member countries.

33.4.1 Opposition

A final hurdle to obtaining patent in some countries and under the EPC is known as the opposition proceeding. Under these procedures, the patent office, after determining that the invention is patentable, publishes the patent application and allows the public an opportunity to file written reasons opposing the grant of patent. Generally, the reasons for opposition must be filed in the patent office within 2 to 3 months following publication of the approved application. The applicant (inventor) is afforded an opportunity to rebut the opposer's arguments. If the opposer is not entirely successful, the patent application proceeds to grant in the form approved or in the form modified.

33.5 PUBLICATION

While a patent application is in the patent office, it is maintained in confidence. However, in several countries and under the EPC, each patent application is published, or availability thereof published, 18 months from the earliest filing date of the patent application. Thus, in most cases the inventor is well advised to bear in mind that his discovery likely will be published 18 months from the filing date of the first patent application. This feature is an example of the patent system providing for early scientific intercommunication and promotes the progress of science. Countries which publish or permit access to copies of patent applications 18 months after filing include the Republic of South Africa, Japan, France, West Germany and the Netherlands.

33.6   THE APPLICATION

The patent application, a legal document, is one of the most difficult documents to prepare. Bearing in mind that a patent application must be written in a form and content capable of fulfilling the different requirements of all countries in which it might be filed, the following outline is presented as a guide.

33.6.1   Specification — title, field and utility

The outline (Table 33.1) may be considered in connection with a chemical invention. The title and field of invention are analogous to titles and abstracts used in conjunction with the announcement of papers to be presented at a scientific convention. The field of invention or abstract should be informative to the reader and indicative of the more detailed description to follow. Similarly, a concise title should fairly indicate the subject matter of the invention. Following the field of invention should be described the utility of the invention, and in particular, the utility which will later in the description be supported by experimental or test description along with data. A description of the prior art follows, which may be brief or fairly lengthy depending upon the state of the art and complexity of the invention. This part is sometimes referred to as the "background of the invention" and contains description of what is already known in the field of the invention. The summary of the invention is a brief definition or description of the invention. It may be in general or broad terms. Sometimes the summary may closely resemble or be the same as the broadest claim of the patent application.

TABLE 33.1. OUTLINE OF PATENT
APPLICATION

Title, field of invention
Utility
Prior art
Summary of the invention
Materials (general properties)
Conditions of operation
Examples
Claims

Following the summary, there is usually set forth a description of the materials used in practice of the invention. For example, a description of the reactants may be presented in both general and specific terms. The conditions of operation are also described in both general and specific terms. In other words, a general range of conditions and a narrower or preferred range of conditions are given for practicing the invention.

Examples of the practice of the invention are next set forth, and constitute a very important part of the patent application. They are sometimes referred to as "working examples" and should describe to the reader how to carry out the invention. The examples are basic to the consideration underlying the contract between the inventor and the government. Examples which are insufficient to enable practice of the invention represent, in effect, failure of consideration and may cause the patent to be found invalid by the court. As a part of the examples or following the examples, there are usually illustrations or descriptions of how to use the invention. These should correspond to the utility described earlier, and include a description of tests performed.

33.6.2   The claim

The total of the foregoing parts is usually referred to as the "specification" of the patent application. Immediately following the specification appear the claims of the patent application. The claims are the heart of every patent. Only what is claimed is protected. The claims define the metes and bounds of the exclusivity represented by the patent grant. The importance of the claims is underscored by the fact that many patent professionals in preparing a patent application write the claims first and the specification second, using the claims as a foundation.

Underlying the patent application is, of course, the scientific discovery — the invention conception. It is important in obtaining a patent position to reduce the invention conception to writing and communicate it to the patent professional at the earliest date possible. The written description should contain, at a minimum, the following:

(a) Brief description (abstract) of the invention, including field of use (utility — general and specific).

(b) A description of how to practice the invention in sufficient detail to enable your colleagues to carry out the invention from the description given. This is a very important part of the procedure. As an example, description of a new chemical compound must describe a method of synthesis.

(c) A description of the novelty of the invention. This is sometimes referred to as a description of the "background of the invention". In other words, what is (are) the novel feature(s) of the invention as compared to prior knowledge in the field and literature. This is very important for the patent attorney to know in order to assist him in obtaining the strongest patent possible.

33.7 FILING A PATENT

The filing of the patent application in the patent office should be accomplished as soon as possible after realization of the discovery. The filing date of the application is determinative of your priority of patent rights with the rest of the world. Without going into detail, the patent rights to a given invention are awarded to the first to file for patent, except in Canada and the USA. In the latter two countries, patents are awarded on the basis of first to invent.

Filing of the patent application at an early date is also important for safeguarding against loss of novelty. Novelty is an essential universal prerequisite to obtaining patent. Public disclosure of the invention, even though inadvertently or innocently done, destroys the opportunity to protect the invention by patent in many countries. Thus, novelty may be lost by publication of the invention in a journal or by discussion thereof at a seminar or scientific meeting prior to the filing of the patent application. To safeguard against loss of patent rights, early filing of the patent application is encouraged.

The subject of patents or patenting is controversial and patent erosions do occur, especially in developing countries. Industrial or developed countries keep a high standard of patentability and a strong patent system in order to protect and encourage research. Progress is not possible without research. Examples of some patent-erosion trends are: (1) exclusion of certain inventions, e.g. new compounds; (2) exclusion of certain invention categories, e.g. pharmaceuticals; (3) shortening of the term of years of the patent, e.g. India — 5 years, Mexico — 10 years; (4) working requirement — that is production of the invention in the country as a condition to continuance of the patent; (5) compulsory licensing of the patent and (6) free government use of the patent. As research increases in a country, patent-erosion tendencies decrease. A research standard is not possible without a good patent law. Switzerland, Italy and Japan are good examples of countries who have changed to stronger patent laws with increasing industrialization and growing importance of their research.

CHAPTER 34

Patent Laws of Kenya and their Procedure

PRESENTED BY JULIUS MENN

Zoecon Corporation, Palo Alto, California, USA

CONTENTS

34.1 Introduction   549
34.2 Procedure for Obtaining Patent Rights   550
34.3 Transfer of Inventions to Industry   551
34.4 Conclusions   551
34.5 Discussion   552

34.1 INTRODUCTION

Kenya, as is true for Uganda and Tanzania also, provides patent protection through the registration in Kenya of patents granted in the United Kingdom. Duration of patent protection is the same as the life of the United Kingdom patent, which is now 20 years from the filing date of the patent application. Only patents granted in the United Kingdom are registerable.

The scope and form of patent protection in Kenya is the same as the United Kingdom. Thus, novel substances are patentable without regard to the process of manufacture or use.

Patent protection for natural substances may take the form of patenting of the extract; for example, the method for obtaining the extract and the use of the extract. In many countries, the patent laws do not permit patent protection of the extract itself. In the UK, for example, as a general rule the extract is not patentable except when produced synthetically. A patent to the use of the extract, however, is a very effective form of patent protection. Patent covering the method of obtaining the extract may be an effective form of protection, also. Of course, if the active component(s) of the extract are

549

identified and novel, patent protection of the active component(s) as a new compound(s) or mixture of compounds may be obtained.

34.2   PROCEDURE FOR OBTAINING PATENT RIGHTS

File a provisional patent specification in the United Kingdom. Within one year from the date of filing of the provisional file a complete specification in the UK and also file application in other countries in order to obtain the advantage of the filing date of the provisional for priority of invention.

The provisional is more in the nature of a brief scientific description and does not require the legal drafting of claims. The complete specification should be written with much more care than the provisional. In the complete specification, the objective is to define the boundaries or scope of the invention. The complete specification requires the drafting of claims — that is, defining the invention so that the reader knows precisely the boundaries of exclusivity claimed. Drafting of claims is the most difficult part of patent application preparation and should be done by an experienced patent attorney or, in the case of the UK, by an experienced Chartered Patent Agent.

It is recommended that both the provisional and complete specifications be done by or with the guidance of an experienced patent professional.

Procedure for obtaining patent rights starts with preparation of a written description of invention, which should include:

(a) Brief description (abstract) of the invention, including field of use (utility — general and specific).

(b) A description of how to practice the invention in sufficient detail to enable your colleagues to carry out the invention from the description given. This is a very important part of the procedure. As an example, if a new compound, then this description must describe a method of synthesis and should include identifying physical data of the new compound. For some countries, such as the USA, this description must also include a description of "how to use" the new compound.

(c) A description of the inventor's opinion of the novelty of the invention. This is sometimes referred to as a description of the "background of the invention". In other words, what is (are) the novel feature(s) of the invention as compared to prior knowledge in the field and literature? This is very important for the patent attorney to know in order to guide and obtain the best patent possible.

Steps (a) to (c) could form the provisional specification referred to above and be filed at the UK Patent Office.

It is recommended, however, that steps (a) to (c) represent your communication to a patent professional who in turn would prepare the provisional specification with your guidance as to technical accuracy. Both the provisional and complete specifications are legal documents whose care is best left with the patent professional. A good patent document is one of the most difficult legal documents to prepare.

It is highly recommended that the filing of the provisional specification be accomplished as soon as possible after realization of the discovery. This filing date is determinative of your priority of patent rights with the rest of the world. Without going into detail, the patent rights to a given invention are awarded to the first to file, except in Canada and the USA. In the latter two countries, patents are awarded on the basis of first to invent. Thus, it is important to keep written records containing points (a) and (b) above.

34.3   TRANSFER OF INVENTIONS TO INDUSTRY

Industry is more likely to be receptive and enthusiastic to the communication of new ideas (inventions) from an institute if the institute has first filed an application for patent. That is, at least the provisional specification. This procedure is also in the best interest of the institute in that it protects the institute's rights.

The initial communication can be in general terms with the advice that patent has been applied for. If the company is interested in the field, a copy of the patent specification should be provided along with a stipulated time period in which the company may consider it. Negotiations can then proceed to agreement for development and royalties to the institute.

Economy may be achieved in communicating with industry at an early date after filing of the provisional specification. If industry is interested in the invention, an agreement could include a provision that industry pay for the filing of the complete specification in the UK as well as filing in other countries. It can easily run to several thousand dollars to file in only a few countries.

34.4   CONCLUSION

I would encourage the institute to designate a patent liaison person. The person should be technically competent. Analogy would be a patent liaison or patent technicians that I understand are used by large companies such as Stauffer and Shell. This type of person, after a certain amount of experience, could become competent in preparation of provisional specifications which would be an economy for the institute.

For the UK, I would recommend, from personal professional experience,

David C. Harrison, whose address is Mewburn Ellis & Company, 70 & 72 Chancery Lane, London WC2A 1AD, England.

If I can be of assistance, I would be glad to communicate with an institute representative for guidance.

34.5   DISCUSSION

Johnson: Further to the nice and useful presentation you have given us, I would like to add that in the UK no patent would be granted to a person who had published a paper on the patentable material first.

Bowers: In the US, however, you have a year to file a patent application after disclosure of the material you wish to patent.

Menn: I think that even in the UK, one can go ahead and prepare a paper for publication immediately after filing a patent application.

Johnson: For those working for institutes like ICIPE, it depends on the prevailing policy of that organization with regards to matters of patenting. Usually, the period between filing an application for a patent and preparing the patentable material for publication is too long, and for those who regard application as a vehicle for promotion, this could be frustrating. Therefore, the director of the institute has to approach the issue tactfully.

Menn: Provided there is a good legal counsel available to the institute concerned, preparation of the material for the publication and patenting can be done simultaneously. This would reduce the period between publication and patent application to "Patent application to 3 or 4 months".

Johnson: Besides taking inventions direct to Industry for consideration as Dr. Menn has suggested, there is a research corporation in the US which will also take patents and approach firms likely to be interested in the invention on behalf of the inventor. I understand that some ICIPE patents have been handled this way.

Menn: Involving third parties like the research corporation of US, etc., would be expensive and time consuming. I think it is logical and better for an institution like ICIPE to establish direct relations with a number of companies and work with them. A research corporation would, in my opinion, be suitable for an individual working as a freelance agent.

Bowers: If one had an invention but had doubts about its real potential, it might be worthwhile to take it to a research corporation for placement to firms for assessment. However, if one was fully persuaded about the market potential of his invention, it is much better to forward it to the big firms direct. In the event, it is imperative to disclose your protection to the firm. It is also crucially important to give the firm 90 days to reply to your communication.

Szabo: I would like to make a few observations regarding the principal concept of what to patent and how and where to patent it. First, with specific reference to an institution like the ICIPE, even extracts from plant or animal sources which display activity with potential for commercial exploitation can be patented as compositions provided the biological and chemical characters of the extract are fully understood. Further, a process patent can be filed on the experimental procedure pertinent to obtaining the extract.

Secondly, with regards to the procedure to follow while seeking commercialization of an invention, our experience with the research corporation of the US mentioned by the previous speakers has been disappointing. This organization does not seem experienced and fast enough to act competently as a middleman between the inventor and the developer of an invention. In my view, it is therefore most advisable to write to a few companies systematically, on a confidential basis, fully disclosing your protection and clearly specifying the period the companies must reply.

Most companies will be interested in signing a secrecy agreement on top of the patent which will enable them to claim a share of the benefits of modifications to patent specifications which might arise from the development effort leading to the commercialization of the invention. Therefore, one needs to be extra careful when one deals with the companies. It is important, for example, to make sure that you have a perfect claim to the use and composition of your invention before making a move to a company.

Finally, patent rules and requirements vary from one country to another. In the US, for example, it is important to note that patents are granted on the basis of the first to invent. Therefore, information intended for patenting in the US must be withheld for publication until patented first. Further, in the US, professional notes or notebooks from foreign countries are unacceptable as evidence in support of patent claims. These circumstances must be evaluated thoroughly before making a move to patent in the States. All in all, it is much more beneficial to have full patent protection in the domicile country.

Arunga: I would like to make an observation on what Dr. Szabo has just said on the issue of patent protection in the country of domicile. It seems strange to me that after independence, none of the three East African countries of Kenya, Uganda and Tanzania has established any patent laws. Would you advance a comment on this? Further I wish to know whether it is possible to obtain a patent on a mixture of known compounds.

Menn: It is not strange that the three East African countries do not have patent laws of their own. It is a matter of social economic priorities. I think that when you establish a country there are certain basic needs that are much more important to fulfil than establish a patent agency or laws. Under the circumstances, I think that what these countries have adopted is very practical and logical. The present arrangements which these countries have with UK are in my opinion satisfactory until they will be able to make comprehensive patent laws of their own. Regarding your second question, it is possible to get a patent for a mixture, provided some unexpected activity can be demonstrated to confer novelty to the mixture.

Bowers: Suppose ICIPE came up with an active compound or extract with good prospects for commercialization and approached a company for further screening. Further, suppose that the company did the tests and expressed an interest in developing the compound, would ICIPE be entitled to any royalty payments?

Gubler: In this case, since the relationship between ICIPE and the company regarding the nature of screening or exploitation of patentable material would not have been formalized, the company would be under no obligation to advance royalty payments to the ICIPE.

Menn: Yes, that is right, I think the best procedure under the circumstances would be to exchange a letter of understanding with the company on the nature of screening and agree that should further commercially exploitable activity be found by the firm, both parties would then discuss the nature of further collaboration and spell out mechanisms for sharing the benefits. Under the terms of such letter, the company would guarantee protection to the inventor during initial screening programmes. The same terms would forbid the inventor from contacting another firm or firms on any respect of the same invention. Such formal arrangement does not necessarily constitute a secrecy agreement. It may be entered into on a goodwill basis.

Kezdy: To what extent do the funding agencies restrict ICIPE's freedom to patent potentially useful inventions?

Bowers: I do not know for sure.

PART IV

The Need for Trained Scientists and Technologists

CHAPTER 35

Training of Scientists and Technologists in the Management of Insect Pests and Vectors in Tropical Countries

ANTONY YOUDEOWEI

Department of Agricultural Biology, University of Ibadan, Nigeria

CONTENTS

35.1 Introduction 557
35.2 Diagnosis of the Problem 558
35.3 Training Programmes and Training Needs 561
    35.3.1 Professional training 563
    35.3.2 Field operators 565
    35.3.3 In-service training 565
35.4 Some Modest Proposals 567
    35.4.1 Research and training centres 567
    35.4.2 National commitments 569
    35.4.3 Assistance from industry 570
    35.4.4 International aid 570
35.5 Summary 570
35.6 References 571
35.7 Discussion 571

35.1 INTRODUCTION

Perhaps one of the crucially important pressing problems facing developing countries of the tropics is the availability of adequately trained manpower, in particular, scientists and technologists for national development programmes. On a much broader perspective, this problem in Africa was highlighted by the following extract from the memorandum presented by the Economic Commission for Africa (ECA) to the Vice-Chancellors of African Universities at their second annual conference in Zaria (Nigeria) in 1966:

557

"There is an acute shortage of trained indigenous personnel to undertake feasibility studies, evaluate development projects, determine what projects would best further their country's development, formulate viable development programmes and manage and supervise the implementation of projects. In a number of professional technical occupations, Africans are poorly represented to the extent that their countries rely heavily on external sources of recruitment, or in order to carry out even the most modest activities' in important development sectors ..."

The memo goes on:

> "... African countries experience trained manpower shortages because their educational systems have not yet been devised to meet the challenge of rapid transition from a traditional economy to an industrial one. Educational programmes in both objectives and course content are lacking in those elements that most rapidly foster innovation and economic growth."

Although this statment was made over 12 years ago, it is still quite valid even today. The views so vividly expressed in these statements are particularly relevant to the manpower requirements for the effective management of insect pests and disease vectors whose activities largely determine the state of feeding, nutrition and well-being of the human populations in tropical countries.

In this paper, an attempt will be made to examine very briefly the insect-pest and vector problems of developing countries, followed by a discussion of some issues connected with insect-pest and disease vector-management training. Proposals for a deliberate training strategy for rapid production of scientists and technologists are made, together with an indication of the roles expected to be played by national governments, industry and international agencies in assisting tropical developing countries to train indigenous scientists and technologists for the management of insect pests and disease vectors.

35.2   DIAGNOSIS OF THE PROBLEM

The need for developing countries to train scientists and technologists can be best appreciated against the background of the numerous development problems which face these countries, especially those in Africa. Fortunately the papers by R. Kumar and F. Lambrecht have given details of the insect problems associated with crops and with human and animal health in the tropics. All that I wish to do therefore is to present a general overview of three major areas of concern to illustrate the magnitude and diversity of the problems challenging the tropical countries.

First is the question of food. Developing countries do not produce enough food for their teeming populations. While they produce only 30% of world food, they account for over half of the world population. Growth rate of food production is estimated at 1% per year with an average annual growth rate of the human population of between 2.0 to 2.5%. There is thus a serious food gap between food production and the food needs and the gulf between starvation and adequate feeding continues to widen. The significant point which must be stressed here is that this depressingly unfortunate situation occurs in the tropics, which has the greatest share of the total world land resources still available for agricultural production. For example, out of the remaining 1.8 billion hectares of cultivatable land in the world, 1.1 billion hectares (i.e. 61%) is to be found in the tropics (Odhiambo, 1976). This land has great potential for high agricultural production because of the availability of high temperatures, rainfall, abundant solar radiation and almost continuous favourable biological growth conditions, if only these resources are scientifically and efficiently exploited for food production. But among the environmental factors which reduce crop yields, pests are considered of major importance. May (1977) concluded that 48% of world's agriculture production is lost to pests and disease organisms in the field and in storage. The pest problem is particularly serious in the tropics where conditions favourable for the development of pests seem to prevail all the year round and detailed biological knowledge of the pest complexes and their economic effects is grossly lacking.

Second, in the prevalence of vector-borne diseases in these countries, malaria claims the lives of at least 1 million people, mostly children, every year in Africa alone. Service (1979) points out that in India there were an estimated 2.8 million cases of malaria in 1974 and that the present estimate of malaria sufferers is about 10 million or more. He attributed this deplorable situation to many "human and political" reasons including "poor surveillance and lack of trained personnel". In West Africa, onchocerciasis, transmitted by the blackfly of the *Simulium damnosum* complex, renders the occupancy of many agriculturally productive river valleys impossible. The situation in West Africa is well illustrated by Service (1979) as follows:

> "the highest endemnity occurs in West Africa and the worst foci are in Ghana and upper Volta. In heavily infested communities some 20 per cent of the population, mostly males over the age of 20, may have serious degrees of impared vision or total blindness. They become a very serious burden to the community which is already living at subsistence level, and on top of this the younger people start moving away which leads to a general deterioration in productivity of the community. Finally, everyone may eventually abandon the fertile but fly infested valleys for less fertile areas away from rivers, which then tend to be overgrazed and the soil exhausted by intensive farming."

Similar gloomy stories can be told of other tropical diseases such as sleeping sickness, schistosomiasis, leishmaniasis, yellow fever, filariasis and others which are widespread in tropical countries. An understanding of the biology and control of the vectors of these diseases is urgent and important if these countries must free themselves of these sources of human suffering and waste. Third is the socio-economic issue of industrial development and urbanization characterized by the rapid build up of shanty squatter, satellite settlements in huge slums around the crowded cities. Sanitation here is of the lowest standards imaginable. Water shortage, inadequate waste-disposal systems and overcrowding are typical of these peri-urban slums. These conditions favour the rapid development and transmission of arthropod-borne diseases which affect the health and productivity of the human populations residing there. Clearly then the developing countries are saddled with the challenge of insect pests and disease vectors which need immediate confrontation. The crucial issue to which these countries must address themselves is how to organize their financial, human and material resources to manage insect pests and disease vectors in order to drastically reduce, and possibly eliminate altogether, their effects on development activities. The management approach is believed to offer the most flexible, efficient and permanent solution to pest and vector problems.

Management is used here to include:

(a) Complete understanding through scientific research, of the ecosystem and the dimension of pest and disease vector problems of the area.
(b) Accumulation of knowledge about the identity, biology, behaviour and ecology of the insect pests and disease vectors.
(c) The development of environmentally sound and acceptable technologies to reduce the ravages of pests and disease vectors to economically tolerable levels.

In order to organize effective management, there must be a cadre of indigenous scientists and technologists in the developing countries since they have a key role to play in the development programmes of their countries. But the sad fact of the matter is that most developing countries lack the necessary scientific manpower capabilities to meet research and development needs. Quite apart from the mere shortage of trained scientists, two other aspects of the manpower situation seem to compound the problem. In many developing countries, particularly in Africa, young scientists and technologists who have been trained and who should remain on the bench as practising professionals soon become saddled with administrative duties which take them completely away from research; in some cases they are forced, because of lack of enough people, to be appointed to jobs for which they were not originally trained. Thus their expertise is lost to scientific research and training. This is not to say that their contributions in administration are not valuable; the point being made is that there is a kind of local "scientific and technological brain drain" which reduces the number of people available for active scientific and technological research on the identified problems of their environment.

The second aspect is that of national recognition. Scientists and technologists in developing countries do not seem to be encouraged by the societies in which they live, to develop an interest in the science and technology which is relevant and will directly benefit their countries and especially the poor in their midst. Unlike priests, lawyers and medical practitioners, they do not yet enjoy a high social status in the community and therefore they are forced to depend upon international sources for scientific recognition. (It may well be that developing societies do not quite understand what role scientists and technologists could and should play in the community.) In frustration, scientists and technologists fashion their science after Western patterns, and pursue prestige scientific research of the sort which will lead to publications in internationally reputable scientific journals and thereby receive international recognition to the neglect and detriment of research into the local pressing problems of hunger, malnutrition, poverty and diseases. Very low priority is given to funding and development of relevant basic research which will enable them to evaluate accurately and develop the technology that is really needed by the community for rapid industrial and socio-economic development.

Because of the diversity and magnitude of the development problems in tropical countries, the pressure and demand for well-trained scientists and technologists is extremely high. These specialists either have to develop their science and technology relevant to the problems of their countries or, like Japan did during the 1950s, acquire the latest and most productive technologies already developed in industrialized countries and modify these for introduction into their own systems. They may even do both. Indeed experience has shown that successful development and adoption of appropriate technologies for any country can be best achieved by indigenous scientists and technologists who are naturally equipped to appreciate and understand the socio-cultural background and economic panorama of their communities, and who must have the "mental freedom to explore and elucidate their own scientific problems". The late Homi Bhabha of India strongly believed that the developing countries must have indigenous scientific capacities equivalent to that of industrialized countries for successful acquisition and transfer of technology.

The case is, therefore, hopefully made for developing countries in the tropics to embark on deliberate programmes to train large numbers of indigenous scientists and technologists for the management of insect pests and disease vectors which constitute impediments to rapid socio-economic development.

35.3 TRAINING PROGRAMMES AND TRAINING NEEDS

National institutions exist in most developing countries for training professional staff in aspects of insect pests and disease vector biology and control at various levels. These range from undergraduate and postgraduate programmes in universities to diploma and certificate course in polytechnics and advanced colleges of agriculture and animal health schools. The first two years of the normal 4-year university undergraduate programme is usually occupied with courses in the basic sciences such as botany, zoology, biology, chemistry, physics, mathematics, climatology and biochemistry. This solid science background is highly desirable for an understanding of advanced courses in agricultural, biological or veterinary sciences. The last two years are devoted to general agriculture or veterinary sciences, in some cases with specializations in applied entomology, plant pathology, ecology or plant breeding. From the faculties of science, graduates in biology, zoology or botany are produced. At higher levels of masters and doctorate degrees, further specialization occurs and the products are employed in research institutes to continue with full-time research or they go into university teaching and research. A large number are absorbed into the civil service in ministries of agriculture and health or in the agro-based industries of the public and private sector.

When these programmes are examined critically, the following picture emerges: at the bachelor degree level, a generalist is produced while a specialist is the product of the higher degree programme. Hardly any institutions provide any kind of multidisciplinary training in integrated insect pest and disease vector management which is a real need of developing countries.

Unfortunately most of the institutions of higher education in developing countries are poorly staffed, especially with trained indigeneous professionals. In many cases, departments are poorly equipped and the necessary logistic support for teaching and research are not available so that the departments are unable to admit and train a respectable number of students to meet the needs of their countries for research and development. The few professional staff present are overburdened with teaching and administrative duties and therefore have very limited time for meaningful research into the development problems of their environments. One consequence of these training inadequacies is that scientists and technologists from developing countries, flood the industrialized countries like England, the United States of America, Germany and France to study for various degrees (Table 35.1). There, they are involved in sophisticated research and acquire expensive scientific and technological tastes which may be largely irrelevant to the peculiar development problems of their home countries. On returning home, they grapple with readjustment difficulties, become frustrated by grossly inadequate organizational and other arrangements which prevent them from performing satisfactorily and utilizing their training for research towards national development. I have no doubts, whatsoever, about the value and significance for scientists and technologists in developing countries to be

TABLE 35.1. NUMBER OF TRAINEES AND STUDENTS FROM DEVELOP-
ING COUNTRIES ASSISTED BY OFFICIAL AID (MODIFIED AFTER
ODHIAMBO, 1977)

Trainees	1965	% of total	1973	% of total
In donor country	29,072	93.8	40,773	86.5
In country of origin	335	1.1	1149	2.4
In third country	1584	5.1	5233	11.1
Total	30,991	100	47,155	100

Students	1965	% of total	1973	% of total
In donor country	27,842	91.1	44,633	87.7
In country of origin	985	3.2	4274	8.4
In third country	1743	5.7	1961	3.9
Total	30,570	100	50,868	100

Note: In both years and for both trainees and students, over 85% of them were trained in the donor country compared with less than 10% training within the country of origin of the trainees or students. Thus training of people in developed countries is higher than training in developing countries by a factor of 8.

regularly exposed to recent advances in scientific research and development and to earn some of the new and novel technologies which can be suitably adapted to their needs. The entire exercise should, however, not degenerate as often happens, into exact replication, in developing countries of what has been seen in the industrialized countries.

In effect most of the present training institutions and programmes are inadequate in providing the right kind of orientation needed for development-oriented research. In addition, there is a lack of real coordination and collaboration of training programmes at regional or international levels. In some cases there is even no national coordination of training programmes. The advantages of regional and international co-operation in training for developing countries are numerous but of utmost importance is the maximization in the use of available limited financial and other resources. But what are the areas of training that need extensive development? Training for insect pest and disease vector management can be conveniently grouped into three major categories as follows:

35.3.1 Professional training

This category includes training of scientists and technologists at university level. Trainees undergo the usual undergraduate and postgraduate program-mes and on graduation become involved in fundamental research in order to provide a sound basis for developing long-term management systems for insect pests and disease vectors. The major component of this category of training is postgraduate research which provides a variety of high-level scientific and technological manpower needed by tropical countries. This is an important area of training in which the International Centre of Insect Physiology and Ecology (ICIPE) in Nairobi, Kenya, has developed expertise and has concentrated in recent times (Table 35.2.)

TABLE 35.2. NUMBER OF SCIENTISTS TRAINED OR IN TRAINING BY THE ICIPE RESEARCH CENTRE DURING A 6-YEAR PERIOD (1974–1979)

Mode of training	Trained/in training
Professional graduate training	25
Research associateship scheme	3
Group training in pest- and vector-management systems	75
Science bursars scheme	55
Technical training programme	133

Besides these traditional undergraduate and postgraduate programmes, there is a further need to introduce programmes which would provide training in integrated pest and vector management for the tropical communities. This kind of programme in crop protection has already been designed (Youdeowei and Adeniji, 1978) with particular reference to the pest problems of the peasant, small-scale African farmer. The project was supported by the FAO Plant Protection Service and the Rockefeller Foundation under the auspices of the Association of Faculties of Agriculture in Africa (AFAA). In this programme, African institutions of higher agricultural education are encouraged to introduce a two-year postgraduate degree course leading to a master of science in crop protection, following a flexible model already approved by the FAO and AFAA in 1978. This course was designed specifically for the tropical African environment, with special emphasis on the broad range of the pest problems of peasant, small-scale farmer mixed cropping systems. The ultimate goal is to produce a cadre of highly trained pest-management specialists in the area of integrated pest management who can function as agricultural extension specialists, pest-management consultants, research scientists and technologists.

In short, professional training programmes will build up the manpower who will not only be primarily concerned with research and development of appropriate technologies for pest and vector management but also will participate fully in the field application and evaluation of the various management systems developed.

I have advocated elsewhere (Youdeowei, 1979) that this concept of training for pest and vector management leads to the emergence of a new breed of professionals called Pest and Vector Managers in the scientific communities of developing countries. I wish to emphasize that there is an

...

urgent need for the international scientific community to accept this concept of training for pest and vector management and to recognize this new career in tropical countries.

35.3.2 Field operators

Money and effort spent on research and development in pest and vector management will be wasted unless the results of these efforts are put into practical use and this requires efficient and well-trained men and efficient extension services to operate at field levels. The field operators here therefore include the extension specialists who according to O. C. Onazi "must be mature and experienced persons who are not only well trained in their professional area (technical subject) but also have knowledge and experience in extension methodology". The extension services in most developing countries suffer the greatest shortage of trained manpower at all levels. Surveillance personnel, spraymen, technical support personnel and other field staff are also included in the category of Field Operators that need adequate training. The recently established UNDP/FAO Tsetse Applied Research and Training Project in Lusaka, Zambia hopes to train field operators in tsetse-fly control.

35.3.3 In-service training

In-service training is the system in which professionals and other personnel who are actually in employment receive instructions in order to improve their competence and skills on their jobs. This is sometimes called "continuing education". Seminars, workshops, field courses, postdoctoral research programmes are all included in this category, and all levels of staff can benefit from in-service training. This category of training in relation to integrated pest and vector management has received very limited attention within the tropics. There is a need to organize international in-service training programmes which will be multidisciplinary in concept and approach and which will combine both crop-pest and disease-vector-management systems since these fields are usually interdependent in many respects. An initial attempt at this kind of training has been made. In 1977 the ICIPE in Nairobi, in collaboration with the United Nations Environment Programme (UNEP), initiated an International Group Training Course on the Components Essential for Ecologically Sound Pest and Vector Management Systems. The objectives of this course were defined as follows:

(a) to provide a sound basis for scientific, modern and clear understanding of pest and vector management problems taking into account the preservation of environmental quality;

(b) to establish a foundation for field scientists who are concerned with or who will be assigned the responsibility for pest and vector management in developing countries;

(c) to provide trainees with information on modern trends and approaches in integrated pest and vector management, stressing all the components essential for ecologically sound management systems;

(d) to provide an opportunity for exchange of ideas and experiences between pest and vector control operators from different geographical and ecological zones in the developing world; and

(e) to develop personal and international co-operation between scientists in the area of practical pest and vector management and to encourage further consultation and advice regarding specific national and international pest and vector management programmes.

This in-service course was designed for young scientists in the field of agriculture, public health or livestock health who are actively engaged in pest and vector management or who are starting their professional careers in these fields. Trainees also included research scientists and university lecturers. During the first 4 years of organizing this course, a total of 100 scientists from 22 developing countries have been trained successfully. They are in turn expected to train middle-level personnel in their home countries on new and up-to-date pest and vector-management methods and thereby help to build up the pest and vector management capabilities of their countries. A follow-up evaluation of this training, 12 months after the trainees had returned home showed that this course has proved to be highly relevant and particularly useful to the trainees in their pest- and vector-control programmes. They have also been able to establish international links with their colleagues in other Third World countries. One area of in-service training which has not been sufficiently recognized and explored is what the ICIPE calls the Postdoctoral Research Fellowship and the Research Associate Schemes. The Postdoctoral Research Fellowship Scheme provides young scientists who have just completed their doctorate programme the opportunities to expand and deepen their research experience in a different tropical environment for 1–3 years. The Research Associateship scheme enables mature scientists from a developing country to undertake research in a research centre in another developing country for short periods (3–6 months) every year for up to 4 years. This scheme promotes intellectual intercourse between research scientists in developing countries.

The ICIPE Research Centre in Nairobi has been in the forefront in pioneering efforts in these areas of training (Table 35.2) and efforts should be made to encourage an expansion of these schemes. Another aspect of in-service training often overlooked is the education of administrators and politicians on the level of priority which should be given to research and training in pest and vector management in developing countries. As funding for research and training depends on the goodwill of the politicians and administrators, it is essential that these people are constantly made aware of the existing problems and the financial implications and level of commitments required to tackle them. Seminars and workshops organized for politicians, administrators, policy-makers and those in charge in government on pest and vector problems can greatly influence policies once these people are fully aware of the economic effects of insect pests and disease vectors and the relevance of new approaches for their management. In other words, the pest and vector problems must be made political issues in developing countries in order to give them the priority rating which they deserve.

35.4 SOME MODEST PROPOSALS

35.4.1 Research and training centres

It is clear then that a need arises for the establishment of some form of institutional arrangement within the developing countries to develop and provide these training courses. One practical way to achieve this arrangement is to identify a few specialized Regional Research and Training Centres located within the developing countries.

The functional organization of such a centre is presented in the following flow chart (see Fig. 35.1). The centres would be funded and fully supported by national governments, industry, international foundations, United Nations agencies and foreign donor organizations. Two distinct functions will be carried out, namely fundamental research on insect pests and disease vectors and training. Research will be specifically development-oriented for application to pest and vector management. For both functions, the centre will collaborate with national universities and other higher training institutions which have authority to award diplomas and degrees; the centre providing the research base for professional training of scientists and technologists. This can be readily achieved by means of collaborative agreements between the centre and the educational institutions. The ICIPE has concluded collaborative agreements with the Universities of Ibadan (Nigeria) and the University of Nairobi (Kenya) for professional training of scientists and technologists at the postgraduate levels. Apart from professional training, the centre will train field operators and organize in-service training in the form of formal lectures, seminars, study workshops, symposia, group training and field courses for all levels of staff, involving participants from different countries.

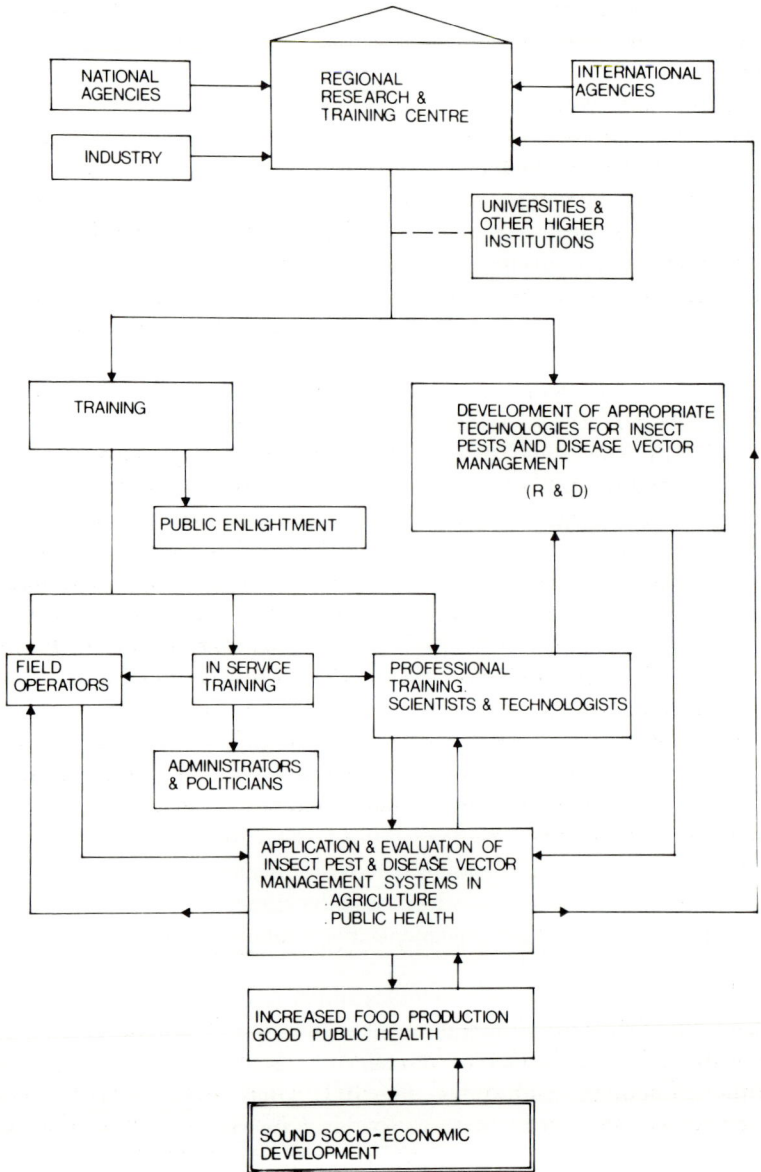

FIG. 35.1. The functional organization of a model Research and Training Centre for Insect Pest and Disease Vector Management.

To foster community participation which is especially important for vector-management programmes the centre would develop public enlightenment activities. Special seminars for administrators, politicians and policy-makers in governments would also be arranged as part of the training programme of the centre. These functions will lead to the development, application and evaluation of insect-pest and disease-vector-management systems in agriculture and public health and eventually to increased food production and a healthy human population. A feed-back mechanism will be established for referring problems of pest and vector management in the field to the Regional Research and Training Centre for investigations and development of appropriate technologies to solve them.

The efficient execution of all these activities would help to achieve sound socio-economic development. The unique factor in the activities of a few well-supported and organized regional research and training centres is international and regional co-operation, exchange and cross-fertilization of information and ideas on pest and vector management.

The ICIPE Research Centre in Nairobi is already pursuing such activities vigorously and it typifies this model of an "activity centre" for research and training for insect-pest and disease-vector management in the developing world. The outstanding training successes achieved by the ICIPE within such a short period in spite of her limited resources has clearly demonstrated that such research and training capabilities can be achieved within the developing countries. I suggest that the ICIPE be now identified as the first and the major Research and Training Centre for Pest and Disease Vector Management in the developing world and that the developing countries, international agencies and other donor organizations (including the Organization of African Unity) provide adequate financial and other supporting facilities to enable the ICIPE to expand her training activities in the field of ecologically sound pest and vector management systems.

35.4.2   National commitments

The establishment of Regional Research and Training Centres have implications for the developing countries themselves. Apart from providing adequate financial support the governments of these countries need to recognize pest and vector management as a matter of utmost priority and commit the necessary resources of manpower for training and the materials for the successful implementation of management programmes. It is their responsibility to determine the priority areas of training needed and to make accurate plans and projections of their manpower requirements for socio-economic development. Adequate recognition, satisfactory job and career opportunities must be created for scientists and technologists so that these specialists can be retained to perform the jobs for which they will be trained.

35.4.3   Assistance from industry

Research and development activities will require the production of materials for field application against pests and vectors from time to time. This is where industry inputs become relevant especially in integrated rural development schemes. It is therefore essential for industry to become involved at the research stage in order to hasten the build up of appropriate technologies. UNIDO's interest in the ICIPE activities is therefore a very welcome development.

35.4.4   International aid

The poor developing countries need considerable assistance from International Foundations, United Nations Agencies and other donor bodies to effectively develop research and training programmes and to enable them achieve self-reliance in these matters. Funding and expertise are essential but the philosophy of the aid is even more important; and this is where problems have arisen in the past. I believe that one reason for the failure of many development aid programmes in developing countries is the erroneous assumption that the donor knew what was best for the recipient country and proceeded to provide these only to discover, too late, that the wrong assumptions had been made. It has been stressed in this paper that the indigenous scientists and technologists of a country are the people who are best equipped to evaluate and determine their countries' needs and therefore international development aid should adopt the system of working in partnership with the personnel in the poor countries.

For international aid to succeed in helping the development of insect pest and disease vector management in tropical countries, I do not see any aid philosophy better than the concepts conveyed in these words of Albert Camus when he said:

> Do not walk in front of me — I may not follow.
> Do not walk behind me — as I may not lead.
> Walk *beside* me — and be my friend.

35.5   SUMMARY

The desire to drastically increase agricultural production and improve the life style of the people in the developing countries of the tropical world has become a matter of international concern. Substantial food shortages and the prevalence of insect vectors of human and animal diseases which are characteristic of these countries need the immediate attention and positive action of all scientists and technologists in the developing countries. Extensive scientific research into pest and vector problems and the development of appropriate technologies can provide environmentally acceptable and long-term management systems for these pests and vectors. Such an action plan emphasizes the urgency for developing countries to build up a cadre of well-trained indigenous scientists and technologists — pest and vector managers, to meet these development challenges.

The introduction of development-oriented research, specific training programmes for professional scientists and technologists, yield operators and regular in-service training for these specialists, including training of administrators and politicians, is essential for the achievement of national capabilities in pest and vector management. A generalized model of Regional Research and Training Centres, working in close collaboration with national universities and other institutions of higher education, is proposed to meet the training needs of tropical countries.

35.6 REFERENCES

May R. M. (1977) Food lost to pests. *Nature* **267**, 669–670.

Odhiambo T. R. (1976) *National Scientific Capabilities*. ICIPE 2nd Annual Public Lecture. ICIPE Nairobi Kenya. 20 pp.

Service M. W. (1979) *Socio-economic Considerations in the Control of Vectors of Human and Animal Diseases*. ICIPE/UNEP Group Training Course in Pest/Vector Management Systems Lecture. 15 pp.

Youdeowei, A. (1979) Training pest and vector managers for African agriculture. *Popular Technology, London* (in press).

Youdeowei A. & Adeniji M. O. (1978) *Master of Science Programme in Crop Protection*. Consultancy Report, AFAA. 75 pp.

35.7 DISCUSSION

Johnson: Can you give any idea of the number of potential technicians in developing countries who would like training?

Youdeowei: I do not have the figures for all developing countries. However, in Nigeria, we have many people who desire to have training, but our training facilities are inadequate. For example, at least for every candidate admitted, 100 qualified potential technicians are left out. I hope this gives you an idea of the magnitude of the problem we face in the developing countries.

Patel: I would elaborate on this issue further by citing our own experience at the Engineering School of the University of Nariobi. About 600 to 700 qualified people apply for admission into our school. Of these, only 175 are admitted because the facilities we have can not allow us to take more. We thus have to turn away a large number of young people aspiring to receive training in the engineering profession. Most developing countries experience closely related problems.

Bowers: I think that the incorporation of a well-organized and effective extension service into the education system of the developing countries would be highly beneficial to their training efforts.

Youdeowei: I agree, but first, we would need a sufficient number of technical experts in order to execute a novel proposal like this. Unfortunately, the few highly skilled people we have do not wish to go into extension service, and this shortcoming has tended to perpetuate the shortage of skilled manpower in the area of extension service.

Kahumbura: Would you say that most developing countries who have been independent for 15 years have made satisfactory progress in their attempts to provide training facilities and institutions to their nationals?

Youdeowei: I would say that most developing countries have made remarkable progress in providing training for their nationals despite lack of adequate funds, training facilities, institutions and skilled manpower.

Taylor: In Nigeria, the Stored Products Research Institute has been in existence for a very long time, and I find it strange that it does not yet have a training function. Many trainees of this institute still go to the UK for specialist courses. Could Professor Youdeowei tell us why NISPRI has not developed an effective training function?

Youdeowei: The Nigerian Stored Products Research Institute has not developed the capacity necessary to run an effective training function. It has insufficient funds and lacks highly competent manpower. It is considering starting this function as soon as funds become available.

Taylor: The ICIPE has an effective and impressive training function. Could you please elaborate on how you evaluate student performance especially those who attend the ICIPE/UNEP Group Training Course?

Youdeowei: We give students attending the ICIPE/UNEP Group Training Course questionnaire forms to complete at the end of their training, and from the entries, we are properly advised on the extent to which the students have absorbed the subject matter of the course. In order to determine the student's baseline knowledge on pest management, we have this year introduced a simple objective test at the beginning and at the end of the course. The average score in this test was 55 %, at the end of the course which is an encouraging indication of how the course contents were assimilated. Further to this method of evaluation, we send out a letter to each participant 12 months after the end of the course. In the letter, we ask how useful the training had been and how it helped improve the individual's work and effectiveness.

Taylor: Do you personally visit the students in their area of work after the training period to actually observe whether they have improved their performance?

Youdeowei: Not yet. We have not been doing this in the past due to lack of funds. It is, however, our intention to do so in the future when funds become available.

Abe: We have made one trip to the ICRISAT to assess the performance of one of ICRISAT employees who was one of our first trainees, and found that his performance and effectiveness has tremendously improved after he attended the ICIPE/UNEP Group Training Course.

Gebreyesus: Professor Youdeowei, given a limited amount of funds to train people in pest and vector management, would it be wise to train few people to the highest sophistication or provide low-level training to a large number of people or do both?

Youdeowei: Public enlightenment and the education of the politicians and the administrators is as important in pest- and vector-management campaigns as the skilled personnel. Both components are essential for the success of vector- and pest-management strategies. Under the circumstances specified in your question, I would advise an approach which incorporates these two components provided financial capabilities permit.

Jondiko: One effective instrument for enlightening people on the pest and vector problems is the local mass media. Have you thought of informing the population about pest and vector control methods via this process?

Youdeowei: Yes, I have. First, I have started a newspaper called AGRI-SCOPE which will inform general readers in agriculture, forestry and veterinary science about insect pests and disease vectors. Secondly, I have recommended that one of the functions of the proposed Regional Research and Training Centre be public education in the general field of vectors and insect pests. I agree that it would be beneficial to our efforts in combating insect pests and disease vectors, if scientists and technologists in the developing countries wrote in simple language about insect pests and their management in the local newspapers for the non-specialist public. Unfortunately, only a few of the scientists in these countries consider it worthwhile to do this.

Kumar: There is inadequate supply of well-trained scientists and agriculturalists in the developing countries especially those in the African continent. Furthermore, there is also an acute shortage of scientific books and training material written by indigenous scientists. I think this is one of the sad situations which frustrate our efforts in combating insect pests and disease vectors in these countries.

Kabaara: I think that intensified training of high-level manpower will improve this situation advantageously. However, it is equally important to plan such training carefully, and to made adequate provision for the employment of trained people when they come out of the training institutions.

Musoke: I would like to stress this point further by adding that poor coordination between the authorities responsible for training and those planning the manpower requirements for the different sectors of the economy leads to unemployment and encourages brain drain thus stagnating the development process in developing countries.

Youdeowei: Yes, close collaboration of these sectors is needed so that training institutions can train people who will be immediately absorbed into jobs. Nonetheless, there are two important features which have a direct bearing on the unavailability of scientists in the developing countries. First and foremost, is the lack of adequate funds to provide enough training institutions and associated facilities, second is the inability of the economies of these countries to provide the intellectual atmosphere necessary for indigeneous scientists to achieve their professional satisfaction and feel just as capable as their colleagues in the industrialized countries.

Young: Mr. Chairman, I was wondering whether Professor Odhiambo would now comment on what has so far been said on the subject of training scientists?

Odhiambo: I do not think I have much more to add to what Professor Youdeowei has just said. However, let me make two important observations. In the first place, I believe that in order to retain the few trained scientists we have, we must develop an intellectual environment in which the scientist can enlarge and enrich his mental capacity and satisfy his professional ambition. In a small way, and despite limitations on our facilities, we have ensured this intellectual atmosphere at the ICIPE; and, as a result, we have not lost any of our scientists from a developing country to another.

In the second place, a scientist working in a frontier area of research must have very clear objectives of his mission. In contrast to senior scientists in government institutions, who are not fully clear of what to do or what is expected of them, the scientists at the ICIPE have very clear and well-defined objectives. Further, they are fully aware of what is expected of them. I think that this is very important, as it gives the scientist the necessary stimulation and direction in his effort to pin down the research problem. We can develop the desirable intellectual environment only in centres of excellence such as the ICIPE.

Kumar: Another essential component of the intellectual atmosphere necessary for a scientists' work is forums for the discussion and exchange of ideas beneficial to the scientists' work. At the ICIPE, several of these forums are organized from time to time.

Bowers: Taken together, the conditions which have been stated by Professors Odhiambo and Kumar have led to the remarkable success for which the ICIPE is well known in the entire world. Research Institutions in the tropical world might consider it worth while to emulate the ICIPE's example.

Index

Abate   395
Acacia   312
Acaricide chlordimeform (CDM)   14, 22
Acceptability testing requirements   530
Acceptable daily intake (ADI)   530, 532
Adzuki bean weevil   372, 377, 378
Aedes aegypti   358, 363, 364
Aedes mosquitoes   392
Ageratum   355
Ageratum houstonianum   52, 357
Agricultural exports   505–6
Agrochemical industry   7, 488
Agro-climatology   431
Ajugarin   194
Alarm pheromones   66
Aldicarb   7
Alkaloids   189, 222
Alkylating agents   344
Allethrin   98, 112, 116, 118, 129, 131–3
Alloboletic acids   76
Allomones   67
Allround screening   37–8
Allylisothiocyanate   257–8
Allylthiocyanate   173
Altosid   20
American trypanosomiasis   473
Amino acids   74, 84–5, 274–7
Amoebiasis   387, 403
Anabasine   313
Angelica pubescens   305–7
Angelicin   170
Anopheline mosquitoes   390
Antiallatal agent   325
Anti-allatotropins   64–5, 324
Antibiosis   209
Antifeedants   191–4, 217, 223–36, 311
  botanical occurrence   227–8
  choice of insects for testing   228–9
  discussion   269–71
  field persistence   266
  field trials   265–7
  from *Clausena anisata* (Willd.) Hook F.
      Ex Benth (Rutaceae)   237–42
  in crop pest management   259–71
  in waste materials   228
  logistics   267
  natural occurrence   259–60
  preliminary field tests   229

prospects   235
research   226–31
search for   260–7
sprayed   264, 266
Antifungal antibiotics   74
Anti-fungal compounds   74, 81, 82, 279–97
  naturally occurring   291–2
Anti-fungal defence   80–4
Anti-juvenile hormones   52, 57, 316, 324,
      358, 364
Appropriate technology (equipment)   513–8
Armyworm   423
Arrestants   244–7
trans-Asarone   255
Association of Faculties of Agriculture in
      Africa (AFAA)   562
Ateleia herbert smithii   84–5, 88
Atherigona soccata   357–68
Attractants   192, 199, 217, 244–9, 251, 481
Azadirachtin   16, 194, 261, 262, 264
Azinphos-methyl   118

Bacillus thuringiensis   496, 497
Bacterial infections   387
Balsam fir   315
Bartinia   354
Bay region theory   339
Bean varieties resistant and susceptible to
      Mexican bean beetle   273–8
Bedbugs   397
Behaviour-controlling chemicals   470–1
3-Benzylbenzyl ester   136
4-Benzylbenzyl ester   135
Benzyl chrysanthemates   133–4
BHC   26, 398, 473, 474, 480
Bifforin   202, 214
Bilharzia   401
Bioallethrins   131
Bioassay   262–4
Biochemical design   39
Biochemical profile   225
Biochemically inspired modifications of
      insecticides   17–20
Biodegradable insecticide, discovery
      approaches   11–22
Biological control agents   497
Biological control methods   496

Biologically active chemicals   463–72
Biologically active natural products   74–85
Biorational pest control   20–2, 251
Bioresmethrin   61, 134–6
Black Death   399
Blackfly   388, 395, 557
Blindness   557
Boletic acids   74–9
Boll weevil   423
Bollworm   353, 422
Borelia recurrentis   399
D-Bornyl acetate   317
Botanicals   451–60
BRAVO   46
Brazil   473
Breeding of resistant plants   250–1
Brown Plant Hopper   124
Bufencarb   7
Bugs   386, 388, 397, 442–3
Burning of stubs   429

Cabbage fly   257–8
Cadalene   212
Cadinene   212
L-Canaline   192
L-Canavanine   192
Cancer risk from precocenes   342–4
Capillin   210
Carbamates   436
Carbaryl   7, 399
Carbofuran   7, 19
Carcinogenicity   399, 343, 348, 353
Cardenolides   214
Carthamus tinctorius   301–3, 309
Cassava   425
Cedrene   197
Celastrol   203
Cereals   424, 427
CGA-13353   39
Chagas' disease   162, 386, 473
Chebulin acid   208
Chemical control agents   496, 497
Chemical defenses   167–73, 175, 178
Chemical modifications   40
Chemical pesticides   485
Chemosensory mechanism   224
Chemosterilants   456–8
Chickpea   439, 445–6
Chitin synthesis inhibitors   21
Chlordane   404
Chlordimeform   34, 37
Chlorinated hydrocarbons   7, 313, 509
Chloroform   256
Chlorophenyl isovalerates   114
Chlorophorin   205
Chlorophrifos   400

Chlorophyll   274, 277
Chlorphoxim   395
Chlorpyrifos   395, 475–7
Chromanocoumarins   242
Chromene derivatives   323
Chromenes   21
Chrysanthemic acid   92–5
(1*R*)-*trans*-Chrysanthemic acid   93
Chrysanthemum dicarboxylic acid   94
CIMMYT   441
Cinergrin   256
Cinerins   60
Cinerolone   93
Clausena anisata (Rutaceae)   237–42
CO_2 anaesthesia   359
Cockroach   403, 479
Cocoa   419, 420, 422, 426, 436–8
Coconuts   419, 423, 424, 429
Codex Alimentarius Commission   534, 536
Coevolution of plants and insects   165–82
      chemical aspects   178
      criticisms of   166–7
      diffuse   166, 175–8
Coffee   213, 417, 420, 426
Colchicine   457
Colorado beetle   235
Compositae   354
Conjunctivitis   387
Controlled droplet application (CDA)   466–7
Controlled release   468–70
Co-operation between countries   489
Copaiferic acid   206
Corpora allata (CA)   323–48, 354–5, 358
Cosmopolitan pests   426
Cost/benefit problem   505–11
Cotton   417, 422, 427
Cotton leaf worm   423
Cotton stainer   318
Coumarins   182, 198, 199, 239, 242
Cowpea   425, 430
Crop damage   484–5
Crop loss   431, 484
Crop protection   90, 311, 484
      chemicals   99–101
      techniques   432
Crop rotation   429
Cruciferous plants   167–9
Crude plant materials   229
L-α-cubabene   68
Cube   55
Culex mosquitoes   391
Culicoides   396
Cultural practices in tropical
      agriculture   429–30, 433
α-Cyanobenzyl esters   139
α-Cyanoesters   140–1

Cyclopentenolones 130
Cyclopropylcarboxylic acid 295
Cypermethrin 117

Dalbergin 203
Dalbergiones 203
Daphne odora Thumb 303–5
DDT 6, 11, 26, 34, 36, 43, 48–50, 67, 71, 72,
   117, 120, 123, 395, 397, 399, 400, 404,
   406, 428, 436, 452, 480, 489, 509, 538
Decamethrin 115–7, 140–1, 143, 150, 162
Defence substances
  formed as response to insect
      infestation 206–7
  in plant resistance 200–6, 217
Deguelin 190
Delany clause 537
Delayed release 469, 470
N-Demethyl CDM 22
Derris 453
Deterrence chemistry in non-host
      plants 225–6
Deterrents 225, 244, 252, 261, 263
Developing countries 29, 48, 483–511, 533
  training requirements 555–73
Development problems 559
Development programmes 558
Diazinon 397
Dibenzuron 38
Dichlobenil 15, 37
N-(2,6-Dichlorobenzoyl)-*N*-(4-chlorophenyl)-
      urea 37
2,2-Dichloro-3,3-dimethyl cyclopropane car-
      boxylic acid 80
4,4′-Dichlorodiphenyl-sulphone 36
Dichlorvos 18
Dieldrin 403, 404
2,4-D 27, 489
2,4-Dienoic ester 20
Diflubenzuron 34, 37, 38
Dihalovinyl cyclopropane carboxylates 137–
      8
Dihydropipercide 370, 371
  insecticidal action 377
  structure determination 374
  synthesis of 377
DIMBOA (2,4,-dihydroxy-7-methoxy-1,4-
      benzoxazin-3-one) 288
Diptera 117
Disease resistance 279–97
Disease vector-control chemicals 99–101
Diterpene 194, 206
Diuretic hormones 65
Diuron 15
DNA 241, 337
Drug plants 420, 426

Du-19111 15
Dutch Elm Disease 67
Dynamic devices 514–6

Earhead bug 442–3
Ecdysones 51, 197, 213
Ecological factors in tropical agriculture 413
Ecological time 176
Economic Commission for Africa
      (ECA) 555
Economic factors in tropical agriculture 417
Efficacy testing 528
Efficiency of utilization 464–5
Electrophysiological recordings 246
Elicitors 283
Embelin 213
Encapsulation, natural pyrethrum 156–9
Endemic relapsing fever 386, 398, 399
Endocrine regulation 51
Environment factors in tropical
      agriculture 431
Environmental chemistry data
      requirements 523–8
Environmental Protection Agency 48, 536,
      537, 538
Ephemeral plants 171–3
Epidemic relapsing fever 386
Epieudesmin 194
Epilachna varivestis 363
Epoxidase activity 90
Epoxides 335–8, 353–5
Equipment development 513–8
Ethylene dibromide 437
Eudesmin 194
European Patent Convention (EPC) 541,
      543
Evaporation losses 468, 470
Evaporation potential 468
Evaporation rates 468
Evolutionary time 176
Extension services 433

Falcarindiols 310
trans-β-Farnesene 318
(E)-β-Farnesene 319
Fatty acids, nematocidal 305
FDA 536, 537
Feeding stimulants 245, 250, 251
Fenthion 400
Fenvalerate 117, 118
Fibre crops 422–4, 427
Field operators 563
Filariasis 386
Flavanic units 193
Flavanoids 203

Flavoquinones   309
Fleas   387, 388, 399–400
Flies   386, 388, 393–7
Foliage of trees   173
Food additives   537
Food and Agriculture Organization   534, 562
Food production   506–7, 510, 557
Food Standard Regulations   534
Forecasting of pest attacks   433
Fourth generation pesticides   47–58, 324
Fumigants   8
Fungal diseases   74, 80, 200, 207
Fungal growth control   285–90
Fungal infections   387
Fungicides   76
Furamenthrin   116
Furanocoumarins   170–1, 177, 178, 241, 244
Furocoumarins   239

Gallotannins   207
Gene-for-gene concept   281, 295
Germacrene A   319
Germacrene D   317, 318
Glabrescia   191
Glabrescione A   191
Glossina. *See* Tsetse fly
Glossina austeni   407
Glucans   283
Glucogallin   208
Glucosinolates   168–9, 177
Glutathion (GSH)   18
Glycolipid antibiotics   87
Glycoprotein   283
Glycosides   222, 286–7
Grain legumes   425, 430
Grapevines   77–80
Groundnuts   425, 439, 446–8
Grounds   430
Guineensine   370, 371
    insecticidal action   377
    structure determination   374
    synthesis of   377

Hamamelitannin   208
Hammett's constant   42
Hansch equations   42
Haplophyton cimicidum   453
Hardwick acid   206
Harrissonin   194
Harvesting regimes   431
Hazard evaluation   536
    humans and domestic animals   526–8
    wildlife and aquatic organisms   525
Head-bug   448
Hellebore   453
Herbaceous plants   183
Herbicides   48, 403

Homomevalonate   63
Hordatin   212
Hormonal control   50–2
Hormone-like substances   191, 197, 217
Horse flies   388
Host-parasite interactions   284
Host-parasite recognition   280–4
Host receptors   284
Host resistance   280
Housefly   372, 403
Human testing   536
Hydroquinone   202
4-Hydroxy stilbenes   82
Hypericin   199

ICIPE   28, 30, 38, 45, 54, 55, 69, 71, 194,
    214, 228, 550–2, 562–5, 567, 570–3
ICRISAT   439–49, 571
Identification techniques   246–7
Imperatorin   239
Inhibitory biochemical profile   225, 226,
    228–30
Insect growth regulators (IGRs)   20, 21,
    314–6, 324
Insect-plant relationships   243–58
Insecticidal dusts   437
Insecticide chemistry   33–43
Insecticide design   60–8
Insecticide development   6, 9–11, 28
    history of 49–50
Insecticide discovery by guided
    empiricism   12
Insecticide production   8
Insecticides   389, 452–5
    biochemically inspired modifications   17–
    20
    history   6
    insect resistance to 149
    plant-derived   313
Integrated Pest Management   7, 176
International Sorghum Pest Resistance
    Testing Programme (ISPRTP)
    445–6
Investment-return balance   503
Ipolamiide   194
Iridomyrmecin   215
Iris japonica thumb   305
Irradiation effects   457
Isocordoin   191
Isolation techniques   246–7
Isothiocyanate   173

Jasmololone   94
Jassids   447

Jute 424, 427
Juvenile hormone 20, 62–4, 191–2, 197, 213,
  314–6, 324, 326, 345, 347, 367, 474–5,
  481, 496

Kairomones 67
Kaurenoic acids 205
Kenya 511
  patent laws 547–52
  pesticide production 507–9
Kinoprene 21
Knockdown agents 116
Knockdown resistance 118

Label development 528–9
Labelling of chemicals 244
Lapachol 201
Lapachone 201
Larvacides 389, 391
Lead finding 35
Lead optimization 35
Lead structure 35, 37
Leaf-hoppers 118
Lectins 295, 296
Legal protection of industrial property. *See*
  Patents and patent protection
Lignans 194
Legume crops 429
Leishmaniasis 386
Leptospirosis 387
Lice 386, 388, 399
Lindane 397, 399, 403, 404
Lipoferolide 194
Locusta migratoria 325–30, 335, 342, 344, 348
Locusts 257
Lonchocarpin 191
Longicidines 89
Longistylines 205
Lycopersicon hirsutum 453

Maize 429, 435
Malaria 386, 387, 557
Malathion 118, 397, 399, 403, 404, 475–7
Mammea americana 495
Mammein 453
Mandulea sericea 495
Mangoflies 397
Mansonia mosquitoes 393
Mansonones 201, 202, 214
Mass trapping 250, 258
Mathematical models 41
Max Planck Institute for Biochemistry 446
Mechanical insect vectors 387, 403–4
Metabolic detoxification 119

Metabolic experiments 536
Metabolic studies 526, 537
Metabolic tests 525–6
Metabolites 116
Metamorphosis 315
2,4-Methano-glutamic acid 85
2,4-Methanoproline 85
Methomyl 7
Methoprene 20, 21, 39
Methoxychlor 404
6-Methoxy-7-geranylcoumarin 198
Methropreme 62
2-Methyl-buten-2-ol 216
2-Methyl-3-buten-2-ol 216
Methyl carbamate 7, 55
N-(2-Methyl-4-chlorophenyl)-*N'*-
  dimethylformamidine 37
Methylchloropyrofos 395
Methylene chloride extracts 75
4-Methyl-3-heptanol 68
Methyl parathion 7, 124, 154, 162, 489
Mexican bean beetle 273–8
Microanalytical techniques 532
Microencapsulation 151–61, 469, 472
  advantages of 153–5
  comparative activity 154
  comparative study 158, 159, 160
  field tests 155
  materials and methods 156
  preparation of formulations 156
  rationale for 152–3
  results and discussion 157
Midges 396, 442, 448
Minimum effective dose (MED) 157
Mites 386, 397–9
Mitin FF 6, 7
Molluscicides 403, 405
Molt-disrupting compounds 13
Monocroptophos 18
Monoculture of tropical and subtropical
  crops 413
Monoterpenes 213
Mosquito coil 103, 117
Mosquito larva 372
Mosquitoes 386, 388–93, 455, 479
Moulting hormones 197
Multistriatin 68
Muristerone 213

Nairobi 570
Naphthoquinone derivatives 201
Natural compound models 38
Natural insecticides 312–6
Natural products 59–70, 235, 490–6
  biologically active 74–85
  chemicals 59–70

development of synthetic
  insecticides   109–25
in plant-insects and plant-fungi
  interaction   187–220
models   14–7
nematocidal   299–310
research   85–6, 86–7
structure of   213–6
Natural resistance   209–11
Natural sources   73–90
Nematocidal activity, screening for   301
Nematocidal crude extract   303
Nematocidal fatty acids   305
Nematocidal natural products   299–310
Nematocidal polyacetylenes from *Carthamus
  tinctorius* L.   301–3
Nematocidal substance from *Angelica
  pubescens*   305–7
Nematodes   299
Nereistoxin   17, 38, 310
Nerve-insensitivity factor   117
Nicandra physaloides   495
Nicandrenone   194
Nicotine   53, 313, 457, 493–4
Nigeria   569, 570
Non-host plants   223, 250
Non-host resistance   280
Non-preference type of plant resistance   230
No-observable-effect level (NOEL)   532
Number mean diameter (NMD)   466

Odoracin   303–5, 310
Odoratrin   303–5, 310
Oil of citronella   455
Oil palm   424, 429
Oleoresins   206, 371
Oligophagous insects   224
Onchocerca volvulus   396
Onchocerciasis   386, 387, 557
Oncopeltus fasciatus   325, 330, 342, 357, 358,
  363
Organochlorids   150
Organochlorines   55, 452, 455
Organoleptic evaluation   537
Organophosphates   150, 452, 455,
  475–8
Organophosphorous esters   7, 11, 18, 27
Organophosphorous pesticides   118
Ornithodorus moubata   398
Ovicides   37
Oviposition stimulants   244–6, 251
Oxaphenalene derivatives   202
Oxydemetonmethyl   17

Pachyrrhizus erosus   495
Paint   480

Palms   419, 423
Pantropical pests   426
Parasitic diseases   387
Patent Cooperation Treaty (PCT)   541, 543
Patents and patent protection   539–46
  application   544–6, 550
  claims   545
  filing   546
  flow of procedure   541–2
  Kenya   547–52
  opposition proceeding   543
  outline of application   544
  overview of system   540–1
  procedure   551
  publication   543
  specification - title, field and utility   544–5
  towards international uniformity   541–2
  transfer of inventions to industry   549, 550
  unity of invention   542–3
Pearl millet   439, 444, 448
Pellitorine   370
Peltogynol   203
Penncap-M^R 155
Pepper plant   369
Periplaneta americana   335, 358
Periplanone B   318
Permethrin   62, 189, 456, 480
Peruvian ground cherry   495
Pest control, new approaches   496–7
Pesticide application   513, 514
Pesticide chemistry   33–43
Pesticide development   10, 28
Pesticide discovery   10
Pesticide establishment, advantages and
  difficulties   487–8
Pesticide formulation   497–500
Pesticide hazards   509
Pesticide overview   6–7
Pesticide production   507–9
  in developing countries   486–90
Pesticide production data   6–7
Pesticide registration   521–38
  data requirements   523
  international considerations   533–4
  procedure for   522
Pesticide regulations   522
Pesticide regulatory, advisory and testing
  body   502
Pesticide role   505–7
Pesticide tolerances   522–3, 529–33, 537, 538
  evaluation   532–3
  international considerations   533–4
Pesticides   428, 454
  economic benefits from   500
  industrial production of   505–11
  insect resistance to   149
  world market growth   487

PH60-41   14
pH effects   286–7, 336–7
Phagostimulants   199, 217, 276
Phaseolunatin   274
Phenoldienones   203
Phenolic compounds   226, 274
3-Phenoxybenzyl alcohols   138
3-Phenoxybenzyl chrysanthemates   139
3-Phenoxybenzyl dichlorovinyl esters   138
3-Phenoxybenzyl esters   135
Phenylformamidines   37
Phenyl ureas   37
Pheromones   66–8, 70, 163, 215, 246, 316–9, 471, 481, 497
Phlebotomus   396
Phosphate insecticides   40
Photochemical (Z)-(E) isomerization   93, 95
Photochemical rearrangements   92–4
Photodegradation   114
Phoxim   395
Phyroecdysones   191
Phytoalexins   77–80, 83, 84, 209, 210, 211, 218, 285, 288–9, 292, 294, 296
Phytochemical action on insect morphogenesis, reproduction and behaviour   311–20
Phytochemicals   478–9
Phytolacea dodecandra   405
Phytotoxicity   264
Phytotoxins   207
Pigeonpea   439, 444, 448
Piper guineense Schum   374
Piper nigrum L.   370, 371
Piper trichostanchyon C. DC.   375
Piperaceae amides   369–82
Pipercide   370, 371
   insecticidal action   377
   structure determination   372
   synthesis of   375
Piperenone   196
Piperine   370, 372
Piperlongumnine   371
Piperolein   371
Piperonyl butoxide   162
Piperstachine   370
Plague   387, 399
Plant apparency   173–5
Plant defense   173–5
Plant deterrency   182
Plant escape   173
Plant extracts   454
Plant-fungi interaction   207–13, 218
Plant-insect interaction   188–207, 218
Plant protection *See* Crop protection
Plant resistance   173, 279–97, 439–49
   defence substances in   200–6
   non-preference type of   230

Podfly   448
Polynuclear aromatic hydrocarbons (PAHs)   337, 339, 343, 344
Post-harvest technology centre   437
Precocene-epoxides   335–8, 353–5
Precocenes   21, 53–8, 64, 65, 192, 197, 316, 323–5, 357–68, 478
   cancer risk from   342–4
   lethal bioactivation of   325–30
   morphogenetic and antigonadotropic effects   358
   structure/activity optimization   338–42
   tissue and species specificity   330-5
   toxicity of   367
Precocious metamorphosis   317
Pre-formed inhibitors   285–8
Preharvest interval requirement   531
Presumed safe levels   530
Primate testing   536
Primin   213
Pristimerin   203
Pro-allatocidins   323–55
Pro-carcinogen   325
Product chemistry data requirements   523
Propionate   63
Propoxur   400
Proprietary know-how   539
Purified compounds   229
Pyrethrins   15, 60, 91–102, 109–25, 132, 189, 404, 435, 472, 479, 490
   acid components   136–8
   actions on insects and insect nerves   116–7
   biodegradability   116
   biology   112–4
   bioresmethrin   190
   chemistry of   112–4
   compared with pyrethroids   120
   impurities in   112
   instability of   100, 103, 104, 112–4
   mixtures of photoproducts   116
   natural   128, 145
   resistance to   117
   toxicities of   114
Pyrethroid acid and alcohol combinations   142–3
Pyrethroids   15, 16, 30, 34, 60–2, 72, 104, 109–25, 354, 472
   action in mammals   118
   actions on insects and insect nerves   116–7
   biodegradability   116
   case history of   120
   characteristics of   145
   developments in chemistry and action of   127–48
   enhanced photostability   110
   environmental persistance   123

impurities in   112
insect resistance to   149
metabolism rates of   115
mode of action   125
persistence of   115
pest resistance to   117–8
photostable   138–9, 141–2
potency   119
present situation of   143–5
resistance to   118, 436
stabilizing   114–5
standard   131
structure   119
synthetic compounds   129–38
toxicosities of   114
types of   117
types of action   119
Pyrethrolone   96, 114, 130
Pyrethrum   49, 55, 189, 312–3, 424, 425, 429,
    451, 454, 455, 457, 490–3
  insect pests   103
  natural, microencapsulated   151–61
  production practices   492–3
  residual repellency of   152
Pyrethrum Board of Kenya   102–4
Pyrethrum clones   104
Pyrethrum extracts
  composition of   110–2
  need for   110
Pyrethrum flowers   99–100, 104, 110
Pyrethrum industry   102

Quantitative structure-activity relationship
    (QSAR)   41
Quassia   453, 495
Quinones   201, 202, 213, 220, 221

Rain forests   183
Random screening   36
Raw materials   505–6
Registration. *See* Pesticide registration
Repellency tests   156, 161
Repellents   192, 196, 217, 244, 252, 271, 311,
    455–6
Research centres   565–7
Research programmes   569
Reserpine   457
Residual deposits   467–8
Residue intake analysis   533
Residue testing   531–2
Resmethrin   61, 116, 118, 134–6, 138, 400
Resveratrol   81, 82, 93–4, 98–9, 113
Rethrolone   92–4, 96–8
Rhinoceros beetle   424
Rhodnius prolixus   317

Rice   80–4, 124, 419, 423, 429
Rice blast pathogen   280
Rice pests   72
Rice weevil   372
Rickettsiosis   386, 398
Rodenticides   404
Rodents   387, 404–5
Root crops   425
Rotenoids   190
Rotenone   53, 55, 190, 313, 457, 494
Ryamia speciosa   453
Ryanodine   190, 494–5

Sabadilla   453, 495
Safety aspects   235–6, 535
Salithion   17
Salmonella typhimurium   344
Sandflies   388, 396
Sapindus saponaria   495
Saponins   495
Schistocerca gregaria   335, 366
Schistosomiasis   387, 401, 405, 495
Scitophylus zea mays   435
Sclareol   211
Scolytidae   67
Screening procedures   21–2, 35, 301, 452,
    507
Scrub typhus   386, 398
Secrecy agreement   551
Seed beetle   447
Seretonin   457
Sesamin   314
Sesamolin   314
Sex pheromones   66, 67
Shell Biosciences Laboratory   85–6
Shootfly *See* Sorghum shootfly
Shootfly plant   495
Silica   278
Simulium   395
Skythantine   215
Snails   387, 400–3, 495
Soap berry   405
Solanum mammosum   312
Sorghum   439–43
Sorghum shootfly   357–68
  effect of topical application of
  pupae   362
  fumigation effects   359–62, 365–8
  mortality   360
Specificity factors   283
Spraying   264, 266, 390, 394, 465–7, 514
Static devices   515–6
Stem borer   441–2, 448
Sterile Insect Technique (SIT)   456
Sterility control   457
Steroid derivatives   214

Stilbenes 204
Stimulants 244–7, 251, 276
Storage pests 435
Stored products protection 437
Stress substances 211–3, 218
Structure activity relations (SAR) 11
Sugarcane 419, 422, 424, 427
Suppressors 283
Synergism 234–5, 314, 381
Synthesis concepts 36

Tabanids 396, 405
Tannins 193, 203, 207, 211
Tanzania 547, 551, 552
Target site sensitivity 119
Tarsal chemoreceptors 256
Tea 417, 420, 421, 426
Tenebrio molitor 335
Termites 448
Terpenes 199, 204
Terpenoids 197
Tetramethrin 116
Tetramethylsilane 78
Thallium sulphate 404
Theoretical maximum residue contribution
    (TMRC) 532
Thermal-acid-catalysis 94–9
Thiocarbamate herbicides 19
Thiocarbamate sulfoxides 19–20
Third-generation insecticides 62
Thujopsene 197
Ticks 157, 159, 160, 241, 386, 397–9, 456
Tingenone 203
Token stimuli 223, 224
Tomatine 286, 292
Toxaphene 7
Toxic substances 189–90, 217
Toxicants 311, 314
Toxoplasmosis 387
Trachylobanoic acids 205
Trade secrets 539, 540
Training 555–73
    assistance from industry 568
    diagnosis of the problem 556–9
    discussion 569–73
    in-service 563
    international aid 568
    international recognition 559
    national commitments 567–8
    national recognition 559
    problems and recommendations 569
    professional 561–3
    proposals 565–8
Training centres 565–7
Training needs 559–65

Training programmes 559–65, 569
Trap crops 429
Tree foliage 173
Triatomid bugs 397
Triatomine control agents 473–81
Trichomes 278
Trichonine 371
Trichostachine 371
2-Tridecanol 220
3,5,6-Tridecanol 220
2-Tridecanone 201
Triggering effect 284–5
Tripterygium wilfordii 453
Tropical agriculture 409–38
    control strategies for crop pests
        430–1
    controlled use of pesticides 428
    crop diversity 410–1
    crop resistance 428
    cultural practices 429–30, 433
    current methods of pest control
        426–7
    development needs for pest control 432–3
    discussion 435–8
    ecological factors 413–7
    economic factors in 417
    environment factors in 431
    features of 410–1
    host/natural enemy relationship 416
    intra- and interspecific competition 416
    introduction of pests to new
        environments 416
    major crops and pests 417–20
    management and economic aspects 431
    monoculture 413
    pest information required 431
    pest status variation 412–3
    quality and quantity of food 416
    reasons for crop diversity 411
Tropical disease vectors 385–407
Tropical diseases 558
Tri-*o*-tolyl phosphate (TOCP) 17
Trypanosoma congolense 393
Trypanosoma gambiense 393
Trypanosoma rhodesiense 393
Trypanosoma vivax 393
Trypanosomiasis 386, 473
Tsetse fly 150, 156, 160, 388, 393, 406, 456
Tuber crops 425
Two-spotted mites 118
Type I effects 118
Type II effects 118
Type I symptoms 117
Type II symptoms 117
Typhoid fevers 387
Typhus 387

Uganda   547, 551, 552
Umbelliferae   167
Umbelliferous plants   169–71
UNIDO   504
United Kingdom, patent laws   547–50
United Nations Environment Programme
      (UNEP)   563, 570, 571
United States   522, 538
   patent rules   551
Unity of invention   542–3
USDA   228

Vegetable oil crops   424, 429
Viniferins   77–81
Vinylstilbenes   88
Viruses   386, 387
Volta River Basin Project   395
Volume mean diameter (VMD)   466

Wisanidine   371
Wisanine   371

WL 28325   82–4, 292
World Health organization (WHO)   389,
      534
Wortman's Yield Equation   484
Wyerone   211

Xanthotoxin   170
Xanthoxyletin   239–41
Xenopsylla cheopsis   400
Xylomollin   194

Yams   425–6
Yellow fever   386
Yersinia (Pasteurella) pestis   399

Zinc phosphide   404
Zonocerus variegatus   266–7